STRUCTURES

Daniel L. Schodek

Harvard University

STRUCTURES

PRENTICE-HALL, INC., *Englewood Cliffs, New Jersey* 07632

Library of Congress Cataloging in Publication Data

Schodek, Daniel L
   Structures.

   Includes index.
   1. Structures, Theory of. 2. Structural design.
I. Title.
TA645.S37   624'.17   79-15529
ISBN 0-13-855304-1

Editorial/production supervision and interior design
    by Barbara A. Cassel
Cover design by Edsal Enterprises
Manufacturing buyer: Gordon Osbourne

Printed in the United States of America

10 9 8 7 6 5 4 3 2

PRENTICE-HALL INTERNATIONAL, INC., *London*
PRENTICE-HALL OF AUSTRALIA PTY. LIMITED, *Sydney*
PRENTICE-HALL OF CANADA, LTD., *Toronto*
PRENTICE-HALL OF INDIA PRIVATE LIMITED, *New Delhi*
PRENTICE-HALL OF JAPAN, INC., *Tokyo*
PRENTICE-HALL OF SOUTHEAST ASIA PTE. LTD., *Singapore*
WHITEHALL BOOKS LIMITED, *Wellington, New Zealand*

CONTENTS

2 Principles of Mechanics   38

3 Introduction to Structural Analysis and Design   84

7   Members in Compression: Columns   283

8   Continuous Structures: Beams   309

IRON AND STEEL IN LARGE BUILDINGS

Our first picture (The Palace Hotel, Denver, Colorado) affords a vivid representation of the manner in which large, high buildings, in all the principal cities, are now erected. The contrast it presents to the old method of building, with wooden posts, beams, joists, and stringers, the structure all supported by the walls, is very great. The revolution in building construction which this change represents may be said to have commenced about 1850. . . . In all similar structures up to a very recent period, the walls were depended upon to furnish the principal support of the several floors and give the necessary strength and stability to the building. Such dependence upon the walls alone has been found to be increasingly difficult and vastly more expensive with every addition to the height of the building; and where it was necessary to make the walls, at the first story, four or five feet or more thick as has often been the case in eight- or nine-story buildings, a large proportion of the most valuable room was thus taken up. The modern method of building obviates this difficulty and enables the architect to put up structures twenty or more stories high having every desired element of strength and stability, but with the walls forming only a mere shell enclosing the building, and in no way depended upon for its support. . . . In this way of building the walls are only intended to support their own weight, serving such purposes of ornamentation or embellishment as may be sought, the openings for the admission of light and air to the interior being largely increased, or, as has been followed in some cases, the exterior may be formed almost entirely of glass. (Scientific American, May 21, 1892)

PREFACE

When the fact that a structure of one type or another is or has been present in every building ever built by man is considered, any recent emergence of something really new in the way of structural forms is actually rather remarkable. Yet new structural approaches are continually evolving. The structural system described as a "revolution in building construction" in the accompanying excerpt from an 1892 edition of *Scientific American* is one such development. Since this relatively recent date, however, the system described has achieved the dubious status of becoming a commonplace type of structural system that is even regarded by many as being at best antiquated and at worst inefficient and hopelessly noncreative.

Despite the fact that rapid changes of the type described are occurring in what structures are used in buildings, there exist an invariant set of physical principles founded in the field of mechanics that can be used by designers as aids to understanding the behavior of existing structural forms and in devising new approaches. The development of these principles has flowered during the past three centuries to the extent that they are amazingly well established and documented. Some new understandings, of course, are continuing to occur and will hopefully always occur. Still, the analytical tools already available to the designer are extensive and enormously powerful. This is true to the extent that the real challenge in the field of structures lies not so much in developing new analytical tools but in bringing those currently in existence to bear in the designing and formulation of creative structural solutions with the intent of making better buildings.

This book discusses in an introductory way the nature of the invariant physical principles that underlie the behavior of structures under load. The primary goal of the book, however, is not to simply teach analytical techniques but to more generally explore their role in the design of structures in a building context. Because of this larger goal, the book covers material discussed not only in specialized engineering curricula but also, to some extent, that covered in architectural curricula as well. The traditional hard boundaries between subdisciplines in engineering, (e.g., statics and strength of materials) have also been deliberately softened and a more integrative approach taken.

The book is divided into three major parts. Part I is an introduction to the subject and to fundamental concepts of analysis and design. Part II introduces the reader to most of the primary structural elements used in buildings and discusses their analysis and design. Each of the chapters in this part is divided into sections that (1) introduce the element considered and explain its role in building, (2) discuss its behavior under load in qualitative terms (an "intuitive" approach), (3) examine its behavior under load in quantitative terms, and (4) discuss methods for designing (rather than just analyzing) the element. Part III contains a unique discussion of the logic of structural design. The various chapters take the reader from early design stages (where the designer is still manipulating functional spaces and the structural design issues faced hinge upon the characteristics of different structural grids or patterns) all the way through more detailed design development stages where the nature of the structure is resolved in greater detail. The Appendix generally discusses more advanced principles of structural analysis.

The book is largely intended as a resource for students and instructors wishing to design their own curriculum. For those wanting to adopt a strictly qualitative approach to the subject, for example, it is possible to read only Chapter 1 in Part I, the sections entitled "Introduction" and "General Principles" in each of the chapters in Part II, and all of Part III. This coverage will provide a brief qualitative overview of the field with a special emphasis on design rather than analysis. For those students already having a background in the analytical aspects of structures, Part III contains summary information useful in a design context. Part III can be read independently by such students.

Within Parts I, II, and III there is a certain redundancy in the way analytical topics are covered so that students or instructors can integrate the material in the order seen best fit. Shear and moment diagrams, for example, are first introduced in an abstract way in Chapter 2. They are reintroduced in connection with the analysis of a specific structural element—the truss. *Where* the different presentations are introduced, if at all, may be varied by the instructor. The author, for example, typically chooses to introduce shear and moment diagrams initially as a part of truss analysis and then follow up with the more abstract development of shear and moment diagrams in Chapter 2 before going into beam analysis and design. Other instructors may choose to approach the subject material differently. The book is designed to have sufficient flexibility to support different approaches. In any event, however, the material is presented in such a way that a direct cover-to-cover reading is also appropriate.

The author is, of course, indebted to a vast number of people in either a direct or indirect way for the approach taken in this book. The general works and ideas of other authors or educators in the field—Mario Salvadori, Robert Heller, Henry Cowan, Peter McCleary, and Waclaw Zalewski, to name but a few—have undoubtedly had an impact on the content and approach that has been taken. Architects such as Edward Baum and Urs Gauchat, who have had a long and abiding interest in the teaching of structures, have also been influential. The contributions of David Wright, Richard Rauh, Elinore Charlton, Rick Cureton, and Denys Purcell are also acknowledged. Kay and Ned must also not go without notice, especially since Ned firmly believes that the book was written just for him. The endless patience and contributions of several years of first-year students in the Graduate School of Design at Harvard who have suffered through courses involving the material contained herein is also greatly appreciated. So here it is; have fun with it.

DANIEL L. SCHODEK
Cambridge, Massachusetts

STRUCTURES

part I

INTRODUCTORY
CONCEPTS

The three chapters in Part I provide an overview and introduction to structures and their use in buildings. The first chapter is a self-contained overview of the field and discusses different ways of classifying structural elements and systems. The second chapter reviews certain fundamental principles of mechanics that are generally applicable to the analysis of any structure. The third chapter considers the loads that structures must be designed to carry and generally discusses the structural analysis and design process as it occurs in a building context.

chapter 1

STRUCTURES: AN OVERVIEW

1.1 INTRODUCTION

Definitions are a time-honored way of opening any book. A simple definition of a *structure* in a building context is that a structure is a device for channeling loads that result from the use and/or presence of the building to the ground. Important in the study of structures are many widely varying concerns. The study of structures certainly involves coming to understand the basic principles that define and characterize the behavior of physical objects subjected to forces. More fundamentally, it even involves defining what a force itself is, since this familiar term represents a fairly abstract concept. The study of structures in a building context also involves dealing with much broader issues of space and dimensionality. The words "size," "scale," "form," "proportion," and "morphology" are all terms commonly found in the vocabulary of a structural designer.

As a way of getting into the study of structures, it is useful to reconsider the first definition of a structure given above. Although valuable in the sense that it defines the purpose of a structure, the original definition unfortunately provides no insight into the makeup or characteristics of a structure: What *is* this device that channels loads to the ground? To adopt the complex and exacting style of a dictionary writer, a *structure* could be defined as a physical entity having a unitary character that can be conceived of as an organization of positioned constituent elements in space in which the character of the whole dominates the interrelationship of the parts. Its purpose would be defined as before.

Although it might be very hard to believe, a contorted and relatively abstract definition of this type which is almost laughably academic in tone actually has some merit. It first states that a structure is a *real* physical object, not an abstract idea or interesting issue. A structure is not a matter of debate, it is something that is built. The implication is that a structure must be dealt with accordingly. Merely verbally postulating that a structure can carry a certain type of load or function in a certain way, for example, is inadequate. A physical device must be provided for accomplishing the desired ends that conforms to basic principles governing the behavior of physical objects. Devising such a structure is the role of the designer.

The expanded definition also makes the point that a structure functions as a whole. This is a point of fundamental importance and one easily forgotten when confronted with a typical building composed of a seemingly endless array of individual beams and columns. There is in such cases an immediate tendency to think of the structure only as an assembly of individual small elements in which each element performs a separate function. In actuality, all structures are, and must be, primarily designed to function as a whole unit and only secondarily as an array of discrete elements. In line with the latter part of the expanded definition, these elements are invariably so positioned and interrelated as to enable the overall structure to function as a whole in carrying either vertically or horizontally acting loads to the ground. No matter how some individual elements are located and attached to one another, if the resultant configuration and interrelation of all elements does not function as a whole unit in channeling loads of all anticipated types to the ground, the configuration cannot be said to be a structure. The reference to anticipated types of loads in the previous statement is included to bring up the important fact that structures are normally devised in response to a specific set of loading conditions and function as structures only with respect to these conditions. They are often relatively fragile with respect to unanticipated loads. A typical building having a structure capable of carrying normally encountered occupancy and environmental loads cannot, for example, be simply picked up by a corner and transported through space. It would simply fall apart, since its structure would not have been designed to carry the unique loadings involved. So much for Superman carrying buildings around.

While on the subject of formal definitions, the act of designing a structure can be defined in language at least as complex as that used previously to define a structure, but again the result is of some value. Designing a structure is the act of positioning constituent elements and formulating interrelations with the objective of imparting a desired character to the resultant structural entity. The notions that elements are positioned and that relationships exist among these elements are basic to the concept of designing a structure. Figure 1-1(a) illustrates, for example, a simple structure with columns and a beam positioned to carry a vertical load. The beam merely rests on top of the columns and is not rigidly affixed to them (this type of connection defines one particular type of relationship between members). If this same assembly were suddenly called upon to carry lateral forces as might be associated with winds acting on the side of the building, then this assembly no longer functions as a structure, in that it cannot carry the lateral load to the ground. It would collapse, as illustrated in Figure

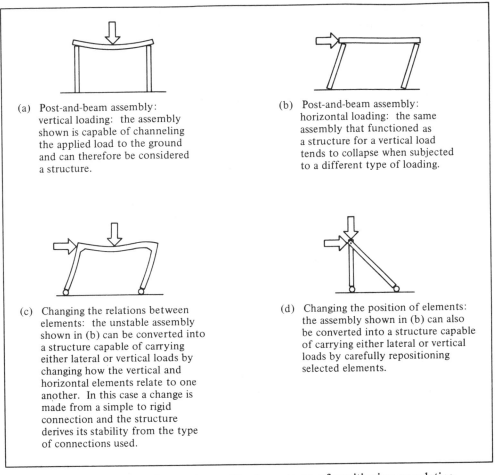

(a) Post-and-beam assembly: vertical loading: the assembly shown is capable of channeling the applied load to the ground and can therefore be considered a structure.

(b) Post-and-beam assembly: horizontal loading: the same assembly that functioned as a structure for a vertical load tends to collapse when subjected to a different type of loading.

(c) Changing the relations between elements: the unstable assembly shown in (b) can be converted into a structure capable of carrying either lateral or vertical loads by changing how the vertical and horizontal elements relate to one another. In this case a change is made from a simple to rigid connection and the structure derives its stability from the type of connections used.

(d) Changing the position of elements: the assembly shown in (b) can also be converted into a structure capable of carrying either lateral or vertical loads by carefully repositioning selected elements.

FIGURE 1-1 Basic structures: The importance of positioning or relating elemental parts.

1-1(b). From a design viewpoint, the difficulty with the assembly is either that the elements used are incorrectly positioned, incorrectly related, or perhaps both. The assembly can be redesigned into a viable structure for the lateral load by either changing the relationships that exist between the elements and/or repositioning them. An example of changing the relations that exist between elements would be to use a rigid rather than simple connection between the elements [see Figure 1-1(c)]. A rigid joint behaves essentially like a monolithic unit. The assembly then gains stability with respect to lateral loads by virtue of these connections in much the same way that a table derives its stability from the rigid connections that exist between the table top and legs. Alternatively, the elements of the assembly could be repositioned in a way shown in Figure 1-1(d), such that one of the elements serves as a brace which transfers the lateral load to the ground.

There are many ways that elements can be positioned to carry loads. Similarly, there are many types of relations that may exist. Many of the ways in which constituent elements can be positioned and related will be explored in this book. The physical nature of the constituent parts of a structure are also important, since this influences the attributes of the whole structure. The nature of many elements will also be explored in this book.

1.2 GENERAL CLASSES OF STRUCTURES

1.2.1 Primary Types of Structures

CLASSIFICATIONS. Fundamental to an understanding of any field is a knowledge of the way type groups within the field are systematically distinguished, ordered, and named. A knowledge of the criteria or presumed relationships that form the basis for classifications of this type is similarly important. This section introduces one general method for classifying structural elements and systems, which is simply according to their shape and basic physical properties of construction (see Figure 1-2). Since this classification scheme implies that complex structures are the result only of additive aggregations of elements, it is inherently simplistic. Of significance in aggregations is only the additive nature of the elements. Of significance in structures is that the elements are also positioned and related with the intent of giving the structure certain load-carrying attributes. Still, the simpler classification approach illustrated in Figure 1-2 is useful as an introduction.

As Figure 1-2 indicates, the general geometrical form of a typical structure is either one of the basic geometrical shapes illustrated in the left portion of the figure or is derived from some combination or aggregation of these shapes. Corresponding to these basic shapes are a series of primary structural elements having certain physical properties. This discussion focuses on these elements.

GEOMETRY. In terms of their basic geometries, the structural forms indicated to the left in Figure 1-2 can generally be classified as either *line-forming elements* (or composed of line elements) or as *surface-forming elements*. Line-forming elements can be further distinguished as straight or curved. Surface-forming elements are either planar or curved. Curved-surface elements can be of either single or double curvature.

Strictly speaking, there is, of course, no such thing as a line or surface element, since all structural elements have thicknesses. For classification purposes, however, it is still useful to classify any long, slender element (such as a column whose cross-sectional dimensions are small with respect to its length) as a line element. Similarly, surface elements also have thicknesses, but this thickness is again small with respect to length dimensions. The terms "line" and "surface" are used for convenience only.

Closely coupled with whether an element is linear or surface-forming is the material and/or method of construction used. Many materials are naturally line-forming. Timber, for example, is inherently line-forming simply because of way in which timber is grown. It is possible, however, to make minor surface-forming elements directly from timber (as is evidenced by common plywood) or larger surface-forming structures by aggregating more elemental pieces. Other materials, such as concrete, can be line-forming or surface-forming with equal ease. Steel is primarily line-forming, but it is also possible to make directly minor surface-forming elements (e.g., steel decking).

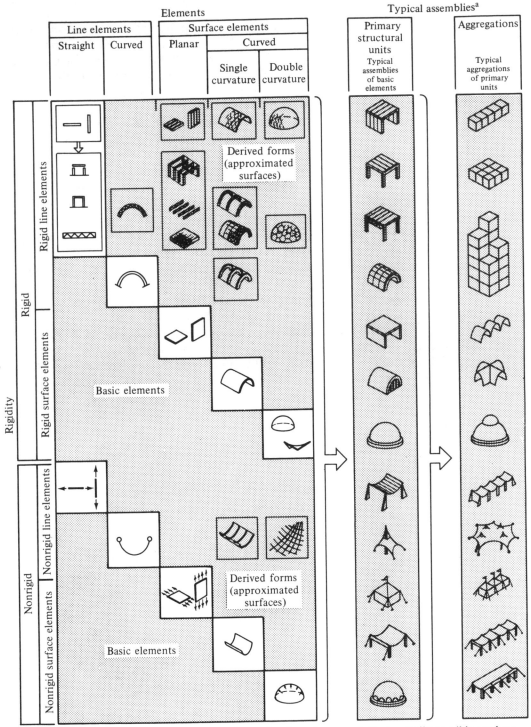

FIGURE 1-2   Classification of basic structural elements according to geometry and primary physical characteristics. Typical primary structural units and other aggregations are also illustrated.

STIFFNESS. Figure 1-2 also illustrates a second fundamental classification, which is according to the stiffness characteristics of the structural element. The primary distinction here is whether the element is rigid or flexible. *Rigid elements,* such as typical beams, are those that do not undergo *appreciable* shape changes under the action of a load or changing loads [see Figure 1.3(a)]. They are, however, usually bent or bowed to a small degree by the action of the load.

Flexible or *nonrigid elements,* such as cables, are those in which there is a tendency for the element to assume one shape under one loading condition and to change shape drastically when the nature of the loading changes [see Figure 1-3(b)]. Flexible structures maintain their physical integrity, however, no matter what shape they assume.

For either type of structure, the general effect of a loading is to cause deformations to occur in the structure. In general, deformations can either be elongations or shortenings. In flexible structures that are subjected to tension forces only, the deformations are invariably elongations. When the action of a load on a rigid structure is such to cause a bowing or bending in the member considered, both elongations and shortenings can occur in the same cross section of the member. The importance of the latter observation will become increasingly significant as the behavior of structures is looked at in greater detail in the remainder of the book.

Often associated with whether an element is rigid or flexible is the material of construction used. Many materials, such as timber, are inherently rigid. Other materials, such as steel, can be used to make either rigid or flexible members. A good example of a rigid steel member is the typical beam (an element that does not undergo appreciable change in shape under changing loads). A steel cable or chain, however, is clearly a flexible member since the shape that such elements assume under loading

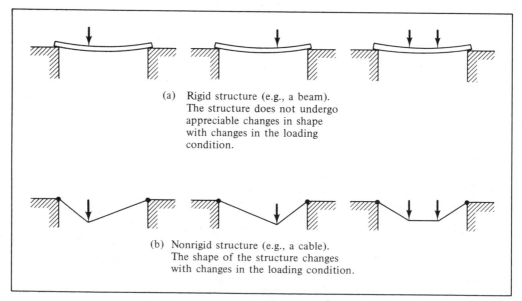

(a) Rigid structure (e.g., a beam). The structure does not undergo appreciable changes in shape with changes in the loading condition.

(b) Nonrigid structure (e.g., a cable). The shape of the structure changes with changes in the loading condition.

FIGURE 1-3 Nonrigid and rigid structures.

is a function of the exact pattern and magnitudes of the load carried. A steel cable thus changes shape with changing loads. Whether a structure is rigid or flexible therefore depends *either* upon the inherent characteristics of the material used or on the amount and microorganization of the material in the element.

Many specific structures that are usually classified as rigid are in actuality rigid only under certain given loading conditions or under minor variations of the given loading condition. When the loading changes dramatically, structures of this type become unstable and tend to collapse. Structures made by aggregating smaller rigid elements (e.g., blocks) into larger shapes are often in this category. The rigidity of structures of this type will be considered in more detail in a later section that discusses **arches**.

MATERIALS. A very common classification approach to structures is simply by the type of material used (e.g., wood, steel, or reinforced concrete structures). A strict classification by materials, however, is somewhat misleading and is not adopted here since the principles governing the behavior of similar elements composed of different materials (e.g., a timber and steel beam) are invariant, and the actual differences present are superficial. More general descriptions, such as rigid versus nonrigid elements, are of more intrinsic value at this stage.

As one begins taking a more detailed look at structures, however, the importance of materials will increase. One reason for this is that there is a close relationship between the nature of the deformations induced in a structure by the action of the external loading and the material and method of construction that is most appropriate for use in the structure. Steel can be used under virtually all conditions. Plain concrete can be used only in situations where the structure is only compressed or shortened under the action of the load. Concrete will crack and fail when subjected to forces that tend to elongate the material. The material (and structure) is thus rigid only under one type of loading. Concrete reinforced with steel, however, can be used in situations where elongating forces are present, since the steel can be designed to carry these forces. These and other considerations will be studied in more detail later in this book.

1.2.2 Primary Structural Elements

ELEMENT NAMES. Common rigid elements include beams, columns, arches, flat plates, singly curved plates and shells having a variety of different curvatures. Elements that are nonrigid or flexible include cables (straight and draped) and membranes (planar, singly curved, and doubly curved). In addition, there are a number of other types of structures that are derived from these elements (e.g., frames, trusses, geodesic domes, nets, etc.). Assigning a specific name to refer to an element having certain geometrical and rigidity characteristics is done as a matter of convenience only and has its basis in historical tradition. Naming elements in this way can, however, be misleading, since it is too easy to fall into believing that if two elements have different names, the way they carry loads must also be different. This is not necessarily so. Indeed, one of the basic principles of structures that later portions of this book will clarify and elaborate on is that the fundamental load-carrying mechanism for all structures is the same. At this point, however, it is still useful to retain and use the traditional names as a way of gaining familiarity with the subject.

BEAMS AND COLUMNS. Structures formed by resting rigid horizontal elements on top of rigid vertical elements are commonplace. The horizontal elements (*beams*) pick up loads that are applied transversely to their lengths and transfer them to the supporting vertical columns or posts. The *columns*, loaded axially by the beams, transfer the loads to the ground. Since the beams are bowed or bent as a consequence of the transverse loading carried [see Figure 1-4(a)], they are often said to carry loads by *bending*. The idea of bending in structural elements is an important one and will be explored in detail in later parts of this book. The columns in a beam and column assembly are not bent or bowed, since they are subjected to axial compressive forces only. In a building, the absolute length of individual beams and columns that is possible is rather limited

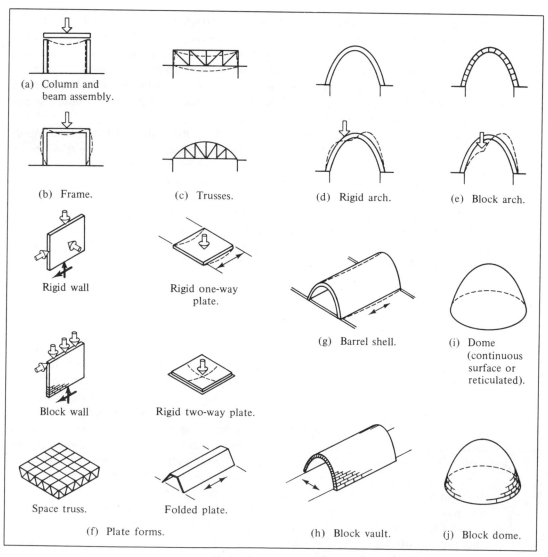

(a) Column and beam assembly.

(b) Frame.   (c) Trusses.   (d) Rigid arch.   (e) Block arch.

Rigid wall

Rigid one-way plate.

(g) Barrel shell.

(i) Dome (continuous surface or reticulated).

Block wall

Rigid two-way plate.

Space truss.   Folded plate.

(f) Plate forms.   (h) Block vault.   (j) Block dome.

FIGURE 1-4 Typical rigid structures.

as compared to some other structural elements (e.g., cables). Beams and columns are therefore typically used in a repetitive pattern. Simple single-span beams are discussed more extensively in Chapter 6 and columns in Chapter 7. Continuous beams that bear on multiple support points are discussed in Chapter 8. Continuous beams often exhibit more advantageous structural properties than simpler single-span beams supported only at two points.

FRAMES. The relatively recently developed *frame*, illustrated in Figure 1-4(b), is similar in appearance to the post-and-beam type of structure but has a different structural action because of the rigid joints that are made between vertical and horizontal members. This joint rigidity imparts a measure of stability against lateral forces that is not present in the post-and-beam system. As previously discussed, this joint ridigity is an illustration of a different type of relation that can exist between members than is present in the post-and-beam system, where beams are nonrigidly rested on the columns. In a framed system both beams and columns are bent or bowed as a consequence of the action of the load on the structure [see Figure 1-4(b)]. As with the post-and-beam structure, the lengths of individual elements in a frame structure that are possible are limited. Consequently, members are typically formed into a repetitive pattern when used in a building. Chapter 9 discusses frames in detail.

TRUSSES. *Trusses* are structural members made by assembling short, straight members into triangulated patterns [see Figure 1-4(c)]. The resultant structure is rigid as a result of the exact way the individual line elements are positioned relative to one another. Some patterns (e.g., a pattern of squares rather than triangles) do not necessarily yield a structure that is rigid (unless joints are treated in the same way as in framed structures). Pattern implications will be explored in detail in Chapter 4. It should be noted that a truss composed of discrete elements is bent or bowed in an overall way under the action of an applied transverse loading in much the same way that a beam is bent or bowed. Individual truss members, however, are not subject to bending but are only either compressed or pulled upon. The internal action of trusses will also be explored in depth in Chapter 4.

ARCHES. An *arch* is a curved, line-forming structural member that spans between two points. The common image of an arch is that of a structure composed of separate wedge-shaped pieces that retain their position by mutual pressure induced by the load. The exact shape of the curve and the nature of the loading are critical determinants as to whether the resultant assembly is stable. When shapes are formed by simply stacking rigid block elements, the resultant structure is functional and stable only when the action of the load is to induce in-plane forces that cause the structure to compress uniformly. Structures of this type cannot carry loads that induce elongations or any pronounced type of bowing in the member (the blocks simply pull apart and failure occurs). Block structures can be very strong, however, when used properly, as their extensive historical usage attests. The strength of a block structure is due exclusively to the *positioning* of individual elements, since blocks are typically either simply rested one on another or mortared together (the mortar does not appreciably increase the strength of the structure). The positioning is, in turn, dependent on the exact type of loading involved. The shape of an arch made of blocks, for example, is not an arbitrary curve but a very specific one appropriate for the loading involved.

The resultant structure is thus rigid only under very particular circumstances. These issues will be discussed more extensively in Chapter 5.

Less known but frequently used in modern building is the *rigid arch*, which is curved similarly to block arches but is made of one continuous piece of deformed rigid material [Figure 1-4(d)]. If rigid arches are properly shaped, they can carry a load to supports while being subject only to axial compression, and no bowing or bending occurs. As will be discussed later, this is an advantageous and efficient way of carrying loads. The rigid arch is better able to carry variations in the design loading than is its block counterpart made of individual pieces. Arches of this type will be discussed in more detail in Chapter 5.

FLAT PLATES. A *flat plate* is a planar surface-forming structural element capable of carrying bending that can be used horizontally, vertically, or in an inclined way. When used horizontally, such as for a floor surface, it typically picks up loads that act transversely to its surface and transfers them horizontally by bending to the supports of the plate. When used vertically, it typically carries in-plane loadings. The surface-forming nature of flat-plate elements makes them very useful in situations where it is desirable that the enclosure surface of a building and its structure be one and the same.

Plates can be made in a variety of ways. Reinforced concrete can be used to form either vertical or horizontal surfaces. Vertical surfaces designed to carry vertical loads can also be made by stacking rigid blocks. Assemblies of this type (which are typically called *load-bearing walls*) are not, strictly speaking, plates since they can carry in-plane loads only and cannot be used in any situation where appreciable bending or bowing is induced in the structure by the external loading.

Horizontal plates can also be made by assembling patterns of short, rigid line elements. Three-dimensional triangulation schemes are used to impart stiffness to the resultant assembly. Plate structures will be explored in more detail in Chapter 10.

Long, narrow rigid plates can also be joined along their long edges and used to span horizontally in beamlike fashion. These structures, called *folded plates*, have the potential for spanning fairly large distances. Figure 1-4(f) illustrates a typical folded-plate structure. Folded plates are explored in detail in Chapter 10.

CYLINDRICAL SHELLS AND VAULTS. Cylindrical barrel shells and vaults are examples of *singly curved-plate* structures [see Figure 1-4(g)]. A barrel shell spans longitudinally such that the curve is perpendicular to the direction of the span. When fairly long, a barrel shell behaves much like a beam with a curved cross section. Barrel shells are invariably made of rigid materials (e.g., reinforced concrete or steel). A *vault*, by contrast, is a singly curved structure that spans transversely. A vault can be conceived of as basically a continuous arch. Vaults are typically made in much the same way as block arches and function similarly.

SPHERICAL SHELLS AND DOMES. A wide variety of doubly curved surface structures are in use. These include structures that are portions of spheres and those that form warped surfaces (e.g., the hyperbolic paraboloid). The number of shapes possible is actually boundless. Probably the most common doubly curved structure is the spherical shell made of reinforced concrete (a continuous rigid material). It is convenient to think of this structure as a rotated arch. This analogy, however, is actually misleading with respect to how the structure actually carries loads because of the fact that loadings induce circumferential forces in spherical shells which do not exist in arches. Exact differences will be discussed in Chapter 12. Domed structures can be made of stacked blocks instead of a continuous rigid material. Shells and domes are very efficient structures capable of spanning large distances using a minimum of material. Dome-shaped structures can also be made by forming short rigid line elements into repetitive patterns. The *geodesic dome* is a structure of this type.

CABLES. *Cables* are flexible structural elements (see Figure 1-5). The shape they assume under a loading depends on the nature and magnitude of the load. When a cable is simply pulled on at either end, it assumes a straight shape. This type of cable is often called a *tie-rod*. When a cable is used to span between two points and carry an external point load or series of point loads, it deforms into a shape made up of a series of straight-line segments. When a continuous load is carried, the cable deforms into a continuously curving shape. The self-weight of the cable itself produces such a

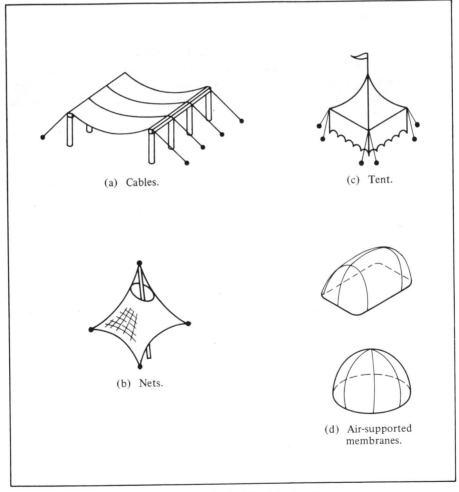

(a) Cables.

(c) Tent.

(b) Nets.

(d) Air-supported membranes.

FIGURE 1-5 Typical flexible structures.

curve (typically called a *catenary curve*). Other continuous loads produce curves that are similar in appearance but not exactly the same as the catenary curve.

There is a close relationship between the shape of a cable and the shape of an arch. When the loadings are similar, the shape that the cable assumes under the loading is the appropriate shape for the arch. One shape is an inversion of the other. Tension is developed in the cable and compression in the arch. Cables can be used to span extremely large distances. They are quite often used for bridges, where they support a road deck, which in turn carries the traffic loading. Since moving traffic loads would ordinarily cause the primary support cable to undergo shape changes as load positions changed, and with this changes in the shape of the road surface would occur, the horizontal bridge deck is made into a continuous rigid structure so that the road surface remains essentially flat and the load transferred to the primary support cables remains essentially constant. Cables are also used to support roof surfaces in buildings, particularly in long-span situations. Care must be taken, however, to keep the roof surface from fluttering because of alternating pressures and suctions caused by wind blowing over the roof (see Sections 3.2.3 and 5.2). Cable structures are discussed more extensively in Chapter 5.

MEMBRANES, TENTS, AND NETS.   A *membrane* is a thin, flexible sheet. A common *tent* is made of membrane surfaces. Both simple and complex forms can be created using membranes. For surfaces of double curvature, such as a spherical surface, however, the actual surface would have to be made as an assembly of much smaller segments, since most membrances are typically available only in flat sheets (a spherical surface is, of course, not developable). A further implication of using a flexible membrane to create the surface is that it either has to be suspended with the convex side pointing downward or, if used with the convex side pointed upward, supplemented by some mechanism to maintain its shape. Pneumatic or air-inflated structures are of the latter type. The shape of the membrane is maintained by the internal air pressure inside the structure. Another mechanism is to apply external jacking forces that stretch the membrane into the desired shape. Various stressed-skin structures are of this general type. The need to pretension the skin, however, imposes various limitations on the shape that can be formed. Spherical surfaces, for example, are very difficult to pretension by external jacking forces, while others, such as the hyperbolic paraboloid, can be handled with comparative ease.

Nets are three-dimensional surfaces made up of a series of crossed curved cables. Nets are very analogous to membrane skins. By allowing the mesh opening to vary as needed, a wide variety of surface shapes can be formed. An advantage of using crossed cables is that the positioning of the cables prevents the roof from fluttering due to wind suctions and pressures. In addition, tension forces are typically induced into the cables by jacking devices, so that the whole surface is turned into a type of stretched skin. This also gives the roof stability and resistance to flutter. Membranes and nets are more extensively discussed in Chapter 11.

1.2.3 Primary Structural Units

While many of the basic elements discussed in the previous section can indeed function in isolation as load-carrying structures, it is evident that some of them must be combined with others if the intent is to create a structure that encloses or forms a volume. It is in this respect that structures used in buildings are often distinct from those used for other purposes. Building structures are invariably volume-forming in nature; others are not necessarily so. Bridge structures, for example, are typically used to form or support linear surfaces.

In this context it is useful to introduce the general notion of a *primary structural unit*, which is simply a discrete volume-forming structural element or assembly of structural elements (see Fig. 1-2). It is a minimum structure appropriate for use in a building context and one that can be used either individually or in a repetitive way. Four columns supporting a rigid planar surface at its corners, for example, form a primary unit. Units of this type can be stacked and/or placed side by side to form a connected series of volumetric units. When placed side by side, columns are typically shared between units. Primary units are often an intermediate step between a series of discrete elements (e.g., beams and columns) and an entire building complex. The way discrete elements can be conceptually assembled into units and then aggregated often, but certainly not always, reflects the way building complexes are actually constructed.

The importance of thinking about structures in terms of units of this type is most apparent in preliminary design stages. The usefulness of the idea stems from the fact that the dimensions of a unit must invariably be related to the programmatic requirements of the building considered. Many buildings, for example, can be considered to be made up of a cellular aggregation of volumetric units of sizes related to the intended occupancy. Housing is an illustration of such a building type (see Figure 1-6). In this case the dimensions of the primary structural unit would be directly related to the functional dimensions of the housing units themselves. The primary structural unit, however, could be larger and encompass several functional units. It could not be smaller than the minimum functional subdivision of a unit. The point herein is simply that the dimensions of the primary structural unit are either the same as, or a multiple of, the critical functional dimensions associated with the building occupancy. This is a very simple but immensely valuable concept useful in early design stages that will be explored in more detail in Part III of the book.

Building types other than housing would have critical dimensions for the unit that are appropriate for the building type considered. In some cases, it should be noted, the building can be defined as essentially consisting of one large functional unit (e.g., a skating rink) and not consisting of an aggregation of cellular volumes. In this case, the primary structural unit must be equivalent to the entire structure.

As Figure 1-2 indicates, many of the basic structural elements are already volume forming in nature and could thus constitute a primary structural unit, while others are not. Line elements, for example, must always be assembled in some way to make volume-forming structures. Rigid surface elements must also be typically used in conjunction with these line elements if the structure is to be useful in a building context. Even when the basic element is already volume-forming, the making of a viable structure typically requires the use of other elements. As will be discussed in Chapter 12, for example, a domed structure (which is obviously volume-enclosing) cannot normally be used alone but requires the presence of a surrounding tension ring to contain thrusts generated by the action of loads on the structure. These facets will be explored in more detail later. The structural behavior of such combined units depends upon the interactive behavior of the discrete elements.

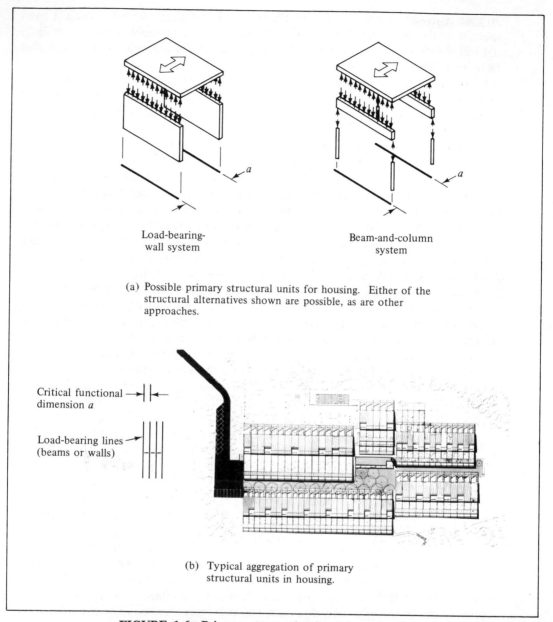

Load-bearing-
wall system

Beam-and-column
system

(a) Possible primary structural units for housing. Either of the
structural alternatives shown are possible, as are other
approaches.

Critical functional
dimension a

Load-bearing lines
(beams or walls)

(b) Typical aggregation of primary
structural units in housing.

FIGURE 1-6 Primary structural units and aggregations in housing.

Although a topic outside this coverage, it should be noted that there are
general approaches to the way elements are assembled to form units that could be
stated and classified in an explicit way. At this point, however, it is only the idea that
is of primary importance. Figure 1-2 illustrates several primary structural units.
Clearly, this is a sampling only, since there are a huge number of ways that the slate
of basic elements could be combined to form units.

1.2.4 Aggregations

As noted above, primary structural units can be combined to form larger complexes that have essentially a repetitive nature. In this context it is crucial to note that the nature of the aggregation possible is strongly influenced by the physical characteristics, including the geometry and rigidity, of the primary structural unit used.

As is evident from Figure 1-2, rectilinear units formed from rigid pieces are highly versatile. Aggregations of virtually any character can be made from units of this type. It should be noted that there are ways of characterizing and classifying these aggregations (e.g., linear aggregations, planar aggregations). The character of other units, however, limits the ways in which aggregations can be made. Geometrical considerations, in particular, strongly influence the ways units involving curved surfaces can be aggregated. Vertical aggregations involving curved surfaces, for example, tend to be very difficult. The same is true for units involving flexible surface elements. These units can only be assembled in specific ways if the result is to have any practical applicability as a building structure. As a consequence of these limitations, aggregations involving flexible structural elements typically occur only in a horizontal plane, with the flexible elements providing roof surfaces only. In cases of this type it is often more efficient to enclose an area by using one large structural unit than to aggregate several smaller ones (as will be discussed later, flexible structures often have long-span capabilities). Figure 1-2 illustrates a sampling of several types of aggregations. Figure 1-6(b) illustrates a series of rectilinear units aggregated to form the structure for a housing development.

1.3 BASIC ISSUES IN THE ANALYSIS AND DESIGN OF STRUCTURES

1.3.1 Introduction

The method for classifying structures discussed in the previous section was founded on basic considerations relating to the geometrical form of structural elements or assemblies and to their primary physical properties. Critically, this descriptive classification system does not address or reflect any of the relations that must exist between the constituent parts of a structural assembly so that the whole structure functions as an entity. Structures were reflected simply as aggregations of elements. A real structure, however, must function as a whole in carrying loads to the ground. There is nothing implicit in any aggregation of elements that ensures that the structure is capable of doing this. The need for a structure to function as an entity in carrying loads is best illustrated through an investigation of the basic *stability* of a structure under load. The idea of stability is addressed in the following section.

Another limiting aspect of the descriptive classification discussed in the previous section is that it does not reflect many of the distinctions that are often of primary importance in designing a structure. With respect to the structures described, for example, the classification does not address distinctions that might (and do) exist in the basic internal load-carrying mechanisms involved. Considerations of exactly how a structure or structural element carries a load are among the basic tools used by a designer in developing a structure for use in a specific context or by the analyst concerned with determining if a given structure is acceptable for use in a particular situation. The issue of stability alluded to earlier forms part of these considerations, but a host of other issues are also involved. The sections following the one on stability provide an introduction to the fundamental ways a structure or structural element carries a load. Another classification method for structures based on concepts developed in these sections will then be discussed.

1.3.2 Structural Stability

A fundamental consideration in designing a structure is that of assuring its *stability* under any type of possible loading condition. All structures undergo some shape changes under load. In a stable structure the deformations induced by the load are typically small, and internal forces are generated in the structure by the action of the load that tend to restore the structure to its original shape after the load has been removed. In an unstable structure, the deformations induced by a load are typically massive and often tend to continue to increase as long as the load is applied. An unstable structure does not generate internal forces that tend to restore the structure to its original configuration. Unstable structures quite often collapse completely and instantaneously as a load is applied to them. It is the fundamental responsibility of the structural designer to assure that a proposed structure does indeed form a stable configuration.

Stability is very often a crucial issue in the design of structures that are assemblies of discrete elements. In theory, for example, the post-and-beam structure illustrated in Figure 1-7(a) is apparently stable. Any slight change in the loading condition such that a horizontal force on the structure is developed, however, tends to cause deformations of the type indicated in Figure 1-7(b). Clearly, the structure has no capacity to resist horizontal loads, nor does it have any mechanism that tends to restore it to its initial shape after the horizontal load is removed. The large changes in angle that occur between members characterize an unstable structure that is beginning to collapse. This particular structure will collapse almost instantaneously under load. Consequently, this particular pattern of members is referred to as a *collapse mechanism*. Other patterns and relations between members will also form collapse mechanisms.

There are really only a few fundamental ways of converting a self-standing structure of the general type shown in Figure 1-7(b) from an unstable to a stable configuration. These are illustrated in Figure 1-7(d). The first is to add a *diagonal member* to the structure. The structure cannot now undergo the parallelograming indicated in Figure 1-7(b) without a dramatic increase in the length of the diagonal member (this would not occur if the diagonal were adequately sized to take the forces involved). Another method used to assure stability is through *shear walls*. These are rigid planar surface elements that inherently resist shape changes of the type illustrated. A reinforced concrete or masonry wall can be used as a shear wall. Either a full or a partial wall can be used (the required extent of a partial wall depends on the magnitudes of the forces involved). A final method used to achieve stability is through stopping the large angular changes between members that is associated with collapse by assuring that the nature of the connections between members is such that their angular rela-

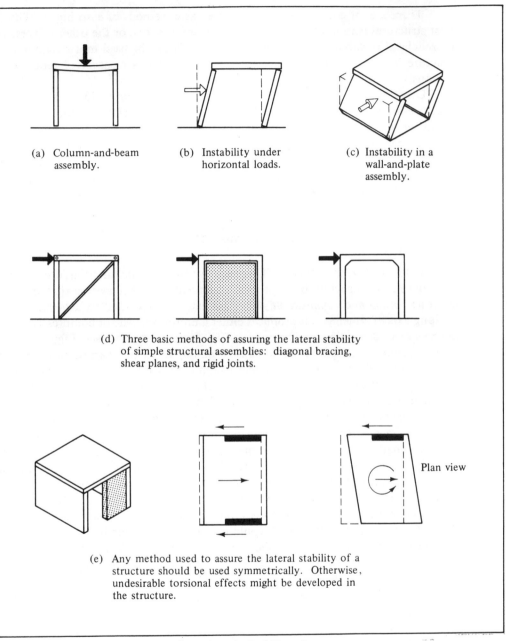

(a) Column-and-beam assembly.

(b) Instability under horizontal loads.

(c) Instability in a wall-and-plate assembly.

(d) Three basic methods of assuring the lateral stability of simple structural assemblies: diagonal bracing, shear planes, and rigid joints.

Plan view

(e) Any method used to assure the lateral stability of a structure should be used symmetrically. Otherwise, undesirable torsional effects might be developed in the structure.

FIGURE 1-7 Stability of structures.

tionship remains a constant value under any loading. This is done by making a *rigid joint* between members. This is a very common form of joint. A typical table, for example, is a stable structure because there is a rigid joint between each table leg and the top that maintains a constant angular relationship between the elements. Structures using rigid joints to assure stability are often referred to as *frames*.

There are, of course, variants on these basic methods of assuring stability. Still most structures composed of discrete elements rely on one or the other of these basic approaches for stability. More than one approach can be used in a structure (e.g., a structure having both rigid joints and a diagonal), but a measure of redundancy is obviously involved.

Questions of stability arise in other situations than simply in connection with a structure made up of an assembly of pieces and are often related to the behavior of a specific element under load. As will be discussed in the next section, a long compressive member can be potentially unstable under load. Similarly, part of a shell surface (e.g., a structure that consists of a portion of a sphere) can also be potentially unstable under load (the whole surface can suddenly snap through). These and other issues related to the stability of a structure under load will be elaborated upon in following chapters.

1.3.3 Internal Force States: Tension, Compression, and Bending

FUNDAMENTAL FORCE STATES.   There are two fundamental internal force states that are induced in a structure by the action of an external force system that are of basic interest: *tension* and *compression*. When an external force system acts exactly along the long axis of a member, it produces either a uniform tensile or compressive force in the member, depending on the sense of the external forces (see Figure 1-8). The general action of these forces is to cause a pulling apart or crushing of the material, depending on whether tension or compression is present. The load-carrying capacity of a tension member generally depends on the type of material used and the cross-sectional area of the member. The same factors are important vis-à-vis the load-carrying capacity of a compression member. In addition, however, the load-carrying capacity of a relatively long compression member tends to decrease with increasing member lengths. Long compression members tend to become unstable under the action of the load and to snap out suddenly from under the load at a certain critical load level. This sudden inability to carry additional load typically occurs without evident material distress. Once it occurs, however, the structure remains in a deformed state, since it cannot generate internal forces that tend to restore the structure to its initial shape. Continued increase in load would finally cause the member to fail by bending. The phenomenon described is called *buckling*. Because of this phenomenon, long compression members are not capable of carrying very high loads. This contrasts markedly with the behavior of members in tension. The length of a tension member does not significantly influence its load-carrying capacity. These and other concepts are explored in more depth in Chapter 7.

There is another type of force state involving a combination of internal tensile and compressive forces that is of basic interest. When a member carries external loads that are applied transversely to the long axis of the member (rather than along the axis of the member), the action of the external forces is to tend to cause the member to bow. As the member is bowed under the action of the loads, it deforms in the manner illustrated in Figure 1-8(d). In order for this type of deformation to be possible, it is evident that some fibers in the member must stretch, while others must compress. In Figure 1-8(d) it is also evident that both stretching and compressing of

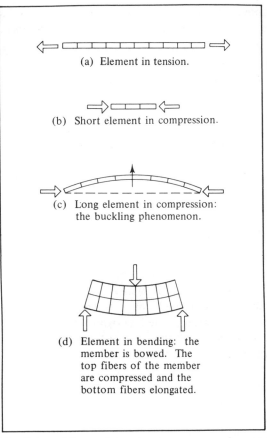

(a) Element in tension.

(b) Short element in compression.

(c) Long element in compression: the buckling phenomenon.

(d) Element in bending: the member is bowed. The top fibers of the member are compressed and the bottom fibers elongated.

FIGURE 1-8 Internal force states: tension, compression, and bending.

member fibers occur at the same cross sections. Associated with this phenomenon are internal tensile and compressive forces. A member that carries transverse forces, undergoes a bowing because of them, and consequently develops internal tensile and compressive forces at the same cross section is said to be in *bending*. The common beam is typically subject to bending.

TENSION, COMPRESSION, AND BENDING: A COMPARISON. In the preceding discussions it was not stressed that members are more sensitive to some force states than to others. A member in pure tension, for example, can carry an enormous tensile load relative to the size of the member. A short compressive member can also carry high loads relative to its size. As noted, however, the load-carrying capacity of a compressive member is reduced as its length is increased. A member subject to bending can carry only a surprisingly small amount of load relative to the size of the member and in comparison to a member carrying purely tensile forces. At this stage these relative load-carrying capacities cannot be quantified, but the differences should be intuitively obvious. A small, slender stick, for example, can quite easily be broken by hand by snapping it (loading it transversely to its axis). It can also be broken by hand by compressing it, but this is usually more difficult, depending on the length of the stick. It is almost impossible, however, to simply pull a stick of any reasonable size apart by hand. The analogy can be applied to structural members as well.

GENERAL DESIGN PRINCIPLES. Taking cognizance of differences in the load-carrying capacities of members subject to tension, compression, or bending is fundamental to designing an efficient structure. A general structural design objective is often to minimize bending in structures. Techniques for doing this vary, but the principle is always the same. Another design objective is to minimize the use of long compressive elements. Preferred elements are pure tension members or short compressive members. Yet another design principle is the appropriate matching of the type of force state present with the proper choice of materials so that the inherent characteristics of the material can be fully utilized.

1.3.4 One-Way and Two-Way Systems

A very basic way of distinguishing among structures is according to the spatial organization of the system of support used and the relation of the structure to the points of support available. The two primary cases of importance here are one-way and two-way systems. In a *one-way system* the basic load-transfer mechanism of the structure for channeling external loads to the ground acts in one direction only. In a *two-way system* the direction of the load-transfer mechanism is more complex but always involves at least two directions. A linear beam spanning between two support points is an example of a one-way system [see Figure 1-9(a)]. A system of two crossed elements resting on two sets of support points not lying on the same line and in which both elements share in carrying any external load is an example of a two-way system. A square, flat rigid plate resting on four continuous edge supports is also a two-way system. An external load cannot be simplistically assumed to travel to a pair of the supports in one direction only.

The distinction between one-way and two-way structural action is of primary importance in a design context. As will be discussed in more detail later, there are situations typically involving certain patterns in the support system used that often lead to certain advantages (in terms of the efficient use of materials) in using a two-way system as compared to a one-way system. Other patterns in the support system, however, often lead to the converse result. For this reason it is useful even at this early stage to begin distinguishing between one-way and two-way systems.

1.4 STRUCTURAL ASSEMBLIES

1.4.1 Cellular Structures

Many complete structures can be characterized as primarily carrying loads by tension, compression, or bending. More typically, however, most whole structures are composed of elements involving various combinations of these force states.

Figure 1-10 illustrates several of the many structural options possible for forming a simple single volumetric cell. The horizontal spanning system can be composed of a series of one-way beams (or trusses) and planar elements arranged in any

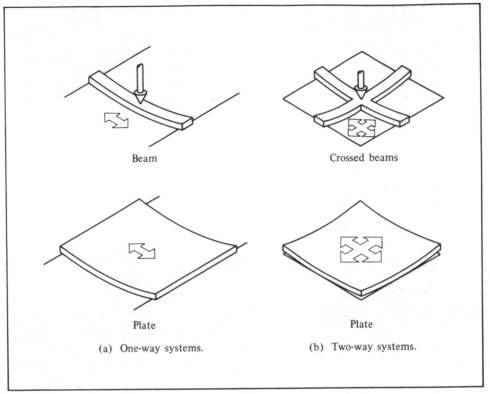

Beam	Crossed beams
Plate	Plate
(a) One-way systems.	(b) Two-way systems.

FIGURE 1-9   One-way and two-way structural systems.

one of several ways or, alternatively, some type of rigid one- or two-way plate element. A hierarchical arrangement is usually present when one-way beam and planar elements are used. The horizontal spanning systems shown are in a state of bending when vertical loads act on the members.

The vertical support system for the same cell can be either load-bearing walls or some type of beam-and-column arrangement. Columns and load-bearing wall elements are primarily in compression, although some bending may also be present under some circumstances. It is absolutely necessary that one of the options previously discussed be used to assure that lateral stability is maintained (see Figure 1-10). The options selected may influence whether a particular member is or is not subject to bending. Frame action, for example, invariably introduces bending in all members (see Chapter 9).

Structures more complex than shown can usually be decomposed into a series of more basic elements subjected to one or more of the primary force states discussed. Techniques for making decompositions of this type will be discussed in Chapter 3.

1.4.2 Design Trade-Offs

Designing or selecting an appropriate structural option for a given case can often involve many trade-offs. Assume that some type of diagonal system was to be designed to be used to provide lateral stability for the cellular structure previously discussed

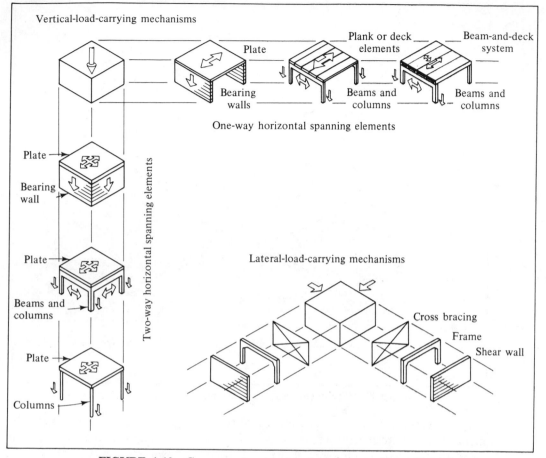

FIGURE 1-10 Common structural options for a typical volumetric unit.

(see Figures 1-10 and 1-11). As Figure 1-11 indicates, whether tension or compression forces are developed in any diagonal element used depends upon its orientation. The designer can opt to orient the diagonal such that only tension forces are developed in the member. A cable could therefore be used. Alternatively, the diagonal could be oriented such that compressive forces develop in the member. A rigid element capable of carrying compression would therefore be needed. In line with the general design principles discussed in Section 1.3.3, it would seem that the first alternative of using a cable is preferable. The designer must be careful, however, to make sure that the element functions as a stabilizer under all possible loading conditions. Evidently, if the applied load were reversed, the structure would collapse if only a single cable were used, since the cable could not develop the compressive force resistance needed to stabilize the structure under the new loading. A crossed-cable system of the type illustrated in Figure 1-11 would be needed if cables are to be used at all in situations involving the possiblity of load reversals. Alternatively, the designer could opt to use a single rigid diagonal capable of developing either tension or compression forces, which could thus stabilize the structure under any type of loading. The designer is

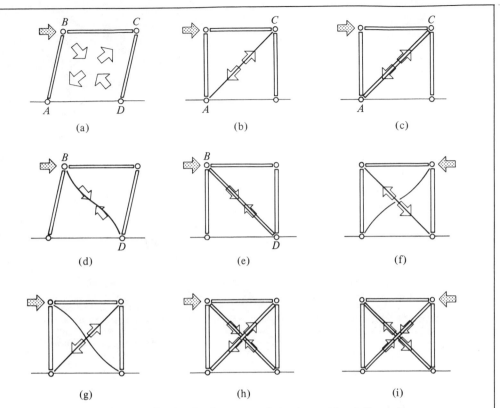

(a) Deformed shape of a simple pin-connected structure without diagonals. The original distance between points A and C tends to be increased and that between B and D tends to be decreased.

(b) A cable placed between points A and C will have tension forces developed in it because it resists the tendency of the two points to separate. The cable will stabilize the structure and keep it from collapsing.

(c) A rigid element placed between A and C will serve the same function as a cable. Tension forces will be developed in the member.

(d) Placing a cable between points B and D is useless in keeping the structure from collapsing. The points move toward one another. A cable placed between these two points would simply buckle out of the way. The same thing would happen to the cable in (b) if the direction of the load were reversed.

(e) Placing a rigid element between points B and D can stabilize the structure. Compressive forces would be developed in the member.

(f) In order to completely stabilize the structure with respect to loads
(g) from either direction, by using cables, it is necessary to use a crossed-cable system. Under a specific loading, one cable will operate effectively in stabilizing the structure, while the other simply buckles out of the way. Reversing the loading causes a reversal in which member operates effectively.

(h) Crossed rigid elements can also be used, but a certain degree of
(i) redundancy is involved, since a single diagonal is capable of stabilizing the structure alone with respect to loads in either direction.

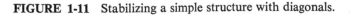

FIGURE 1-11 Stabilizing a simple structure with diagonals.

Funicular line

(a) Structure in tension:
the shape assumed by
a flexible cable under
an applied loading is
called the funicular
shape for that loading.

(b) Structure in tension:
Funicular shape for a
uniformly distributed
loading.

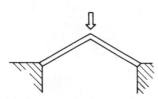

(c) Structure in compression:
inverting the shape assumed
by a flexible cable under an
applied loading yields a shape
that carries the load by axial
compression only.

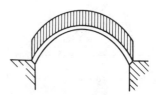

(d) Structure in compression:
inverting the tension funicular
yields a shape that carries loads
by axial compression. The
common "arch," shown above
is thus one type of funicular
structure.

(e) Structure in bending:
a structure that is shaped
to be funicular for a concentrated load at midspan becomes
subject to bending when the
loading changes.

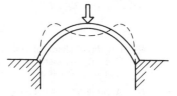

(f) Structure in bending:
a funicular structure for a
uniformly distributed load
becomes subject to bending
when the loading changes.

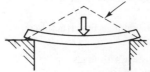

Funicular line

(g) Structure in bending:
when the shape of the structure
does not correspond to the funicular shape for the loading,
bending invariably develops.

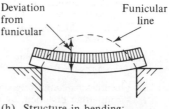

Deviation
from
funicular

Funicular
line

(h) Structure in bending:
the deviation at any point
between the structure and the
funicular line for the loading
generally reflects the amount
of bending present in the
structure at that point.

FIGURE 1-12 Funicular and nonfunicular structural shapes.

consequently faced with a trade-off of using either a single rigid diagonal or two crossed cables. It is not immediately clear which solution is preferable. Only by carrying the designs to a further stage of development and evaluating them according to some prespecified criterion (e.g., volume of material required or cost) could a statement of preferability be made. Trade-offs of this type are commonly encountered in structural design.

1.5 FUNICULAR STRUCTURES

1.5.1 Basic Characteristics

Many whole structures can be characterized as being primarily in a state of pure tension or compression. These are interesting structures deserving special treatment. Consider a simple flexible cable spanning between two points and carrying a load. This structure must be exclusively in a state of tension, since a flexible cable can withstand neither compression nor bending. A cable carrying a concentrated load at midspan would deform as indicated in Figure 1-12(a). The whole structure is in tension. If this exact shape were simply inverted and loaded precisely in the same way, it is evident by analogy that the resultant structure would be in a state of pure compression. If the loading condition is changed to a continuous load, a flexible cable carrying this load would naturally deform into the parabolic shape indicated in Figure 1-12(b). Again, the whole structure is in tension. If this exact shape were inverted and loaded with the same continuous load, the resultant structure would be in a state of compression [Figure 1-11(d)]. The common arch is predominantly a structure of this type.

Structures wherein only a state of tension or compression is induced by the loading are referred to as *funicular structures*. It is interesting to note that despite the fact that loads are applied transversely to the length of the members, as typically occurs in a beam, only tension or compression exists in these structures—not bending.

Why a linear beam is in bending and a cable is not is, of course, partly a function of the material used. The cable cannot withstand bending; therefore, it deforms under the action of the load. The rigid linear beam can take bending and therefore resists being deformed. Material alone, however, does not explain why some structures are funicular and others are not. The arch, which is predominantly in a state of compression, is often made from rigid materials.

Obviously, the *shape* of the structure is the prime determinant of whether a structure is in pure tension or compression or is subject to bending.

The importance of shape is clear from looking at Figure 1-12. The peaked linear form that was in a state of pure compression when carrying the single concentrated load becomes subject to bending when the loading is changed to a continuous one [see Figure 1-12(e)]. Similarly, the parabolic form that was in pure compression under a continuous loading becomes subject to bending when a concentrated load is applied to the structure [see Figure 1-12(f)]. Evidently, there must be a fixed relationship between shape and loading if the structure is to be a funicular one and carrying loads by either pure tension or compression.

The easiest way to determine the funicular response for a particular loading condition is by determining the exact shape a flexible string would deform to under the load. This is the tension funicular. Inverting this shape exactly yields a compression funicular. There is only one funicular shape for a given loading condition. Bending would develop in any structure whose shape deviates from the funicular one for the given loading. Figure 1-13 illustrates an early, but nonetheless latter-day, analysis of a well-known structure that is based on the idea that an arch can be conceived of as an inverted catenary.

Some more general principles can be extracted from the simple examples considered thus far. Note that in a funicular structure, the shape of the funicular structure *always changes beneath an external load.* Where the structure is not loaded, the structure remains straight. The funicular shape appropriate for a continuous load must therefore change continuously. By a similar token, if the shape of the structure changes when there is no load change present, bending will be present. It is also interesting to note that if the shape of a funicular structure is simply imagined and superimposed on the actual structure considered, the amount of deviation of the actual structure from the funicular shape generally reflects the severity of the bending present in the

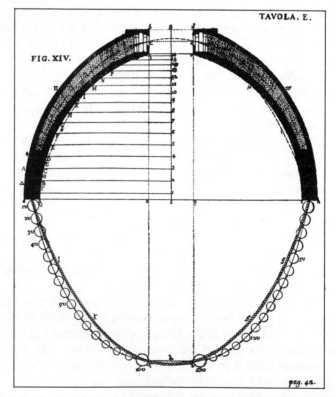

FIGURE 1-13 St. Peter's Dome, Rome. Construction of the thrust line as an inverted catenary. (From Poleni, *Memorie istoriche della Gran Cupola del Tempio Vaticano*, 1748.)

actual structure [see Figure 1-12(g) and (h)]. These points will be returned to a
elaborated upon in later chapters.

While there is only one general shape of structure that is funicular for a giv
loading, there is invariably a family of structures having the same general shape f
any given condition. All the structures illustrated in Figure 1-14, for example, a
funicular for the loadings indicated. Clearly, all structures in a group have the san
shape, but the physical dimensions are different. Within a family, the relative pro
portions of all the shapes are identical. For Figure 1-14 it is obvious that such a famil
could be obtained simply by using a series of flexible cables of different lengths. A
would deform in a similar way under the action of the load, but the actual amoun
that the structure would sag would be different. The magnitude of the internal force
developed in the members varies with the structural depth present (large forces are
developed in shallow shapes, and vice versa).

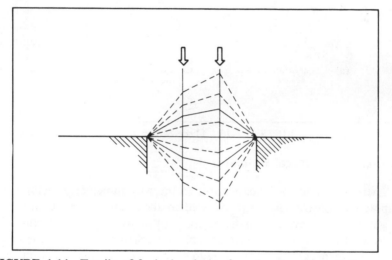

FIGURE 1-14 Family of funicular shapes for a loading. The relative proportions of all shapes are similar.

It is also evident that funiculars need not be only two-dimensional structures,
but can be three-dimensional as well (see Figure 1-15.) The shape a flexible membrane
assumes under a uniformly distributed load is of special interest. Since the funicular
response is clearly not spherical, but rather parabolic, it follows that the ubiquitous
spherical shell often commonly thought ideal as a structural form for this type of
loading is, in fact, not exactly a funicular response for the vertical loads which commonly act on buildings. It is, however, for the type of loading indicated in Figure
1-15(c). This does not, however, limit its usefulness, since there are other mechanisms
at work (circumferential forces) in a shell structure of this type that still cause the
structure to be highly efficient. These mechanisms will be discussed in Chapter 12.

Figure 1-16 illustrates a well-known example of the application of the idea of
three-dimensional funicular shapes as the basis for determining the form of a structure.

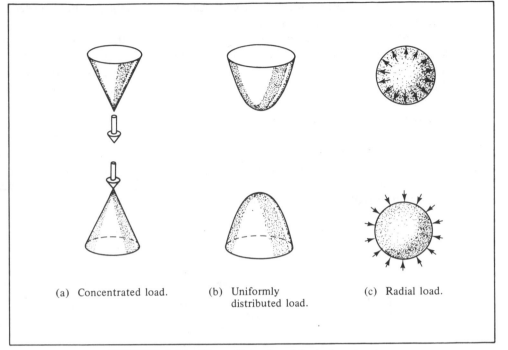

(a) Concentrated load.   (b) Uniformly distributed load.   (c) Radial load.

FIGURE 1-15 Three-dimensional funicular structures.

1.5.2 Structural Behavior

There are many variants possible in the way funicular structures are used in actual practice. Clearly, they can be used in their pure forms as arches or cables. More generally, however, the basic behavior of many different structural forms can be described in terms of funicular action. These structures can consequently be viewed as special forms of funicular structures. Some basic transformations of simple funicular shapes into other structural types are illustrated in Figure 1-17.

Figure 1-17(a) illustrates basic funicular responses for a concentrated load and a uniformly distributed loading. Figure 1-17(b) illustrates the type of movement that can be expected at the support points if the structure were not constrained at these points. A tension funicular typically wants to pull inward and downward. The support or foundation system must apply an outward and upward force on the structure at each end to maintain its shape and to keep it from collapsing. These forces are reactive in nature and are thus commonly referred to simply as *reactions* [see Figure 1-17(c)]. Note that the reactions exert an equal and opposite pull on the structure. The applied forces are thus balanced by equal and opposite reactive forces. A compression funicular produces a converse effect. The structure wants to move outward and downward. There is thus an outward and downward force exerted on the foundation which must in turn exert an equal and opposite set of forces on the structure if the shape of a structure is to be maintained. The combinations of applied forces acting on the foundation are commonly called thrusts. The foundation must contain these thrusts.

The final or resultant direction of the thrusts associated with a funicular struc-

(a) Inverted photograph of the funicular model for the Colonia Guel chapel by Antonio Gaudi.

(b) Sketch of the exterior of the chapel drawn from an inverted photograph of the model.

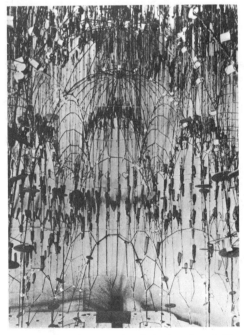

(c) Inverted photograph of the interior of the model.

(d) Companion sketch for the interior of the chapel.

FIGURE 1-16 Use of three-dimensional funicular models. The suspended weights are carefully modeled to represent the actual weights involved.

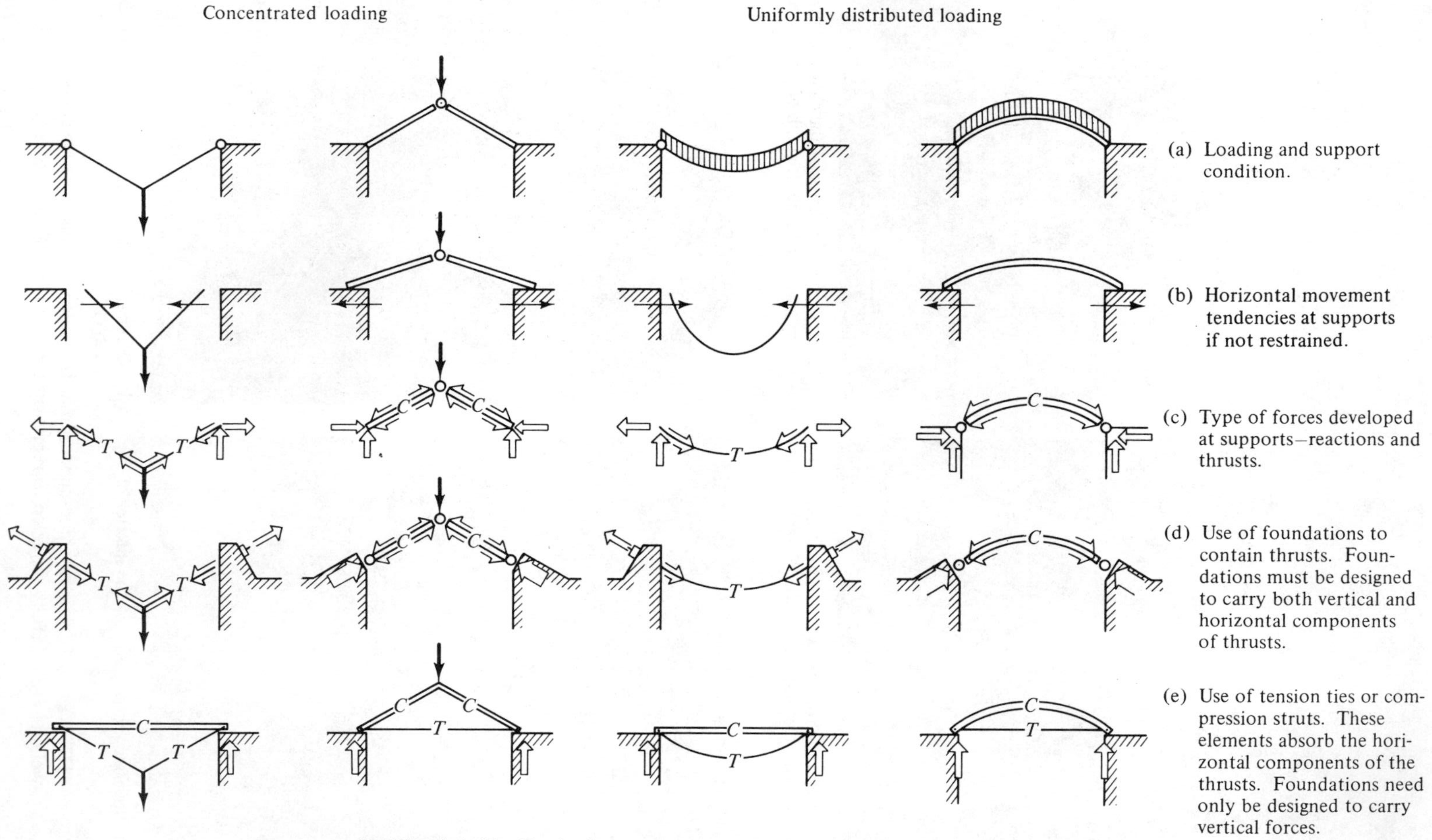

Uniformly distributed loading

Concentrated loading

(a) Loading and support condition.

(b) Horizontal movement tendencies at supports if not restrained.

(c) Type of forces developed at supports—reactions and thrusts.

(d) Use of foundations to contain thrusts. Foundations must be designed to carry both vertical and horizontal components of thrusts.

(e) Use of tension ties or compression struts. These elements absorb the horizontal components of the thrusts. Foundations need only be designed to carry vertical forces.

FIGURE 1-17. Funicular structures. Transformations derived from basic shapes.

32

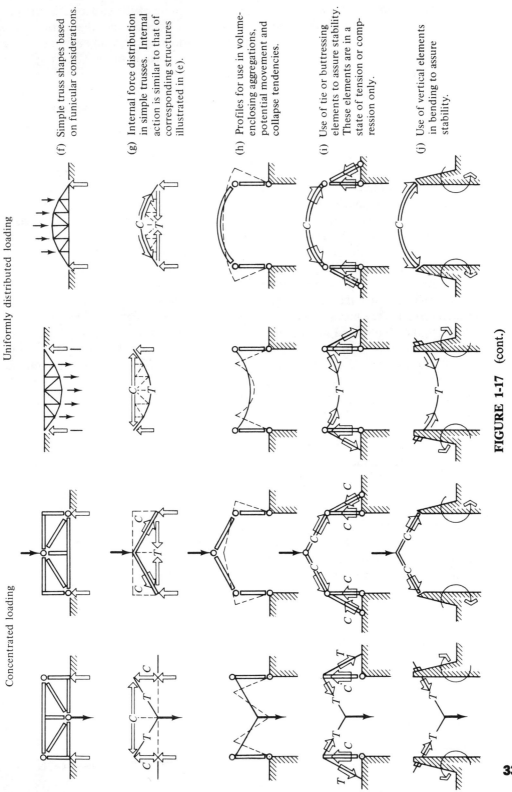

Uniformly distributed loading

Concentrated loading

(f) Simple truss shapes based on funicular considerations.

(g) Internal force distribution in simple trusses. Internal action is similar to that of corresponding structures illustrated in (e).

(h) Profiles for use in volume-enclosing aggregations, potential movement and collapse tendencies.

(i) Use of tie or buttressing elements to assure stability. These elements are in a state of tension or compression only.

(j) Use of vertical elements in bending to assure stability.

FIGURE 1-17 (cont.)

33

ture can be shown to be along the tangent to the slope of the structure at the point where the structure meets the support. This follows from the fact that since bending is not present in the structure, all internal forces are directed axially along the length of the member. As Figure 1-17(d) indicates, this can have important bearings on the design and shaping of the foundation structure.

The use of built-in massive foundations is not the only way of handling the thrusts developed in a funicular structure. Part of the thrusts can be absorbed into the structure itself by adding an additional member. Figure 1-17(e) illustrates doing this for several structures. In tension funiculars, the tendency of the two ends to move inward can be restrained by simply introducing a linear member capable of carrying a compressive force (a strut) between the two points. This would have the net result of relieving the foundation of the necessity of providing restraint to the inward pulling forces of the structure. Hence, the foundation could be designed to carry only vertical forces. The same is true for compressive funicular structures (e.g., arches). By simply tying the two ends of an arch together, the outward-spreading tendency of the structure is eliminated and the foundation can be designed to carry vertical forces only. A tension force would develop in the tie-rod connecting the ends of the arch.

There are several points of interest in connection with the preceding discussion. One is that the force that must be carried by the compressive strut or tension tie-rod is exactly equal to the horizontal component of the total force developed by the structure at the foundation. Using struts or ties does not eliminate this force but merely handles it in a different way. In some cricumstances it might be preferable to use a strut or tie rather than absorbing the horizontal thrust by means of a massive foundation. Designing and building foundations capable of handling horizontal forces is no easy matter. It is relatively easy, however, to design and build a foundation capable of carrying vertical forces only. For this reason ties or struts are frequently used. Ties, in particular, are quite an efficient way of taking up the horizontal component of the thrust in a compression funicular, since they can be simply long tension members. Struts for cable structures are less desirable, since the element is a long member in compression and thus potentially susceptible to buckling.

The remaining illustrations in Figure 1-17 indicate further evolution of the basic funicular shapes considered into other forms. While it is not the intent to explore these transformations in detail, some general observations will be made.

In some triangulated configurations of linear members (i.e., trusses), the primary action of the structure can also be discussed in terms of funicular shapes. In the truss shown to the left in Figure 1-17(f), a close inspection reveals that the center diagonals function exactly like a cable carrying a concentrated load, and the member across the top as a compression element serving the function previously described. The two end verticals simply translate the whole assembly vertically upward and are thus in compression. Other truss configurations can be discussed in similar ways. The analogy, however, does not extend to all conceivable shapes of triangulated bar networks. More sophisticated methods of analysis for any type of truss will be discussed in Chapter 4.

Figures 1-17(h), 1-17(i), and 1-17(j) illustrate other transformations of the basic funicular structure into forms that could be aggregated to enclose volumes. Obviously, the method of taking up the horizontal thrusts generated by the structure is a crucial determinant of the exact characteristics of the structures developed. The assembly shown in Figure 1-17(i), which carries a uniformly distributed load and is arch-shaped, is a time-honored way of taking care of horizontal thrusts by simply adding buttressing elements. More preferable than the case shown would be lining up the buttressing elements with the slope of primary structure at the point of connection.

There are thus a wide variety of structures that are apparently different but are actually related in terms of their internal structural behavior.

1.6 OTHER CLASSIFICATIONS

As a way of summarizing some of the concepts previously discussed, a classification of structures that tends to reflect their internal structural behavior is illustrated in Figure 1-18. The nature of the loading and boundary conditions are clearly important, whereas they were not at all reflected in the classification previously discussed and illustrated in Figure 1-2. There are other differences, but also overlaps, as a close comparison of the two schemes will indicate.

The level of detail associated with the classification scheme presented in Figure 1-18 is clearly limited. Many of the entries could be expanded into a categorization of their possible transformations of the general type previously discussed and illustrated in Figure 1-17. Still, it is useful as a conceptual device.

The two methods of classification thus far discussed are not the only ways structures could be classified. By selecting other criteria for classification, a wide range of schemes could be developed. Another might be based on the relative complexity of the internal force states in the structure. Curiously, many of the structures having the longest span capability, such as cables, have internal force states far easier to describe and predict than do many ubiquitous short-span elements, (e.g., a wood-joist flooring system). This type of ordering would involve distinctions between funicular actions and bending as before, but be organized toward a different objective.

Other classification schemes could be developed on the basis of span or load-carrying capabilities. Those factors will be considered later in this text and comparisons drawn in the final chapters. A highly interesting way of classifying structures would be according to energy considerations. Although a topic not yet discussed, energy is stored in a structure as work is done on it (in the form of applied loads). Considerations of this type would be used to characterize structures into categories such as low-energy structures or high-energy structures.

These and other classification schemes are well worth reviewing or delving into as a way of developing a general appreciation for both the wide variety of structures in use and the ways in which they are interrelated. Many of the more important relations, however, are based on a detailed understanding of the force distributions present in structures. In Chapter 2 we begin to develop an understanding of the forces in structures by first simply looking at the characteristics of forces in general. Later chapters will demonstrate how the ideas developed in Chapter 2 can be applied as tools for analyzing and designing structures.

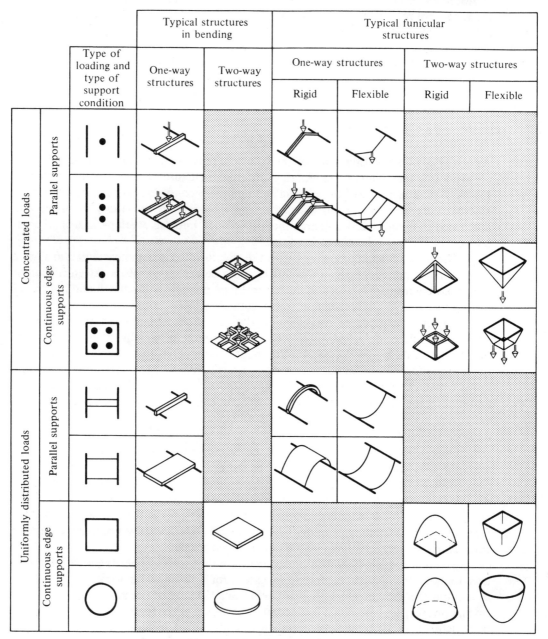

FIGURE 1-18 Classification of typical structures according to their basic load-carrying actions.

QUESTIONS

1-1. Consider the structure shown in Figure 1-19, which consists of a heavy rigid plate supported by four columns. Assume that all connections are nonrigid. If the diagonal members $A'B$, $B'C$, $C'D$, and $D'A$ are rigid members capable of carrying either tension or compression, is the assembly a stable configuration of elements? Why? If the diagonal members are all cable elements rather than rods, is the assembly stable? Why? If not, can additional cables be added to convert the assembly into a stable structure? If so, where? In all of the above, assume that the structure can be subjected to horizontal loads acting at any point on the rigid plate and in any direction.

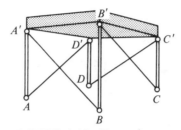

FIGURE 1-19 Example.

1-2. Determine how the load-carrying capacity of a long slender column varies with length; for example, as the length of a column is doubled, what happens to its load-carrying capacity? Do this by measuring the load required to buckle a series of slender compression elements of similar cross sections but varying lengths. Square basswood pieces ($\frac{1}{8}$ in. $\times \frac{1}{8}$ in.) are recommended. Simply support the ends without restraining rotations. Concentrate on long pieces only (as pieces get shorter, the failure load is more difficult to measure). Graphically plot your results (buckling load versus column length).

1-3. Take a thin sheet of rubber and inscribe a grid on it. Subject the sheet to different loading conditions and support types. Minimally, do the following:
(a) Take a square sheet, continuously support its edges, and pile sand on it.
(b) Repeat, except support the sheet at corners only.
(c) Repeat (a) and (b), using a concentrated load at midspan instead.
Note how the surface deforms in each case by studying how elemental parts of the grid deform. Identify the most highly deformed areas in each case (regions of highest stress). Sketch the deflected shapes obtained.

1-4. Consider Figure 1-16, which illustrates the hanging-weight model used by Gaudi for the Colonia Güell chapel. Were similar model analyses used for other Gaudi works? Was the structure actually built? Is the chapel illustrated a singular example of Gaudi's concern for structure, or are there other examples? Consult your library.

chapter 2

PRINCIPLES
OF MECHANICS

2.1 INTRODUCTION

Mechanics is the branch of applied science dealing with forces and motions. Fundamental to the field is the notion of *equilibrium*, the condition existing when a system of forces acting on a body is in a state of balance. The term *statics* is used to describe the part of mechanics specifically concerned with relations between forces acting on rigid bodies that are in equilibrium and at rest. The term *dynamics* refers to the part of mechanics dealing with rigid bodies in motion. If correctly placed inertial forces are taken into account, bodies in motion can also be considered to be in equilibrium. The field of study usually referred to as *strength of materials* is an extension of mechanics that addresses the relationship between applied or external forces acting on a body and the internal effects produced by these forces in the body. The study of the deformations produced in a body by a set of external forces is an integral part of the field of strength of materials.

The distinctions made above reflect the way the study of the general subject area has evolved in the engineering disciplines. In most engineering curricula, statics, dynamics, and strength of materials are normally treated as separate topics presented sequentially under the general umbrella of mechanics. During the process of analyzing and designing structures in buildings, however, professionals freely use as tools ideas and elements from each of these basic fields (as well as others) in a nonsequential manner. The chapters in Part II of this book adopt this more integrative approach. At this point, however, it is pedagogically useful to retain the traditional engineering distinctions.

The first part of this chapter will introduce certain fundamental ideas in statics. The second part will focus on basic elements of strength of materials. The field of dynamics is outside the scope of this text and is not addressed. The following presentation is intended to provide only an overview of the basic issues involved in statics and strength of materials. Topics are therefore succinctly presented. The reader is referred to any of a number of basic texts which treat the subject matter more elaborately.[1]

2.2 FORCES AND MOMENTS

2.2.1 Forces

Fundamental to the field of mechanics is the concept of force and the composition and resolution of forces. A *force* is a directed interaction between bodies. Force interactions have the effect of causing changes in the shape or motion, or both, of the bodies involved. The basic concept of force is undoubtedly familiar to the reader and is probably felt to be intuitively obvious. But viewed from a historical perspective, there was originally nothing obvious about the idea of force and the characterization of a force in terms of magnitude, sense, and direction. The precise formulation of these concepts is really quite a remarkable accomplishment in view of the degree of abstraction involved. Indeed, distinctions between force and weight as a special form of force and the notion of nonvertical forces were only just beginning to be appreciated by scholars in the Middle Ages. The name of Jordanus de Nemore is recurrently connected to the emergence of these concepts. Once force was conceived in vectorial (directional) terms, the problem of force components and the general composition and resolution of forces was addressed by a variety of individuals, including Leonardo da Vinci, Steven, Roberval, and Galilei. This problem, often termed *the* basic problem in statics, was finally solved by Varginon and Newton.

2.2.2 Scalar and Vector Quantities

A distinction is invariably made in the study of mechanics between scalar and vector quantities. *Scalar* quantities can be adequately characterized by magnitude alone. *Vector* quantities must be characterized in terms of both magnitude and direction. Forces are typically vector quantities. Any vector quantity can be represented by a line. The direction of the line with respect to a fixed axis denotes the direction of the quantity. The length of the line, if drawn to scale, represents the magnitude of the quantity [see Figure 2-1(a)].

A force is actually not completely specified by its magnitude and direction alone, since its point of application along its line of action can often have importance. The *line of action* of a force is a line of indefinite length of which the force vector is a

---

[1]For example, Ferdinand P. Beer and E. Russell Johnston, *Vector Mechanics for Engineers: Statics.* McGraw-Hill Book Company, New York, 1972.

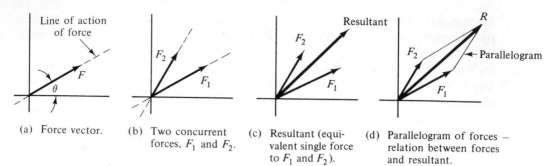

(a) Force vector.

(b) Two concurrent forces, F_1 and F_2.

(c) Resultant (equivalent single force to F_1 and F_2).

(d) Parallelogram of forces — relation between forces and resultant.

FIGURE 2-1 Force vectors, force interactions, resultant forces, and the parallelogram of forces.

segment. Since most structures that will be dealt with in this text are essentially rigid bodies that deform only slightly under the application of a force, it can be assumed that the point of application of a given force may be transferred to any other point on the line of action without altering the translatory or rotational effects of the force on the body. Thus a force applied to a rigid body may be regarded as acting *anywhere* along its line of action.

2.2.3 Parallelogram of Forces

Essential to a study of structural behavior is knowing the net result of the interaction of several vector forces acting on a body. This interaction can be studied in terms of the laws of vector addition. These laws and fundamental postulates are based on experimental observation. Historically, the first method of adding vector quantities was based on the *parallelogram law*. In terms of force vectors, this proposition states that when the lines of action of two forces intersect, there is a single force, or *resultant*, exactly equivalent to these two forces, which can be represented by the diagonal of the parallelogram formed by using the force vectors as sides of the parallelogram [see Figure 2-1(b)–(d)]. In general, a resultant force is the simplest force system to which a more complex set of forces may be reduced and still produce the same effect on the body acted upon.

A graphic technique for finding the resultant force of several force vectors whose lines of action intersect is illustrated in Figure 2-2. The individual vectors, drawn to scale, are joined in tip-to-tail fashion. The order of combination is not important. Unless the resultant force is zero, the force polygon thus formed does not

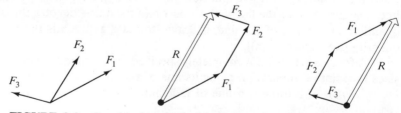

FIGURE 2-2 Graphical method of finding resultant forces for concurrent systems.

form a closed figure. The closure line is identical to the resultant force of the several individual vectors (i.e., the resultant is that vector which would extend from the tail of the first vector to the tip of the last vector in the group). The resultant closes the force polygon. This general technique follows from the parallelogram law. An algebraic method for finding the resultant of several forces acting through a point is discussed in Section 2.2.6.

Although conceptually very simple, graphical approaches to finding the resultants of force systems are extremely powerful as structural analysis aids. Historically, they found wide application because of the ease of their application. Graphical techniques were used extensively by early investigators in their attempts to understand the behavior of complex structures. Figure 2-3 illustrates a latter-day analysis of a gothic structure by graphic techniques. Figure 2-4 illustrates the *sphere model* of an arch that utilizes graphic techniques. The model, based simply on the parallelogram law, is still a very elegant way of looking at arches. Although going into either of these analyses in detail is beyond the scope of this text, it can be noted that the techniques

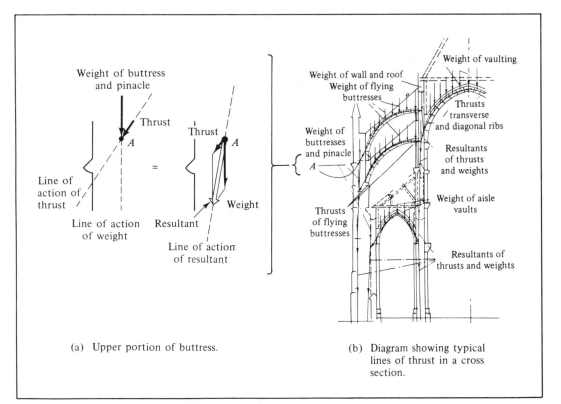

(a) Upper portion of buttress.

(b) Diagram showing typical lines of thrust in a cross section.

FIGURE 2-3 Application of graphic methods to the analysis of a gothic structure (Amiens). The dead weights of the vertical buttresses and pinacle help turn the thrusts of the flying buttresses downward through the middle portions of the buttresses themselves. The vertical buttresses are stable and not prone to overturning or cracking when the force resultants pass through the middle portions of these buttresses.

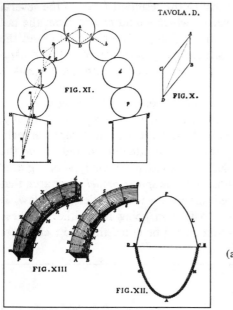

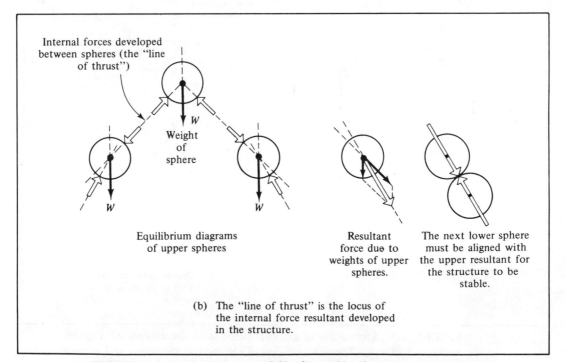

(a) From Poleni, *Memorie istoriche della Gran Cupola del Tempio Vaticano, 1748*: The spheres are arranged in accordance with the "line of thrust."

Internal forces developed between spheres (the "line of thrust")

W
Weight of sphere

W

W

Equilibrium diagrams of upper spheres

Resultant force due to weights of upper spheres.

The next lower sphere must be aligned with the upper resultant for the structure to be stable.

(b) The "line of thrust" is the locus of the internal force resultant developed in the structure.

FIGURE 2-4 Early "sphere model" of an arch: if a series of spheres are stacked as illustrated, the assembly obtained is stable. Note that the shape is the inversion of a freely hanging chain made of similar spheres. The assembly will collapse if either the loading or positioning of the spheres is changed.

used are extremely simple and based on the concepts discussed in this chapter. Graphical techniques are no longer extensively used because of the advent of improved quantitative methods and calculation aids such as the computer. Still, graphic techniques remain a very elegant way of looking at structures and are very useful in developing an intuitive feeling for the flow of forces in a structure.

2.2.4 Resolution and Composition of Forces

A process that follows directly from the fundamental proposition of the parallelogram of forces is that of breaking up a single force into two or more separate forces that form a force system equivalent to the initial force. This process is usually referred to as *resolving a force into components*. The number of components that a single force can be resolved into is limitless (see Figure 2-5). In structural analysis it is often most

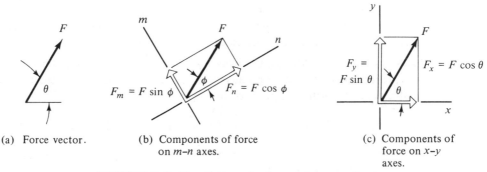

(a) Force vector.   (b) Components of force on *m–n* axes.   (c) Components of force on *x–y* axes.

FIGURE 2-5 Resolution of a force into components.

convenient to resolve a force into rectangular, or cartesian, components. By utilizing right angles, components can be found by using simple trigonometric functions. When a force F is resolved into components on the x and y axes, the components become $F_x = F(\cos \theta)$ and $F_y = F(\sin \theta)$. The process can be reversed if F_x and F_y are given and it is desired to know the resultant force F:

$$F = \sqrt{F_x^2 + F_y^2} \quad \text{and} \quad \theta = \tan^{-1}\left(\frac{F_y}{F_x}\right)$$

It should be stressed that using cartesian components is a matter of convenience only. A right angle is nothing more than a special form of a parallelogram. Figure 2-5 illustrates several manipulations with components. Components obviously could be found in three dimensions as well as two.

2.2.5 Moments

MOMENT OF A FORCE. A force applied to a body tends to cause the body to translate in the direction of the force. Depending on the point of application of the force on the body, the force may also tend to cause the body to rotate. This tendency to produce rotation is called the *moment* of the force (see Figure 2-6). With respect to a point or line, the magnitude of this turning or rotational tendency is equal to the product of the magnitude of the force and the perpendicular distance from the *line of action*

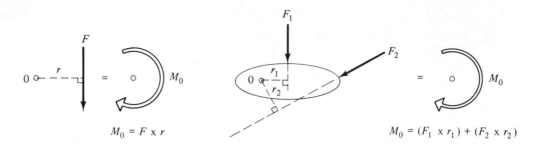

(a) Moment of a force about a point
(moment = force x distance).

(b) Moment of several forces about a point.

FIGURE 2-6 Moments.

of the force to the point or line under consideration. The moment M of a force F about a point 0 is thus simply $M_0 = F \times r$ where r is the perpendicular distance from the line of action of F to point 0. r is often called the *moment arm of the force*. A moment has the units of force times distance (e.g., ft-lb or N-m).

MOMENTS DUE TO MULTIPLE FORCES. The total rotational effect produced by several forces about the same point or line is merely the algebraic sum of their individual moments about that point or line. Thus,

$$M_0 = (F_1 \times r_1) + (F_2 \times r_2) + \ldots + (F_n \times r_n)$$

MOMENTS ABOUT A LINE. The rotational effect on a rigid body caused by multiple forces acting about a line, but not in the same plane, is the same as that which would result if all of the forces were acting in the same plane.

MOMENT OF A COUPLE. A *couple* is a force system made up of two forces equal in magnitude but opposite in sense and having parallel lines of action that are not on the same straight line ($\rightleftharpoons$). A couple tends to cause only rotational effects on bodies and does not cause translation. The moment of a couple is simply the product of one of the forces times the perpendicular distance between the two forces. It can be shown, however, that the moment of a couple is actually independent of the reference point selected as a moment center. The magnitude of the rotational effect produced by a couple on a body is also independent of the point of application of the couple on the body.

MOMENT OF A DISTRIBUTED LOAD. In structural analysis it is quite often necessary to determine the moment of some sort of continuous or distributed load acting on a body. Consider the continuous load which has a constant magnitude of w lb/ft or kN/m that is illustrated in Figure 2-7. An elemental portion of the load, $w \, dx$, produces a moment about point 0 of $(x)w \, dx$. The total moment of the entire load about point 0 is thus

$$M_0 = \int_0^L wx \, dx = \frac{wL^2}{2}$$

Note that the same moment is obtained by imagining that the continuous load is

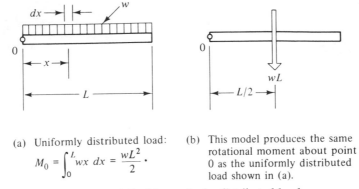

(a) Uniformly distributed load:
$$M_0 = \int_0^L wx \; dx = \frac{wL^2}{2}.$$

(b) This model produces the same rotational moment about point 0 as the uniformly distributed load shown in (a).

FIGURE 2-7 Moment of a distributed load.

replaced by an equivalent concentrated load acting at the center of mass of the load. For the case shown the equivalent concentrated load is wL [i.e., $(w \text{ lb/ft})(L \text{ ft}) = wL \text{ lb}$], which can be considered to act at $L/2$. The moment of this equivalent force system is $M_0 = (wL)(L/2) = wL^2/2$, which is the same moment obtained from $M_0 = \int_0^L wx \; dx$. This technique of modeling a continuous load as a point load is a useful one for finding reactions in complex structures and is often used in structural analysis.

2.2.6 Statically Equivalent Systems

Implicit in the discussions in the previous sections is the notion of static equivalency. When a system of forces applied to a body can be replaced by another system of forces applied to the same body without causing any net change in translational or rotational effects on the body, the two force systems are said to be *statically equivalent*. A resultant force, for example, is statically equivalent to the force system from which it was derived. The series of diagrams in Figure 2-8 illustrate one process for determining the statically equivalent single resultant force for a series of coplanar concurrent forces. *Concurrent forces* act through the same point and thus do not tend to produce rotational effects about that point (the moment arms are zero). The process illustrated depends on resolving each force into components (F_x and F_y forces) and algebraically summing components acting in similar directions. The resultant force is then given by $R = \sqrt{(\sum F_x)^2 + (\sum F_y)^2}$ and its orientation by $\theta_x = \tan^{-1}(\sum F_y / \sum F_x)$.

Since the components of a force are statically equivalent to the force itself, it follows that the total moment produced by the components of a force acting about a point is identical to that produced by the force itself about the same point [e.g., $M = F(r) = F_x(r_x) + F_y(r_y)$].

The method just described can be used as an aid in finding the resultant of nonconcurrent force systems (in which forces do not intersect at a common point). In general, the single resultant force of a nonconcurrent force system is given by $R = \sqrt{(\sum F_x)^2 + (\sum F_y)^2}$, its orientation by $\theta_x = \tan^{-1}(\sum F_y / \sum F_x)$, and its location by $a_0 = \sum M_0 / R$, where $\sum M_0$ is the sum of the moments of the forces about a point 0 and a_0 is the moment arm of R about 0 (see Figure 2-9).

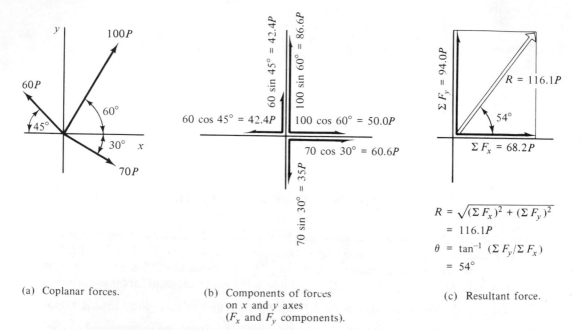

(a) Coplanar forces.

(b) Components of forces
on x and y axes
(F_x and F_y components).

(c) Resultant force.

FIGURE 2-8 Concurrent force systems—resultant and equilibrating forces.

2.3 EQUILIBRIUM

2.3.1 Equilibrium of a Particle

A body is in equilibrium when the force system acting on the body tends to produce no net translation or rotation of the body. It is in a state of balance. Equilibrium exists in concurrent force systems which act on a point or particle when the resultant of the force system equals zero. A concurrent force system having a resultant force may be put in equilibrium by applying another force (typically called an *equilibrant*) equal in magnitude, of the same line of action, but of opposite sense (see Figure 2-10).

The resultant of a system of concurrent forces can be found by considering force components and using $R = \sqrt{(\sum F_x)^2 + (\sum F_y)^2}$. If the force system is in equilibrium, its net resultant must be zero (i.e. $R = 0$), and it follows that $\sum F_x = 0$ and $\sum F_y = 0$. Thus, the algebraic sum of all the components of the forces applied to a particle in the x direction must be zero. Likewise for the y direction. x and y need not be horizontal and vertical. The statement is true for any orthogonal set of axes, no matter what their orientation.

More generally, the conditions $\sum F_x = 0$, $\sum F_y = 0$, and $\sum F_z = 0$ are necessary and sufficient to ensure equilibrium in a concurrent force system. A force system satisfying these conditions will not cause the particle to translate (rotation is not a problem, since all forces act through the same point in concurrent force systems).

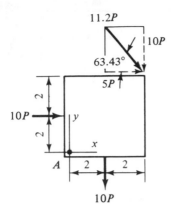

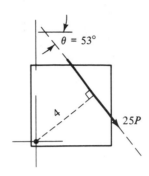

(a) Coplanar nonparallel force.

(b) Statically equivalent force.

Force	F_x	F_y	Moment of F_x about A	Moment of F_y about A
10	$10\,P$	0	$(2 \times 10) = 20\,P$	0
10	0	$10\,P$	0	$(2 \times 10) = 20\,P$
11.2	$5\,P$	$10\,P$	$(4 \times 5) = 20\,P$	$(4 \times 10) = 40\,P$

$\Sigma F_x = 15\,P \quad \Sigma F_y = 20\,P \quad \Sigma M_A = 40\,P + 60\,P = 100\,P$

$$R = \sqrt{(\Sigma F_x) + (\Sigma F_y)} \qquad a = \frac{\Sigma M}{R}$$
$$= 25\,P$$
$$\theta = \tan^{-1}\frac{\Sigma F_y}{\Sigma F_x} = 53° \qquad = \frac{100}{25} = 4$$

FIGURE 2-9 Nonparallel forces—finding statically equivalent forces.

2.3.2 Equilibrium of a Rigid Body

GENERAL EQUILIBRIUM CONDITIONS. When a non-concurrent-force system acts on a rigid body, the potential for both translation and rotation is present. For the rigid body to be in equilibrium, neither may occur. With respect to translation, this implies that, as in the case of concurrent-force systems, the resultant of the force system must be zero. With respect to rotation, the net rotational moment of all forces must be zero. The conditions for equilibrium of a rigid body are, therefore,

$$\Sigma F_x = 0 \qquad \Sigma F_y = 0 \qquad \Sigma F_z = 0$$
$$\Sigma M_x = 0 \qquad \Sigma M_y = 0 \qquad \Sigma M_z = 0$$

When working with general force systems, sign conventions are invariably problematic. For structural analysis and design purposes, it is often most convenient

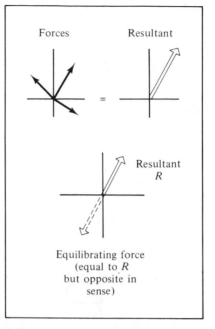

FIGURE 2-10   Relation among forces, the resultant force, and the equilibrating force.

to work with forces, or components of forces, acting along x and y axes. In this text, forces acting in the positive x or y directions will be considered positive. Moments that tend to produce counter-clockwise rotations on a body will be considered positive. These conventions are for equilibrium calculations only. Other conventions will be developed later when forces and moments internal to a structure are considered.

Before looking at the general application of these general equilibrium statements, it is useful to review two special cases (two-force members and three-force members) first.

TWO-FORCE MEMBERS.   When a rigid member is subjected to only two forces, the forces cannot have arbitrary magnitudes and lines of action if the member is to be in equilibrium. Consider the member shown in Figure 2-11. By summing moments about point A it can be seen that there is no way that the structure can be in rotational equilibrium *unless* the line of action of the force at B passes through point A. In a similar way, it is necessary that the line of action of the force at A must pass through point B if the structure is to be in rotational equilibrium about point B. Thus, the two forces *must* be collinear. They must also be equal in magnitude but opposite in sense. This result is particularly important in the analysis of trusses.

THREE-FORCE MEMBERS.   As with two-force members, three forces acting on a member cannot have random orientations and magnitudes if the member is to be in equilibrium. This can be seen with reference to Figure 2-11. For the member to be in rotational equilibrium, it is evident that the lines of action of all three forces must pass through a common point.

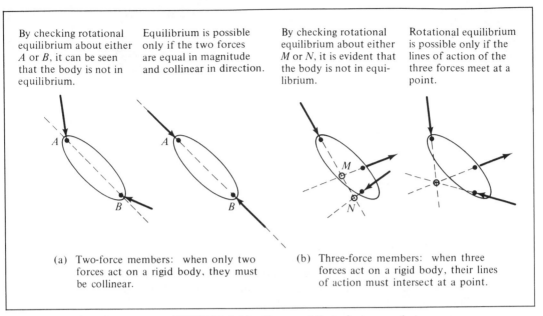

By checking rotational equilibrium about either A or B, it can be seen that the body is not in equilibrium.

Equilibrium is possible only if the two forces are equal in magnitude and collinear in direction.

By checking rotational equilibrium about either M or N, it is evident that the body is not in equilibrium.

Rotational equilibrium is possible only if the lines of action of the three forces meet at a point.

(a) Two-force members: when only two forces act on a rigid body, they must be collinear.

(b) Three-force members: when three forces act on a rigid body, their lines of action must intersect at a point.

FIGURE 2-11 Two- and three-force members.

Figure 2-12 shows a latter-day analysis of a gothic structure which illustrates the principle that the lines of action of three-force members must pass through the same point.

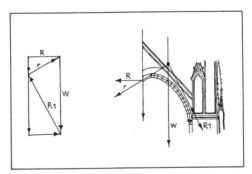

FIGURE 2-12 Application of graphical methods to the analysis of a gothic structure. The weight of the buttress and its reactions form a three-force system. Consequently, all forces meet at a single point.

GENERAL FORCE SYSTEMS. When more complex force systems are considered, it is evident that there must be relationships among forces and lines of action such that the general conditions of $\sum F = 0$ and $\sum M = 0$ are satisfied.

EXAMPLE

Determine the forces F_{A_y}, F_{B_y}, and F_{B_z} necessary to put the rigid body shown in Figure 2-13 in equilibrium.

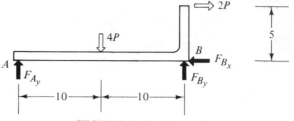

FIGURE 2-13 Example.

Solution:

Force equilibrium in the vertical direction: $\sum F_y = 0$:

$$F_{A_y} + F_{B_y} - 4P = 0$$

This expression obviously cannot be solved for either of the forces since there is only one equation and two unknowns. The next step is to find another independent equation involving the same two unknowns. $\sum M = 0$ will yield another independent equation.

Any point can be selected for a moment center. By selecting a moment center located on the line of action of one of the unknown forces, however, the solution of the problem is facilitated.

Moment equilibrium about point A: $\sum M_A = 0$:

$$(F_{A_y} \times 0) - (4P \times 10) + (F_{B_y} \times 20) + (F_{B_z} \times 0) - (2P \times 5) = 0$$

or

$$F_{B_y} = 2.5P \uparrow$$

From $\sum F_y = 0$:

$$F_{A_y} + 2.5P - 4P = 0 \qquad \text{or} \qquad F_{A_y} = 1.5P \uparrow$$

Force equilibrium in the horizontal direction: $\sum F_x = 0$:

$$-F_{B_z} + 2P = 0 \qquad \text{or} \qquad F_{B_z} = 2P \leftarrow$$

Check.

While all forces sought for are now known, the results found can be checked by arbitrarily choosing another moment center and verifying that rotational equilibrium obtains. Summing if moments about point *B*:

$$-(1.5P \times 20) + (4P \times 10) + (2.5P \times 0) + (2P \times 0) - (2P \times 5) = 0$$

or

$$0 = 0$$

The rigid body is indeed in equilibrium.

2.4 EXTERNAL AND INTERNAL FORCES AND MOMENTS

Up to this point there has been no distinction made in the types of forces or moments acting on a body. Forces and moments, however, can be either external or internal. Forces or moments that are applied to a structure are described as *external* (e.g., a weight attached to the end of a rope). Forces and moments that are developed within a structure in response to the external force system present in the structure are described as *internal* (e.g., the tension in a rope resulting from the pull of an attached weight).

2.4.1 External Force Systems

APPLIED AND REACTIVE FORCES. Forces and moments that act on a rigid body can be divided into two primary types: applied and reactive. In common engineering usage, *applied forces* are those that act directly on a structure (e.g., snow). *Reactive forces* are those generated by the action of one body on another and hence typically occur at connections or supports. The existence of reactive forces follows from *Newton's third law,* which generally states that to every action there is an equal and opposite reaction. More precisely, the law states that whenever one body exerts a force on another, the second always exerts on the first a force which is equal in magnitude, opposite in direction, and has the same line of action. In Figure 2-14(b), the force on the beam causes downward forces on the foundation and upward reactive forces are consequently developed. A pair of action and reaction forces thus exist at each interface between the beam and its foundations. In some cases, moments form part of the reaction system as well [see Figure 2-14(c)].

The diagrams in Figure 2-14, which show the complete system of applied and reactive forces acting on a body, are called *free-body diagrams.*

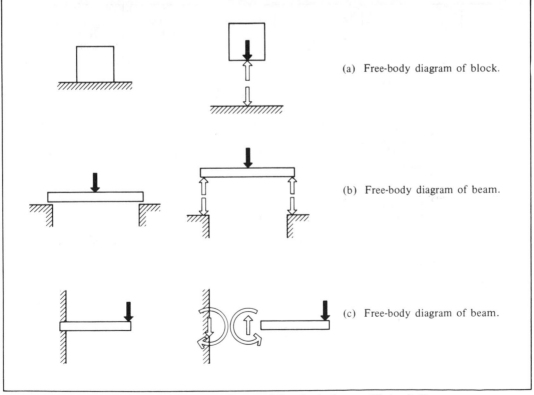

(a) Free-body diagram of block.

(b) Free-body diagram of beam.

(c) Free-body diagram of beam.

FIGURE 2-14 Reactions and free-body (or equilibrium) diagrams.

chapter 2 / PRINCIPLES OF MECHANICS

If a body (such as any of those illustrated) is indeed in a state of equilibrium, it is clear that the general conditions of equilibrium for a rigid body that were stated in the previous section must be satisfied. The magnitude and direction of any reactive forces developed must be such that equilibrium is maintained and are thus necessarily dependent on the characteristics of the applied force system. The whole system of applied and reactive forces acting on a body (as represented by the free-body diagram) must be in a state of equilibrium. Free-body diagrams are consequently often called *equilibrium diagrams*. Drawing equilibrium diagrams and finding reactions for loaded structural members is a common first step in a complete structural analysis.

SUPPORT CONDITIONS. The nature of the reactive forces developed on a loaded body depends on the exact way in which the body is either supported or connected to other bodies. Figure 2-15 illustrates relations between the type of support condition present and the type of reactive forces developed. Several basic types of support conditions are indicated; others are possible. Of primary importance are pinned connections, roller connections, and fixed connections. In *pinned connections*, the joint allows attached members to rotate freely but does not allow translations to occur in any direction. Consequently, the joint cannot provide moment resistance but can provide force resistance in any direction. A *roller connection* also allows rotations to occur freely. It resists translations, however, only in the direction perpendicular to the face of the support (either into *or* away from the surface). It does not provide any force

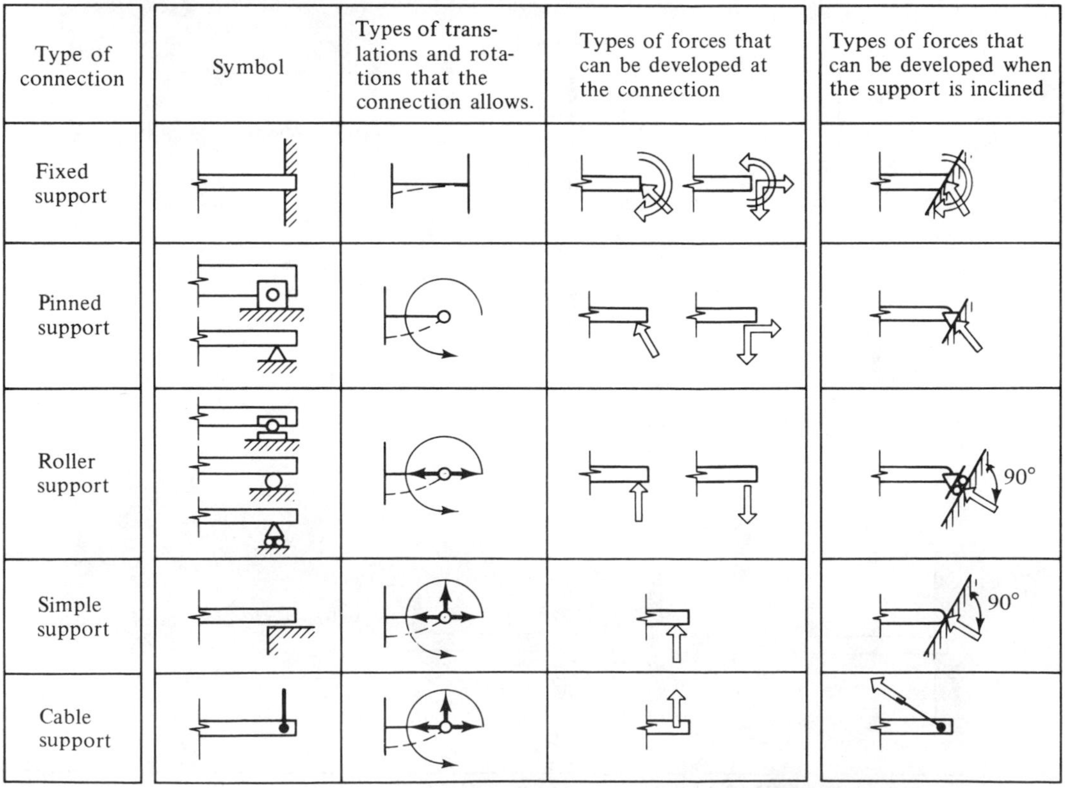

Type of connection	Symbol	Types of translations and rotations that the connection allows.	Types of forces that can be developed at the connection	Types of forces that can be developed when the support is inclined
Fixed support				
Pinned support				
Roller support				90°
Simple support				90°
Cable support				

FIGURE 2-15 Types of support conditions: idealized models.

resistance parallel to the surface of the support. A *fixed joint* completely restrains rotations and translations in any direction. Consequently, it can provide moment resistance and force resistance in any direction. Other types of supports include a *cable support* and a *simple support*. These are similar to the roller connection except that they can provide force resistance in one direction only. The connections shown are, of course, idealized. The relation between these idealized connections and those actually present in building structures is discussed in Section 3.3.2.

For a structure to be stable there must be a specific minimum number of force restraints provided by the supports. For a simple beam loaded with both downward and horizontal forces there must be three (this corresponds to the fact that there are three conditions of equilibrium for this type of structure: $\sum F_x = 0$, $\sum F_y = 0$, and $\sum M = 0$). One way of meeting this requirement is to use a fixed connection. Another is to use a pinned connection on one end and a roller connection on the other. It is, of course, possible to use connections that offer more degrees of restraint than the minimum required. Two pinned joints, for example, could be used to support a beam (see Section 3.3.2).

Structures having connections or supports that provide more than the minimum needed for stability are referred to as *statically indeterminate*. Since there are more unknown constraining forces than there are equations of equilibrium for solution, it is not possible to solve for the magnitudes of these constraining forces by statics alone. Other techniques must be used. These will be discussed in succeeding chapters.

EXAMPLE

Determine the reactions for the structure shown in Figure 2-16.

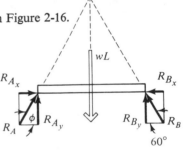
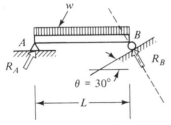

FIGURE 2-16 Model used to find reactions.

Solution:

Equilibrium in the vertical direction: $\sum F_y = 0$:

$$R_{A_y} + R_{B_y} - wL = 0$$

The total load acting downward due to the distributed loading is simply the load per unit length multiplied by the length over which the load acts. Note that the direction of the unknown reaction R_B is initially known because roller joints can transmit loads only perpendicular to the surfaces on which they roll. Hence, R_{B_y} must be $R_B \sin 60°$. The direction of R_A is not known a priori.

Moment equilibrium about point A: $\sum M_A = 0$:

$$+R_{B_y}(L) - (wL)\left(\frac{L}{2}\right) = 0 \qquad \text{or} \qquad R_{B_y} = \frac{wL}{2}$$

Hence, $R_{A_y} = wL/2$ from $\sum F_y = 0$. All other unknown components of the two reactive forces pass through the moment center and consequently have zero moments arms and drop out of the equation. The moment produced by the uniformly distributed load was found by imagining it to be concentrated at its center of mass and finding the moment produced by this concentrated load about point A (see Section 2.2.5). Modeling uniform loads in this way is frequently done for the purpose of finding reactions. It is not a valid technique, however, for other purposes such as finding beam sizes.

Final reactions:

Since R_{B_y} is now known, R_B can be calculated next. Thus, $R_{B_y} = R_B \sin 60°$ or $R_B = R_{B_y}/\sin 60° = (wL/2)/\sin 60° = 0.58wL$. R_{B_z} is simply $R_B \cos 60°$ or $0.29wL$. From $\sum F_x = 0$ it can be seen that $R_{B_z} = R_{A_z}$. Hence, $R_{A_z} = 0.29wL$. Since $R_{A_y} = wL/2$, it follows that $R_A = 0.58wL$.

EXAMPLE

Determine the reactions for the structure shown in Figure 2-17.

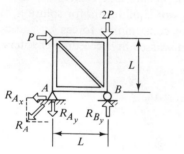 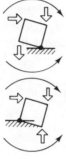

Moment about B produced by external forces

Resisting moment that must be developed by the reaction

Moment about A produced by external forces

Resisting moment that must be developed by the reaction

The directions of the reactions can be determined by inspecting how the external forces tend to rotate the body

FIGURE 2-17 Example.

Solution:

Moment equilibrium about point A: $\sum M = 0 \;\circlearrowright$:

$$-P(L) - 2P(L) + R_{B_y}(L) = 0 \qquad \text{or} \qquad R_{B_y} = 3P \uparrow$$

Equilibrium in the vertical direction: $\sum F_y = 0 \uparrow+$:

$$-R_{A_y} + R_{B_y} - 2P = 0; \qquad \text{hence,} \quad R_{A_y} = P \downarrow$$

Equilibrium in the horizontal direction: $\sum F_x = 0 \overset{+}{\rightarrow}$:

$$-R_{A_x} + P = 0 \qquad \text{or} \qquad R_{A_x} = P \leftarrow$$

2.4.2 Internal Stresses, Forces, and Moments.

TYPES OF INTERNAL FORCE SYSTEMS. Internal forces and moments are developed within a structure due to the action of the external force system acting on the structure and serve to hold together, or maintain the equilibrium of, the constituent particles or elements of the structure. Two general types of internal forces and moments will be considered here: (1) forces and moments developed at the points of connection of elemental parts of a larger structural assembly, and which are internal to the connection, and (2) forces and moments developed within the actual fabric of a structural element or point.

FORCES AT CONNECTIONS.   Forces and moments developed at the point of connection between two parts of a structural assembly are not conceptually different from the reactive forces discussed in the previous section. Indeed, a distinction is made as a matter of convenience only. Any elemental part of a structure has reactions that serve to maintain the equilibrium of the part, just as the larger assembly has reactions at its supports which serve to maintain the equilibrium of the whole structural assembly.

The forces and moments developed at points of connection are such that the reactive forces and moments acting on one attached member are equal and opposite in sense to those acting on the other attached member. This follows from Newton's third law that was previously discussed.

Forces at connections are explored in more detail in Sections 4.3 and 5.3 which deal with the analysis of trusses and three-hinged arches, respectively.

TENSION AND COMPRESSION.   As noted earlier, internal forces and moments can be developed *within* the fabric of a body subjected to an external force system. This section begins a consideration of internal tension and compression forces in a member.

Consider the system shown in Figure 2-18. It is intuitively obvious that there is a tension force developed in the cable supporting the block which has a magnitude equal to the weight of the block. More formally, equilibrium diagrams can be drawn as illustrated. As the diagram illustrates, the equilibrium of the block is maintained by the development of an internal force, F_t, in the cable, which in this case is simply equal to the weight of the block. It is also evident that the system could be decomposed in several different ways. The relation between this resultant internal force F_t, and the unit intensity of force within the material of the tension member, will be discussed later.

Internal compressive forces are obviously similar in character but opposite in sense. Figure 2-18 illustrates a member in which the internal compressive force varies due to the character of the external force system present. The magnitude and direction of the internal forces developed are such that *all parts* of the structure are in a state of equilibrium, as they must be no matter what part of the structure is considered.

Tension and compression forces that are collinear with the long axis of a member are often called *axial* forces or sometimes *normal* forces. Diagrams of the type illustrated in Figure 2-18 are frequently drawn to display graphically the variation in the internal axial force present in a member.

INTERNAL STRESSES IN TENSION MEMBERS.   In the previous section internal tension or compression forces present at a cross section within a structure were considered. This section considers the internal forces developed at a specific point within the cross section. The idea of fundamental importance introduced is the concept of *stress* or the *force intensity per unit area.*

Consider a simple tension member of the type illustrated in Figure 2-19. It is intuitively evident that the internal tension present is not actually concentrated at a specific spot (as the arrows symbolizing internal force in Figure 2-19 would seem to indicate) within the cross section of the member, but is rather distributed over the entire cross section. The total internal force necessary to equilibrate the external force acting on the member is in actuality the resultant of all distributed forces, or stresses, acting at the cross section.

With respect to a simple element carrying a tension force, it is reasonable to assume that if the external force acts along the axis of the member and at the centroid or point of symmetry of the cross section, the stresses developed at the cross section are of uniform intensity. Their resultant would have the same line of action as the external force present. When stresses are uniformly distributed, their magnitude is given by

$$\text{stress} = \frac{\text{force}}{\text{area}} \quad \text{or} \quad f = \frac{P}{A}$$

where f is the stress (force intensity per unit area), P is the axial force present, and A is the area of the cross section considered. Stresses of this type are often called *axial* or *normal* stresses. While the above was developed in connection with axial or normal tension forces, the stresses developed in a member loaded in direct compression can be similarly described.

The assumption that stresses associated with axial loads are uniformly distributed across a cross section is a reasonable one when the load is applied in a purely axial way and the member involved is straight and of uniform cross section. At unique points, such as where the external load is applied or at discontinuities in the member, more complex stress patterns may occur. The assumption, however, is generally quite good and perfectly acceptable as a working hypothesis for preliminary design purposes.

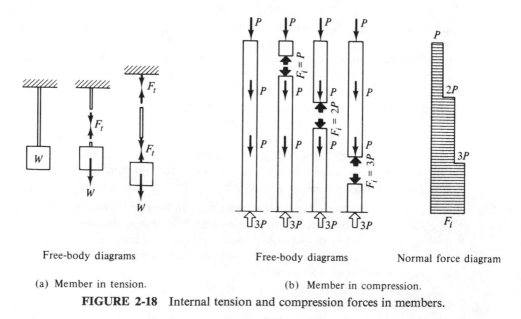

Free-body diagrams   Free-body diagrams   Normal force diagram

(a) Member in tension.   (b) Member in compression.

FIGURE 2-18 Internal tension and compression forces in members.

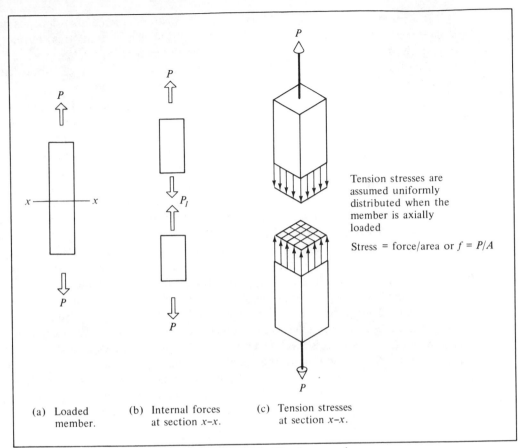

(a) Loaded member.

(b) Internal forces at section x–x.

(c) Tension stresses at section x–x.

Tension stresses are assumed uniformly distributed when the member is axially loaded

Stress = force/area or $f = P/A$

FIGURE 2-19 Tension members.

EXAMPLE

A member having a square cross section that measures 2 in. $\times$ 2 in. (50.8 $\times$ 50.8 mm) carries a tension load of 24,000 lb (106.75 kN). What is the stress level present at a typical cross section?

Solution:

Assuming that the internal stress are uniformly distributed:

$$\text{stress} = \frac{\text{force}}{\text{area}} \quad \text{or} \quad f = \frac{P}{A} = \frac{24{,}000 \text{ lb}}{2 \text{ in.} \times 2 \text{ in.}} = 6000 \text{ lb/in.}^2$$

or

$$f = \frac{106.75 \text{ kN}}{50.8 \times 50.8 \text{ mm}}$$

$$= 0.0414 \text{ kN/mm}^2 = 41.4 \text{ N/mm}^2$$

It is important to note that the axial or normal stress in a member depends only on the force applied and cross-sectional area involved. The actual stress developed does *not* depend on the type of material of which the member is made.

Bending, shear, and other types of stresses are nonuniformly distributed and *not* described by $f = P/A$. These stresses are considered in Chapter 6.

STRENGTH OF TENSION MEMBERS. Consider a simple axially loaded member carrying a tension force. The strength of an element of this type depends on a number of factors. The probable mode of failure is a pulling apart of the member at the weakest location along its length. The load-carrying capacity of a member subjected to pure tension is independent of the length of the member if the member has a uniform cross section throughout (in terms of both area and material). The presence of a single weak spot, such as a point of reduced cross-sectional area, determines the capacity of the whole member (at such locations the actual stress level as given by $f = P/A$ is higher than elsewhere because of the smaller value of A).

When a member is axially loaded, stress levels defined by $f = P/A$ develop. If the material used can sustain this stress intensity, the member will carry the load. As stress levels increase, due to increasing loads, for example, there comes a point where the stress intensity present exceeds the capability of the material used to withstand the pulling apart. This is the *failure stress level* of the material. It is a property of the material used. Failure stress levels are experimentally determined for various materials. When the *actual* stress level, given by $f = P/A$, in a member exceeds the failure stress level for the material used, the member will pull apart.

For steel, experiments have shown that the apparent tension stress level associated with the *beginning* of the material pulling apart or *yielding* is approximately $F_{yield} = 36,000 \text{ lb/in.}^2$ (248 N/mm² or 248 MPa). Section 2.5 explores material properties in greater detail.

ALLOWABLE STRESSES. From the point of view of determining if a member is of adequate size to support a given tension load, it is necessary first to determine the actual stress level in the member ($f = P/A$). If this actual stress level is less than the experimentally determined failure stress level for the material, the member can support the loading involved. From a design point of view, however, it is desirable that a statement that a member is adequately sized involve a factor of safety of some type. Loads and failure stresses can never be predicted with absolute certainty and a conservative note should be introduced when designing members.

One way of introducing a factor of safety into the design process is to use an *allowable stress* (F_t) rather than failure or yield stresses as the measure against which actual stresses are compared. Thus,

$$F_{t_{allowable}} = \frac{F_{yield}}{\text{factor of safety}}$$

For steel, for example, a factor of safety of 1.5 is often used. Thus, the allowable stress level in tension for a steel member is given by $F_t = (36,000 \text{ lb/in.}^2)/1.5 = 24,000 \text{ lb/in.}^2$ or $F_t = (248 \text{ N/mm}^2)/1.5 = 165 \text{ N/mm}^2$. This value can be used in evaluating the acceptability of a given member or as a tool for determining the required size of a new member, given the load it must carry. In sizing a new member, the required cross-sectional area for a member in tension carrying a load P is given by

$$A_{\text{required}} = \frac{\text{force}}{\text{allowable stress}} = \frac{P}{F_t}$$

EXAMPLE

What diameter of steel rod is required to support a tension load of 10,000 lb (44.5 kN)? Assume that the allowable stress for the steel is $F_t = 24,000$ lb/in.^{2} (165 MPa or 165 N/mm^2).

Solution:

Area required:

$$A = \frac{P}{F_t} = \frac{10,000 \text{ lb}}{24,000 \text{ lb/in.}^2} = 0.416 \text{ in.}^2$$

$$= \frac{44,500 \text{ N}}{165 \text{ N/mm}^2} = 270 \text{ mm}^2$$

Diameter:

$$\frac{\pi d^2}{4} = 0.416 \text{ in}^2. \quad \text{or} \quad d = 0.73 \text{ in.}$$

$$= 270 \text{ mm}^2 \quad \text{or} \quad d = 18.5 \text{ mm}$$

2.4.3 Shear and Moment

BASIC PHENOMENA. This section begins a study of the type of internal forces and moments generated in a member carrying an external force system that acts transversely to the axis of the member. The concepts of shear and moment in structures introduced here are of fundamental importance to an understanding of the behavior of structures under load. They also provide the basis for developing tools for designing structures. This section introduces shear and moment considerations in a direct but abstract way. These same concepts are later developed in the context of looking at a particular structural type (trusses). Succeeding chapters draw out and elaborate on the theme of shear and moment in connection with other structural elements.

Consider the loaded cantilever member illustrated in Figure 2-20. There are two primary ways in which the member might fail as a result of the applied load. One

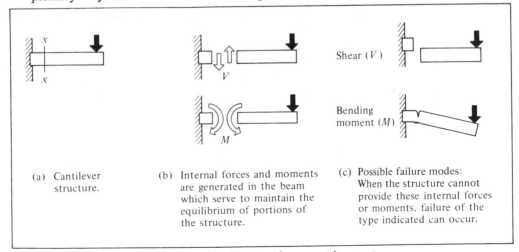

(a) Cantilever structure.

(b) Internal forces and moments are generated in the beam which serve to maintain the equilibrium of portions of the structure.

(c) Possible failure modes: When the structure cannot provide these internal forces or moments, failure of the type indicated can occur.

FIGURE 2-20 Shear and moment in structures.

potential type of failure is for the load to cause two contiguous parts of the member to slide relative to each other in a direction parallel to their plane of contact. This is called a *shear failure*. The internal force developed in the member that is associated with this phenomenon is called an *internal shear force*. This internal shear force is developed in response to the components of the external force system that act transversely to the long axis of the member and tend to cause the type of transverse sliding indicated in Figure 2-20(c). These internal shear forces resist or balance the net external shearing force tending to cause the sliding. Failure of the type illustrated occurs when the member can no longer provide an equilibrating internal force of this type.

The second possible mode of failure is that illustrated in Figure 2-20(c). This type of failure is obviously associated with the tendency of the transverse external forces to cause part of the structure to rotate, or *bend*. Since free rotation cannot occur in a rigid body (unless a pin connection is present), an internal resisting or balancing moment equal and opposite to the applied moment (the moment associated with the rotational tendencies of the external forces) must be developed within the structure. Failure of the type shown occurs when the structure can no longer provide a resisting moment equal to the applied moment.

At any section of the loaded member, internal shear forces and moments are developed simultaneously. If the member is decomposed at this point into two parts, the forces and moments developed internally serve to maintain the translational and rotational equilibrium of each part. They also represent the internal actions and reactions of one part of the member on the other part. As will be seen later, one of the objectives of the structural design process is to create a structural configuration capable of providing these internal shears and moments in an efficient way and with factors of safety sufficient to prevent shear and moment failures.

Internal shear forces and moments of this type described above are developed in *any* structure carrying transverse loads. Determining the magnitude of these quantities is a straightforward process based on the fundamental proposition that any structure, or any part of any structure, must be in a state of equilibrium under the action of the complete force system (including internal as well as external forces and moments) acting on it. The external force system is typically known. Parts of the force system not initially known, such as reactions at supports, can be readily calculated by methods previously discussed. If the equilibrium of an isolated portion of a structure is considered, the unknown internal shears and moments that must be present at the point of decomposition can be determined through equilibrium considerations involving known parts of the external force system.

The member shown in Figure 2-21 is in equilibrium. Assume that it is desired to know the nature of the internal forces at some arbitrary location, such as section x–x located at midspan. To find these internal forces, the structure is decomposed into two parts at this location. Consider the left portion of the structure [Figure 2-21(b)]. As is evident, this part of the structure is not in equilibrium if only the effects of the external force system acting on this part are considered. The net effect of the portion of the external force system acting on this part of the structure is to cause it to translate vertically downward and rotate in a clockwise direction. The net downward force is

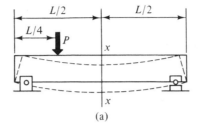

(a)

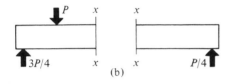

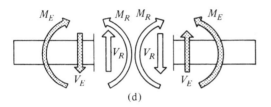

(b)

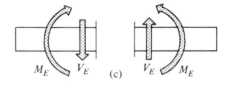

(c)

(d)

(e)

(a) Loaded structure beam and deflected shape.

(b) Separating the beam into two parts. As shown, the parts, are not in equilibrium.

(c) Net effects of the external force system on each part of the beam about section x–x.

Left part: $V_E = +3P/4 - P = P/4 \downarrow$

$M_E = (3P/4)(L/2) - (P)(L/4) = PL/8$

Right part: $V_E = P/4 \uparrow$

$M_E = (P/4)(L/2) = PL/8$

Note that V_E and M_E are of the same magnitude on the two parts of the beam but are of opposite sense.

(d) For each part of the beam to be in equilibrium, an internal resisting shear force, V_R, and internal resisting moment, M_R, must be provided by the structure. For equilibrium, $V_R = V_E$ and $M_R = M_E$.

(e) Final free-body diagram of each part showing the external force system acting on each part and the internal resisting shears and moments. Note that the internal shears and moments are of equal magnitude but opposite sense on the two parts of the structure. The internal shears and moments represent the actions and reactions of one part of the beam on the other part.

FIGURE 2-21 Internal shear and moment at a section of a beam.

called the *external shear force* (V_E) present at the section. In this case it is given by

$$V_E = \frac{+3P}{4} - P = \frac{P}{4} \downarrow$$

For this part of the structure to be in equilibrium, it is evident that the structure must somehow provide an *internal resisting shear force* (V_R) equal in magnitude but opposite in sense. Thus, $V_E = V_R$. Hence, $V_R = P/4\uparrow$.

The net rotational effect associated with the portion of the external force system acting on the part of the structure considered (in this case the portion to the left of midspan) is called the *external bending moment* (M_E). If sign conventions are temporarily ignored (a new convention to be described in the next section is used), the external bending moment is simply given by

$$M_E = \left(\frac{3P}{4}\right)\left(\frac{L}{2}\right) - (P)\left(\frac{L}{4}\right) = \frac{PL}{8}$$

For equilibrium to be maintained, it is evident that the structure must provide an *internal resisting moment* (M_R) equal in magnitude but opposite in sense at this location. Since $M_R = M_E$, $M_R = PL/8$.

A more direct approach is to write immediately the equations of equilibrium for the complete free-body diagram of the part of the structure considered [see Figure 2-21(e)]. Thus,

$\Sigma F_y = 0$:

$$\frac{+3P}{4} - P + V_R = 0 \qquad \therefore V_R = \frac{P}{4} \uparrow$$

$\Sigma M_x = 0$:

$$+\left(\frac{3P}{4}\right)\left(\frac{L}{2}\right) - (P)\left(\frac{L}{4}\right) - M_R = 0 \qquad \therefore M_R = \frac{PL}{8}$$

Obviously, this is not a conceptually different approach. It does not, however, stress that the external force system creates a net translatory force and applied moment at the section which must be balanced by numerically similar internal forces and moments acting in the opposite senses.

These internal shear and moment values could have also been found by considering the equilibrium of the right rather than left portion of the structure. The external shear force V_E for this part is given by $V_E = P/4\uparrow$. Since $V_E = V_R$, $V_R = P/4\downarrow$. Numerically, this is the same result as found previously. The sense of the force, however, is opposite.

In a like vein, the external moment on the right part is given by

$$M_E = \left(\frac{P}{4}\right)\left(\frac{L}{2}\right) = \frac{PL}{8}$$

Since $M_E = M_R$, $M_R = PL/8$. This is numerically equal to the moment found previously, but again of opposite sense.

SIGN CONVENTIONS. That the shears and moments found by considering alternate parts of the structure should be numerically equal but opposite in sense is reasonable in view of the fact that the forces found are actually internal to the structure and represent the *action and reaction* of one part of the structure on the other. This does bring up a difficulty, however, in terms of sign conventions. Designating the internal shear or moment as either positive or negative by its direction only, as was done for calculating reactions, is misleading because of the fact that with reference to one part of the structure a value would be positive and with reference to the other part the value would be negative. The same shear forces and moment, however, are involved. A different convention is therefore adopted. The general convention usually employed is based on the physical effects of forces and moments or structures. With respect to moments, note that the tendency of either the external force system acting on the left part or the external force system acting on the right part is to cause the structure to deform or bend, as illustrated to the left in Figure 2-22(a). When the external force system causes this type of *concave* curvature to develop at a section in the member, the moment present at the section is said to be *positive*. An equivalent convention is to say that upward-acting external forces tend to cause positive moments to develop at a section. If the external force system tends to produce a *convex* curvature (concave downward) at a section, the moment present is said to be *negative*. Equivalently, downward-acting forces tend to produce negative moments at a section. Note that conventions of this type give the same sign to the moment developed at a section, regardless of whether the right or left part of a structure is considered. Relating

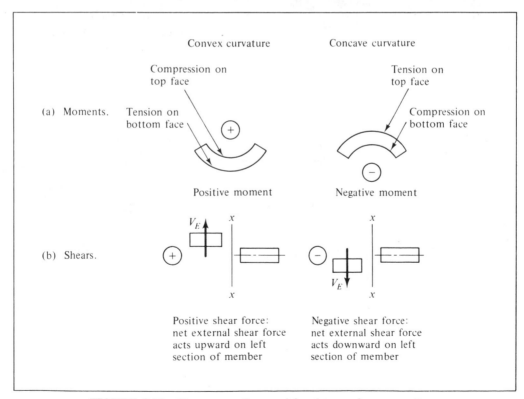

FIGURE 2-22 Sign convention used for shear and moment diagrams.

moment to curvature is also a way of looking at structures that is useful in developing a feeling for the physical behavior of structures.

For shear, the convention typically used is that when the external forces acting on the left part of a structure have a net resultant that acts vertically upward, the shear is said to be positive. Positive shear is associated with the tendency of the external forces to produce relative movements of a section of the type illustrated to the left in Figure 2-22(b).

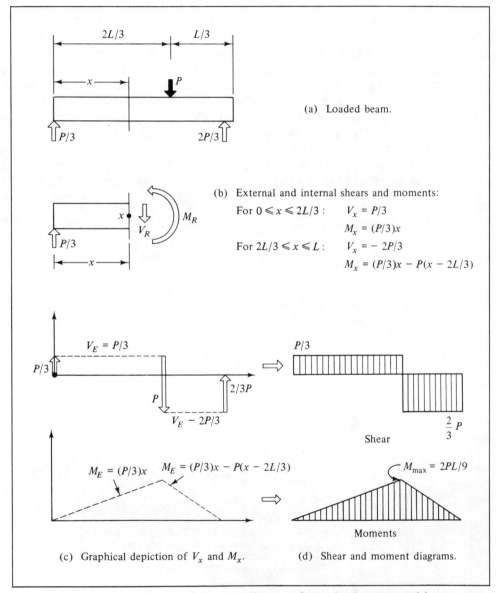

(a) Loaded beam.

(b) External and internal shears and moments:

For $0 \leqslant x \leqslant 2L/3$: $V_x = P/3$

$M_x = (P/3)x$

For $2L/3 \leqslant x \leqslant L$: $V_x = -2P/3$

$M_x = (P/3)x - P(x - 2L/3)$

$V_E = P/3$

$P/3$

$2/3P$

P

$V_E - 2P/3$

Shear

$\frac{2}{3}P$

$M_E = (P/3)x$ $M_E = (P/3)x - P(x - 2L/3)$

$M_{max} = 2PL/9$

Moments

(c) Graphical depiction of V_x and M_x.   (d) Shear and moment diagrams.

FIGURE 2-23 Shear and moment diagrams for a simply supported beam carrying a concentrated load.

2.4.4 Distribution of Shears and Moments

SHEAR AND MOMENT DIAGRAMS. As would be intuitively expected, there is commonly a variation in the magnitude (and often the sense) of the shears and moments present at different sections in a structure. The distribution of these shears and moments in a structure can be found by considering in turn the equilibrium of different elemental portions of the structure and calculating the shear and moment sets for each elemental portion. As an aid in visualizing the distribution of these shears and moments, the values thus found can be plotted graphically to produce what are called *shear and moment diagrams*.

Consider the beam shown in Figure 2-23(a) and the free-body diagram of an elemental portion of the beam (having a length x) to the left of the load. As shown in the figure, the rotational effect of the set of external forces acting on the left portion of the beam about the cross section defined by the distance x is simply given by $M_E = (P/3)(x)$. The net translatory effect is given by $V_E = P/3$. Thus, if $x = 0$, $M_E = 0$ and $V_E = P/3$. If $x = 2L/3$, $M_E = 2PL/9$ and $V_E = P/3$ (if the section is infinitesimally to the left of the load). Once x moves to the right of the load, new equations for shears and moments are needed, since a new force now acts on the elemental portion of the beam considered. Thus, $M_E = (P/3)(x) - P[x - \frac{2}{3}(L)]$ and $V_E = +(P/3) - P = (-\frac{2}{3})P$. Graphically, these shears and moments can be plotted as illustrated in Figure 2-23(c).

The example above illustrates the general procedure for finding the distributions of shears and moments in a structure. Shear and moment diagrams are invaluable aids in the analysis and design of structures, and a working knowledge of what they mean and how to draw them quickly is of permanent importance. It should be emphasized that their use is not solely restricted to beams. They represent a graphic way of looking at the effects of an external force system acting on any type of structure.

EXAMPLE

Draw shear and moment diagrams for the structure illustrated in Figure 2-24.

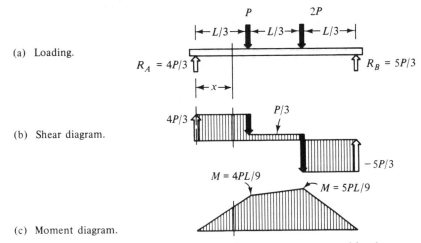

(a) Loading.

(b) Shear diagram.

(c) Moment diagram.

FIGURE 2-24 Shear and moment diagrams: concentrated loads.

Solution:

Equations for shear and moment at a distance of x from the left reaction are determined by considering the equilibrium of the portion of the structure to the left of x.

For $0 < x < L/3$:

$$V_E = \frac{4P}{3}$$

$$M_E = \left(\frac{4P}{3}\right)x$$

For $L/3 < x < 2L/3$:

$$V_E = \frac{4P}{3} - P = \frac{P}{3}$$

$$M_E = \left(\frac{4P}{3}\right)x - P\left(x - \frac{L}{3}\right)$$

For $2L/3 < x < L$:

$$V_E = \frac{4P}{3} - P - 2P = -\frac{5P}{3}$$

$$M_E = \left(\frac{4P}{3}\right)x - P\left(x - \frac{L}{3}\right) - 2P\left(x - \frac{2L}{3}\right)$$

These expressions can be plotted to obtain shear and moment diagrams as illustrated in Figure 2-27. The shear forces are constant between loads while the moments vary linearly. Note that critical values of the moment occur at $x = L/3$, where $M = 4PL/9$, and $x = 2L/3$, where $M = 5PL/9$. Major changes in the shear present also occur at these points. It is also interesting to note that the maximum value of the moment does not occur at midspan but at a point where the shear diagram passes through zero.

In structures carrying concentrated loads, critical moment and shear values invariably occur directly beneath the concentrated load application points. By utilizing this observation, the moment diagram can be drawn more easily by simply considering sections beneath the load points. Thus, the moment at $x = L/3$ (the first load point) is simply given by $M_E = (4P/3)(L/3) = 4PL/9$. That at $x = 2L/3$ (the second load point) is given by $M_E = (4P/3)(2L/3) - (P)(L/3) = 5PL/9$. The latter is even easier to find if the equilibrium of a portion of the structure to the right instead of left of the section studied is considered. Thus, $M_E = (5P/3)(L/3) = 5PL/9$. The values of $4PL/9$ and $5PL/9$ define critical points on the moment diagram. Straight lines can be drawn between them, since by inspection it is evident that the moments vary linearly between critical points.

EXAMPLE

Draw shear and moment diagrams for the cantilever structure shown in Figure 2-25.

Solution:

Shears and moments at any section defined by a distance x from the left end of the member are determined through considering the equilibrium of the structure to the left of x. The moment of the uniform load to the left of the section considered is found by replacing the uniformly distributed load by a statically equivalent concentrated load acting at the center of mass of the portion of the load considered (see Section 2.2.5).

For $0 < x < L$:

$$V_E = -wx$$

$$M_E = -wx\left(\frac{x}{2}\right) = -\frac{wx^2}{2}$$

Critical values occur at $x = L$, where $V_E = -wL$ and $M_E = -wL^2/2$. Note that the shear forces vary linearly according to the first power of x and that the moments vary according to the second power of x.

EXAMPLE

Draw shear and moment diagrams for the uniformly loaded simply supported structure shown in Figure 2-26.

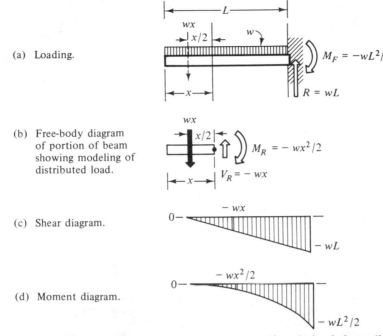

(a) Loading.

$M_F = -wL^2/2$

$R = wL$

(b) Free-body diagram of portion of beam showing modeling of distributed load.

$M_R = -wx^2/2$

$V_R = -wx$

(c) Shear diagram.

$-wx$

$-wL$

(d) Moment diagram.

$-wx^2/2$

$-wL^2/2$

FIGURE 2-25   Shear and moment diagrams: uniformly loaded cantilever structure.

Solution:

As before, the net effect of the external force system acting on an elemental portion of the structure is to produce at the section considered a translational shear force, V_E and rotational moment, M_E. These external shears and moments must be balanced by an internal resisting shear force, V_R, and internal resisting moment, M_R, provided by (or developed in) the structure if equilibrium is to be maintained.

For $0 \lessgtr x \lessgtr L$:

$$V_E = \frac{wL}{2} - wx$$

$$M_E = \left(\frac{wL}{2}\right)x - (wx)\left(\frac{x}{2}\right)$$

Note that the shear diagram varies linearly and has maximum values at $x = 0$ and $x = L$ where $V_E = wL/2$ and $V_E = -wL/2$, respectively. The moment diagram varies according to the second power of x and reaches a maximum at midspan (where the shear is zero). The maximum moment present is $M = wL^2/8$. This is one of the most frequently encountered expressions in structural theory. It is commonly used in the analysis and design of simple beams but crops up in other less expected places as well.

SHAPES OF DIAGRAMS.   Some general observations can be made about the shapes of shear and moment diagrams. Concentrated loads generally produce shears that are constant in magnitude along sections of a structure between concentrated loads, and the shear diagram consequently consists of a series of horizontal lines. Moments in such structures vary linearly (according to the first power of x) between concentrated

(a) Loading.

(b) Segment of structure of length x: external force system and equivalent external shear force and bending moment.
$V_{E_x} = wL/2 - wx$
$M_{E_x} = (wL/2)x - wx(x/2)$

(c) Equilibrium diagrams: an internal resisting shear force and bending moment is developed in the structure to maintain the equilibrium of the segment.

(d) Expressions for the shear and moment as a function of x and their graphical depictions.

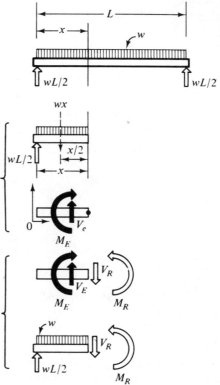

$V_E = wL/2 - wx$

$x = 0$
$V_E = wL/2$

$x = L$
$V_E = -wL/2$

Slope $= -w$

$x = L/2$
$M_E = wL^2/8$

$M_E = (wL/2)x - wx^2/2$

FIGURE 2-26 Shear and moment diagrams: uniformly loaded simply supported structure.

loads. Moment diagrams are consequently composed of sloped lines. Uniformly distributed loads tend to produce linearly varying shear forces. Shear diagrams correspondingly consist of a sloped line or a series of sloped lines. Uniformly distributed loads produce parabolically varying moments. Moment diagrams are correspondingly curved. Combined loads produce combined shapes. While not rigorously proveable at this point, it is also generally true that a point of zero shear corresponds to a location of a maximum or minimum moment. This latter observation is particularly valuable for locating critical design moments in structures having unusual loading conditions.

EXAMPLE

Draw shear and moment diagrams for the partially loaded beam shown in Figure 2-27.

Solution:

The equations for shear and moment are determined and plotted as before.

$$\text{For } 0 \lessgtr x \lessgtr 12: \qquad \text{For } 12 \lessgtr x \lessgtr 16:$$

$$V_E = 15 - 2x \qquad\qquad V_E = 15 - 2(12)$$

$$M_E = 15x - 2x\left(\frac{x}{2}\right) \qquad M_E = 15x - 2(12)(x - 6)$$

Note that where uniform loads are actually on the member, the shear diagram varies linearly and the moment diagram parabolically. Determining the maximum moment present in the structure could be a tedious job since it is *not* necessarily obvious where this maximum moment occurs. A trial-and-error process of assuming different values of x would work, but a more direct approach making use of the fact that the moment has a critical value where the shear is zero is much easier.

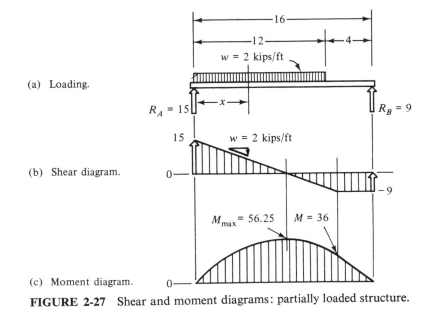

(a) Loading.

(b) Shear diagram.

(c) Moment diagram.

FIGURE 2-27 Shear and moment diagrams: partially loaded structure.

If a correct but less rigorous approach to constructing the shear diagram is adopted, it can be seen that the shear diagram can be easily drawn by noting that the left reaction, so to speak, pushes the shear diagram initially up to a level corresponding to the left reaction (indicating that a shear force exists in the beam immediately to the right of the reaction of a value equal to that of the reaction—a result that would have been found by passing a section through that point). The effect of the uniform load is to cause the shear diagram to decrease at a rate per unit length equal to the load per unit length (this is also evident by looking at the shear equation noted above). The shear diagram thus has a slope equal to the uniform load itself. This information can then be used to find where the shear diagram passes through zero. Hence, $15 = 2x$ or $x = 7.5$. This is equivalent to setting $V_E = 0$ in the original shear equation (i.e., $0 = 15 - 2x$). By noting that the moment is generally a critical value at points of zero shear, the moment can be calculated at $x = 7.5$ to yield the maximum moment present. Thus, $M_E = 15(7.5) - 2(7.5)(7.5/2) = 56.25$.

RELATION OF MOMENT DIAGRAM TO DEFLECTED SHAPE OF STRUCTURE. As previously discussed, transverse loads produce a bowing in the structure. Depending on the way the loads act, the bowing produced can be either concave upward or the reverse. The overall deflected shape of the structure is intimately related to the nature of the bowing present, and, consequently, to the moments present.

Consider the member shown in Figure 2-28. The loads on the ends of the projecting cantilevers produce a bowing that is concave downward in the left and right

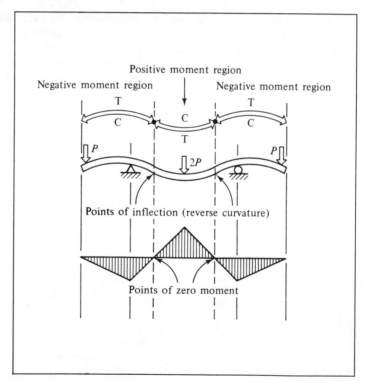

FIGURE 2-28 Relationship between the moment diagram and the deflected shape of a structure.

regions of the member. The moment diagram is negative in these zones, which are commonly called *negative moment regions*. Associated with the concave downward bowing is a stretching of the upper fibers of the structure, which puts them in a state of tension, and a shortening of the lower fibers, which are consequently put in a state of compression. Converse pheonomena occur in the middle portion of the member, in the *positive moment region*. Between positive and negative moment regions transition points must obviously occur. Because of the reversal of curvature that takes place, it is evident that no bowing or bending is present at the transition points. Consequently, they are points of zero moment. These points are typically called *points of inflection*. Quite often it is possible, first, to sketch the deflected shape of a structure and then to infer the shape of the moment diagram from the sketch. Such relations are also useful for checking the accuracy of a diagram obtained through calculation.

2.4.5 Relations Among Load, Shear, and Moment in Structures

Although a topic generally outside the scope of this text, it can be shown that relationships of the following type exist among the load, shear, and moment in a structure:

$w = dV/dx, \quad V = dM/dx, \quad V_B - V_A = \int_A^B w\, dx, \quad \text{and} \quad M_B - M_A = \int_A^B V\, dx.$[2] Hence, the value of the load at a point is equal to the slope of the shear diagram at the same point. The value of the shear at a point is equal to the slope of the moment diagram at the same point. The *change* in moment between points A and B on a structure is represented by the area under the shear diagram between the same two points (points of reference and boundary conditions must be carefully considered). Techniques of this kind are highly useful in checking the accuracy of shear and moment diagrams. They clearly point out that the maximum/minimum value of the moment in a structure occurs when the shear diagram passes through zero (since $dM/dx = 0$ at this point).

2.5 MECHANICAL PROPERTIES OF MATERIALS

2.5.1 Introduction

This section presents a brief coverage of the various properties of materials that are of interest from a structural design viewpoint. Of primary interest are the strength and load-deformation properties of a material. In the final analysis, these properties are really explicable only in terms of the internal forces that act between the constituent parts of the material (i.e., the molecules, or, in some cases, the atoms). This extent of investigation is, however, beyond the scope of this text. Instead, the coverage will be primarily descriptive in tone and will not dwell on the reasons underlying particular material behavior.

[2]F. L. Singer, *Strength of Materials*, 2nd ed. Harper & Row, New York, 1962.

2.5.2 General Load-Deformation Properties of Materials

The application of a load to a member invariably produces dimensional changes in the member. The member under load undergoes changes in size or shape or both. For many common materials, such as steel, the dimensional changes produced can be roughly categorized into one of two general types, elastic and plastic deformations, that occur sequentially with increasing loads, When a member is first loaded, deformations are in the *elastic range* of the material. In the elastic range, the member generally returns to its original dimensions if the applied loads are removed (the behavior is similar to that of a spring). Deformations in the elastic range depend directly on the magnitude of the stress level present in the member. As loads increase, deformations in many materials move into the *plastic range*. This occurs when the stress in the material reaches a sufficiently high level to cause a permanent change (i.e., a breaking down) in the internal structure of the material. Once these internal changes have occurred, the original state cannot be exactly regained, even if the load is removed completely. Consequently, when a material moves into the plastic range, it undergoes irreversible dimensional changes and a permanent set exists even if all external loads are completely removed. The load or stress levels associated with the plastic range are invariably higher than those associated with the elastic range. In the plastic range, deformations are nonlinearly dependent on the load or stress level present. Deformations in the plastic range are very large relative to those in the elastic range. Quite often material in the plastic range simply pulls apart (i.e., undergoes massive deformations) under a relatively constant load.

As will be discussed in more detail later, not all materials demonstrate both elastic and plastic behavior under increasing loads. Steel does; plain concrete does not.

2.5.3 Elasticity

ELASTIC BEHAVIOR. This section looks more closely at the behavior of materials within their elastic range, wherein the material returns to its original size and shape if the applied stress is removed. The concept of stress has already been dwelt on extensively. The primary way of describing changes in size or shape is through the concept of *strain* (ϵ). Strain is generally defined as the ratio of the change in size or shape of an element subjected to stress to the original size or shape (S) of the element (i.e., $\epsilon = [\Delta S/(S + \Delta S)] \doteq \Delta S/S$). As a ratio, it has no physical dimensions. There is a general relationship between stress and strain in elastic materials that was first postulated by Robert Hooke (1635–1703) and which is known as *Hooke's law*. Hooke's law states that for an elastic body, the ratio of the stress present on an element to the strain produced is a constant. Thus,

$$\frac{\text{stress}}{\text{strain}} = \text{constant for a material}$$

$$= \text{modulus of elasticity} = E$$

The magnitude of the constant is a property of the material involved and, as noted above, is usually referred to as the *modulus of elasticity*. The units for this constant are the same as those for stress (i.e., force per unit area) since strain is a dimensionless quantity. The relationship between stress and strain shown above implies that strain in a member is linearly dependent on the stress level present. The constant relating stresses and strains (the modulus of elasticity) is determined experimentally.

When a member is subjected to a simple tensile force, it will elongate a certain amount (Figure 2-29). If L is the original length of the member and ΔL is the change in length, the strain, ϵ, present in the member is

$$\text{strain} = \frac{\text{increase in length}}{\text{original length}} \quad \text{or} \quad \epsilon = \frac{\Delta L}{L}$$

Strain is actually dimensionless, as previously noted. It may be useful, however, to think of strain as the amount of deformation per unit of length. In these terms, the dimensions of strain would be millimeters/millimeter or inches/inch.

A typical way of determining the modulus of elasticity (E) of a material is to take a member of known length and cross-sectional area, subject it to a known load, and measure the amount of elongation (ΔL). Since the stress present can be readily calculated by using the relation $f = P/A$ and the strain found from the relation $\epsilon = \Delta L/L$, the modulus of elasticity can be determined, or $E = f/\epsilon$. Moduli of elasticity (E) of many different materials can be found by a general procedure of this type. For steel, $E_s = 29.6 \times 10^6$ lb/in.^{2} (204,000 N/mm^2 or 204,000 MPa), and for aluminum, $E_A = 11.3 \times 10^6$ lb/in.^{2} (77,900 N/mm^2). A typical value for concrete is $E_c = 3 \times 10^6$ lb/in.^{2} (20,700 N/mm^2) and for timber is $E_T = 1.6 \times 10^6$ lb/in.^{2} (11,000 N/mm^2). The values for concrete and timber vary widely according to the precise characteristics of the concrete mix tested or the grade and species of wood used.

Once the value of E for a material is known, it can be used as a constant to predict deformations in the material under different conditions of stress.

With respect to Figure 2-30, the modulus of elasticity is mothing more than the *slope* of the stress–strain curve within the elastic range of the material. As the stress level present in a member continues to increase, a point is reached where the strains developed are no longer linearly dependent on the stress. This is the transition point between the elastic and plastic range for the material. This point is termed the *proportional limit* of the material. After this point is passed, the concept of a constant modulus of elasticity is not valid. For some materials, such as steel, the magnitude of the deformations that occur in the plastic range are literally enormous as compared to those in the elastic range.

It should be noted that some materials, such as aluminum, do not have pronounced proportional limits (see Figure 2-31). Other materials, such as cast iron, exhibit virtually no plastic deformations at all. Different materials thus behave in widely differing ways under load.

LATERAL DEFORMATIONS IN THE ELASTIC RANGE. As illustrated in Figure 2-29, a member subjected to an axial force undergoes elastic changes in its lateral dimensions as well as in the direction of the applied load. Lateral dimensions in a member decrease when the member is subjected to a tensile force and increase when the member is subjected to a compressive load. A constant relation exists between these lateral changes and those that occur in the longitudinal direction. This relation is usually called *Poisson's ratio* (v) and is defined by $v = -\epsilon_y/\epsilon_x$. For steel, Poisson's ratio is about 0.3.

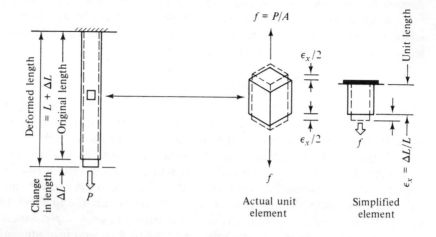

$f = P/A$

$\epsilon_x/2$

$\epsilon_x/2$

f

f

Actual unit element

Simplified element

(a) Axially loaded tension member: an applied tension force produces a longitudinal elongation whose magnitude depends on the cross sectional area of the member, the length of the member, the material the member is made of and the magnitude of the applied force.

(c) "Strain" in a tension member: strain is the amount of deformation per unit length that occurs in a loaded member.

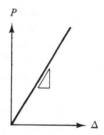

(b) Elastic behavior: for members of a given size made of "linearly elastic materials," (e.g., steel), the amount of deformation that takes place is proportional to the magnitude of the applied load. This elastic relationship holds true until the load is of sufficient magnitude to cause a breakdown to occur in the material.

(d) Relation between stress and strain: in linearly elastic materials there is a constant relationship between the magnitude of the stress present and the resultant strain, i.e., stress/strain = constant = modulus of elasticity (E).

FIGURE 2-29 Stresses, strains, and elongations in a simple tension element.

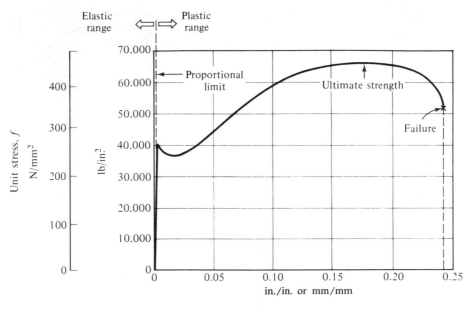

(a) Stress–strain diagram.

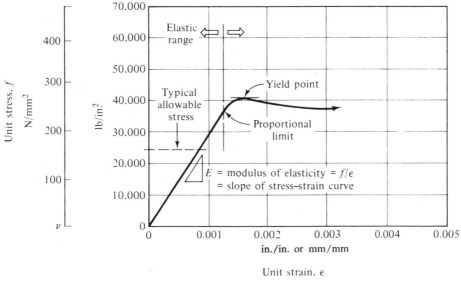

(b) Enlarged portion of elastic range.

FIGURE 2-30 Typical stress–strain diagrams for structural steel tested in tension.

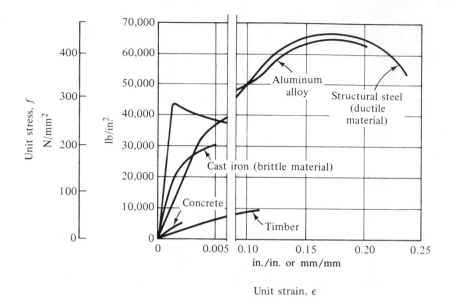

FIGURE 2-31 Typical stress–strain diagrams for different materials.

2.5.4 Strength

The term *strength* is often used with reference to the load-carrying capacity of a material. As noted above, however, materials often exhibit relatively complex behaviors under load which tend to necessitate a more precise definition of the term "strength." Many materials, for example, can continue to carry additional loads even after the proportional limit of the material has been passed. Steel, for example, continues to be able to carry increasing stress levels past the proportional limit, but it does so with greater amounts of deformation for given increases in unit stress than occurred within the elastic range. A critical point, termed the *yield point*, is reached when the steel deforms without any increase in stress at all. In fact, when steel is being experimentally studied in a load-testing machine (which typically applies deformations and measures stresses or loads rather than vice versa), an actual decrease in the stress level appears to occur (see Figure 2-30). When a load is directly applied (instead of deformations), the yield point simply marks a sudden increase in deformations. The material then undergoes massive deformations (in the plastic range) at a relatively constant stress level. As deformations increase, however, the steel begins to be able to carry small additional loads and the stress level present increases again. A level is reached that marks the maximum stress the steel can carry. This is called the *ultimate strength* of the material. After this level is reached, the steel deforms extremely rapidly, reduces in cross-sectional area (i.e., *necks*) and finally pulls apart (*ruptures*).

2.5.5 Ductile Versus Brittle Materials

DUCTILE AND BRITTLE BEHAVIOR. Materials that undergo large plastic deformations of the type described above before rupture occurs are generally referred to as *ductile* materials. Steel is a classic example of a ductile material. Conversely, if a material does

part I / INTRODUCTORY CONCEPTS

not exhibit plastic behavior under load but rather ruptures under load with little evident deformation, the material is said to be *brittle*. Cast iron is a brittle material, as is plain concrete. The stress–strain curves shown in Figure 2-31 illustrate the differences in behavior under load between the two general types of material.

The amount of ductility or brittleness present in materials such as steel can actually be controlled by altering their constituency or method of processing. Increasing the carbon content of steel, for example, reduces ductility. Alternatively, a steel exhibiting little ductility can be made more ductile by annealing it (heating it to a high temperature and allowing it to cool slowly).

IMPLICATIONS OF DUCTILITY FOR STRUCTURAL DESIGN.   From a structural design viewpoint, materials such as steel that exhibit the ductile or plastic behavior described above are very desirable, since the plastic region (with its slightly increased load-carrying capacity after the yield point) represents a measure of reserve strength. Design stress levels, or *allowable* stress levels, are invariably set below the yield stress of the material and are well within the elastic range of the material. A steel beam, for example, would be designed to have its actual stress level equal to or less than the allowable stress value. A certain level of elastic beam deflection corresponds to this stress level and its associated elastic strain. If loads on the beam increase beyond the anticipated design levels, bending stress levels and strains also increase until the yield point of the material is reached. At this point, the steel yields, but does not physically rupture, and the beam begins undergoing the massive deflections associated with the move of the material into the plastic range. These deflections are very visibly evident compared to design deflections, and can serve as a visual warning of impending failure. Because of the increase in stress required to reach the ultimate strength of the material, the beam can still carry slightly increased loads even when these massive deflections are evident. Only after the ultimate strength of the material is reached does the member fail. Since this phenomenon is coupled with an increased load-carrying capacity in the beam due to the plastic stress redistribution that occurs (see Section 6.3.9), the beam has a significant reserve load-carrying capacity. The plasticity of a material is consequently a very useful and desirable property.

Brittle materials do not exhibit plastic behavior. Structural members using brittle materials, such as cast iron beams, do not visibly deflect to any great degree prior to failure and thus give no advance warning of impending collapse. Such members are dangerous and are not used.

Concrete is also a brittle material, but when used in conjunction with ductile steel the overall member can be designed to have a measure of ductility (see Section 6.3.10).

2.5.6 Other Mechanical Properties

STRAIN RATE EFFECTS.   As the rate of load application on a structure increases, many normally ductile materials begin behaving in a brittle manner (less plastic deformation is present). Proportional limits and yield points often increase with increasing strain rates.

TEMPERATURE EFFECTS.   Low temperatures often cause many materials that are normally ductile, such as steel, to begin to exhibit brittle behavior. In many senses, the effects of low temperatures on materials are equivalent to those of high strain rates.

CREEP EFFECTS. The term "creep" refers to the continued deformation with time of a material under a constant stress level. Plastics and plain concrete, for example, tend to creep greatly under load. Steel does not. Long-term deflections in structures due to creep effects can be significant. Creep can also cause unfavorable stress redistributions to occur in reinforced concrete members.

FATIGUE EFFECTS. A material subjected to alternating stress cycles may fail at a relatively low stress level (even less than the elastic strength of the material). The endurance limit of a material is the maximum unit stress that the material can withstand for an indefinite number of cycles without failure. Most ferrous materials, such as steel, have well-defined endurance limits. Nonferrous materials, e.g., aluminum, do not. Fatigue is not generally a problem in buildings because of the absence of continuing sources of fatigue. Most vibrations simply do not last long enough to be a problem.

EFFECTS OF LOCALIZED STRESSES, CRACKS, AND FLAWS. In any structure it is likely that micro-cracks or flaws exist. At these points stresses often develop that are very high over a small area. These are called *stress concentrations*. When brittle materials are used, it is likely that at these stress concentrations cracks will develop and continue to propagate until the member fails. When ductile materials are used, the material deforms locally a slight amount at these stress concentrations, allowing a relieving redistribution in stresses to occur (the tip of the crack becomes, so to speak, blunted). Cracks are less likely to propagate than in brittle materials. Consequently, minor cracks or flaws in a structure such as a typical wide-flange steel beam are not particularly serious and do not significantly affect the load-carrying capacity of the member. This is not true with brittle members.

2.6 DEFORMATIONS IN TENSION AND COMPRESSION MEMBERS

It is often very important to be able to quantitatively describe the deformations caused in a member due to the application of an external load. For axially loaded tension or compression members in which internal stresses are uniformly distributed at a cross section, the elongation or shortening that occurs depends on the magnitude of the applied load, the cross-sectional area of the member, the length of the member, and the type of material of which the member is made.

For materials exhibiting elastic behavior, such as steel, the deformations that take place in axially stressed materials are proportional to the stress level present, assuming that other parameters such as length are held constant. For a steel tension member of a fixed length and cross-sectional area, for example, a specified axial load will produce a certain total elongation in the member. If the load on the member is doubled, the elongation present will double since the stress level doubles.

The deformations present in an axially loaded member can be evaluated by using the fact that for any elastic material the ratio of the stress (f) present to the strain (ϵ) caused is a constant value [i.e., stress/strain = constant = modulus of elasticity (E)]. The elongation in a tension member can be found by simply determining the strain associated with the stress level present ($\epsilon = f/E$) and then using this ratio to find the total deformation present ($\Delta L = \epsilon L$).

EXAMPLE

Consider a simple tension member that carries an axial load of $P = 5000$ lb (22,240 N). Find the total elongation in the member due to the load shown. Assume that the member is made of steel which has a modulus of elasticity of $E = 29.6 \times 10^6$ lb/in.² (204,000 N/mm²). Also, assume that the member is 120 in. (3048 mm) long and has a cross-sectional area of 2 in.² (1290 mm²).

Solution:

$$\text{load} = P = 5000 \text{ lb} = 22,240 \text{ N}$$

$$\text{area} = 2.0 \text{ in.}^2 = 1290 \text{ mm}^2$$

$$\text{stress} = f = \frac{P}{A} = \frac{5000 \text{ lb}}{2.0 \text{ in.}^2} = 2500 \text{ lb/in.}^2$$

$$= \frac{22,240 \text{ N}}{1290 \text{ mm}^2} = 17.24 \text{ N/mm}^2$$

$$\text{strain} = \epsilon = \frac{f}{E} = \frac{2500 \text{ lb/in.}^2}{29.6 \times 10^6 \text{ lb/in.}^2} = 0.0000845 \text{ in./in.}$$

$$= \frac{17.24 \text{ N/mm}^2}{204,000 \text{ N/mm}^2} = 0.0000845 \text{ mm/mm}$$

$$\text{elongation} = \Delta L = \epsilon L = 0.0000845(120 \text{ in.}) = 0.0101 \text{ in.}$$

$$= 0.0000845(3048 \text{ mm}) = 0.257 \text{ mm}$$

Note that the magnitude of both the strain and the total elongation are relatively small numbers. If the same member were made of aluminum having a modulus of elasticity of $E = 11.3 \times 10^6$ lb/in.² (77,900 N/mm²), the strains and elongations would be increased. Thus,

$$\text{strain} = \epsilon = \frac{f}{E} = \frac{2500 \text{ lb/in.}^2}{11.3 \times 10^6 \text{ lb/in.}^2} = 0.000221 \text{ in./in.}$$

$$= \frac{17.24 \text{ N/mm}^2}{77,900 \text{ N/mm}^2} = 0.000221 \text{ mm/mm}$$

$$\text{elongation} = \Delta L = \epsilon L = 0.000221(120 \text{ in.}) = 0.0265 \text{ in.}$$

$$= 0.000221 (3048 \text{ mm}) = 0.67 \text{ mm}$$

Instead of finding stresses and strains first and then calculating elongations, it is possible to determine a single expression for the elongation in a member in terms of the load P, member length L, cross-sectional area A, and modulus of elasticity E:

$$\Delta L = \epsilon L = \left(\frac{f}{E}\right)L = \left(\frac{P/A}{E}\right)L$$

$$= \frac{PL}{AE}$$

This is an often-used expression in structural analysis and design. It clearly shows that elongations in an axially loaded member are linearly dependent on the load present and the length of the member, and inversely dependent on the cross-sectional area and modulus of elasticity.

EXAMPLE

Find the elongation caused by a tensile force of 7500 lb (33,360 N) in a steel member 144 in. (3658 mm) long having a cross-sectional area of 0.785 in.^{2} (506.5 mm^2). Note that $E = 29.6 \times 10^6$ lb/in. (204,000 N/mm^2).

Solution:

$$\Delta L = \frac{PL}{AE} = \frac{(7500 \text{ lb})(144 \text{ in.})}{(0.785 \text{ in.}^2)(29.6 \times 10^6 \text{ lb})}$$

$$= 0.046 \text{ in}$$

$$= \frac{(33,360 \text{ N})(3658 \text{ mm})}{(506.5 \text{ mm}^2)(204,000 \text{ N/mm}^2)}$$

$$= 1.18 \text{ mm}$$

QUESTIONS

2-1. A force of P defined by the angle $\theta_x = 75°$ to the horizontal acts through a point. What are the components of this force on the x and y axes?
Answer: $P_x = 0.26P$, $P_y = 0.97P$.

2-2. The components of a force on the x and y axes are $0.50P$ and $1.50P$, respectively. What is the magnitude and direction of the force?
Answer: $1.58P$ at $\theta_x = 71.5°$.

2-3. The following three forces act concurrectly through a point: a force P acting to the right at $\theta_x = 30°$ to the horizontal, a force P acting to the right at $\theta_x = 45°$ to the horizontal, and a force P acting to the right at $\theta_x = 60°$ to the horizontal. Find the single resultant force that is equivalent to this three-force system.
Answer: $2.93P$ at $45°$.

2-4. The following three forces act through a point: P at $\theta_x = 45°$, $2P$ at $\theta_x = 180°$, and P at $\theta_x = 270°$. Find the equivalent resultant force.
Answer: $1.33P$ at $192.8°$.

2-5. Determine the reactions for the structure shown in Figure 2-32(a).
Answer: $R_{Av} = P/2$ ↑, $R_{Bv} = 5P/2$ ↑.

2-6. Determine the reactions for the structure shown in Figure 2-32(b).
Answer: $R_{Av} = 10P/3$ ↑, $R_{Bv} = 8P/3$ ↑.

2-7. Determine the reactions for the structure shown in Figure 2-32(c).
Answer: $R_{Av} = 5wL/18$ ↑, $R_{Bv} = wL/18$ ↑.

2-8. Determine the reactions for the structure shown in Figure 2-32(d).
Answer: $R_{Av} = wL/2$ ↑, $R_{AH} = wL/2$ →, $R_{Bv} = wL/2$ ↑, $R_{BH} = wL/2$ ←.

2-9. Determine the reactions for the four beams shown in Figure 2-32(e). Notice that the three inclined members are identical except for the type of end conditions present.
Answer: $R_A = P/2$ ↑, $R_B = P/2$ ↑; $R_{Av} = P/2$ ↑, $R_{AH} = 0$, $R_{Bv} = P/2$ ↑, $R_{BH} = 0$; $R_{Av} = 0$, $R_{AH} = P/2$ →, $R_{Bv} = P$ ↑, $R_{BH} = P/2$ ←; $R_{Av} = P/2$ ↑, $R_{AH} = 0$, $R_{Bv} = P/2$ ↑, $R_{BH} = 0$.

part I / INTRODUCTORY CONCEPTS

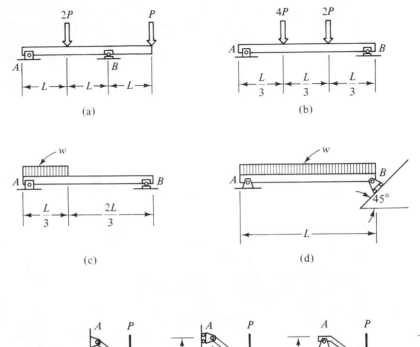

(a)   (b)   (c)   (d)

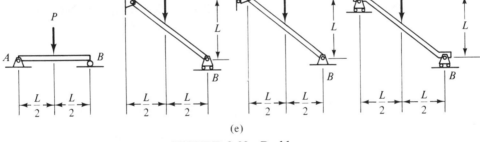

(e)

FIGURE 2-32   Problems.

2-10. Draw shear and moment diagrams for the same beam analyzed in Question 2-5. What is the maximum shear force present? What is the maximum bending moment present? Answer: $V = -3P/2$, $M = -PL$.

2-11. Draw shear and moment diagrams for the beam analyzed in Question 2-6. What is the maximum shear force present? What is the maximum bending moment present? Answer: $V = +10P/3$, $M = 10PL/9$.

2-12. Draw shear and moment diagrams for the beam analyzed in Question 2-7. What is the maximum shear force present? What is the maximum bending moment present? Answer: $V = +5wL/18$, $M = 25wL^2/648$.

2-13. Draw shear and moment diagrams for the beam analyzed in Question 2-8. What is the maximum shear force present? What is the maximum bending moment present? Answer: $V = \pm wL/2$, $M = wL^2/8$.

2-14. Draw shear and moment diagrams for the four beams analyzed in Question 2-9. For the inclined members, the shear and moment diagrams should be drawn with respect to the longitudinal axes of the members. Transverse components of the applied and reactive forces should thus be considered in determining shears and moments. Compare the maximum moments developed in all four beams. Discuss.

2-15. Determine by reviewing manufacturer's literature or other sources the moduli of elasticity for at least two different types of plastics (e.g., plexiglass and polyvinylchloride). Consult your library.

2-16. What is the unit strain present in an aluminum specimen loaded to 10,000 lb/in.²? Assume that $E_a = 11.3 \times 10^6$ lb/in.²
Answer: 0.000885 in./in.

2-17. What is the unit strain present in a steel specimen loaded to 24,000 lb/in.²? Assume that $E_s = 29.6 \times 10^6$ lb/in.²
Answer: 0.000811 in./in.

2-18. A steel bar that is 2 in. × 2 in. square is 20 ft long and carries a tension force of 16,000 lb. How much does the bar elongate? Assume that $E_s = 29.6 \times 10^6$ lb/in.²
Answer: 0.032 in.

2-19. A steel bar that is 20 mm in diameter is 5 m long and carries a tension force of 20 kN. How much does the bar elongate? Assume that $E_s = 0.204 \times 10^6$ N/mm².
Answer: 1.56 mm.

2-20. Conduct a detailed study of how the technique of graphic statics can be used to analyze the forces in a gothic structure. A highly sophisticated example of such an analysis is shown in Figure 2-33. Consult your library. See, for example, the following references: Jacques Heyman, "The Stone Skeleton," *International Journal of Solids and Structures*, Vol. 2, 1966, 249–279; and William Wolfe, *Graphical Analysis, A Text Book on Graphic Statics*, 1st ed., McGraw-Hill Book Company, Inc., New York, 1921.

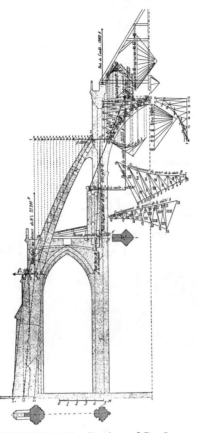

FIGURE 2-33 Section of St. Ouen.

chapter 3

INTRODUCTION TO STRUCTURAL ANALYSIS AND DESIGN

3.1 ANALYSIS AND DESIGN CRITERIA

This chapter provides an introduction to the analysis and design of structures in a building context. In order to analyze or design a structure, it is necessary to establish criteria that can be used as either measures for determining whether a given structure is acceptable for use in a specified circumstance or for use directly as design objectives that must be met. The more important criteria are discussed below.

3.1.1 Serviceability

The structure must be able to carry the design load safely—without excessive material distress and with deformations within an acceptable range. The ability of a structure to carry loads safely and without material distress is achieved by using safety factors in the design of the element. By altering the size, shape, and choice of materials, stress levels in a structure can be maintained at a level considered acceptable for life safety and such that material distress (e.g., cracking) does not occur. This is basically a *strength* criterion and is of fundamental importance.

Another aspect of the serviceability of the structure is the nature and extent of the deformations caused by the load and whether these deformations are acceptable or not. Excessive deformations could result in the structure's interfering with or causing distress to some other building element. They could also simply be visually undesirable. It should be noted that because a structure appears to be deforming excessively does not necessarily mean that the structure is unsafe. Large deflections or deformations may be associated with an unsafe structure, but not necessarily (consider the large deflections associated with diving boards). Deformations are controlled by varying the *stiffness* of a structure. Stiffness depends largely on the type, amount, and distribution of material in a structure. Often it is necessary to use a larger member to achieve necessary stiffness than is required to achieve the necessary structural strength.

Associated with deformations, but not the same phenomena, are *movements* in structures. In some situations, the actual accelerations and velocities of structures carrying dynamic loads are perceptible to building occupants and can cause discomfort. The movements associated with multistory buildings under the action of wind are a case in point. Criteria thus relate to limiting accelerations and velocities. Control is achieved through manipulations involving the stiffness of the structure and its damping characteristics.

3.1.2 Efficiency

This criterion involves the relative economy with which a structure achieves its design objective. A measure often used is the amount of material required to support a given loading in space under the conditions and constraints specified. It is quite possible that several different alternative structural responses to a given loading situation will be equally serviceable. This does not mean, however, that each will require exactly the same amount of structural material to achieve the same level of serviceability. Some solutions may require less material than others. The use of minimum volume as a criterion is one that has intrinsic conceptual appeal to a large number of architects and engineers and is one that will be extensively used throughout this text. Other measures of efficiency, however, could obviously be developed and used.

3.1.3 Construction

Construction considerations are often influential in choosing a structural response. It is entirely possible that a structural assembly that is highly efficient in terms of the way material is used will not be easy to fabricate and assemble. Construction criteria are diverse and include such considerations as the amount and type of effort or manpower required to construct a given facility, the type and extent of equipment required, and the total amount of time necessary to complete construction.

A general factor influencing the ease of construction of a facility is the complexity of the facility as reflected in the number of different pieces involved and the relative degree of effort necessary to assemble the pieces into a whole. The size, shape, and weight of the pieces are also important since the type of equipment necessary for construction is often dictated by these factors. In general, an assembly consisting of pieces of a size and shape manageable with easily available construction equipment and that can be assembled with either few or highly repetitive operations will be relatively easy to construct. Although highly important, construction ease will not be dealt with extensively in this book.

3.1.4 Costs

Costs are invariably an influential factor in the choice of most structures. The concept of cost cannot be dissociated from the two considerations discussed immediately above, material economy and ease of construction. Total costs of a structure are primarily dependent on the amount and cost of material used, the amount and cost of labor required to construct the facility, and the cost of equipment needed during construction. Obviously, a highly efficient structure that is not difficult to construct will probably be an economical one. This text will not deal, however, with costs in any comprehensive way.

3.1.5 Other

There are, of course, a host of other factors that are influential in selecting a structure. In contrast to the relatively objective and measurable criteria discussed above, many of these additional factors are subjective in nature. Not the least of these is the role of structure as it forms a part of the larger image of the building as envisioned by the architect (e.g., the role of structure as a space definer, etc.). Although obviously a subject of interest and depth, these considerations are largely beyond the scope of this text.

3.1.6 Multiple Criteria

With respect to the preceding list of criteria, it is rare that a structure is strictly responsive to only one or another. The concept of serviceability and the involved notion of safety, however, are typically common to all structures. Indeed, this is the fundamental responsibility of the structural designer. Other criteria may also be involved, but serviceability should *always* be involved.

3.2 LOADS ON STRUCTURES

3.2.1 Introduction

In analyzing or designing a structure, it is necessary to have a clear picture of the nature and magnitude of the loads applied to the structure. Figure 3-1 illustrates in a diagrammatic way the primary loads that must be considered and ways of describing and characterizing them.

Of fundamental importance is the distinction between static and dynamic loads. *Static* forces are those that are slowly applied to a structure and are steady-state in character. Resultant deformations in the structure associated with these forces also slowly develop, are steady-state in character, and reach a peak when the static force is maximum. *Dynamic* forces are those that are suddenly applied to a structure. Often they are not steady-state forces and are characterized by rapid changes in

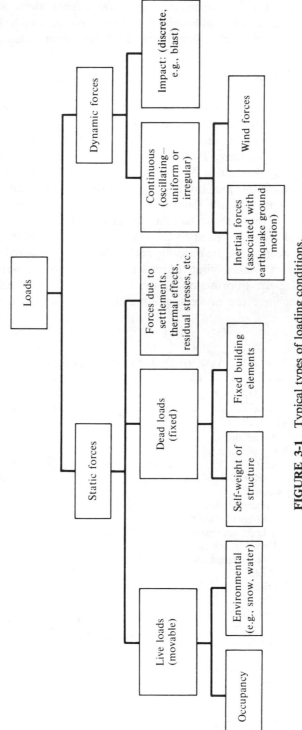

FIGURE 3-1 Typical types of loading conditions.

magnitude and location. Associated structural deformations also vary rapidly. Dynamic forces can cause oscillations to occur in structures such that peak deformations do not necessarily occur when the applied forces are maximum.

3.2.2 Static Forces

DEAD LOADS. Static forces are typically subclassified into dead loads, live loads, and forces due to settlements or thermal effects.

Dead loads are those forces acting vertically downward on a structure that are relatively fixed in character. The self-weight of the structure itself is a dead load. The weights of any permanent building elements, such as floor surfaces, mechanical equipment, nonmoveable partitions, and so on, are also dead loads. The exact weights of these elements are typically known (or can be easily determined) to a high degree of accuracy. All methods for calculating the dead load of an element are based on a consideration of the unit weight of the material involved and on the volume of the element. Unit weights are empirically determined and tabulated for easy use in a number of sources. Determining volumes is a straightforward but occasionally tedious process. Tabular information of the type illustrated in Tables 3-1 and 3-2 is available to facilitate the process of obtaining dead loads for common building elements.

Table 3-1. AVERAGE DENSITIES OF MATERIALS

Material	lb/ft^3	kg/m^3
Metals		
Aluminum, cast	165	2,643
Copper, cast	556	8,907
Lead	710	11,374
Steel, rolled	490	7,849
Masonry		
Concrete, plain		
Stone aggregate	144	2,307
Light aggregate	75–110	1,201–1,762
Concrete, reinforced		
Stone aggregate	150	2,402
Brick	100–130	1,602–2,083
Earth		
Clay, dry	63	1,009
Clay, damp	110	1,762
Earth, dry	75–95	1,201–1,521
Earth, damp	80–100	1,281–1,602
Other		
Glass, common	156	2,499
Glass, plate	161	2,579
Pitch	69	1,105
Tar	75	1,202

Table 3-2. AVERAGE LOADS OF CONSTRUCTION COMPONENTS

Component	lb/ft^2	kg/m^{2a}	N/m^{2b}
Roofs			
Asphalt shingles	2.0	9.8	95.8
Cement asbestos shingles	4.0	19.5	191.5
Composition roofing			
3-ply and gravel	5.5	26.8	263.3
5-ply and gravel	6.5	31.8	311.2
Corrugated metal			
20 U.S. std. gage	1.7	8.3	81.4
28 U.S. std. gage	0.8	3.9	38.4
Concrete plank decking			
(per inch of thickness)	6.5	31.8	311.2
Insulation, rigid fiberglas	1.5	7.3	71.8
Lumber sheathing (1 in., 2.54 cm)	2.5	12.2	119.7
Plywood sheathing (1 in., 2.54 cm)	3.0	14.6	143.6
Floors			
Concrete slab, stone aggregate			
(per inch of thickness)	12.0	58.6	574.6
Hardwood (1 in., 2.54 cm)	4.0	19.5	191.5
Hollow-core concrete planks			
(6 in., 15.2 cm, thick)	45–50	220–244	2155–2394
Linoleum ($\frac{1}{4}$ in., 0.635 cm)	1.0	4.9	47.9
Plywood (1 in., 2.54 cm)	3.0	14.6	143.6
Steel decking	2–10	10–49	96–480
Walls			
Brick			
(4 in., 10.2 cm)	40	195.3	1915.2
(8 in., 20.4 cm)	80	390.6	3830.4
Hollow concrete block, heavy			
Heavy aggregate			
(4 in., 10.2 cm)	30	146.5	1436.4
(8 in., 20.4 cm)	55	268.5	2633.4
Light aggregate			
(8 in., 20.4 cm)	38	185.5	1819.4
Wood studs (2 × 4, 16 in. oc)			
Window assembly	8	39.1	383.0

[a]Conventional metric system.
[b]SI system.

LIVE LOADS. Live loads are those forces that may or may not be present and acting upon a structure at any given point in time. Although movable, live loads are still typically applied slowly to a structure. Use or occupancy loads are live loads. Occupancy loads include personnel, furniture, stored materials, and other similar items. Snow loads are also live loads. All live loads are characterized by their movability. They typically act vertically downwards but occasionally can act horizontally as well.

Wind loads and earthquake forces, considered later, are special forms of live loads typically considered separately because of their dynamic aspects. It is probably obvious that predicting exactly what magnitudes and distributions characterize live loads is a highly difficult task. An empirical approach is consequently taken in which measurements have been made of the nature and magnitudes of the loads associated with different occupancy types. From these statistical data, equivalent loads typically expressed in terms of a load per unit area have been derived for design purposes. Concentrated loads are occasionally used also. These equivalent loads generally reflect in one figure the net effects of all movable vertically acting loads associated with a particular occupancy type. They generally represent the maximum expected combination of occupancy-related forces that the structure would possibly have to carry. As such they are founded on certain assumptions about what loads can be reasonably expected on a structure and the frequency of their occurrence.

The actual live loads on a structure at any particular point in time are typically less than the loads a structure actually must be designed to carry. At some point, however, there is a high probability that the structure will indeed have to carry the design loading. It should also be evident that a structure initially designed to carry loads derived from one occupancy type should be carefully checked before being subjected to loads from other occupancy types. A structure initially designed to carry loads derived for apartment houses would be inadequate, for example, if the building were converted into use as an office building or storage warehouse.

Table 3-3 illustrates equivalent live loads recommended for several different occupancy types. Provisions in building regulations typically require that structures be designed according to loadings of this general type. The loadings shown are illustrative only and should not be used for design purposes (local building code recommendations should be used instead).

SNOW LOADS. Snow loads on roofs vary very widely and depend on such factors as elevation, latitude, wind frequency, duration of snow fall, site exposure, roof size, geometry, and inclination. As a rule of thumb, the weight of snow is about 0.5 to 0.6 lb/ft² per inch of snow depth.

These figures vary widely, however, depending on snow density. Most design snow loads for typical urban areas range from 20 to 60 lb/ft² (958 to 2873 N/m²) (see Table 3-4). Local experience should always be checked, however, since in particular areas loads can be much higher than these figures indicate. This is particularly true in mountainous regions. Snow loads in some inhabited areas have been as high as 250 to 300 lb/ft² and even higher in uninhabited areas.

3.2.3 Wind Loads

STATIC EFFECTS OF WIND.   A structure in the path of wind causes wind to be deflected, or in some cases, stopped. As a consequence, the kinetic energy of the wind is transformed into the potential energy of pressure or suction. The magnitude of the pressure or suction caused by the wind at a point on a structure depends on the velocity of the wind, the mass density of the air, the geometrical shape, dimensions and orienta-

Table 3-3. TYPICAL UNIFORMLY DISTRIBUTED LIVE LOADS

Occupancy or use	lb/ft²	kg/m²	N/m²
Apartments, multifamily housing:			
Corridors	80	391	3830
Private apartments	40	195	1915
Public rooms	100	488	4788
Assembly halls			
Fixed seats	60	293	2873
Movable seats	100	488	4788
Dance halls	100	488	4788
Garages, public	100	488	4788
Gymnasiums	100	488	4788
Hotels			
Corridors	100	488	4788
Guest rooms	40	195	1915
Public rooms	100	488	4788
Libraries			
Reading rooms	60	293	2873
Stacks	150	732	7182
Manufacturing	125	610	5985
Office buildings			
Offices	60	293	2873
Lobbies	100	488	4788
Residential dwellings			
First floors	40	195	1915
Second floors	30	146	1436
Schools			
Classrooms	40	195	1915
Corridors	80	391	3830
Stairs	100	488	4788
Stores, retail			
First floor	100	488	4788
Upper floors	75	366	3591
Theaters			
Aisles, corridors	100	488	4788
Balconies	60	293	2873
Floors	60	293	2873
Stage floors	150	732	7182

Table 3-4. TYPICAL SNOW LOADS

Region	lb/ft²	N/m²
Southern states	0–15	0–718
Central states	25–35	1197–1676
Northern states	30–50	1436–2394
Great Lakes, New England and mountain areas	30–80	1436–3830

tion of the structure, the exact location of the point considered on the structure, the nature of the surface that the wind acts on, and the overall stiffness of the structure.

As any fluid such as air flows around an immersed object, a complex flow pattern is generated around the object. The nature and complexity of the flow pattern depend upon the shape of the object. Flows can be either smooth or turbulent. The forces acting on the object as a result of the impinging air flow can be either pressure forces or suction forces. The more an object is streamlined, the less is the total reactive force exerted by the structure in opposing the motion of the air. Figure 3-2 illustrates air flows around some typical shapes in a diagrammatic way.

The magnitude of the forces involved as air moves around a shape is dependent on the velocity of the wind, among other factors. Design wind velocities for different geographical locations are determined from empirical observations. They range from lows of 60 mph (96 km/hr) in some inland regions to highs of 100 mph (161 km/hr) in other inland regions and 120 mph (193 km/hr) in some coastal zones. Design velocities are usually based on a 50-year mean recurrence interval. Since wind velocities increase with height above ground, design values are accordingly increased. Allowance is also made for whether the building site is in an urban or rural setting. More sophisticated analyses also include gust-response factors, which are a function of the size and height of the structure, the surface roughness, and the obstacles present in the surrounding terrain. Local building codes should be consulted for exact design loadings or velocities.

Once the design wind velocity is known, it is possible to determine the dynamic pressure arising from wind action and to express it as an equivalent static force. The dynamic pressure of wind can be determined by application of the well-documented *Bernoulli equation* for fluid flow, which yields the expression $q = \frac{1}{2}\rho V^2$, where q is the dynamic pressure, ρ the mass density of air, and V the velocity of wind. For standard air (usually defined as 0.0765 lb/ft³), the velocity pressure in pounds per square foot at a height h is given by $q_h = 0.00256 V_h^2$, where V_h is expressed in miles per hour. The conversion of this force into an equivalent static force involves the use of a pressure coefficient, C_D, that depends on the geometry of the body that the wind impinges upon. The reactive force, F_D, exerted by the structure in opposing the motion of the wind can be found by using this pressure coefficient and the dynamic

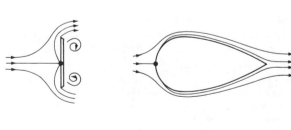

(a)   (b)   (c)

FIGURE 3-2   Flow about different shapes.

part I / INTRODUCTORY CONCEPTS

pressure expression. Thus, $F_D = C_D q_h A$, where C_D is the pressure coefficient for the shape involved, q_h the velocity pressure at height h, and A the exposed area of the building surface normal to the wind. A more exact expression would also involve the use of a gust factor, G_F. Thus, $F_D = C_D q_h A G_F$. Using the simpler expression, the *average* static pressure, p, is given by $p = F_D/A$ or $p = C_D q_h$. The latter expression is frequently used.

The *pressure coefficient*, C_D, depends on the building form. C_D is also often referred to as the *shape coefficient*. Pressure coefficients for different shapes generally reflect the relative amount of obstruction caused by the shape to an impinging air flow. Streamlined shapes have pressure coefficients smaller than nonstreamlined shapes. When the flow of air is completely stopped by the shape, the pressure coefficient is equal to unity ($C_D = 1$). In actual building shapes, the flow of air is never actually stopped, but is deflected. Flow patterns around actual buildings are extremely complex. For this reason, pressure coefficients for a number of different building shapes and configurations have been developed and tabulated. Since on any given building the wind produces both pressure and suction (depending on the location of the point considered), there are specific pressure coefficients tabulated for different locations on a building. Typical coefficients are shown in Figure 3-3.

These coefficients are a sampling only. The literature contains detailed information for a wide variety of situations and building configurations.[1] Typical resultant design wind loads are on the order of 20 to 40 lb/ft² (960 to 1920 N/m²).

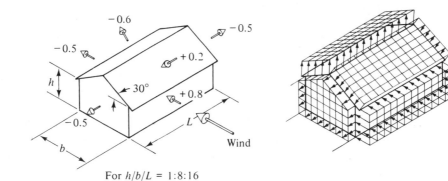

For $h/b/L = 1:8:16$

Wind force $= C_d q_h A$

C_d = pressure coefficient
(+ pressure − suction)

q_h = velocity pressure $= \dfrac{1}{2}\rho v^2$

A = exposed surface area

FIGURE 3-3 Typical wind-force coefficients (C_D) for a gabled building (pressure coefficients actually depend on exact dimensions of building).

[1] See "Wind Forces on Structures," Paper 3269, Final Report of the Task Committee on Wind Forces of the Committee on Loads and Stresses of the Structural Division, American Society of Civil Engineers; and *Timber Construction Manual*, American Institute of Timber Construction, pp. 3–23 to 3–29.

DYNAMIC EFFECTS OF WIND. In our previous discussion, the static nature of wind forces was stressed. The dynamic nature of wind forces, however, is also extremely important. Dynamic effects can arise in several ways. One is simply that wind is rarely a steady-state phenomenon. Buildings are therefore subjected to alternating forces. This is particularly true for buildings in an urban setting. As is evident in Figure 3-2(c), the flow pattern of air around a building is not smooth. If there is a group of closely spaced buildings, the wind pattern is even more complex, since some buildings are in the turbulent wake of others. The action of the wind becomes that of a buffeting. The result is that buildings can begin swaying to and fro due to these alternating load effects.

Winds can produce a dynamic response in buildings even when the wind is at a relatively steady-state velocity. This is particularly true in structures that are relatively flexible, as is the case in many cable-supported roof structures. The wind causes varying force distributions on the roof surface, which in turn cause shape changes, either major or minor, to occur. The new shape causes different pressure or suction distributions, which again lead to new shape changes. The result is a constant movement or flutter in the roof. The problem of roof flutter is a major one in the design of flexible structures. Techniques for controlling roof flutter invariably have major design implications. The phenomenon of roof flutter in flexible structures is discussed in much greater detail in connection with cable design (see Section 5.2).

The dynamic effects of wind can also be problematic in high-rise structures by causing undesirable swaying (see Section 14.3.3).

3.2.4 Earthquake Forces

Earthquakes are vibratory phenomena associated with shock loadings on the earth's crust. While these shock loads can result from a number of causes, one of the primary reasons is the sudden slippage that frequently occurs between adjacent crust plates that make up the earth's surface. Locations where this slippage occurs are called *fault zones*. The shock associated with this slippage is propagated in the form of waves. These waves cause the earth's surface and any buildings resting on it to vibrate. As the building is vibrated, forces are developed in the structure of the building because of the tendency of the mass of the building to resist the motion. The forces developed are consequently inertial in character. The magnitude of these forces depends on many factors. The mass of the building is clearly important, since the forces involved are inertial. Other factors include the way the mass is distributed, the stiffness of the structure, the stiffness of the soil, the type of foundation, the presence of damping mechanisms in the building, and, of course, the nature and magnitude of the vibratory motions themselves. The latter are difficult to describe precisely since they can be almost random in nature. Some discernible patterns are present, however. The motions generated are three-dimensional in nature. The horizontal ground movements are usually the most important from a structural design viewpoint.

The mass and rigidity of a structure, and its related natural period of vibration, are the most important of all the factors that affect the overall response of a structure to motion and the specific nature and magnitude of the forces developed as a consequence of the motion. As a way of understanding some of the phenomena involved, it is useful to consider first how a completely rigid structure responds to a simple vibratory motion. Structures having flexibility, as is common in all building structures, will be considered next.

When a completely rigid structure, such as a solid block, is subjected to base vibrations, the structure simply moves as a unit along with the forcing motions. Since there is a natural physical tendency for any mass initially at rest to remain at rest, there are internal forces set up in the rigid body due to the ground motion that are inertial in character (Figure 3-4). The upper mass of the body tends to remain at rest while the base tends to be translated due to the ground motion. These different tendencies result in the development of internal forces. For a simple rigid block of weight W subjected to a horizontal ground motion characterized by an acceleration a, the inertial force F_i that is developed is simply $F_i = (W/g)a$, where g is the acceleration due to gravity. This expression follows directly from Newton's first law, discussed earlier. For equilibrium, a shearing force (V) is developed at the base of the structure such that $V = F_i = (a/g)W$. This is also the force that the structure must be capable of withstanding. Quite simply, then, when ground accelerations are known or can be estimated, the shear force that a totally rigid structure must resist at its base can be found as a percentage of the weight of the building (say, 5 to 20%). Structural elements whould have to be designed to carry this total shearing force.

The preceding analysis indicates that the forces generated in a body are very sensitive to the magnitudes of the accelerations involved. Note that if a body is very slowly displaced so that the accelerations involved are essentially zero, or if it is moved a constant velocity (again a case of zero accelerations), then there are no inertial forces generated in the body at all (happily, the earth spins and moves through space at a constant velocity or we would all be in what could only be described as deep trouble). Consequently, the aspects of earthquake motions that are of most importance are the magnitudes of the accelerations involved.

Although the concepts described above formed the basis for many of the earliest

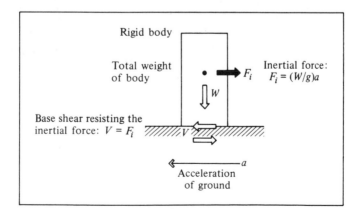

FIGURE 3-4 Inertial forces due to ground motion in a rigid body.

approaches to earthquake design, the model is undoubtedly simplistic. If the actual flexibility that a structure possesses is taken into account, a more sophisticated model is needed to predict the exact forces generated in a structure due to ground accelerations. One very important reason for going to a more complex model is that the forces generated in a flexible structure depend not only on the magnitude of the ground accelerations involved but on the relative stiffness and vibratory-motion characteristics of the structure itself. An interaction can take place which can lead to forces much higher than those predicted by the simpler static model.

An aspect of primary importance in considering the behavior of flexible structures subjected to ground accelerations is the *natural period of vibration* of the structure. Consider the simple structure illustrated in Figure 3-5. If the top of the structure is horizontally displaced and then released, it is obvious that the top of the structure would oscillate back and forth with slowly decreasing amplitudes until the structure finally came to rest. Of interest here is that these free vibratory movements are not random at all but vary in precise ways. The vibrations are said to be harmonic, since the displacement varies in a sinusoidal way with the time after release. The time required for the motion to go through one complete back-and-forth cycle is

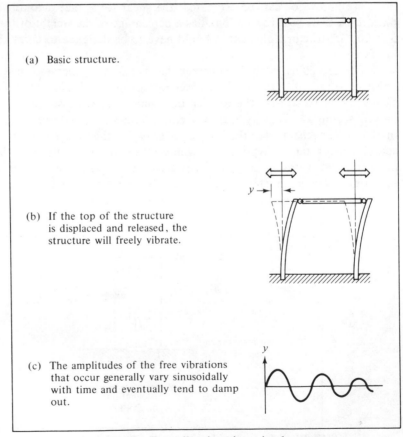

(a) Basic structure.

(b) If the top of the structure is displaced and released, the structure will freely vibrate.

(c) The amplitudes of the free vibrations that occur generally vary sinusoidally with time and eventually tend to damp out.

FIGURE 3-5 Free vibrations in a simple structure.

termed the natural period of vibration T. The frequency f of this vibration is defined as the number of cycles that occur per unit time. Note that $f = 1/T$.

The frequency of free vibration that occurs in a structure of the type described depends directly on the mass of the structure and the relative stiffness of the columns with respect to horizontal forces. If the columns were of low stiffness, they would provide relatively little in the way of restoring forces which would tend to cause the deflected structure to resume its original shape after release. Consequently, the top of the structure would lumber back and forth at a relatively slow pace until the oscillations died down. A structure of this type is said to have a long natural period of vibration. If, on the other hand, the columns were very rigid, they would provide significant restoring forces, tending to cause the structure to resume its original shape after release. Consequently, the structure would still oscillate back and forth, but this time would do so quite rapidly. A structure of this type is said to have a short natural period of vibration.

The fact that the oscillations in a structure of either long or short natural periods die out with time is due to the presence of damping mechanisms in the structure. Damping mechanisms that absorb energy are present in all building structures. In the structure shown in Figure 3-5, frictional forces that are bound to exist in the pinned connections between the beam and columns, for example, would cause damping to occur. In larger structures other damping mechanisms, such as cracking of partitions, also exist.

The situation described above is one where free vibrations occur due to a release of an imposed displacement at the top of the structure. If the base of the structure is instead suddenly translated a short distance horizontally, the mass of the structure (particularly the beam) would initially tend to resist translation because of inertial tendencies. The structure would consequently be set in vibration (a situation converse to the initial one would be created). If the base of the structure were translated back and forth continuously, as would be the case in an earthquake, the structure would obviously continue to vibrate as long as the ground motion is present. In this case, however, the actual vibratory motion that occurs can potentially be strongly influenced by the precise characteristics of the ground motions and is not simply a free vibration. In cases when the frequency of the ground movements is completely different from that of the natural frequency of free vibration associated with the structure, the consequence is that no real interactive effect occurs. In cases when the frequency of the ground movement is approximately the same as that of the natural frequency of vibration of the structure, however, an interactive effect can occur which tends to cause the amplitudes of the vibrations to increase above the magnitudes associated with the imposed ground displacements.

The phenomenon described above, called *resonance*, is a crucial one in structural design. One way of visualizing the phenomena is to think in terms of how a weight suspended from a simple spring behaves (see Figure 3.6). There are some very direct analogies, as will be discussed shortly, between a spring system of this type and the frame discussed above. If the suspended weight is displaced and released, a free vibration similar to that described for the frame will occur. The natural period of this vibration is given by the familiar $T = 2\pi\sqrt{W/gk}$ where W is the weight involved, g is the acceleration due to gravity, and k is the spring constant defining the load-

(a) Free vibrations: if
    initially displaced and
    released, the weight
    freely vibrates at a
    certain natural
    frequency.

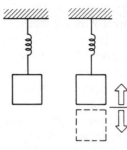

(b) Force vibrations: if the
    system is subjected to a
    oscillatory motion of a
    frequency very much lower
    than the natural frequency
    of the system, the weight
    simply translates up and
    down with the applied
    motion (no elongations or
    contractions occur in the
    spring due to the applied
    motion).

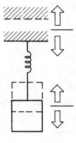

(c) Forced vibrations: when
    the frequency of the applied
    oscillations is very much
    greater than that of the natural
    frequency of the system, the
    inertial tendencies of the weight
    preclude it from following the
    rapid applied oscillations and
    the weight remains essentially
    stationary. Spring deformations
    comparable to the amplitude of
    the applied oscillation occur.

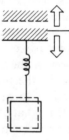

(d) Forced vibrations: when the
    frequency of the applied
    oscillations is similar to that
    of the natural frequency of
    the system, a resonance con-
    dition can develop in which
    the amplitude of vibration of
    the weight exceeds that of the
    applied oscillation. Abnormally
    high forces can consequently
    develop in the spring.

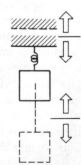

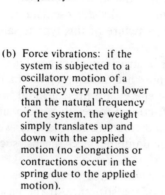

FIGURE 3-6 Free and forced vibrations in a simple spring–weight system.

deformation characteristics of the spring (this expression is discussed in any basic physics textbook). If, on the other hand, the weight were simply suspended from the spring and the end of the spring itself moved up and down, then one of the following phenomena would occur. If the applied oscillatory movement were very slow (i.e., a long period), the suspended weight would simply translate along with the spring at the same rate. The spring would not be elongated but would also simply ride along. On the other hand, if the applied oscillatory motion were extremely rapid, the suspended weight would merely stay relatively stationary in space since the inertial tendency of the weight would prevent it from following the imposed rapid movements. Rapid elongations and relaxations would occur in the spring (which means internal forces develop and relax) but these are not necessarily large and no greater than the amplitude of the applied oscillations.

The critical case arises when the period of the applied oscillations is exactly equal to the natural period of vibration of the spring–weight system. In this event the initial applied oscillations cause the weight to begin vibrating up and down. If the applied oscillations continue, the amplitude of the vibratory movements in the spring–weight system will continue to increase. The relative elongations and contractions of the spring would hence also continue to increase, and along with this so would the magnitude of the force developed in the spring. The amplitudes could easily increase far beyond those of the original applied oscillations. The result can be that the system approaches a point of resonance wherein the relative movements tend to become indefinitely large (and, consequently, so do internal forces in the spring), so that eventually the system self-destructs. All the phenomena described above can be demonstrated by suspending a weight from a rubber band and vibrating it at different speeds.

Real structures can and do behave in much the same way that the simpler system described above does. When the natural frequency of the applied movements exactly equals the natural period of vibration of the system itself, the phenomenon of resonance can occur. The effects can be disastrous as is illustrated by the dramatic failure of the Tacoma Narrows Bridge in Washington State in 1940. A wind caused an initial twisting to develop in the structure, which then began to oscillate in a twisting motion of increasing amplitude (the deck was at one time 45° to the horizontal) until the 2800-ft (840-m) structure collapsed. Fortunately, most buildings inherently possess sufficient damping mechanisms so that dramatic resonances of this type rarely occur in buildings. Still the phenomenon does cause abnormally high displacement amplitudes to occur, which correspond to abnormally high force intensities in structural members. Failure to take into account the dynamic action of an earthquake loading can thus potentially lead to highly undesirable consequences.

The dynamic analysis of a structure more complex than the simple ones discussed can be very involved due to the many modes of vibration that are possible. As will be described in more detail later, a multistory building can vibrate in different modes which can cause floors on different levels to accelerate in different directions at the same time (see Section 14.3.4). Predicting how a multistory structure will vibrate can be very complex. Analytical predictions, however, are possible by modeling the structure as a complex system of point masses (representing building weights), springs (representing the stiffnesses of the structural members), and damping devices (representing the energy-absorbing mechanisms present in a building, such as those formed when partitions crack because of vibratory movements).

Because of the complex dynamic action described, a static model for assessing earthquake forces of the type initially described can be highly misleading. For design purposes, however, a variation of the static model is still often used, owing simply to the complexity of the dynamic analyses required. An expression in a common building code for determining earthquake design forces, for example, is of the form $V = ZIKCSW$. In this expression, V is the total static shear at the base of the structure, W is the total dead load of the building, C is a coefficient that depends on the fundamental period of vibration (T) of the building, Z is a factor depending on the geographical location of the building and the probable seismic activity and intensity of the location, K is a factor depending on the type of structure and construction used (primarily with respect to its relative stiffness and ductility), I is an importance factor depending on the building type (e.g., hospitals) and S is a coefficient depending on (among other things) the relation between the natural period of the building and that of the soil on which the building rests. The fundamental period of vibration (T) on which the factor C depends is often estimated through expressions of the form $T = 0.05H/\sqrt{D}$, where D is the dimension of the structure in a direction parallel to the applied forces and H is the height of the main portion of the building above the base (in feet). The coefficient C then takes the form $C = 15/\sqrt{T} \leq 0.12$. All these expressions and factors are largely empirically determined. The base shear V obtained through the evaluation of these factors is then distributed at the various story levels through prescribed methods to act as applied lateral loads.

An implication of the static equation $V = ZIKCSW$ is that for a given building the design force V is greater when a relatively "stiff" bracing system is used than it is when a more "flexible" one is adopted. This result is important from a design viewpoint (see Section 14.3.4).

The empirical approach briefly outlined above is described in detail in the *Uniform Building Code*.[1] Another approach has recently been developed by the Applied Technology Council, which uses a different formulation.

3.2.5 Load Combinations

As is evident from the preceding sections, a great number of different types of loadings act on a structure. Of critical importance in determining design loads is the question of whether or not all of these loadings act on the structure simultaneously. Dead loads, it should be reemphasized, by definition always act on a structure. The variability is in the live load or combination of live loads. Is it reasonable, for example, to design a structure to sustain at the same time the maximum loads that might result from an earthquake and those associated with a maximum wind force, and a full-occupancy loading condition? The probability that all these loadings will occur with maximum intensity at the same point in time is remote.

The structure *could* be designed to carry the simultaneous maximums of all possible loadings, but the resultant structure would, in all probability, be greatly

---

[1] International Conference of Building Officials, *Uniform Building Code*, Whittier, California, 1976.

oversized for the load combinations that would actually occur during the lifetime of the structure. In recognition of this, many codes or regulations allow specific reductions in design loadings when certain load combinations are present.

With respect to occupancy loads in a multistory building, it is extremely unlikely that all floors of the building will simultaneously carry maximum occupancy loads. For this reason, a reduction is allowed in the design loading for members carrying the contributory effects of live loadings from many floors. A typical recommendation is as follows:

1. No reduction shall be applied to the roof live load.
2. For live loads of 100 lb or less per square foot, the design live load on any member supporting 150 ft² or more may be reduced at the rate of 0.08 %/ft² of area supported by the member, except that no reduction shall be made for garages or for areas to be occupied as places of public assembly. The reduction shall exceed neither R as determined by the following, nor 60%:

$$R = 23\left(1 + \frac{D}{L}\right)$$

in which R = reduction, %
$\qquad D$ = dead load per ft² of the area supported by the member
$\qquad L$ = design live load per ft² of the area supported by the member

For storage live loads exceeding 100 lb/ft² no reduction shall be made except that the design live loads on columns may be reduced 20%.

For combinations of different types of loadings, it is often (but not always) permissible to reduce the total design force by some specified factor. For example, rather than designing for 1.00 [dead load + live load + wind load + earthquake load], many codes allow you to design for 0.75 [dead load + live load + (wind load or earthquake load)]. The expectation is clearly that not all loads will act on the structure at their full value simultaneously. On the other hand, it is usual that the structure is designed for the full dead and live load considered acting simultaneously, or 1.00 [dead load + live load]. Usually multiple types of possible loading conditions must be considered and the structure designed to carry the worst possible combination.

3.3 THE GENERAL ANALYSIS PROCESS

3.3.1 Basic Steps

This section explores the basic steps in analyzing a given structure with the intent of determining its adequacy for a given situation. The general process of analysis typically includes the steps described below:

1. The nature and extent of the structure to be analyzed must first be determined. In some instances, for example, it may be possible and desirable to isolate the constituent elements of the whole building structure and look at them in detail (e.g.,

focusing on a specific beam in a larger assembly of beams and columns). This often facilitates the process of determining whether the specific element considered is adequate to play its part in the overall functioning of the structure. If a structure is decomposed into its more basic elements, it is necessary to *model* the boundary conditions of the element so that its true context is correctly represented. This modeling often assumes the form of forces and moments at the ends of the member. In many cases, this modeling is simple (e.g., in a simply supported beam system). This modeling will be briefly explored in the following section. In other cases, it is more difficult (e.g., a beam in a framed structure having rigid joints), and it may be mandatory to increase the extent of the structure studied. This is often a more complicated analysis process, as will be seen later.

While in some cases the extent of the structure to be considered is obvious, in many cases it is not at all apparent what constitutes the full extent of the structure to be considered. This situation usually arises when the entire structure in a building is considered as a whole as a device for channeling loads to the ground. It is often unclear, for example, whether certain building elements serve a structural function in addition to other functions. An exterior enclosure wall, for example, may serve not only as a weather-control device, but also as a mechanism for assuring the lateral stability of the whole structure. Whether it does or not depends on the relation of the element to the other building elements, particularly to those carrying vertical loads, as typically manifested in the way the elements are joined or connected. The detailing of the connection might be the determinant of whether the enclosure wall serves a structural function or not. Alternatively, the obvious existence of other mechanisms for ensuring lateral stability (e.g., cross bracing) may be an indicator of whether or not the element must serve in this capacity. On the other hand, there might be redundant mechanisms involved. The whole point is simply that each element in the building must be reviewed with respect to its potential contribution as a constituent part of the whole structure. This often cannot be done, however, without a simultaneous consideration of the nature of the loadings associated with the building as a whole.

A final point in connection with defining the extent of the structural system to be considered is that in many cases a distinction should be drawn between the portions of the structure that are *man-made* and those *naturally existing* physical entities that are often incorporated into a structure. A cable structure, for example, in which the cable is tied into foundations is often referred to as being inherently superior as a structural system to a cable structure of the type using a cross strut [see Figure 1-17(e)] to carry the horizontal thrusts involved because it has a simpler load-carrying action and involves fewer elements. This simplicity and apparent efficiency is of course possible only because the foundation, and, indeed, the immediately surrounding earth, are highly loaded and are serving a necessary structural function. Thus, if a system larger than the man-made spanning structure is considered, a free span cable structure tied directly into end foundations does not display any simpler action as compared to a similar cable structure using cross struts. As noted previously, the same total forces must be handled, and the latter merely handles them in a different way than the former by increasing the role of the man-made structure and decreasing the role of the immediately surrounding earth.

2. Another basic step in the analysis process that is generally done concurrently

with the first described is determining and characterizing in a useful way the nature of the external force system acting on the structure considered. This often involves such activities as determining how distributed occupancy loads that act on a surface supported by several structural elements are channeled to the ground. This would need to be established to determine what portion of the total load is carried by each constituent element. The adequacy of each specific element can then be investigated. These activities vary from being relatively simple in structures of limited extent to potentially very difficult when the structure considered is extensive.

3. Once the external force system acting on the structure considered is defined, the next step in the analysis process is typically to determine through application of equilibrium principles the set of reactive forces and moments that are developed at the boundaries of the structure as a consequence of the external loading present. This is often a straightforward process if the structure considered is not extensive, but can be extremely difficult in more involved structures. For *statically determinate structures* where there are typically no more than three unknown reactions, the reactions can be found through simple application of the basic equations of statics ($\sum F_x = 0$, $\sum F_y = 0$, and $\sum M_0 = 0$). A beam simply resting on two supports is a statically determinate structure. In more complex structures there are often more than three unknown reactive forces. These forces cannot be solved for directly by use of the equations of statics, since (1) there are more unknown forces than there are equations for solution, and (2) the reactive forces are dependent on the physical properties of the member (including its shape and material of construction). More involved methods of analysis are thus required. Methods suitable for this type of structure will be discussed in following chapters. Simpler structures whose reactions can be found by statics alone will be considered in the latter part of this chapter.

4. After establishing the nature of the complete force system consisting of both applied and reactive forces acting on the structure, the next step in the general analysis is to determine the nature of the internal forces and moments developed in the structure as a consequence of the action of the external forces. For linear rigid elements such as beams, this typically involves determining the magnitude and distribution of internal resisting shears and moments in the structure. For other structures, more specific measures (e.g., the axial forces in truss members) may alternatively be found. A fundamental premise that will be stressed throughout this text, however, is that the basic load-carrying mechanism for all structures is the same and is best examined with reference to concepts of shear and moment. Emphasis is thus placed on determining these quantities. Specific techniques for determining internal force states are developed in the chapters in Part II that deal with specific structural elements.

5. By knowing the internal force states present in the structure, one can next determine whether the actual member used is adequate to carry these forces without material distress or excessive deformations. Determining this typically involves predicting the internal stress levels associated with the internal force states present and then comparing these stress levels with stress levels known to be safe for the material involved. Predicting actual stress levels involves considerations of the amount and distribution of material in a structure. The allowable stress levels to which these are compared are based on empirical findings for the material involved (see Section 2.4.2). Predicting deformations involves consideration not only of the amount and distribu-

tion of material, but also of the basic load-deformation characteristics of the material. Criteria for what constitutes an acceptable deformation are also empirically based. Techniques for making predictions of this type are discussed in succeeding chapters.

The preceding steps more-or-less describe the process common to analyzing any type of structure.

3.3.2 Modeling the Structure

This section briefly explores how to isolate an elemental part of a whole structure and to model it in a way useful for analysis purposes. Structures are decomposed into more basic elements by conceptually taking them apart, typically at connections between members, and by replacing the action of the adjacent elements acting on them with a set of forces and moments having an equivalent effect. This is often trivially easy, as in the simple structure illustrated in Figure 3-7(a). In this case, the modeling forces are simply reactive forces. At each point of connection, a set of equal and opposite forces exists. On column B, for example, the force R_B acts downward, is equivalent to, and replaces the effect of the entire loaded horizontal beam system acting on the column. The modeling in this case is simple. Note, however, that an assumption has been made. This is that the beams are simply resting one on top of the other and on the column. The joint is assumed to transmit vertical forces only. This is true only under certain conditions, notably when the connection between members is not rigid. Thus, the column does not restrain the free rotation of the end of the beam. If it did, the modeling shown would be incorrect, and a more complex model of the type generally illustrated in Figure 3-7(b) is necessary. This type of structure, a *frame*, is studied in more detail in Chapter 9. The discussion in Section 2.4 is also relevant in this context.

As is evident from the discussion above, effective modeling depends on identifying the exact structural nature of the joints between members. For analytical convenience, connections are modeled as one of the basic types discussed in Section 2.4.1 (e.g., pins, rollers, and rigid joints). Determining what model is most appropriate for an actual connection in practice is no easy matter and involves judgment. Several actual connections and their equivalent models are illustrated in Figure 3-8.

The first step in analyzing a joint is to determine whether the nature of the joint is such that the rotations induced by a load acting on one member are transmitted to the other through the joint. If the joint does *not* transmit rotations, it is usually modeled as a pin or roller. The choice between the two depends on whether the joint can transmit forces in only one direction or in any direction. If forces can be transmitted in any direction, the joint is considered pinned. Pinned joints allow relative rotations to occur between members, but not translations, and can transmit forces in any direction. If forces can be transmitted in one direction only, the joint is considered as a roller. The latter allows not only rotations to occur between members, but also translations in the direction perpendicular to the transmitted force. In many situations, it should be noted, pinned joints are literally made with a pin connecting two members, and a roller connection with actual rollers. Older bridges and buildings were often done this way—hence, the names of the joints (Figure 3-9). Such literal connections are occasionally still made in very large building structures.

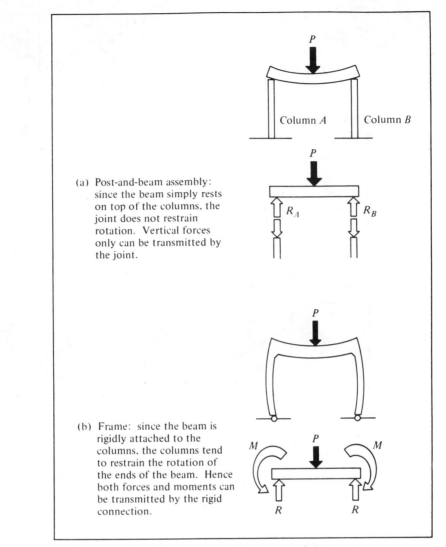

(a) Post-and-beam assembly: since the beam simply rests on top of the columns, the joint does not restrain rotation. Vertical forces only can be transmitted by the joint.

(b) Frame: since the beam is rigidly attached to the columns, the columns tend to restrain the rotation of the ends of the beam. Hence both forces and moments can be transmitted by the rigid connection.

FIGURE 3-7 Isolating structural elements.

If the joint does indeed transmit rotations, hence moments, between members, it is typically considered a *rigid* joint. A rigid joint always maintains a fixed angle between two members. When a rigid joint of the type shown in Figure 3-8(f) is part of a frame, however, it can translate and rotate as a whole unit (see Chapter 9). When a member joins a firm foundation with a connection that does not allow any rotation or translation to occur between the member end and the foundation, the resulting joint is called a *fixed-end* connection [see Figure 3-8(o)]. Forces of any direction can be transmitted by this type of joint.

Quite often the difference between a pinned joint and a rigid one is difficult to determine immediately. Usually, if one member is connected to another at only one point, the joint is pinned. If the member is connected at two widely separated

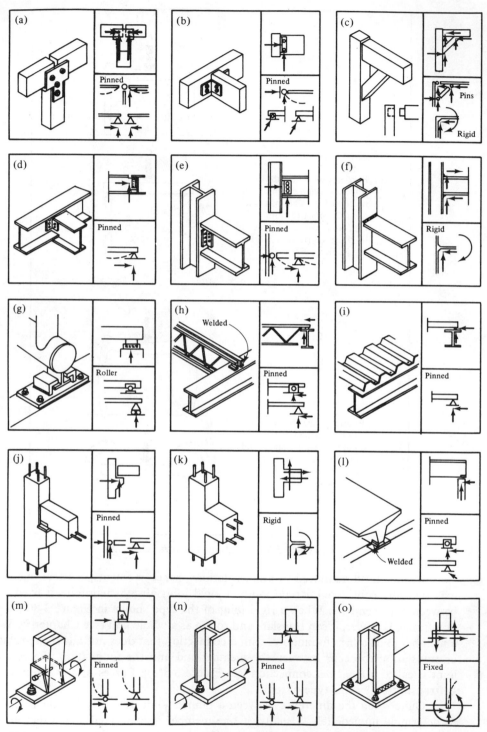

FIGURE 3-8 Types of connections and idealized models.

106

(a) Early "lenticular" bridge located in Lowell, Massachusetts. (b) Typical "pinned" connection.

FIGURE 3-9 Many early structures used connections that were quite literally "pinned." The term is still used to describe conceptually similar connections in modern structures.

points, the joint is typically rigid. It is virtually impossible for rotations in one member to be transmitted to another through a connection that can be idealized as a point, as long as the member is large with respect to the size of this point. Figure 3-8(e) and (f) illustrates two steel wide-flange members connected in these two different ways. Figure 3-8(e) represents a pin connection, since the members are joined essentially at only one point. In Figure 3-8(f), the welding that joins the top and bottom flanges of one member to the other causes this connection to be rigid.

In real structures, roller joints may or may not resist uplift forces. As Figure 3-8(g) illustrates, however, they can be designed to do so. When loadings are known to be only downward, actual joints are sometimes not designed to resist uplifting. In this text it will be assumed, however, that a roller joint can resist either downward or uplifting forces, since if they must resist upward forces, they can be designed to do so.

The previous discussion focused on the nature of various joints. It should also be noted that there are minimum requirements with regard to the number and type of joints connecting the structure to the ground. Sufficient restraints must be provided by the joints so that the basic equations of equilibrium, $\sum F_x = 0$, $\sum F_y = 0$, and $\sum M_0 = 0$ can be satisfied. As illustrated in Figure 3-10(c), for example, a simple beam cannot simply rest on two rollers, since if a horizontal force were applied, the structure would translate horizontally. An alternative way of looking at this is to say that $\sum F_x = 0$ could never be satisfied. Thus, at least one of the two joints must be pinned. Both joints could be pinned as well. Typically, one of the joints is not restrained horizontally, however, since making it into a roller serves the very useful function of allowing expansions and contractions due to thermal effects to occur at will. Pinning both ends would cause a very large buildup of forces to occur as the structure tried to change length with a change in temperature. The forces potentially involved

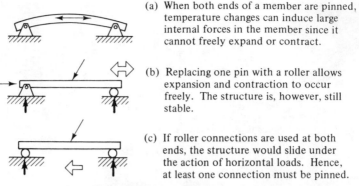

(a) When both ends of a member are pinned, temperature changes can induce large internal forces in the member since it cannot freely expand or contract.

(b) Replacing one pin with a roller allows expansion and contraction to occur freely. The structure is, however, still stable.

(c) If roller connections are used at both ends, the structure would slide under the action of horizontal loads. Hence, at least one connection must be pinned.

FIGURE 3-10   Use of roller connections.

are enormous and can lead to serious failures. Thus, the support condition illustrated in Figure 3-10(b) is typical.

3.3.3   Modeling the External Loads

This section considers methods for characterizing in a way useful to the analysis process the nature of the external force system acting on the structure considered.

As noted previously, actual loadings in a building are typically either concentrated or uniformly distributed over an area. The former need no modeling other than as necessary to characterize them as a force vector. In the latter, however, some modeling is needed when the area considered is actually made up of an assembly of one-way line and surface elements. These elements would pick up different portions of the total load acting over the surface depending on their arrangement.

Consider the simple structural assembly shown in Figure 3-11(a). Since all connections in this illustration are simply supported, the structure can be decomposed as indicated. For the wide plank elements, the reactions are better characterized as *line* reactions [as illustrated in Figure 3.11(b)] than as point reactions. The reactions from all the planks supported by a beam then become loads acting on the beam. Note that these loads form a continuous line load. Loads of this type are expressed in terms of a load or force per unit length (e.g., lb/ft or N/m) and are commonly encountered in the structural analysis process. These loads can be calculated by first determining the reactions of the planks and then considering them as loads on the supporting beam [see Figure 3-11(c)].

Another way of looking at this same loading is to think in terms of *contributory areas*. By a symmetry argument, each of the beams can be considered as supporting an area of the extent indicated in Figure 3-10(d). The width of each area is often called the *load strip*. When the member is looked at in elevation, the load acting over the width of the load strip is considered transferred to the support beam. If the uniformly distributed load is constant and the load strip is of a constant width, the amount of load carried *per unit length* by the support beam is simply the load per unit area multiplied by the width of the load strip. This process is illustrated in Figure 3-11(e). The result is again a continuous line load describable in terms of a load per unit length. This process is valid for symmetrical loads only.

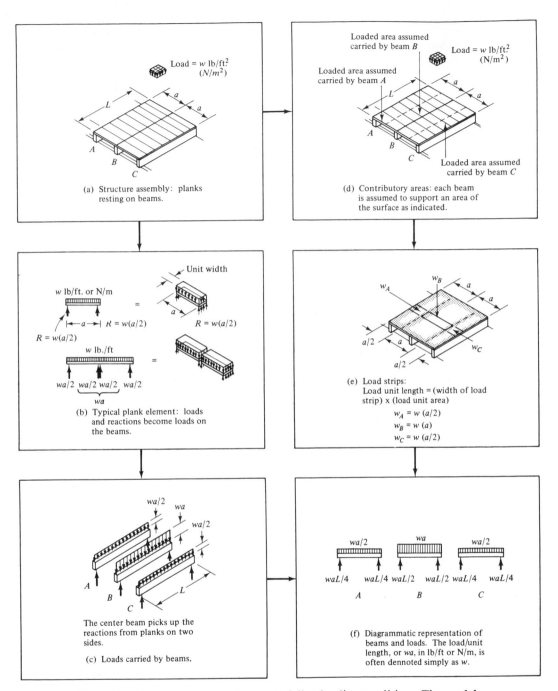

FIGURE 3-11 Two approaches to modeling loading conditions. The model shown on the left most correctly reflects how surface loads are picked up by a beam system. The model on the right, which is based on the concept of contributory areas, is often used for convenience. Both approaches yield the same loading on the beams.

Both of the methods described above should yield the same load on the supporting beams [see Figure 3-11(f)]. Of the two, the former is a more correct model of how the structure works and how loads are carried to the ground. The latter is, however, often more convenient to use and is consequently utilized as a shorthand way of determining loads.

When a collector member is supporting a series of closely spaced joists that in turn support a plank deck, it is evident that each joist carries a uniform load per unit length. This load can be found by multiplying the width of the load strip carried (i.e., the joist spacing) by the load per unit area. This method is typically used whether the planks over the joists are continuous over several joists or simply span from joist to joist. Theoretically, there is some difference in the two cases, but it is not of sufficient consequence to change the model for a simple joist system. Joist reactions can next be found. The collector, in turn, carries the reactions of the joists. The collector thus actually carries a series of closely spaced concentrated loads. For calculation convenience, however, these loads on the collector beam are often replaced with a uniformly distributed load per unit length found by considering the load strip width of the contributory area of the surface supported by the member—a reasonable modeling *if* the concentrated loads are closely spaced. Figure 3-12 illustrates the process.

The model used can be as simple or as sophisticated as the analyst deems necessary. In early design stages, the type of approximation previously discussed is adequate for most purposes. The loading considered should, of course, include both live- and dead-load components. The exact value of the latter can be found only through a detailed consideration of the size and unit weights of the various elements that constitute the whole assembly. Determining these values can be tedious. An alternative is to use an approximate equivalent weight, typically expressed as a force per unit area, to represent the weight of an entire assembly of some complexity. Thus, in Figure 3-12 the average dead load of the complete deck, joist, and beam system can be expressed as (x_2) lb/ft² or (x_1) N/m², where the value of x is estimated or found from empirical data. Since live loads are also expressed in terms of a force per unit area, the calculation process is facilitated, since both loads can be considered

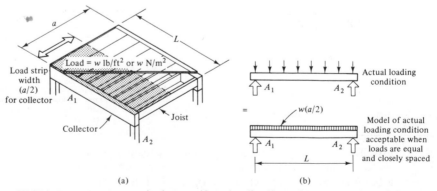

FIGURE 3-12 An equivalent uniformly distributed load can be used to approximate a series of closely spaced concentrated loads.

simultaneously. Procedures of this type are alright for preliminary purposes, but more detailed takeoffs should be made at later stages in the complete analysis process.

The following examples illustrate some of the principles discussed above. A simple example of a floor joist system is first considered, followed by more complex structures.

EXAMPLE

Determine the reactions for a typical interior joist in the floor framing system shown in Figure 3-13(a). Assume that the spans and loads are as shown below.

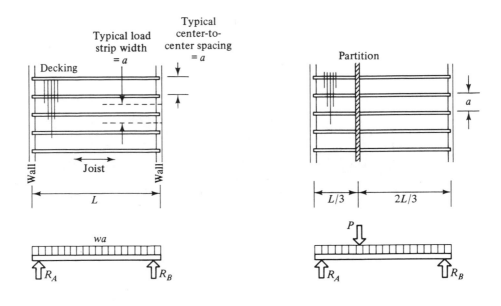

(a) Basic joist system.

(b) Joist system supporting a partition.

FIGURE 3-13 Example floor framing system.

Solution:

Typical joist span $= L = 16.0$ ft $= 4.88$ m

Typical center-to-center spacing $= a = 16$ in. $(1.33$ ft$) = 0.406$ m

Live load $= w_L = 40$ lb/ft² $= 1915$ N/m²

Dead load $= w_D$:

Finished flooring	= 2.5 lb/ft² =	119.7 N/m²
Rough flooring	= 2.5 lb/ft² =	119.7 N/m²
Sheet rock ceiling	= 10.0 lb/ft² =	478.8 N/m²
Joists (estimated)	= 9.0 lb/ft² =	430.9 N/m²
Total dead load	= 24.0 lb/ft² =	1149.1 N/m²

$$\text{Total load} = w_D + w_L = (24.0 + 40.0) = 64.0 \text{ lb/ft}^2$$
$$= (1915 + 1149.1) = 3064.1 \text{ N/m}^2$$

Reactions:

$\sum F_Y = 0:$

$$R_A + R_B - \underbrace{(w_D + w_L)}_{\substack{\text{force per unit} \\ \text{length}}} \underbrace{(a)(L)}_{\substack{\text{joist} \\ \text{length}}} = 0$$

$$\underbrace{}_{\substack{\text{total downward} \\ \text{force}}}$$

$\sum M_A = 0:$

$$0(R_A) + L(R_B) - \underbrace{(w_D + w_L)(a)(L)}_{\substack{\text{total downward} \\ \text{force}}} \underbrace{(L/2)}_{\substack{\text{moment} \\ \text{arm}}} = 0 \qquad R_B = \frac{(w_D + w_L)(a)(L)}{2}$$

From $\sum F_Y = 0$,

$$R_A = (w_D + w_L)(a)(L)/2$$

Thus,

$$R_A = R_B = \frac{(64.0 \text{ lb/ft}^2)(1.33 \text{ ft})(16.0 \text{ ft})}{2} = 682 \text{ lb}$$

$$= \frac{(3064.1 \text{ N/m}^2)(0.406 \text{ m})(4.88 \text{ m})}{2} = 3035 \text{ N}$$

A typical joist reaction is therefore 682 lb (3035 N).

EXAMPLE

Determine the reactions for a typical interior joist in the floor framing system which supports a partition wall as shown in Figure 3-13(b). Assume that all the loads and spans are the same as those used in the preceding example and that the only difference is the addition of the partition wall.

Solution:

Partition:

Assume that the partition is 8.0 ft (2.44 m) high and weighs 20 lb/ft² (958 N/m²). The concentrated force on a joist exerted by the partition is computed as follows. The joists are spaced 1.33 ft (0.406 m) on centers, and, as the partition is 8.0 ft (2.44 m) high, the area of the partition that bears on one joist is

$$8.0 \times 1.33 = 10.67 \text{ ft}^2 \text{ (or } 0.406 \times 2.44 = 0.99 \text{ m}^2)$$

The concentrated force on a single joist is therefore

$$10.67 \text{ ft}^2 \times 20 \text{ lb/ft}^2 = 213.4 \text{ lb (or } 0.99 \text{ m}^2 \times 958 \text{ N/m}^2 = 950 \text{ N)}$$

This force is located as shown in Figure 3.13(b). All other loads (i.e., $w_D + w_L$) are as in the previous example.

Reactions:

$\sum M_A = 0$:

$$0(R_A) + L(R_B) - [(w_D + w_L)(a)(L)]\frac{L}{2} - \frac{L}{3}(P) = 0$$

or

$$16.0(R_B) - [(64.0 \text{ lb/ft}^2)(1.33 \text{ ft})(16.0 \text{ ft})](8.0 \text{ ft}) - (16.0/3 \text{ ft})(213.4 \text{ lb}) = 0 \qquad R_B = 753 \text{ lb}$$

or

$$4.88(R_B) - [(3064.1 \text{ N/m}^2)(0.406 \text{ m})(4.88 \text{ m})](2.44 \text{ m}) - (4.88 \text{ m}/3)(950 \text{ N}) = 0 \qquad R_B = 3352 \text{ N}$$

$\sum F_y = 0$:

$$R_A + R_B - (w_D + w_L)(a)(L) - P \qquad R_A = 824 \text{ lb}$$
$$= 3668 \text{ N}$$

EXAMPLE

Determine the loads on columns M and N in the structure shown in Figure 3-14. Assume that the loading is given by w_T lb/ft² or w_T kg/m² and that this value reflects both live and dead loads.

Solution:

The first step is to determine how the surface load is channeled to the columns. This is best done by drawing free-body diagrams for each element in the structure [see Figure 3-14]. To determine the magnitudes of the column loads it is necessary first to calculate the load carried by each joist, then to calculate the reactions for each joist, and finally to calculate the reactions of the beams that carry the joists. These reactions are the column loads. This process of starting with an analysis of the smallest members actually picking up the loading and tracing it through by considering each collecting member is common in most problems of analysis.

Joist reactions: members A and D

The load/unit length carried by each joist is found by considering the width of the load strip carried by each joist. The end joists, A and D, carry load strips $(a/2)$ in width. If the total load per unit area is w_T, the load per unit length carried by joists A and D is simply $w_T(a/2)$.

$\sum M_{R_{A_1}} = 0$:

$$-\underbrace{[w_T(a/2)(L)]}_{\substack{\text{total load}}}\underbrace{(L/2)}_{\substack{\text{moment} \\ \text{arm}}} + \underbrace{1.0L}_{\substack{\text{moment} \\ \text{arm}}}\underbrace{(R_{A_2})}_{\substack{\text{unknown} \\ \text{reaction}}} = 0 \qquad R_{A_2} = 0.25 w_T a L$$

$\sum F_y = 0$:

$$+\underbrace{R_{A_1}}_{\substack{\text{unknown} \\ \text{reaction}}} + \underbrace{R_{A_2}}_{\substack{\text{calculated} \\ \text{above}}} - \underbrace{w_T(a/2)(L)}_{\substack{\text{total downward} \\ \text{load}}} = 0 \qquad R_{A_1} = 0.25 w_T a L$$

Similarly,

$$R_{D_1} = 0.25 w_T a L \qquad \text{and} \qquad R_{D_2} = 0.25 w_T a L$$

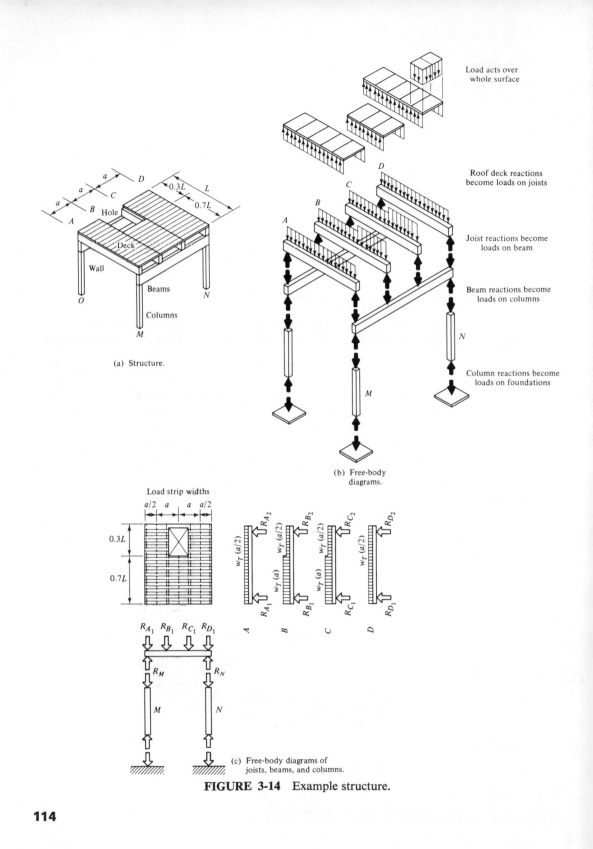

Load acts over
whole surface

Roof deck reactions
become loads on joists

Joist reactions become
loads on beam

Beam reactions become
loads on columns

Column reactions become
loads on foundations

(a) Structure.

(b) Free-body
diagrams.

Load strip widths

(c) Free-body diagrams of
joists, beams, and columns.

FIGURE 3-14 Example structure.

Joist reactions: members B and C

The load/unit length carried by joists B and C is not a constant due to the presence of the opening. In sections where the decking is continuous, the width of the load strip is (a). The load/unit length is consequently $w_T(a)$. In sections where the hole is present, each joist carries a load strip of one-half that in continuous sections, or $(a/2)$. The load/unit length on these sections is $w_T(a/2)$.

$$\sum M_{R_{B_1}} = 0:$$

$$-\underbrace{[w_T(a)(0.7L)]}_{\substack{\text{load}}}\underbrace{\left(\frac{0.7L}{2}\right)}_{\substack{\text{moment} \\ \text{arm}}} - \underbrace{\left[w_T\left(\frac{a}{2}\right)(0.3L)\right]}_{\substack{\text{partial load}}}\underbrace{\left(0.7L + \frac{0.3L}{2}\right)}_{\substack{\text{moment arm}}} + \underbrace{R_{B_2}}_{\substack{\text{reaction}}}\underbrace{(1.0L)}_{\substack{\text{moment} \\ \text{arm}}} = 0$$

$$R_{B_2} = 0.3725 w_T aL$$

$$\sum F_y = 0:$$

$$\underbrace{R_{B_1}}_{\substack{\text{reaction}}} + \underbrace{R_{B_2}}_{\substack{\text{reaction}}} - \underbrace{[w_T(a)(0.7L)]}_{\substack{\text{load}}} - \underbrace{\left[w_T\left(\frac{a}{2}\right)(0.3L)\right]}_{\substack{\text{partial load}}} = 0 \qquad R_{B_1} = 0.4775 w_T aL$$

Similarly,

$$R_{C_1} = 0.4775 w_T aL \qquad \text{and} \qquad R_{C_2} = 0.3725 w_T aL$$

Beam reactions:

The loads on columns M and N are the reactions of the transverse beam-carrying joist loads (reactions) R_{A_1}, R_{B_1}, R_{C_1}, and R_{D_1}.

Beam $\sum M_{R_M} = 0:$

$$-R_{A_1}(0) - R_{B_1}(a) - R_{C_1}(2a) - R_{D_1}(3a) + R_M(0) + R_N(3a) = 0$$

By substituting joist reactions found previously:

$$R_N = 0.7275 w_T aL \qquad \text{and}$$

$$\sum F_Y = 0:$$

$$-R_{A_1} - R_{B_1} - R_{C_1} - R_{D_1} + R_M + R_N = 0$$
$$R_M = 0.7275 w_T aL$$

These are the loads in columns M and N. The loads in the other two columns could be found by a similar process.

If $L = 20$ ft, $a = 6$ ft, and $w_T = 50$ lb/ft^2, the load on column M is given by $R_M = 0.7275 w_T aL = 0.7275(50 \text{ lb/ft}^2)(6 \text{ ft})(20 \text{ ft}) = 4365$ lb. The load on column N is the same.

Alternatively, using SI units, assume that $L = 6.1$ m, $a = 1.83$ m, and $w_T = 244.1$ kg/m^2. The first step is to convert the distributed load into a force measure; thus, $(244.1 \times 9.807) = 2393.9$ N/m^2 = 2.39 kN/m^2. The force on column M is then given by $0.7275(2.39 \text{ kN/m}^2)(1.83 \text{ m})(6.1 \text{ m}) = 19.41$ kN.

EXAMPLE

Determine the forces on a typical interior truss in the structure illustrated in Figure 3-15. Also determine the reactions for the truss analyzed. Assume that the loads and spans are as shown below.

Roof loading—uniformly distributed: w

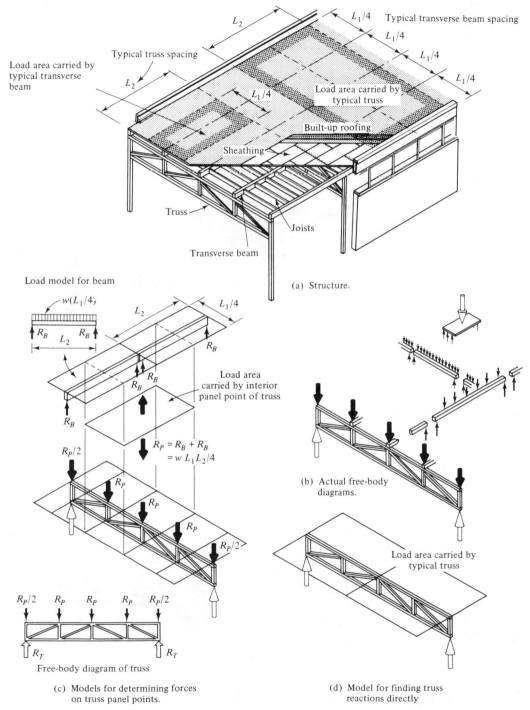

Load area carried by typical transverse beam

Typical truss spacing

L_2

L_2

$L_1/4$

L_2

$L_1/4$

Typical transverse beam spacing

$L_1/4$

$L_1/4$

$L_1/4$

Load area carried by typical truss

Built-up roofing

Sheathing

Truss

Joists

Transverse beam

(a) Structure.

Load model for beam

$w(L_1/4)$

R_B R_B

L_2

R_B

L_2

$L_1/4$

R_B

R_B R_B

R_B

Load area carried by interior panel point of truss

$R_P = R_B + R_B$
$= w L_1 L_2/4$

$R_P/2$

R_P

R_P

R_P

$R_P/2$

(b) Actual free-body diagrams.

$R_P/2$ R_P R_P R_P $R_P/2$

R_T R_T

Free-body diagram of truss

Load area carried by typical truss

(c) Models for determining forces
on truss panel points.

(d) Model for finding truss
reactions directly

FIGURE 3-15 Loading models.

Solution:

Spans:

$$L_1 = 60 \text{ ft} = 18.3 \text{ m}$$

$$L_2 = 25 \text{ ft} = 7.6 \text{ m}$$

Loads:

Assume that the live load is 35 lb/ft² (1676 N/m²). Assume the following dead loads:

Roofing	7.0 lb/ft² =	335 N/m²
Sheathing	2.5 =	120
Joists and beams	8.0 =	383
Truss	4.0 =	191
Total	21.5 lb/ft² =	1029 N/m²

Here the dead weights of the joists, beams, and trusses were estimated in terms of an approximate equivalent distributed load. This is often done in preliminary design stages. More detailed takeoffs are made in successive stages. Using the equivalent distributed load figures, and the assumed live loading, the total distributed load (live plus dead) is given by

$$35 + 21.5 = 56.5 \text{ lb/ft}^2 \quad \text{or} \quad 1676 + 1029 = 2705 \text{ N/m}^2.$$

Forces on truss:

The way a unit area of distributed live load is eventually carried to the supports is illustrated in the series of free-body diagrams shown in Figure 3-15(b). The load is first picked up by the roof decking, which carries it to adjacent joists. The reactions of the decking become forces on the joists. The reactions of the joists in turn become forces on the transverse beams. The reactions of the transverse beams become forces on the panel points of the trusses. The magnitudes of these forces could be determined by calculating each of the reactions involved in turn. This precise method, which accurately reflects how the structure carries loads, is, however, cumbersome. Forces on the panel points of the truss can be determined more quickly by an alternative approach, which involves the models illustrated in Figure 3-15(c). These models are based on the concept of contributory areas. As the illustration suggests, it is possible to determine the force present at a truss panel point by finding the relative portion of the roof which serves as the contributory area for the panel point and then simply multiplying this area by the magnitude of the distributed uniform loading.

$$\text{contributory area for a typical interior panel point} = \left(\frac{L_1}{4}\right)(L_2) = \frac{L_1 L_2}{4}$$

$$\text{interior-panel-point force} = R_p = \frac{w_T L_1 L_2}{4}$$

$$= \frac{(56.5 \text{ lb/ft}^2)(60 \text{ ft})(25 \text{ ft})}{4} = 21{,}187.5 \text{ lb}$$

$$= \frac{(2705 \text{ N/m}^2)(18.3 \text{ m})(7.6 \text{ m})}{4} = 94{,}053 \text{ N}$$

$$\text{contributory area for a typical exterior panel point} = \left(\frac{L_1}{8}\right)(L_2) = \frac{L_1 L_2}{8}$$

$$\text{exterior-panel-point force} = \frac{w_T L_1 L_2}{8} = \frac{R_p}{2}$$

$$= 10{,}593.75 \text{ lb}$$

$$= 47{,}026.5 \text{ N}$$

The final free-body diagram for the truss is shown in Figure 3-14(c).

Reactions:

The reactions of the truss can be obtained by using the free-body diagram shown in Figure 3-15(c) with the forces found above. Because of symmetry, a typical reaction is evidently given by

$$\frac{(R_p/2 + R_p + R_p + R_p + R_p/2)}{2} = 42{,}375 \text{ lb} = 188{,}106 \text{ N}$$

Alternatively, it is possible to find the reactions of the truss directly (without going through the step of finding the panel-point forces) by considering the loading model illustrated in Figure 3-15(d). Figure 3-15(d) illustrates the relative portion of the total roof surface carried by a typical interior truss. This is the contributory load area for the truss. This area can be used to find the total load carried by the truss and then the reactions. The figure illustrates that each reaction picks up one-half of the total contributory load area.

$$\text{contributory area for a typical truss} = L_1 L_2$$

$$\text{total downward force on truss} = w_T L_1 L_2$$

$$= (56.5 \text{ lb/ft}^2)(60 \text{ ft})(25 \text{ ft}) = 84{,}750 \text{ lb}$$

$$= (2705 \text{ N/m}^2)(18.3 \text{ m})(7.6 \text{ m}) = 376{,}212 \text{ N}$$

Since each truss has two reactions that are symmetrically placed with respect to the load, each reaction is simply one-half the total load, or $84{,}750/2 = 42{,}375$ lb or $376{,}212/2 = 188{,}106$ N. These are the same values that were noted before. For finding truss reactions it is evident that this procedure saves time.

It should be emphasized that the use of loading models based on the notion of contributory areas is valid only when there is not any extreme asymmetry in either the loading condition or structure. For partial loading conditions it is necessary to calculate the reactions for each individual element and consider them as forces acting on the elements supporting it. This process then needs to be repeated until the reactions of the final collecting element on the foundation are found.

EXAMPLE

Determine the forces on the structural frames shown in Figure 3-16 due to the action of wind impinging upon the face of the building.

Solution:

The forces can be determined by either tracing through the way specific elements such as girts are loaded and the way their reactions become loads on the frames or by using a contributory-area concept. Both methods are illustrated in Figure 3-16.

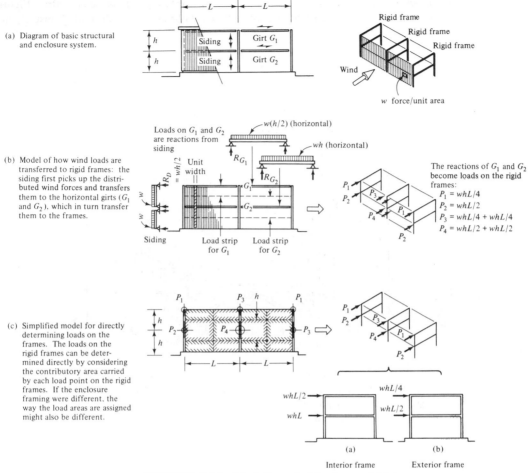

(a) Diagram of basic structural and enclosure system.

(b) Model of how wind loads are transferred to rigid frames: the siding first picks up the distributed wind forces and transfers them to the horizontal girts (G_1 and G_2), which in turn transfer them to the frames.

Loads on G_1 and G_2 are reactions from siding

The reactions of G_1 and G_2 become loads on the rigid frames:

$P_1 = whL/4$
$P_2 = whL/2$
$P_3 = whL/4 + whL/4$
$P_4 = whL/2 + whL/2$

(c) Simplified model for directly determining loads on the frames. The loads on the rigid frames can be determined directly by considering the contributory area carried by each load point on the rigid frames. If the enclosure framing were different, the way the load areas are assigned might also be different.

FIGURE 3-16   Modeling wind loads on rigid frames.

QUESTIONS

3-1. Find two bridges in your local area and identify the type of support conditions present (e.g., hinged, roller). Draw a sketch of the actual supports and a diagram of the symbol used to represent the support.

3-2. Find two different types of trusses used in buildings in your local area and identify the type of support condition present. Draw a sketch of the actual end condition and a diagram of the symbol used to represent the support.

3-3. Consider a typical corridor in the building that you are in or in one nearby. What do you estimate the *actual* live load to be during normal traffic conditions? During fire conditions?

3-4. What are the snow loads and wind loads specifically recommended for buildings in your local area? Consult a copy of the local building code in force.

3-5. Conduct a study of the actual loading conditions on a typical gothic cathedral. Figure 3-17 illustrated how one such structure carried its loads to the ground. Consult your library.

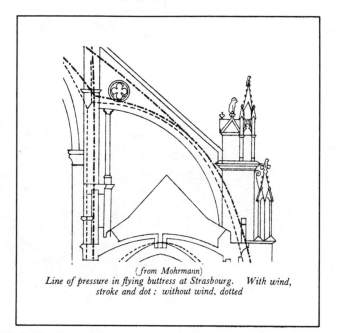

(*from Mohrmann*)
Line of pressure in flying buttress at Strasbourg. With wind, stroke and dot : without wind, dotted

FIGURE 3-17 One of the more remarkable aspects of the way flying buttresses work is their ability to carry forces induced by winds as well as those associated with vertically acting loads.

part II

ANALYSIS AND DESIGN OF STRUCTURAL ELEMENTS

The nine chapters in Part II systematically treat all the major structural elements used in a building context. Each chapter contains initial sections that define the element and discuss its attributes in a qualitative way. The following sections present more detailed analysis of the structural characteristics of the element and discuss principles that are useful in a design context.

chapter 4

TRUSSES

4.1 INTRODUCTION

Although structures made of jointed members have been constructed throughout history, the conscious exploitation of structural advantages inherent when individual linear members are formed into triangulated patterns is of relatively recent origin. Structures of this type, commonly called *trusses*, were indeed built quite early. Simple trusses using relatively few members often appeared in common pitched roofs. More complex trusses were used in isolated instances. A bridge using a form of timber truss, for example, was built across the Danube River by the Romans as early as 500 B.C. Such examples, however, never had any significant impact on the methods of building of the time. The Italian architect Andrea Palladio (1518–1580) gave an illustration of a correctly triangulated truss structure and indicated that he had some knowledge of its potential and the way it carried forces. Trusses were occasionally used afterward in large public buildings such as Independence Hall, Philadelphia, but again without having much impact as a structural innovation. It was the bridge builders of the early nineteenth century who first began systematically to explore and experiment with the potential of the truss. This was in response to the demands of rapidly expanding transportation systems of the time. Emiland Gauthey's *Traité de la Construction des Ponts* (posthumously published between 1809 and 1813 by his nephew, the famous mathematician Louis Navier of the Ecole Polytechnic in Paris), provided a foundation for many subsequent theoretical works in the area. Gauthey's treatise included a discussion of what he termed the principles of *equilibrium of position* and *equilibrium of resistance*. The former was an initial attempt to resolve bridge loads into components in individual members. The latter dealt with material properties and the sizing of truss members. Later important contributions include Squire Whipple's classic of structural engineering, *A Work on Bridge Building*, published in 1847. The development of the truss was thus fostered by a tentative but rapidly expanding body of theoretical knowledge. This contrasts with other structural forms which typically developed slowly over time in a strictly empirical way. The truss soon became a common structural form used in civil engineering structures spanning long distances. The use of trusses in buildings also increased, although more slowly due to different traditions and needs, until they became a common element in modern architecture.

The emergence of the truss as a major structural form has been rapid and its impact significant. The remainder of this chapter explores what a truss is, how it works, and why it is an important structural form. The presentation is an inquiry into a specific structural element. On a more general level, the methods of analysis and design presented for trusses are used to demonstrate principles that are broadly applicable to the analysis and design of a wide range of other structural forms.

4.2 GENERAL PRINCIPLES

4.2.1 Triangulation

A truss is an assemblage of individual linear elements arranged in a triangle or combination of triangles to form a rigid framework that cannot be deformed by the application of external forces without deformation of one or more of its members. The individual elements are typically assumed to be joined at their intersections with *pinned* connections (see Sections 2.4.1 and 3.3.2). Members are customarily arranged so that all loads and reactions occur only at these intersections.

The primary principle underlying the use of the truss as a load-carrying structure is that arranging elements into a triangular configuration results in a stable shape. Consider the two pin-connected structures shown in Figure 4-1(a) and (b). Application of a load to the structure shown in Figure 4-1(a) will cause the massive deformation indicated. This is an unstable structure, which forms a *collapse mechanism* under external loading. A structure of this type may be deformed without a change in length of any individual member. The triangulated configuration of members shown in Figure 4-1(b) clearly could not deform or collapse in a similar manner. This configuration is thus *stable*. Any deformations that occur in this stable structure are relatively minor and are associated with small changes in member length caused by forces generated in the members by the external load. Similarly, the angle formed between any two members remains relatively unchanged under load in a stable configuration of this type. This is in marked contrast to the large angle changes which occur between members in an unstable configuration of elements [see Figure 4-1(a)]. It is also obvious that application of the external force causes forces to be developed in members of the

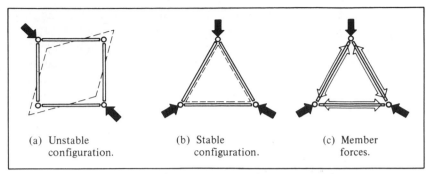

| (a) Unstable configuration. | (b) Stable configuration. | (c) Member forces. |

FIGURE 4-1 Stable and unstable bar assemblies.

stable triangulated structure. As will be explained in detail in Section 4.3.2, the forces developed are either purely tensile or purely compressive. Bending is neither present nor can it be developed.

4.2.2 Configurations

Since a basic triangle of members is a stable form, it follows that any structure made of an assembly of triangulated members is also a rigid, stable structure (Figure 4-2). This idea is the principle underlying the viability and usefulness of the truss in building, since larger rigid forms of any geometry can be created by the aggregation of smaller triangular units. Again, the effect of external loads is to produce a state of either pure tension or compression in the individual members of the assembly. For common trusses with vertically acting loads, compressive forces are usually developed in upper chord members and tensile forces in lower chord members. Either type of force may

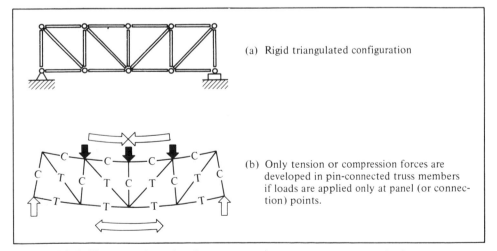

(a) Rigid triangulated configuration

(b) Only tension or compression forces are developed in pin-connected truss members if loads are applied only at panel (or connection) points.

FIGURE 4-2 Triangulated structures: Any structure composed of a series of stable triangular units forms a rigid assembly capable of carrying external loads.

develop in an interstitial member, although there is often an alternating pattern of tensile and compressive forces present.

It is extremely important that trusses be loaded only with concentrated loads that act at joints for truss members to develop only tensile or compressive members. If loads are applied directly onto truss members themselves, bending stresses will also develop in the loaded members in addition to the basic tensile or compressive stresses already present, with the consequence that member design is greatly complicated and the overall efficiency of the truss is reduced.

4.2.3 Member Forces

As will be discussed in following sections, the nature of the forces in the members of any truss can be determined through application of the basic equations of equilibrium. For some simple truss configurations, however, the basic sense (tension, compression, zero) of the forces in many members can be determined by less involved techniques which may prove useful in visualizing how certain trusses carry external loads.

One way of determining the sense of the force in a truss member is to visualize the probable deformed shape of the structure as it would develop if the member considered were *imagined* to be removed. The nature of the force in the member can then be predicted on the basis of an analysis of its role in *preventing* the deformation visualized.

Consider the diagonals shown in truss A in Figure 4-3(a). If the diagonals were imagined removed, the assembly would dramatically deform, as illustrated in Figure 4-3(b), since it is a nontriangulated configuration. In order for the diagonals to keep the type of deformation shown from occurring, it is evident that the left and right diagonals must prevent points *B–F* and points *B–D*, respectively, from drawing apart. Consequently diagonals placed between these points would be pulled upon. Hence, tension forces would develop in the diagonal members.

The diagonals shown in truss B in Figure 4-3 must be in a state of compression since their function is to keep points *A–E* and *C–E* from drawing closer together.

With respect to member *BE* in both trusses, it is fairly easy to imagine what would happen to points *B* and *E* if member *BE* were removed. In truss A, points *B* and *E* would have a tendency to draw together, hence compressive forces develop in any member placed between these two points. In truss B, however, removal of member *BE* leads to *no change* in the gross shape of the structure (since it remains a stable triangulated configuration), hence the member serves no direct role for this loading. It is a *zero-force member*. Note that members *AF*, *FE*, *ED*, and *DC* in truss B could also be removed without altering the basic stability of the remainder of the configuration, and hence are also zero-force members. This is obviously not true for the same members in truss A. Final forces in both truss A and truss B are illustrated in Figure 4-3(c).

Another completely different way of visualizing the forces developed in a truss is to use an *arch* and *cable* analogy. It is evident, for example, that truss A can be conceived of as a cable with supplementary members [see Figure 4-3(d)]. Truss B can be conceived of as a simple linear arch with supplementary members. Diagonals in

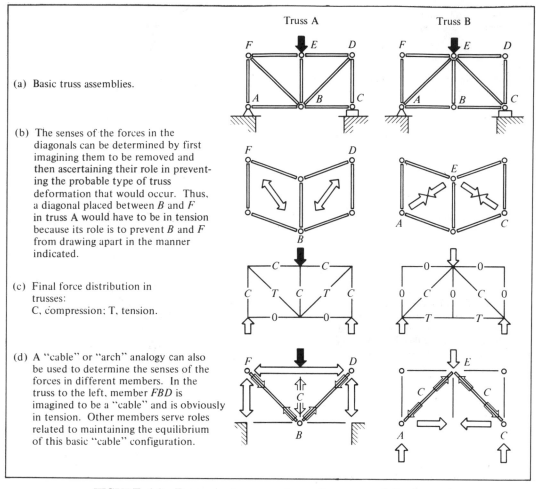

(a) Basic truss assemblies.

(b) The senses of the forces in the diagonals can be determined by first imagining them to be removed and then ascertaining their role in preventing the probable type of truss deformation that would occur. Thus, a diagonal placed between B and F in truss A would have to be in tension because its role is to prevent B and F from drawing apart in the manner indicated.

(c) Final force distribution in trusses:
C, compression; T, tension.

(d) A "cable" or "arch" analogy can also be used to determine the senses of the forces in different members. In the truss to the left, member FBD is imagined to be a "cable" and is obviously in tension. Other members serve roles related to maintaining the equilibrium of this basic "cable" configuration.

FIGURE 4-3 Forces in truss members: the senses of the forces in some simple truss configurations can be determined through intuitive approaches. More complex trusses require quantitative approaches.

truss A are consequently in tension while diagonals in truss B must be in compression. The forces in other members can be determined by analyzing their respective roles in relation to containing arch or cable thrusts or providing load transfer or reactive functions. As Figure 4-4 indicates, special forms of more complex trusses can also be visualized in this same general way.

Obviously, both of the methods discussed above for visualizing forces in trusses become exceedingly difficult or even impossible to apply when more complex triangulation patterns are present. A method of visualizing forces in such trusses on the basis of joint equilibrium considerations is considered in the next section. In general, however, complex truss forms must be mathematically analyzed according to methods to be discussed for correct results to be obtained.

(a) Basic "cable" unit: the diagonal members are obviously in tension.

(b) A simple truss configuration can be formed by placing a horizontal member between the ends of the cable. The horizontal member is in compression since it resists the inwardly directed cable thrusts.

(c) The same configuration can be raised vertically by end compression members.

(d) A more complex truss form can be generated by imagining the entire assembly shown in (c) to be carried by another cable member. Another horizontal compression element is then needed to resist the cable thrusts.

(e) The same process can be repeated to form even more complex trusses. Note that the forces in the vertical and diagonal members increase away from the middle of the truss since increased portions of the external loading are carried. Member sizes could be designed to increase accordingly.

(f) The total force present in the upper chord is greatest at midspan, where the "individual" top chords are actually combined into one member. Chord forces decrease toward either end of the truss. The same is true for the lower tension chord.

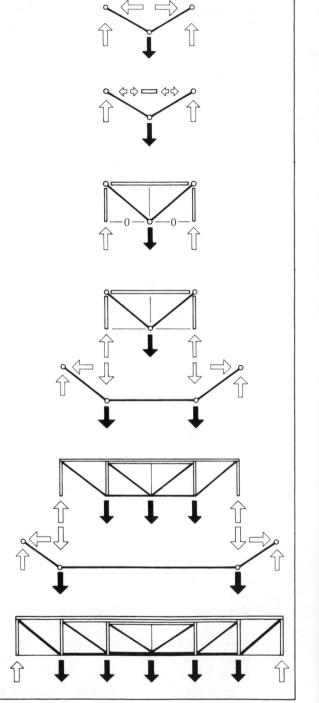

FIGURE 4-4 "Cable" analogy in truss analysis: many complex truss forms can be imagined to be composed of a series of simpler basic cable units. If the directions of the diagonals were all reversed, an "arch" analogy could be used to analyze the structure. This analogy approach, however, is useful for only a few limited truss forms.

4.3 ANALYSIS OF TRUSSES

4.3.1 Stability

The first step in the analysis of a truss is always to determine whether the truss being studied is indeed a stable configuration of members. It is usually possible to tell by inspection whether or not a truss is stable under external loads by considering each joint in turn to determine if the joint will always maintain a fixed relation to other joints under any loading condition applied to the truss. In general, any truss composed of an aggregation of basic triangular shapes will be a stable structure.

The appearance of nontriangular shapes in a bar pattern is an obvious sign that the truss should be looked at carefully. The truss indicated in Figure 4-5(a) is unstable and would collapse under load in the manner illustrated. This truss clearly does not have a sufficient number of bars to maintain a fixed geometrical relationship between joints. Assuming that the remaining truss members are adequately designed for the loads they carry, the addition of a member from *B* to *E* as indicated in Figure 4-5(b) would make this into a stable configuration.

It should be noted that it is possible for a truss to include one or more figures which are not triangles and still be stable. Study Figure 4-6, which illustrates a truss of this type. It is made up of groups of rigid triangulated bar patterns connected to form nontriangular, but still stable figures. The group of triangles between *A* and *C* form a rigid shape, as do those between *B* and *C*. Joint *C* is thus held fixed in relation to joints *A* and *B* in much the same way as occurs in a simpler triangular figure. The group of triangles between *A* and *C* can be thought of as a "member," as can those between *B* and *C*.

In a single truss it is quite possible to have more than the minimum number of bars necessary to achieve a stable structure. One of the two middle diagonal members of the truss shown in Figure 4-7 could be considered redundant. *Either* member *BE* or member *CF* could be removed and the resultant configuration would still be stable. Obviously, removing both would cause the structure to become a collapse mechanism.

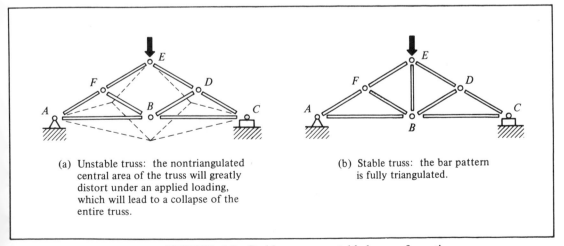

(a) Unstable truss: the nontriangulated central area of the truss will greatly distort under an applied loading, which will lead to a collapse of the entire truss.

(b) Stable truss: the bar pattern is fully triangulated.

FIGURE 4-5 Stable versus unstable bar configurations.

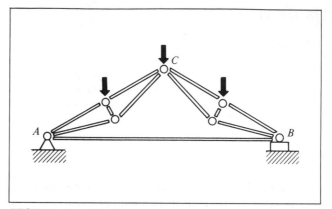

FIGURE 4-6   Stable truss with nontriangular bar pattern.

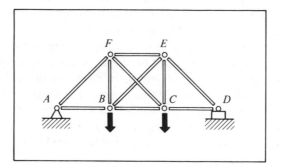

FIGURE 4-7   Stable truss with more than the minimum number of bars necessary for stability.

The importance of determining whether a configuration of bars is stable or unstable cannot be overestimated, since there are few errors more dangerous. Total collapse occurs immediately when an unstable configuration is loaded. Unusual bar patterns often make it difficult to determine the stability of a configuration. As an aid to ascertaining the stability of a planer truss, an algebraic expression has been developed which relates the number of joints present in a truss to the number of bars necessary for stability: If n is the number of necessary members, and j the number of joints, then $n = 2j - 3$.

Testing this expression, which can be determined by inspection, on a few trusses of known stability will indicate its application. Consider the truss shown in Figure 4-8(a). In this case, $j = 5$, so n must be $2(5) - 3$, or 7. This is the minimum number of bars required in a planer truss for stability. Since the truss actually has this number of bars, it is a stable configuration. Generally speaking, fewer bars than are given by this expression will result in an unstable structure; more may indicate a structure with redundant members. The expression, however, is not foolproof and should not be used as a replacement for a careful visual inspection of the truss. The expression is more correctly an indicator of whether or not the internal forces in a structure can be calculated by the equations of statics alone than a predictor of stability. How-

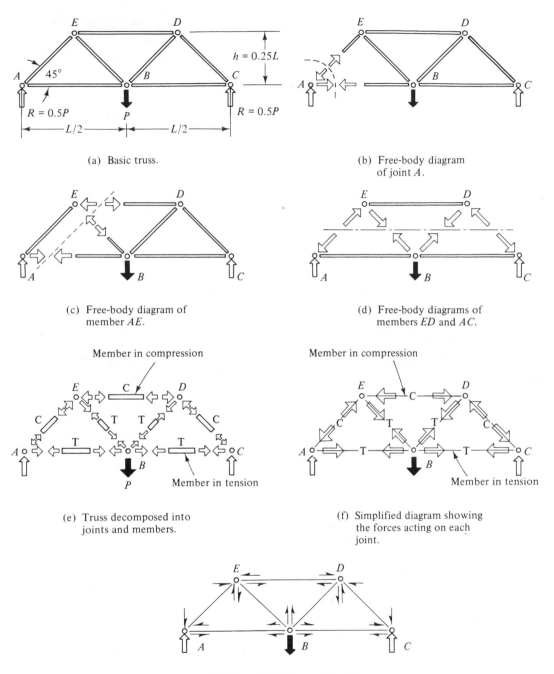

(a) Basic truss.

(b) Free-body diagram of joint A.

(c) Free-body diagram of member AE.

(d) Free-body diagrams of members ED and AC.

(e) Truss decomposed into joints and members.

(f) Simplified diagram showing the forces acting on each joint.

(g) Simplified diagram showing the components of the forces acting on the joints.

FIGURE 4-8 Typical free-body diagrams for elemental truss pieces. These diagrams are based on the fundamental principle that any structure, or any portion of any structure, must be in a state of equilibrium. The free-body diagrams in (e), (f), and (g) are used for solution of bar forces by the method of joints.

ever, it is a useful tool for stability assessments, since it is not possible to calculate the forces in an unstable structure by the equations of statics.

4.3.2 Member Forces: General

This section considers analytical methods for determining the nature and distribution of forces in the members of a truss that has known geometrical characteristics and carries known loads. The methods of analysis used are similar to those developed in Chapter 2 for dealing with the equilibrium of whole structures. The principle underlying the analytical techniques to be developed in this chapter is that any structure, *or any elemental portion of any structure*, must be in a state of equilibrium. This principle is the key to common truss analysis techniques. The first step in analyzing a truss of many members is to isolate an elemental portion of the structure and consider the system of forces acting on the element. If some of these forces are known, it is usually possible to calculate the others by use of the basic equations of statics, since it is known that the element must be in equilibrium. These equations are merely formal statements of the notion that any set of forces, including both those that are externally applied and those internally developed, must form a system whose net result is zero.

The extent of the portion of the structure selected for study is not restricted. A whole segment consisting of several members and joints could be considered, or attention could be limited to a single joint or member. To find the internal forces present at a particular location in a structure, it is necessary that the structure be decomposed at least at that point. Figure 4-8(b)–(e) illustrates free-body diagrams for typical elemental pieces of the truss shown in Figure 4-8(a). Each of the pieces shown must be in a state of equilibrium under the action of the force system present on the piece. The force system considered consists of not only any external loads applied to the piece, but also those which are internal to the structure as a whole. The latter are forces developed in members in response to the external loading on the whole truss, which are necessary to maintain the equilibrium of elemental portions of the truss. In considering the equilibrium of an element, it is often useful to think of these as applied forces. At points where the truss is decomposed, internal forces are shown to be equal in magnitude but opposite in sense in adjacent elements. This follows from the discussion in Section 2.4.1, where we considered Newton's third law: that the forces of action and reaction between elements in contact with each other have the same magnitude and line of action, but are of opposite sense. Since the members of a truss are pin-connected and their ends are free to rotate, only forces, and not moments, can be transmitted from one member to another at the point of connection. Thus, only vector forces are indicated in the free-body diagrams shown. Since only forces and not moments can be transmitted at the connections, the forces at either end of a member must act collinearly (assuming that the member itself is not subjected to external loads) if the member is indeed to be in a state of equilibrium. (See the discussion in Section 2.3.2 on two-force members.) If the member itself is aligned with these forces, it follows that the member is only axially loaded and not subject to bending moments.

4.3.3 Equilibrium of Joints

EQUILIBRIUM OF POINTS WITHIN TRUSSES. The fact that any portion of any structure must be in a state of equilibrium forms the basis for all truss-analysis techniques directed toward finding forces in truss members. In analyzing a truss by the *method of joints*, the truss is considered to be composed of a series of members and joints. Member forces are found by considering the equilibrium of the various joints which are idealized as points. Each of these joints or points must be in a state of equilibrium.

Figure 4-8(e) illustrates a typical truss that has been decomposed into a set of individual linear elements and a set of idealized joints. Free-body diagrams for all the individual members and joints are shown.

By looking at the joints themselves, it can be seen that the system of forces acting on a joint is defined by the bars attached to it and by any external loads that might occur at the location of the joint. As shown in Figure 4-8(e), the forces on a joint are equal and opposite to those on the connecting members. Each joint must be in a state of equilibrium. The system of forces applied to the joint all act through the same point. From an analytical perspective, we are therefore interested in the equilibrium of a point. This requires consideration of translational equilibrium only. Rotational equilibrium is not a concern, since all forces act through a common point and thus produce no rotational effects. This is the key to analysis of trusses by the method of joints. For planar structures, two independent equations of statics exist for a concurrent force system, ($\sum F_x = 0$ and $\sum F_y = 0$). Thus, two unknown forces can be found by application of these equations to the complete system of forces represented in the free-body diagram of a joint. If a joint with a maximum of two unknown forces is considered first, it is possible to calculate these forces. The starting point for an analysis of the forces in a truss is often at a support where the reaction has been determined by considering the rigid-body equilibrium of the whole structure. Once all the forces acting on the initial joint have been found (and thus also the forces in the bars attached to the joint), it is possible to proceed to another joint. Since the bar forces previously found can now be treated as known forces, it is convenient to next consider an adjacent joint. The example illustrated below will clarify the process.

But first a few words should be said about the *arrow convention* used. The arrows illustrate graphically the nature and direction of the forces developed on an element. Thus, the arrows shown on member *DE* in Figure 4-8(e) are used to indicate that forces causing the element to be in a state of compression are developed as a consequence of the loads on the larger structure. Note that the arrows seemingly subject the member to a compressive force. Conversely, the arrows shown on member *BC* in Figure 4-8(e) are used to indicate that a state of tension exists in the element. With respect to the joints, these arrows are shown to be equal and opposite. Thus, the action of member *DE* (in a state of compression) on joint *E* apparently is one of pushing against the joint. In actuality, a reaction is developed. In an analogous way, the action of member *BC* (in a state of tension) is seemingly to pull on joint *C*. It is quite useful to visualize a joint as being in a state of equilibrium when the "pushes" and "pulls" of the members framing into the joint balance each other.

Analysis of bar forces in a truss by the method of joints is generally straight-forward. For the truss shown in Figure 4-8, the first step is preferably to draw a set of free-body diagrams of the type shown in Figure 4-8(e). Alternatively, simplified diagrams of only the forces on the joints themselves may be drawn [see Figure 4-8(f)]. The equations of translational equilibrium ($\sum F_x = 0$ and $\sum F_y = 0$) are then applied in turn to each joint. In drawing the free-body diagrams and writing equilibrium equations, it is necessary to assume that an unknown bar force is either in a state of tension or compression. From a strictly analytical viewpoint, the state of stress assumed can be arbitrary. Whether or not the force is in the state of stress assumed will be evident from the algebraic sign of the force that is found after equilibrium calculations have been made. A positive sign means that the initial assumption was correct, while a negative sign means the converse.

To develop a more intuitive feeling for the force distribution in a structure, it is useful to try to determine correctly whether a member is in tension or compression by a careful qualitative inspection of each joint's equilibrium. Consider joint A in Figure 4-8(e)–(g), where it can be seen qualitatively that the directions assumed are intrinsically reasonable. The reaction is known to act upward. For equilibrium in the vertical direction to obtain, there *must* be a force acting downward. Only force $\overline{AE}$ of the two unknown member forces has a component in the vertical direction and would be capable of providing the downward force necessary to balance the upward reaction. Member AB is horizontal and thus has no component in the vertical direction. Force $\overline{AE}$ must therefore act in the downward direction shown. Thus, member AE must be in a state of compression. If force $\overline{AE}$ acts in the direction shown, it is evident that it has a horizontal component acting toward the left. For the joint to be in horizontal equilibrium, there must be some other force with a horizontal component acting to the right. The reaction acts only vertically and so does not enter into consideration. Force $\overline{AB}$ must therefore act to the right if equilibrium in the horizontal direction is to obtain. Member AB is hence in a state of tension. Looking next at joint E and noting that member AE is in compression, it is evident that member EB must be in tension to provide a downward component necessary to balance the upward one of the force in member AE. Member ED is horizontal and thus can contribute no component in the vertical direction. If forces $\overline{AE}$ and $\overline{EB}$ act in the directions shown, both have a component in the horizontal direction acting to the right. The force in member ED must therefore act to the left to balance the sum of horizontal components of the forces in the two diagonals. Member ED is thus in compression. Since the truss is symmetrical, the state of stress in the remaining members can be found by inspection (member DC must be in compression, BC in tension, and DB in tension). Note that joint B is also apparently in a state of equilibrium. Thus, the state of stress in all the members can be qualitatively determined without resort to calculation. This process is not possible with all trusses, but success is frequent enough to make the attempt worthwhile. This qualitative approach clearly does not yield numerical magnitudes of bar forces. These can be found only by formally writing the equations of equilibrium and solving for the unknown forces. The mathematical process, however, is concep-tually similar to the qualitative one described above. Both are based on the principle that any element of a structure must be in equilibrium. To solve for numerical magnitudes, each joint is considered in turn. In the following, the reference axes used are vertical and horizontal.

EXAMPLE

Determine all the member forces in the truss shown in Figure 4-8.

Solution:

JOINT A

Equilibrium in the vertical direction: $\sum F_y = 0$ ↑+:

$$\underbrace{+0.5P}_{\substack{\text{reaction } R_{A_v} \\ \text{in vertical} \\ \text{direction}}} - \underbrace{\overline{AE}\sin 45°}_{\substack{\text{vertical component of} \\ \text{force in member } AE; \\ \text{the member is assumed to} \\ \text{be in compression.}}} = 0$$

$$\overline{AE} = +0.707P \quad \text{(compression)}$$

Member AE is in compression as assumed, since the sign is positive.

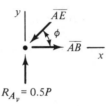

Joint A

Equilibrium in the horizontal direction: $\sum F_x = 0$ →+:

$$\underbrace{-\overline{AE}\cos 45°}_{\substack{\text{horizontal component} \\ \text{of force in member } AE}} + \underbrace{\overline{AB}}_{\substack{\text{Force in member } AB, \\ \text{assumed to be in} \\ \text{tension.}}} = 0$$

Since $\overline{AE}$ is known, $\overline{AB}$ can be found directly:

$$\overline{AB} = +0.5P \quad \text{(tension)}$$

Member AB is in tension as assumed, since the sign is positive. The next step is to proceed to an adjacent joint. Joint B involves three unknown forces for which only two independent equations are available for solution. Joint E, however, involves only two unknown forces. So go to this joint next.

JOINT E

Equilibrium in the vertical direction: $\sum F_y = 0$ ↑+:

$$+\overline{AE}\sin 45° - \overline{EB}\sin 45° = 0$$

Since $\overline{AE}$ is known, $\overline{EB}$ can be solved for:

$$\overline{EB} = +0.707P \quad \text{(tension)}$$

Since these are the only two forces at this joint having components in the vertical direction, the components must be equal. Since the member angles are the same, the member forces must also be the same. Thus, the force in member EB could have been found by inspection.

Joint E

Equilibrium in the horizontal direction: $\sum F_x = 0 \quad \rightarrow +:$

$$+\overline{AE} \cos 45° + \overline{EB} \cos 45° - \overline{ED} = 0$$

or

$$\overline{ED} = 1.0P \quad \text{(compression)}$$

$\overline{ED}$ is merely the sum of the horizontal components in the two diagonals. Next proceed to an adjacent joint (either B or D).

JOINT B

Equilibrium in the vertical direction: $\sum F_y = 0 \quad \uparrow +:$

$$\overline{EB} \sin 45° + \overline{BD} \sin 45° - P = 0$$

Since $\overline{EB}$ is known, $\overline{BD}$ can be solved for:

$$\overline{BD} = 0.707P \quad \text{(tension)}$$

A symmetry argument would yield the same result, that is, $\overline{BD}$ must be similar to $\overline{EB}$ since the truss is geometrically symmetrical and is symmetrically loaded. Otherwise, there is no a priori reason to believe the forces to be the same.

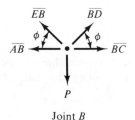

Joint B

Equilibrium in the horizontal direction: $\sum F_x = 0 \quad \rightarrow +:$

$$-\overline{AB} + \overline{BC} - \overline{EB} \cos 45° + \overline{BD} \cos 45° = 0$$

or

$$\overline{BC} = 0.5P \quad \text{(tension)}$$

JOINT D

Equilibrium in the vertical direction: $\sum F_y = 0 \quad \uparrow +:$

$$-\overline{BD} \sin 45° + \overline{DC} \sin 45° = 0$$

or

$$\overline{DC} = 0.707P \quad \text{(compression)}$$

Joint *D*

Equilibrium in the horizontal direction: $\sum F_x = 0 \quad \rightarrow +:$

$$+\overline{ED} - \overline{DB} \cos 45° - \overline{DC} \cos 45° = 0$$

or

$$1.0P - 0.707P \cos 45° - 0.707P \cos 45° = 0$$

All forces are known. The equation sums to zero. This is a good check on the accuracy of calculations.

JOINT C

Equilibrium in the vertical direction: $\sum F_y = 0 \quad \uparrow +:$

$$-\overline{DC} \sin 45° + 0.5P = 0$$

or

$$-0.707P \sin 45° + 0.5P = 0$$

Again, this is a check since all forces are known.

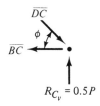

Joint *C*

Equilibrium in the horizontal direction: $\sum F_x = 0 \quad \rightarrow +:$

$$-\overline{BC} + \overline{DC} \cos 45° = 0$$

or

$$-0.5P + 0.707P \cos 45° = 0$$

Again this is a check. All member forces are already known.

EXAMPLE

Determine the forces in members $\overline{FC}$ and $\overline{FG}$ in the truss shown in Figure 4-9.

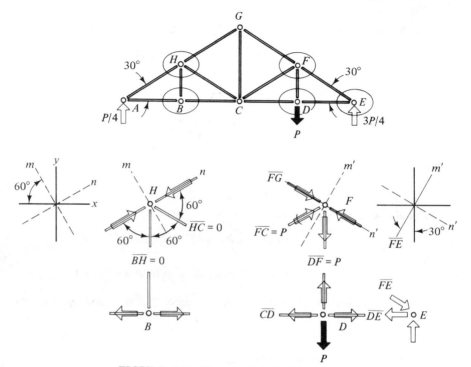

FIGURE 4-9 Use of rotated reference axes.

Solution:

JOINT E

$$\sum F_y = 0 \qquad -\overline{EF}\sin 30° + 3P/4 = 0 \quad \text{or} \quad \overline{EF} = 1.5P \quad \text{(compression)}$$
$$\sum F_x = 0 \qquad \overline{EF}\cos 30° - \overline{DE} = 0 \quad \text{or} \quad \overline{DE} = 1.3P \quad \text{(tension)}$$

JOINT D

$$\sum F_y = 0 \qquad \overline{DE} - \overline{CD} = 0 \quad \text{or} \quad \overline{CD} = 1.3P \quad \text{(tension)}$$
$$\sum F_x = 0 \qquad -P + \overline{FD} = 0 \quad \text{or} \quad \overline{FD} = 1.0P \quad \text{(tension)}$$

JOINT F

$$\sum F_y = 0 \qquad -\overline{FD} + \overline{EF}\sin 30° - \overline{FG}\sin 30° + \overline{FC}\sin 30° = 0$$
$$\sum F_x = 0 \qquad -\overline{EF}\cos 30° + \overline{FG}\cos 30° + \overline{FC}\cos 30° = 0$$

$\overline{FD}$ and $\overline{EF}$ are known. Each equation still involves two unknown forces ($\overline{FG}$ and $\overline{FC}$). Solving the equations simultaneously yields:

$$\overline{FG} = 0.5P \qquad \text{(compression)}$$
$$\overline{FC} = 1.0P \qquad \text{(tension)}$$

The need to solve two equations simultaneously can be obviated by using a rotated reference axis system (m'–n') rather than the traditional vertical and horizontal (x–y) system. Thus, with respect to m'–n':

$$\sum F_{m'} = 0 \qquad -\overline{FD} \sin 60° + \overline{FC} \sin 60° = 0 \qquad\qquad \overline{FC} = 1.0P$$
$$\sum F_{n'} = 0 \qquad \overline{FD} \cos 60° + \overline{FC} \cos 60° - \overline{EF} + \overline{FG} = 0 \qquad \overline{FG} = 0.5P$$

SIMPLIFYING CONDITIONS. Occasionally truss analysis by the method of joints can be facilitated by paying attention to some special conditions that frequently arise. One of these is the appearance of *zero-force* members. Consider the truss shown in Figure 4-10 and refer directly to joint B. By inspecting the equilibrium of this joint in the vertical direction, it is possible to determine by observation that there is no internal force at all developed in member BI. There is no applied external load and members BC and BA are horizontal, having no component in the vertical direction. If there were any force in member BI, the joint would not be in equilibrium; hence, there can be no force present in member BI. It is a zero-force member. The same is true for the force in member HC. By isolating joint J and considering equilibrium in both the horizontal and vertical direction, it can be seen that both members framing into it (JA and JI) are also zero-force members. The same is true of the members framing into joint F.

This approach of picking out selected members and joints for quick determination of forces also can be applied to some joints with external loads applied. Isolating joint D (see Figure 4-10), inspection of equilibrium in the vertical direction indicates

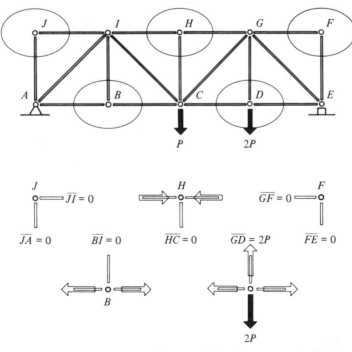

FIGURE 4-10 Simplifying conditions in truss analysis.

that the force developed in member DG must be equal and opposite to the applied load. Thus, member DG carries a force of $2P$ and is in tension.

As a further example of simplifying conditions, consider the truss shown in Figure 4-9 that was previously analyzed. Inspection of joint B indicates that member BH must be a zero-force member. This implies that member HC is also a zero-force member. This can most easily be seen by thinking in terms of the rotated axis system shown in Figure 4-9 and considering equilibrium in the m direction. Member HC cannot have any force in it if joint H is to be in equilibrium in this direction (as it must be in any direction). Inspection of Joint D indicates that DF must carry a tension force of P. This implies that member FC carries a compressive force of P. Again, this is most easily seen by thinking in terms of a rotated axis system, as indicated in Figure 4-9. Since $\overline{DF}$ is shown to be equal to P in tension, the component of this force in the m' direction must be balanced by a similar component in member FC. The members make a similar angle to the m' axis, so it follows that the forces must be numerically the same if the components are the same; thus, $\overline{FC} = P$.

Simplifying conditions of the type described facilitate calculations enormously. But the technique is most valuable in the development of a more intuitive understanding of the force distributions present in a truss.

4.3.4 Equilibrium of Sections

In our discussion of the equilibrium of joints, the elemental portions of the truss defined for equilibrium considerations were the joints themselves. The principle underlying the analytical technique presented next is that *any* portion of a structure must be in a state of equilibrium. It follows that the limits of the portion considered can be extended to a complete subassembly consisting of several joints and members. Consideration of the equilibrium of a larger subassembly can then be the vehicle used for finding unknown bar forces. This concept of considering the equilibrium of an elemental portion of a structure larger than a point is a very powerful one and will form the basis for the analysis and design of many structures other than the truss forms currently under discussion.

Solution of bar forces by this approach is best illustrated by example. Again consider the truss shown in Figure 4-8. Assume that you wish to know the forces in members ED, BD, and BC. The truss is first considered to consist of two subassemblies. Free-body diagrams for these subassemblies are shown in Figure 4-11. The internal set of forces shown is developed as a consequence of the external loading of the whole structure and serves to maintain the equilibrium of the elemental subassemblies of the truss. The complete force system acting on a subassembly, consisting of internal forces and any external forces present, must form a system whose net result is zero. Both translational equilibrium and rotational equilibrium must be considered, since the forces acting on the body form a coplanar, but nonconcurrent and nonparallel force system (see Section 2.3).

Before solving mathematically for the bar forces in the example, by formal application of the equations of statics, it is useful to try to determine by qualitative inspection the senses of the member forces. Study equilibrium requirements in the subassembly shown to the left in Figure 4-11, for example. It can be seen that the force

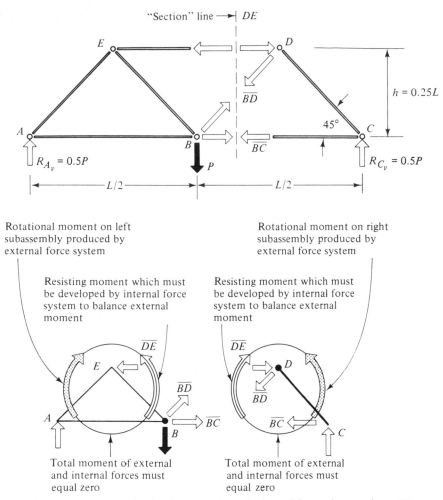

FIGURE 4-11 Free-body diagrams for solution of forces in members *ED*, *BD*, and *BC* by the method of sections.

in member *BD* must act upward as shown to supply the vertical component necessary for balancing the difference between the upward reaction of $0.5P$ and the downward force of P acting on the subassembly (a net of $0.5P$ acting downward). Thus, the member must be in a state of tension. Since the subassembly to the left must be in rotational equilibrium also, the moment produced by the external forces must be exactly balanced by the moment developed by the internal forces (see Figure 4-11). By summing moments about point *B*, it can be seen that the sense of the force in member *DE* must be in the direction shown if moment equilibrium is to occur about point *B*. Remember that the sum of the moments produced by *all* forces must be zero about *any* point. Hence, member *DE* must be in a state of compression. The forces shown on the left subassembly are equal and opposite on the right subassembly. Since the right subassembly must also be in translatory and rotational equilibrium, the sense of the force in member *BC* can be found by summing moments about point

D. For moment equilibrium to obtain about this point, force $\overline{BC}$ must act in the direction shown and so be in a state of tension. Thus, the states of stress in the unknown bar forces can be qualitatively determined. The mathematical process for determining the numerical magnitudes of the forces is conceptually similar to the process described above.

EXAMPLE

Determine the forces in members *DE*, *BD*, and *BC* of the truss shown in Figures 4-8 and 4-11.

Solution:

LEFT SUBASSEMBLY

Translatory equilibrium in the vertical direction: $\sum F_y = 0$ $+\uparrow$:

$$\underset{\substack{\text{reaction } R_{A_v} \\ \text{in vertical} \\ \text{direction.}}}{+0.5P} \quad - \quad \underset{\substack{\text{external} \\ \text{load}}}{P} \quad + \quad \underset{\substack{\text{vertical component of} \\ \text{force in member } BD; \text{ the} \\ \text{member is assumed to be} \\ \text{in tension}}}{\overline{BD} \sin 45°} \quad = 0 \qquad \underset{\substack{\text{No other forces have components} \\ \text{in the vertical } (y) \text{ direction.}}}{}$$

$$\overline{BD} = 0.707P \qquad \text{(tension)}$$

Member *BD* is in tension as assumed, since the sign is positive. We could next try $\sum F_x = 0$, but this would involve two unknown forces ($\overline{ED}$ and $\overline{BC}$) and the equation could not be solved directly. Try to find an equation involving only a single unknown force by considering moment equilibrium about a point. By selecting point *B*, one unknown force, $\overline{BC}$, acts through the moment center and consequently falls out of the moment equation (since its moment arm is zero), leaving an equation involving only the remaining unknown force and known external forces.

Moment equilibrium about point B: $\sum M_B = 0$ ⟲:

$$\underset{\substack{\text{reaction} \\ R_{A_v}}}{-(0.5P} \times \underset{\substack{\text{moment} \\ \text{arm}}}{0.5L)} + (\underset{\substack{\text{member} \\ \text{force}}}{\overline{ED}} \times \underset{\substack{\text{moment} \\ \text{arm}}}{h)} = 0$$

or

$$-0.25PL \quad + \quad \overline{ED}(0.25L) \quad = \quad 0$$

$$\underset{\substack{\text{"applied" moment} \\ \text{produced by ex-} \\ \text{ternal forces}}}{0.25PL} \quad = \quad \underset{\substack{\text{"resisting" moment} \\ \text{developed by in-} \\ \text{ternal forces}}}{\overline{ED}(0.25L)}$$

Thus,

$$\overline{ED} = +P \qquad \text{(compression)}$$

The member is in compression, as assumed. The step where the moment developed by the internal forces is shown equal to that produced by the external forces is not actually necessary. It does, however, represent a fundamentally important way of looking at structural behavior and is important for design purposes. This way of looking at structures will be discussed further in Section 4.3.8. Now that two of the forces acting on the subassembly are known, the remaining unknown force can readily be found by application of the last unused equation of statics: $\sum F_x = 0$.

Translatory equilibrium in the horizontal direction: $\sum F_x = 0$ $\xrightarrow{+}$:

$$-\overline{ED} + \overline{BD}\cos 45° + \overline{BC} = 0$$

or

$$-P + (0.707P)\cos 45° + \overline{BC} = 0$$
$$\overline{BC} = +0.5P \quad \text{(tension)}$$

Member *BC* is in tension, as assumed. All unknown forces acting on the left subassembly have now been found. These forces act in an equal and opposite way on the right subassembly. As is obvious, it should also be in equilibrium. It is good to check this.

RIGHT SUBASSEMBLY

Moment equilibrium about point D: $\sum M_D = 0$ $\curvearrowleft$:

$$-(\overline{BC} \times 0.25L) + (0.5P \times 0.25L) = 0-$$

or

$$-(0.5P \times 0.25L) + (0.5P \times 0.25L) = 0 \quad \text{Check!}$$

Translatory equilibrium in the vertical direction: $\sum F_y = 0$ $+\uparrow$:

$$-\overline{BD}\sin 45° + 0.5P = 0$$

or

$$-0.707P\sin 45° + 0.5P = 0 \quad \text{Check!}$$

Translatory equilibrium in the horizontal direction: $\sum F_x = 0$ $\xrightarrow{+}$:

$$-\overline{BC} - \overline{BD}\cos 45° + \overline{ED} = 0$$

or

$$-0.5P - 0.707P\sin 45° = 0 \quad \text{Check!}$$

The right-hand subassembly is in a state of translational and rotational equilibrium.

The approach illustrated above can be used to find member forces in planar trusses when no more than three unknowns are involved, since there are only three independent equations of statics available for analyzing the equilibrium of planar rigid bodies. Care must be taken in isolating subassemblies so that only three unknowns are present. The decomposition shown in Figure 4-8(d), for example, is perfectly valid, and each subassembly is indeed in a state of translatory and rotational equilibrium. However, the decomposition is not useful from an analytical viewpoint, since there are more unknown forces acting on each segment than could be solved for by using the basic equations of statics.

The descriptive name associated with this approach, the method of sections, stems from one particular way of looking at how subassemblies are separated from the remainder of the truss. An imaginary section line is passed through the truss, dividing it into two segments. In actuality there is no implication that the section line must be straight. The method of joints could quite easily be thought of in terms of passing a section line around each joint [see Figure 4-8(b)]. There is no real difference between these two methods other than the extent of the subassembly considered for equilibrium. Both can be interchangeably used in analyzing trusses. The method of joints is often considered preferable for use when all the bar forces in a truss must be determined. The method of sections is particularly useful when only a limited number of forces must be found.

EXAMPLE

Determine the forces in members *FG* and *FC* in the truss shown in Figure 4-12 (the same truss previously analyzed by the method of joints).

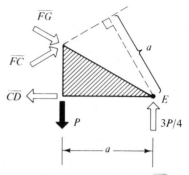

(b) Section for finding $\overline{FC}$.

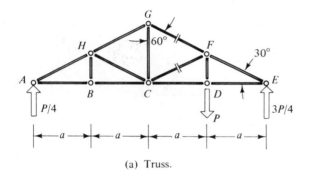

(a) Truss.

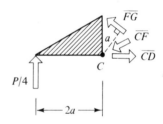

(c) Section for finding $\overline{FG}$.

FIGURE 4-12 Truss analysis by the method of sections.

Solution:

For determination of the force in member *FC*, the truss is decomposed as shown in Figure 4-12(b). Point *E* is selected as the moment center as a matter of convenience, since the unknown forces $\overline{FG}$ and $\overline{CD}$ pass through the point and hence would not enter into the moment equation because their moment arms are zero.

$$\Sigma M_E = 0:$$

$$-\overline{FC}(a) + P(a) = 0 \quad \text{or} \quad \overline{FC} = 1.0P \quad \text{(compression)}$$

To find the force in member *FG*, it is more convenient to decompose the structure as shown in Figure 4-12(c) and use point *C* as a moment center.

$$\Sigma M_c = 0:$$

$$+\overline{FG}(a) - \left(\frac{P}{4}\right)(2a) = 0 \quad \text{or} \quad \overline{FG} = 0.5P \quad \text{(compression)}$$

The results are obviously the same as obtained by the method of joints, but are found in a considerably more direct way.

EXAMPLE

Determine the forces in members MN, ML, and KL of the truss shown in Figure 4-13(a).

(a) Basic truss.

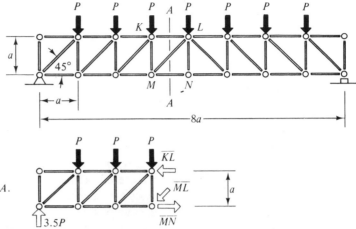

(b) Member forces at section A–A.

FIGURE 4-13 Solution of bar forces by the method of sections.

Solution:

A section line is first passed through these three members and the truss is decomposed into two subassemblies.

LEFT SUBASSEMBLY

$\Sigma F_y = 0 \quad \overrightarrow{+}$:

$$+3.5P - P - P - P - \overline{ML} \sin 45° = 0$$

Thus,

$$\overline{ML} = 0.707P \quad \text{(compression)}$$

$\Sigma M_m = 0 \quad \curvearrowleft$:

$$-(3.5P \times 3a) + P(2a) + (Pa) + \overline{KL}(a) = 0$$

Thus,

$$\overline{KL} = 7.5P \quad \text{(compression)}$$

$\Sigma F_x = 0 \quad \overrightarrow{+}$:

$$-\overline{KL} + \overline{MN} - \overline{ML} \cos 45° = 0$$

or

$$-7.5P + \overline{MN} - 0.707P \cos 45° = 0$$

Thus,

$$\overline{MN} = 8.0P \quad \text{(tension)}$$

chapter 4 / TRUSSES

145

Finding the same forces in the last example by the method of joints would be a long and tedious process, since each joint would have to be considered in turn from one of the ends. The example also illustrates some very important points about forces in a truss. The three member forces found are clearly dependent on the overall geometry and dimensions of the truss and loading condition, and on the local geometry of the three members themselves. Thus, the force in the top chord (member *KL*) is directly related to the height of the truss. If the height were doubled, the member force would be halved, or vice versa. The force in the diagonal is also related to the height. Increasing the truss height would decrease the force in the diagonal, since the sine of the angle involved is increased. The force in the lower chord would also decrease as the truss height is increased. It is very important to note that these forces, which are sensitive to the local geometry at the section considered, are *independent* of many other physical aspects of the truss. Insofar as these three forces are concerned, it makes no difference whether the remainder of the diagonals have the original orientation or an alternative orientation. Indeed, the geometry of the truss at other sections could be drastically altered, and the three forces would not be affected as long as *their* local geometry remained constant. The critical point is that the equilibrium of a whole body is being considered. The internal arrangement of bars making up the body at other locations does not alter the forces in the bars at the cut section that are necessary to maintain its equilibrium.

The discussion above illustrates the power of the method of sections as a conceptual tool. The observations presented could have been made through analysis by the method of joints, but doing so would have been very cumbersome.

4.3.5 Shears and Moments in Trusses

At this point it is useful to introduce a particular way of looking at how trusses carry loads which is of fundamental importance and will prove useful in designing them. This involves studying a truss in terms of sets of externally applied forces and moments and responding sets of internal resisting forces and moments.

Consider the truss shown in Figure 4-14 and the diagram of the portions of the structure to the left of section *A–A*. Note that as yet the diagram in Figure 4-18(b) is *not* an equilibrium diagram, since only the set of *external* forces, consisting of applied loads and the reaction at the support, are shown. Study of this diagram indicates that insofar as translation in the vertical direction is concerned, the *net* effect of this set of forces is to produce an *upward* acting force of 0.5P [see Figure 4-14(c)]. This net force is often referred to as the *external shear force* (V_E). A study of the rotational effects about *A–A* produced by this set of forces reveals that the net effect is a rotational moment of 4Pa [see Figure 4-14(c)]. This net moment is often referred to as the *external bending moment* (M_E) present at the section. The descriptive term "bending" comes from the tendency of this net external moment to produce bowing in the whole truss.

The function of the set of forces developed internally in members of the truss can now be discussed in terms of the external shear force and bending moment present at the section. For vertical equilibrium to obtain in the portion of the structure shown, a set of internal forces must be developed in the structure whose net effect is a resisting force of an equal magnitude but opposite sense to the applied external shear

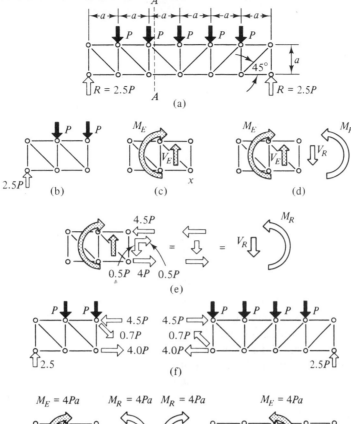

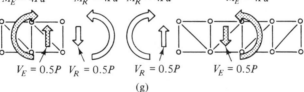

(a) Truss.

(b) External force system acting on left subassembly.

(c) External bending moment and shearing force equivalent to the external force system acting on left subassembly:

$$\Sigma F_y : -P - P + 2.5P = 0.5P$$
External shear (V_E)

$$\Sigma M_x : +2a(2.5P) - a(P) = 4Pa$$
External moment (M_E)

(d) Equilibrium diagram for left subasssembly: an internal resisting shear force (V_R) and internal resisting moment (M_R) is developed in the structure for equilibrium.

(e) The internal resisting shear force (V_R) and internal resisting moment (M_R) are developed through the action of the forces in the bars.

(f) Equilibrium diagrams of left and right subassemblies in terms of internal and external force systems.

(g) Equilibrium diagrams of left and right subassemblies in terms of equivalent internal and external shears and moments. The sum of shear and moment sets on each subassembly must be zero.

FIGURE 4-14 Shear and moment in trusses: These diagrams illustrate the development of shearing forces and moments at a section in a truss through the action of the external force system acting on the truss. A resisting set of shears and moments is developed at the same section through the action of the internal forces developed in the bars.

force. This net force is referred to as the *internal resisting shear force* (V_R) [see Figure 4-14(d)]. In a like manner, the net rotational effect of this set of internal forces must be to provide *an internal resisting moment* (M_R) equal in magnitude but opposite in sense to the external bending moment (or applied moment) present on the portion of the structure considered, so that the *total* rotational moment is zero, as it must be for equilibrium [see Figure 4-14(d)]. Thus $M_E = M_R$ or $M_E - M_R = 0$.

In the structure shown (with respect to equilibrium in the vertical direction), the top and bottom chords cannot provide any of the internal resisting shear force necessary to balance the external shear force because they are horizontal and can have no force components in the vertical direction. The entire resisting shear force must be provided by the vertical component of the force in the diagonal member. This way of looking at the structure could also be used as either a way of calculating the force in the member or checking the force found by some other method. With respect to moment equilibrium, there must be a couple (⇆) developed by the set of internal forces to provide an equilibrating moment to the applied external bending moment. As indicated in Figure 4-14(e), the horizontal forces in the chord members act with the horizontal component of the force in the diagonal to provide the couple which balances the external bending moment.

All the manipulations described above are based on concepts previously discussed in Section 4.3.4. Looking at the behavior of structures this way, it is easier to see that the basic structural function of *any* configuration of bars used in a truss is to provide a resistance to the external shears and moments present. These shears and moments are dependent only on the set of external loads present and their location in space and not on the characteristics of the truss itself. Consider two other similiar structures, indicated in Figure 4-15, in which the external forces are as before, but the bar patterns are different, as are some of the dimensions. The bar forces shown were found by the method of sections (the analyses are not shown). In each case the net result of the forces developed in the members is still to produce at the section shown a net resisting shear force of $0.5P$ and an internal resisting moment of $4Pa$. The *net resultants* of these forces are identical to those in the bars of the structure initially discussed. They must be, since these are the forces necessary for equilibrium.

This, then, is the *thread of similarity* which conceptually ties these and other structures together. The exact way different structures respond to the same loading may vary, but each provides (and *must* provide) the same internal resisting shears and moments. This way of thinking about the behavior of trusses is also important in a design situation. One must determine the structural configuration which most efficiently provides the necessary set of internal resisting shears and moments for given external shear forces and bending moments. The principles involved are of general applicability and not restricted to the study of trusses alone.

4.3.6 Statically Indeterminate Trusses

In all the trusses previously discussed, it was possible to calculate member forces by application of the equations of statics. Trusses of this type are often referred to as being *statically determinate*. There exist other truss structures where this is not possible because of the number of external supports or number of truss bars. Consider

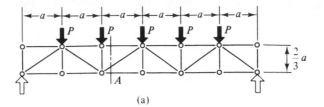

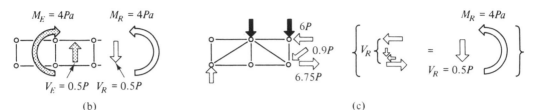

(a)

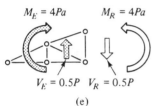

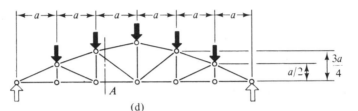

(b)

$M_E = 4Pa$ $M_R = 4Pa$

$V_E = 0.5P$ $V_R = 0.5P$

(c)

$M_R = 4Pa$

$V_R = 0.5P$

(d)

(e)

$M_E = 4Pa$ $M_R = 4Pa$

$V_E = 0.5P$ $V_R = 0.5P$

(f)

$V_R = 0.5P$

$M_R = 4Pa$

(a) Truss carrying the same loading condition illustrated in Figure 4-14.

(b) External bending moment and shearing force equivalent to the external force system acting on left subassembly:

$$\Sigma F_y : -P - P + 2.5P = \underline{0.5P}$$
External shear (V_E)

$$\Sigma M_x : +2a(2.5P) - a(P) = \underline{4Pa}$$
External moment (M_E)

(c) Equilibrium diagram for left subassembly: an internal resisting shear force (V_R) and internal resisting moment (M_R) is developed in the structure to balance V_E and M_E.

(d) Truss carrying the same loading condition illustrated in Figure 4.14

(e) External bending moment and shearing force equivalent to the external force system acting on left subassembly:

$$\Sigma F_y : -P - P + 2.5P = \underline{0.5P}$$
External shear (V_E)

$$\Sigma M_x : +2a(2.5P) - a(P) = \underline{4Pa}$$
External moment (M_E)

(f) Equilibrium diagram for left subassambly: an internal resisting shear force (V_R) and internal resisting moment (M_R) is developed in the structure for equilibrium.

FIGURE 4-15 Shear and moment in trusses: the trusses shown carry the same loading condition as the truss shown in Figure 4-14. The sets of external shears and moments and internal resisting shears and moments present at a section are the same in all three structures. The sets of internal bar forces which provide the resisting shears and moments, however, are different because local bar geometries are different. The net effects of the different internal force systems are the same. All provide an identical internal resisting shear and moment which balances the external shear and moment present at the section.

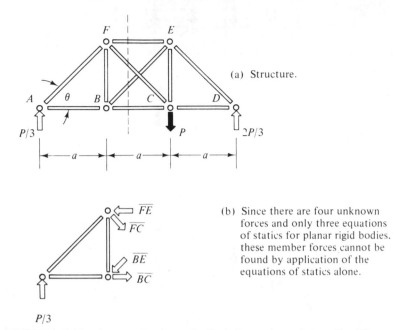

(a) Structure.

(b) Since there are four unknown forces and only three equations of statics for planar rigid bodies, these member forces cannot be found by application of the equations of statics alone.

FIGURE 4-16 A truss that is statically indeterminate internally. The magnitudes of the forces present depend upon the physical properties of the members.

the planar truss shown in Figure 4-16. It is not possible to calculate the forces in the middle members by the equations of statics alone, since there are four unknown bar forces and only three equations of statics available to use in solving for these unknowns. The calculation difficulties are evident when one considers equilibrium in the vertical direction for the left subassembly. Thus:

$$\sum F_y = 0 +\uparrow:$$

$$+\tfrac{1}{3}P - \overline{BE} \sin \theta - \overline{FC} \sin \theta = 0$$

or

$$\overline{BE} \sin \theta + \overline{FC} \sin \theta = \frac{P}{3}$$

Clearly, this equation cannot be solved. Applying the other equations of equilibrium for a planar rigid body, $\sum F_x = 0$, $\sum M_0 = 0$, will not yield equations that could be solved, either alone or simultaneously. Structures of the type described above are referred to as being *statically indeterminate internally*. It should be emphasized that all the principles of statics discussed are still valid and applicable in structures of this type. It is still true that any elemental portion of the truss is in translational and rotational equilibrium ($\sum F_x = 0$, $\sum F_y = 0$, and $\sum M_0 = 0$ still apply) but some other method must be used to calculate member forces. This is a topic generally beyond the scope of this book. One way to interpret the physical meaning of a structure being statically indeterminate, however, is that the senses and magnitudes of the internal forces present in members of the structure are dependent on the actual physical properties of the members themselves.

4.3.7 Space Trusses

The stability inherent in triangulated patterns of bars is also present when extended into the third dimension. Whereas the simple triangle is the basic repetitive element in planar trusses, the tetrahedron illustrated in Figure 4-17 is the basic repetitive element in three-dimensional trusses.

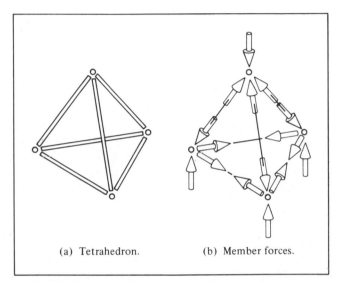

(a) Tetrahedron.   (b) Member forces.

FIGURE 4-17   Three-dimensional bar configuration.

The principles developed for analyzing planar trusses are generally applicable to space trusses. Again, stability is a primary consideration. The presence of non-triangulated patterns is a sign that each joint in turn should be looked at closely to see if it will always maintain the same fixed geometric relationship to other joints when a load is applied to any point on the structure.

The forces developed in the members of a space truss can be found by consideration of the equilibrium in space of elemental portions of the space truss. Obviously, the equations of statics used must be the *full* set of equations developed in Section 2.3 for the equilibrium of a rigid body in three dimensions, rather than the simplified set for the equilibrium of a rigid body in two dimensions used in the previous section on planar truss analysis. The relevant equations are:

$$\sum F_x = 0, \quad \sum F_y = 0, \quad \sum F_z = 0 \quad \text{and} \quad \sum M_x = 0, \quad \sum M_y = 0, \quad \sum M_z = 0$$

While straightforward, application of these equations is usually tedious due to the large number of joints and members typically present in a large three-dimensional truss and will not be undertaken in this text. Figure 4-18, however, illustrates the results of such an analysis for a simple configuration. Computer formulations are usually used to analyze more complex configurations.

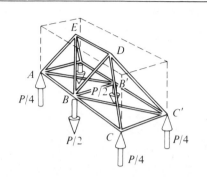

(a) Typical three-dimensional configuration.

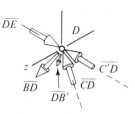

(b) Any joint must be in a state of translatory and rotational equilibrium. Thus, $\Sigma F_x = 0$, $\Sigma F_y = 0$, and $\Sigma F_z = 0$.

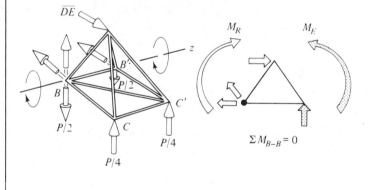

$$\Sigma M_{B-B} = 0$$

(c) Any larger segment must be in a state of translatory and rotational equilibrium. With respect to moment equilibrium, for example, the external moment (M_E) generated at a section must be exactly balanced by an internal resisting moment (M_R) provided by member forces.

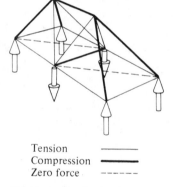

Tension ————
Compression ▬▬▬▬
Zero force ------

(d) Force distribution in truss.

FIGURE 4-18 Three-dimensional bar configurations.

152

4.4 DESIGN OF TRUSSES

4.4.1 Background

As is evident from the examples discussed in the preceding sections, trusses can assume a wide variety of forms. For any given truss, the calculation of member forces is relatively straightforward. Once these forces are known, a physical truss member could be adequately sized and otherwise designed in response to these forces. This is a very important step in designing a truss for use in a building and will be explored in more detail in this section. It is, however, only one step and one that occurs rather late in the complete process of designing a truss for use in a building. Indeed, somewhere along the way in the design of the whole building, an initial decision must have been made to use a truss rather than some other structural element (e.g., a beam or cable) as a horizontal spanning element. Similarly, decisions about the truss span, bar configuration, and overall dimensions must also be made before the process of analyzing member forces and sizing individual elements can take place.

This section begins to explore what is involved in designing a truss once the initial decision to use one has been made. The larger question of whether or not a truss of some type is indeed the preferred structural element for use in a given building context is deferred until Part III. The following discussion will focus on the design of trusses in a simplified context as a way of drawing out general design principles. In actuality, the design of trusses should occur only in the much broader context of the design of the complete building. In the structural system shown in Figure 4-19, for example, not only can the general geometry of the truss and its member properties and proportions be considered design variables, but so can each of the physical dimensions shown ($L_1, L_2, L_3, L_4, L_5, L_6$) in the figure. The spacing of the trusses (L_4), clearly affects the magnitudes of the loads carried by them. The transverse beam spacing (L_5) affects the distribution of the loads on the truss and thus implicitly affects the location of the panel points or joints of the truss. It would not be possible to determine rationally what these spacings optimally would be without considering in detail the rest of the structural system (the transverse beams, roof decking, and columns) and the relation of the structural system to the remainder of the building. A transverse beam spacing which is advantageous with respect to the design of the truss may be completely inappropriate for the roof decking. In general, meaningful design decisions concerning a specific structural element (e.g., a truss) can be made only in a much broader context. The following sections will only begin to explore the issues of importance in design and will do so in a simplified context to make the discussion manageable.

4.4.2 Member Design: General

The selection of physical members for use in a truss is one of the more straightforward aspects of truss design. The crucial issue is predicting what force a member should be designed to carry. In earlier sections it was shown that the nature and magnitude of the force present in a particular member are dependent on the specific loading condition on the whole truss. It is thus necessary to consider the whole spectrum of possible loading conditions and analyze the forces present in each of the members

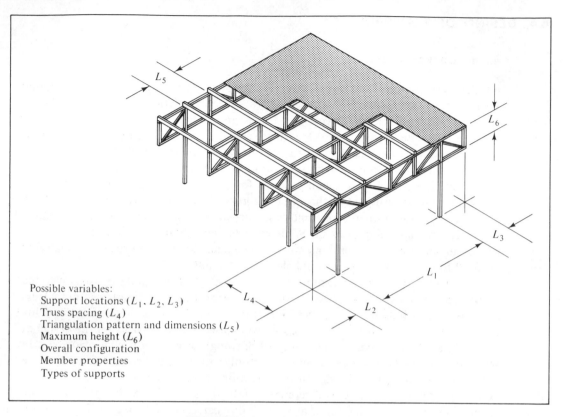

Possible variables:
Support locations (L_1, L_2, L_3)
Truss spacing (L_4)
Triangulation pattern and dimensions (L_5)
Maximum height (L_6)
Overall configuration
Member properties
Types of supports

FIGURE 4-19 Typical truss design variables.

under each of these conditions. A particular member would then be designed in response to the maximum (critical) force possible. Consequently, individual members quite often are designed in response to different loading conditions (e.g., the diagonals of a truss may be designed for one external loading condition and the chord members for another). Only in instances where the loading condition is invariant are all truss members designed in response to the forces generated by the same load.

Once the critical force in a member has been found, the problem becomes one of choosing an appropriate material and a set of cross-sectional dimensions for a member of known length having simple pin-ended connections and subjected to a known tension or compressive force. There is usually no great difficulty in designing the member. Quantitative techniques for designing members of this type are discussed in greater detail in Section 2.4 and Chapter 7. Of primary importance is that members carrying only tension forces can usually be designed to have much smaller cross sections than those carrying compressive forces of a similar magnitude. The cross-sectional area of the former is directly dependent on the magnitude of the force present and the allowable stress of the material used and is independent of member length. Thus: *area required = tension force/allowable stress*. For members in compression it is necessary to take into account the possibility of a buckling failure, which can occur in long members subjected to compressive forces. The load-carrying capacity of the compressive member is inversely proportional to the square of the member's length for long members. If compressive members are relatively short and below a certain maximum length, buckling is not a problem and the cross-sectional area is then directly dependent only on the magnitude of the force involved and the allowable crushing stress of the material and is again independent of the specific length of the element. The general implication of this is that long members (possibly subject to buckling) require a greater cross-sectional area to support a given compressive load than would a short member (not subject to buckling). Exactly what defines a long or short member is discussed extensively in Section 7.3. In truss design, either members in tension or short compressive members are considered more desirable than long compressive members, since less material is required to carry a given force. A truss design objective is often to create a situation where as many members as possible carry only tension forces or are short compressive members, because this often means less total material is required to support the given external loads. Techniques for doing this will be discussed shortly.

The relative sizes of members, designed in response to the forces actually present, in two typical trusses are shown diagrammatically in Figure 4-20. Obviously, the

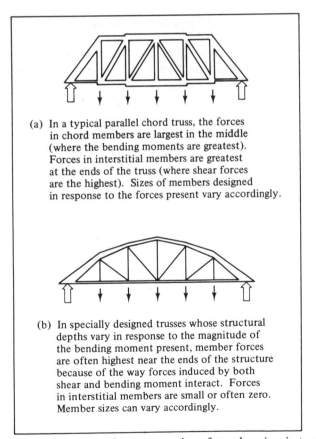

(a) In a typical parallel chord truss, the forces in chord members are largest in the middle (where the bending moments are greatest). Forces in interstitial members are greatest at the ends of the truss (where shear forces are the highest). Sizes of members designed in response to the forces present vary accordingly.

(b) In specially designed trusses whose structural depths vary in response to the magnitude of the bending moment present, member forces are often highest near the ends of the structure because of the way forces induced by both shear and bending moment interact. Forces in interstitial members are small or often zero. Member sizes can vary accordingly.

FIGURE 4-20 Diagrammatic representation of member sizes in two typical trusses.

chapter 4 / TRUSSES

illustration is not meant to represent real members but is intended to convey a graphic sense of material distribution.

Instead of varying the size of each individual member on response to specific forces present, it may be more convenient or less costly to make several pieces (e.g., the entire top or bottom chord) out of constant-cross-section continuous members. Whether or not to do so is a design decision that must be evaluated in light of other factors. When the top chord is designed to be a continuous member of a constant cross section, for example, the cross section used must be designed to carry the maximum force present in the top chord. Since this would usually occur only in one segmental portion of the entire top chord, this same cross section would be excessively large for the remaining segments and so potentially inefficient.

Whether or not to size each truss member in exact response to the forces present or to base all member sizes on certain critical elements is not an easy choice to make. The latter is often done when a large number of trusses are used repetitively, spaced closely together, and designed to carry relatively light loads (e.g., mass-produced bar joists). The members of major trusses carrying large dead loads are often designed more in response to the actual forces carried.

4.4.3 Use of Special Tensile Members: Cables

In all the trusses previously discussed it was tacitly assumed that an individual truss member was capable of carrying the tensile or compressive force generated in the member. Other types of members, however, can be useful. One such member is commonly referred to as a *cable* which is capable of resisting tensile forces only. Physically, members of this type are typically thin steel rods or actual wound cables. Such members cannot resist compressive forces [see Figure 4-21(a)] but are often used when a truss member is known from analysis to always be in a state of tension and need never carry compressive forces. As noted, members carrying only tension forces can have much smaller cross sections than those carrying compressive forces and are often

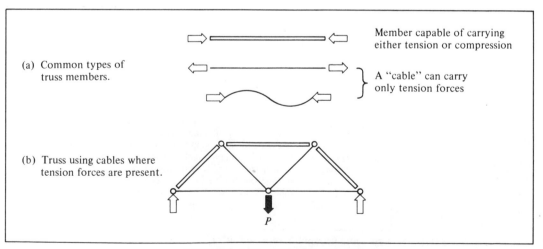

FIGURE 4-21 Use of "cables" in trusses.

considered desirable. For the purposes of this discussion, the important point is
the member has a directional load-carrying capacity. It is easily possible to envisic
counterpart capable of carrying only compression and no tension, but such meml
with this type of directional capacity are rare in actual trusses and will not be c
sidered here. Since a cable could be used wherever analysis has demonstrated tha
tension force is present in a member, the truss analyzed in the first example probl
in Section 4.3.2 could use cables in the manner illustrated in Figure 4-21(b).

One of the crucial issues in using cables as elements in a truss is the question
stability. The repetitive element in the truss is no longer a basic triangular sha
composed of rigid members, inherently stable under any loading condition, but is
special shape, stable only under particular loading conditions. Care must be taken
assure that the cable member, intended to carry tensile forces only, is never expect
to carry compressive forces. *Stress reversals* of this type would usually cause tl
whole truss to become unstable. Stress reversals in members are often caused by
change in the location of the external loads applied to the whole structure. Wit
respect to the truss shown in Figure 4-22(a-1), note that if the load were applied a
joint E rather than joint F, the force distribution in the truss would be as shown i
Figure 4-22(b-1). If an attempt was made to use the cable arrangement shown in Figur
4-22(a-2) to carry forces indicated, the truss would become unstable and collaps
under the second loading condition. This is because member BE, previously in tensior
and thus designed as a cable, would have to be capable of resisting a compressive
force when the truss is subjected to its new loading. Clearly, it cannot, and the member
would "buckle" out of the way. The truss would consequently become unstable and
begin collapsing, as indicated in Figure 4-22(b-2). Potential instability due to stress
reversals is thus a critical issue in using cables in trusses. If cables are used, it is
mandatory to review every possible way that the truss could be loaded, to determine
if a cable member is potentially subjected to compressive forces. If it is, a cable cannot
be used at this location and a member capable of carrying compressive loads must be
used.

In some specific situations, however, an alternative to designing a member capable of carrying either tension or compressive loads is to use additional cables. In the
set of diagrams shown in Figure 4-22 it is evident that if a *crossed-cable* system is used
in the middle [see Figure 4-22(e)], the truss illustrated is stable under varying load
conditions. When the external load is at joint F, the cable between B and E acts in
tension and makes the truss stable, while the cable between F and C merely buckles
(harmlessly) out of the way and carries no load. The converse is true when the external
load is at E. Thus, crossed cables are useful in assuring the stability of trusses under
varying load conditions.

By carefully organizing member orientations, cables can be effectively used in
more complex trusses than that discussed above. Techniques for determining member
orientation will be considered in Section 4.4.4. At this point it can merely be noted
that trusses having bar patterns of the type illustrated in Figure 4-23(a) invariably
have all diagonals in tension (for the loading shown), while trusses having patterns of
the type illustrated in Figure 4-24(a) have all interior vertical members in tension.
The reversal of the direction of diagonals in such trusses is characteristic of designs
for symmetrical loadings.

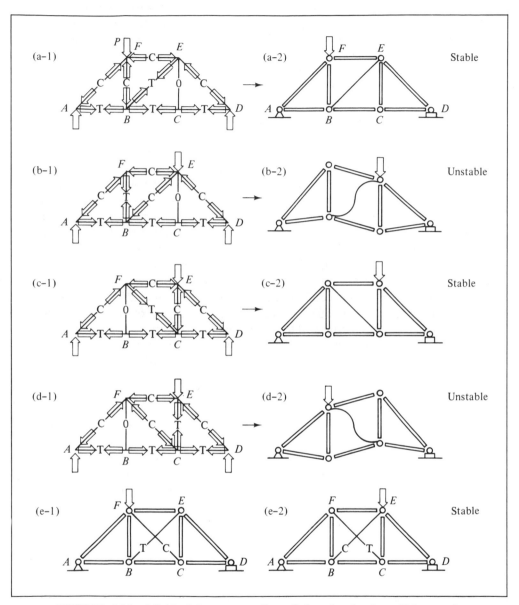

FIGURE 4-22 "Cables" in trusses: effect of changing load conditions and the use of "crossed cables." Parts (a) and (c) illustrate trusses stable under a particular loading condition. Parts (b) and (d) illustrate the instability that could occur if loading conditions are changed. Part (d) illustrates the use of crossed cables as a device for making a truss using cables stable under several loading conditions.

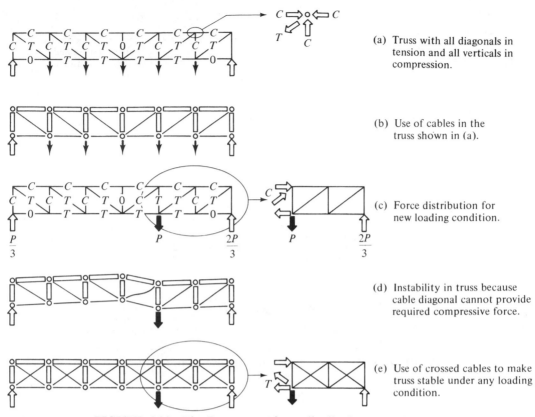

(a) Truss with all diagonals in tension and all verticals in compression.

(b) Use of cables in the truss shown in (a).

(c) Force distribution for new loading condition.

(d) Instability in truss because cable diagonal cannot provide required compressive force.

(e) Use of crossed cables to make truss stable under any loading condition.

FIGURE 4-23 The Pratt truss: force distribution, use of cables, effect of changing load conditions.

Whether cables can be used for tension elements in these trusses depends largely on whether or not the loading condition on the truss is invariant. In the diagrams shown it is obviously suggested that "cables" could be suitable for use as tension elements. If members of this type are used, extreme care should be taken to assure that these tensile members are never expected to carry compressive forces. As previously noted, stress reversals of this type could cause the truss to become unstable. Stress reversals are usually associated with a change in the pattern of external loads carried by the truss. Consider what would happen to the same two trusses under discussion if they were expected to carry the single load shown [see Figures 4-23(c) and 4-24(c)], a case that might occur if the other loads were *simply removed*. Analysis of the member forces due to these loads results in the distributions shown. Clearly, some of the members previously in tension are now in compression. If a cable were used in members previously in tension but now in compression, the trusses would become unstable and collapse as indicated in Figures 4-23(d) and 4-24(d). This illustrates a fundamental design consideration. Each element in a structure must be designed to carry adequately the forces which might develop in it under *any* possible loading condition that *might* occur. This could involve a large number of different load combinations and applies not only to structures using cables, but to all other types of trusses

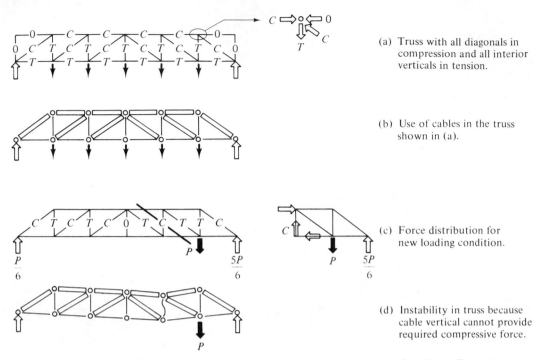

(a) Truss with all diagonals in compression and all interior verticals in tension.

(b) Use of cables in the truss shown in (a).

(c) Force distribution for new loading condition.

(d) Instability in truss because cable vertical cannot provide required compressive force.

FIGURE 4-24 The Howe truss: force distribution, use of cables, effect of changing load conditions.

as well. The trusses shown in Figures 4-23 and 4-24 can be made stable for the new loading condition by using members capable of carrying the compressive forces indicated rather than using cables. Alternatively, for the truss shown in Figure 4-23(c), the addition of crossed cables would cause this truss to be stable. The reader should carefully inspect the truss shown in Figure 4-24(d) to see if the addition of a crossed cable leads to stability. Clearly, it does not. For this reason trusses having this bar pattern rarely use cables, or they are used only when the nature of the external loads are so predictable that stress reversals need not be feared. This is sometimes the case when the dead load carried by a truss significantly exceeds the live load or if there is no live load at all, since the former is usually fixed and precisely determinable while live loads are not.

It is useful to highlight briefly one other general type of bar configuration often encountered in a constant-depth truss. This is a statically indeterminate (internally) configuration shown in Figure 4-25(a). An approximate force distribution for the loading indicated, based on the assumption that the diagonals are of similar stiffnesses and equally share in carrying the shear forces present is also shown. Possible members are diagrammatically illustrated in Figure 4-25(c). The reader should study this truss closely and determine if it is indeed stable under loading conditions other than the one illustrated. (It is, but being able to explain *why* is of crucial importance in gaining a thorough understanding of the structural behavior of trusses.)

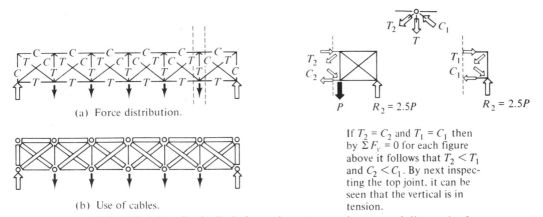

(a) Force distribution.

(b) Use of cables.

If $T_2 = C_2$ and $T_1 = C_1$ then by $\Sigma F_y = 0$ for each figure above it follows that $T_2 < T_1$ and $C_2 < C_1$. By next inspecting the top joint, it can be seen that the vertical is in tension.

FIGURE 4-25 Statically indeterminate truss using crossed diagonals: force distribution, use of cables.

4.4.4 Controlling Force Distributions in Trusses

In order to make the most efficient use of tension or compression elements, it is often desirable to try to control as much as possible the way forces are distributed in a truss. A typical objective in this respect might be to try to cause as many as possible of the longest members of the truss to be in tension rather than compression. This can often be done, to a limited extent, by paying careful attention to how certain members are oriented in the truss. Consider the constant-depth structure shown in Figure 4-26(a) for example, and assume that it is desired to control the states of force in the diagonals that must be added for stability. By studying the equilibrium of different portions of the structure it is possible to determine a bar pattern designed so that each individual diagonal is in either a state of tension or compression, as preferred. In the equilibrium diagram of the structure to the right of section $A–A$, it is evident that with respect to forces in the vertical direction, the net effect of the set of external forces and reactions is to produce an upward acting external shear force of $0.5P$. According to the principles discussed in Section 4.3.5, a set of internal resisting shear forces of an equal total magnitude but opposite sense must develop in the structure in order for the structure to be in *equilibrium* in the vertical direction. In the particular structure shown, the top and bottom chords cannot provide any of this resisting force, since they are horizontal and have no vertical component. The entire resisting shear force must be provided by the introduction of a diagonal member. In this case the resisting shear force would be nothing more than the vertical component of the total force in the diagonal member. The diagonal could be arranged as shown in Figure 4-26(c) or as in Figure 4-26(d). In either case the force arrow must act in the directions illustrated so that the vertical component of the diagonal forces always acts downward, balancing the upward external shear force. The implication of Figure 4-26(c) is that a diagonal placed from K to N is in compression, while the implication in Figure 4-26(d) is that a diagonal placed from L to M is in tension. The designer may now decide whether a tension or compression member is wanted at this location.

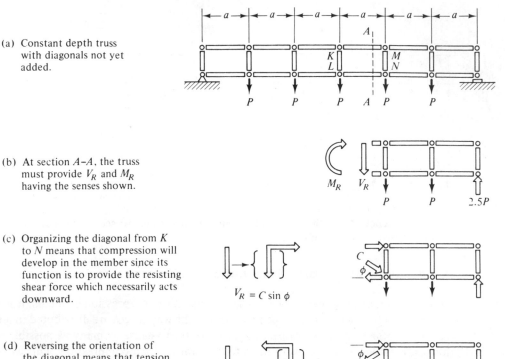

(a) Constant depth truss with diagonals not yet added.

(b) At section $A-A$, the truss must provide V_R and M_R having the senses shown.

(c) Organizing the diagonal from K to N means that compression will develop in the member since its function is to provide the resisting shear force which necessarily acts downward.

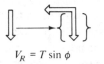

$$V_R = C \sin \phi$$

(d) Reversing the orientation of the diagonal means that tension will develop in the member so that the resisting shear force will be provided.

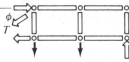

$$V_R = T \sin \phi$$

FIGURE 4-26 Controlling force distributions: The diagonal must provide the internal resisting shear, V_R. The diagonal orientation shown in (c) causes the diagonal to be in compression, while the orientation shown in (d) causes the member to be in tension.

Arguments similar to the above could be made about the members in all the other sections shown in Figure 4-26(a). An orientation that would make the diagonal in each section into either a tension or compressive element can be found.

Figure 4-23(a) illustrates a truss having a bar pattern in which each of the diagonals is in tension, while Figure 4-24(a) illustrates a truss in which all the diagonals are in compression. Note that causing a diagonal to be in either tension or compression implies something about the forces in adjacent members. If a diagonal were placed such that the member is in tension, the adjacent vertical element must be in compression. By a similar token, using a diagonally oriented tension member to be in compression would cause the vertical member to be in tension.

4.4.5 Buckling of Compressive Members; Influence on Triangularization Pattern

The dependency of the load-carrying capacity of a compressive member on its length and the associated design objective of trying to make such members relatively short often influence the pattern of triangulation used. Consider the truss shown in Figure 4-27(a) and the force distribution in its members shown in Figure 4-27(b). Assuming that the members shown in compression are relatively long members, Figure 4-27(a) also indicates how the compression members might buckle (the directions of buckle shown are arbitrary, since there is no way of predicting in which direction the member would actually buckle). The tension members would exhibit no such tendency to deflect in this way. Note that the top chord members are the longest in the truss and carry relatively high compressive forces. Members could undoubtedly be sized to carry these forces, but with a few relatively minor alterations to the bar pattern, it is possible to reduce the total amount of material required in the top chord below what would be required for the configuration shown in Figure 4-27(a). This can be done by locating members to serve as bracing for these top chords. There are four zero-force members in the original configuration. A close look at the truss reveals that these members are not necessary to the stability of the truss (as some zero-force members occasionally are) and could be removed. Or they could be relocated as shown

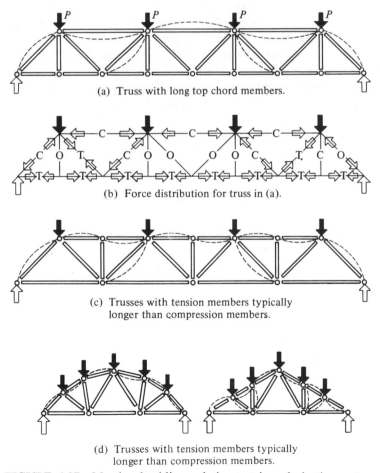

(a) Truss with long top chord members.

(b) Force distribution for truss in (a).

(c) Trusses with tension members typically longer than compression members.

(d) Trusses with tension members typically longer than compression members.

FIGURE 4-27 Member buckling: relation to triangularization pattern.

in Figure 4-27(c). These members would now serve as midpoint bracings, thereby effectively halving the effective lengths of top chord members.

This type of bracing has an enormous influence on the load-carrying capacity of the top chord members. Recall that halving the length of a compressive member increases the load required to cause buckling by a factor of 4 (see Section 7.3). A much smaller member could be used to support the axial load shown than would be the case in the original configuration. Note that the forces in the top chord have not changed, only the member lengths. The configuration shown in Figure 4-27(c) is thus preferable to that shown in Figure 4-27(a). It should also be noted that the longest members in the truss are the lower chords, which are in tension. This is satisfactory, since the cross-sectional areas of tension members required to carry given loads are not dependent on member length. Short compression members and longer tension members often characterize trusses that have been designed with care. This principle is often the design determinant underlying configurations of many trusses commonly found in use.

It might be thought that with the preceding approach the bracing members would have to be very large to serve their function. Usually this is not so, since there is relatively little force exerted on a brace member, so they can be quite small. Sizing these members exactly is a difficult problem beyond the scope of this book. Often they are simply given a certain minimum size empirically known to be adequate or, more likely, those same members carry some relatively large load when the external loading pattern on the truss changes. Sizes for other loads are usually more than adequate when the member serves only a bracing function.

4.4.6 Lateral Buckling of Individual Truss Members

The preceding discussion treated only the behavior of a truss in its own plane. Even in a planar structure, however, the *out-of-plane* direction is extremely important. Consider the planar truss shown in Figure 4-28 and assume that the top chords were made of members symmetrical about both the y and z axes. A square or round member (including a pipe) would exhibit a symmetry of this type. More precisely, the members of interest are those having equal *moments of inertia* about both axes and they need not necessarily be geometrically symmetrical (see Appendix 3). If a top chord of this type carries an axial compressive load, it is equally apt to buckle in the horizontal x–z plane. This implies that a crucial design issue is to assure that this mode of buckling does not adversely affect the load-carrying capacity of the truss.

Assume that transverse beams were used to carry the loads from the decking to the truss as shown in Figure 4-28(b). These beams inherently brace the top chord in the horizontal plane at the points of attachment. Since the chords are braced only at these points, it is still possible for the top chords to buckle in the horizontal plane, as shown in Figure 4-28(b). The effective lengths of the top chord members, insofar as their resistance to buckling is concerned, is $2a$, not just a. Without transverse bracings in the horizontal plane at other points, minimizing top chord length in the vertical plane to gain design advantages is an exercise in futility, as the members will merely buckle in the horizontal plane. Members in the vertical plane do nothing to prevent

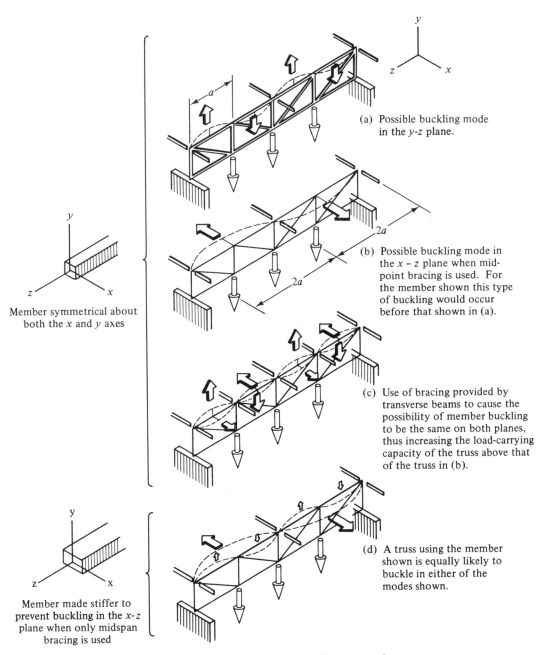

(a) Possible buckling mode in the y-z plane.

(b) Possible buckling mode in the x – z plane when mid-point bracing is used. For the member shown this type of buckling would occur before that shown in (a).

(c) Use of bracing provided by transverse beams to cause the possibility of member buckling to be the same on both planes, thus increasing the load-carrying capacity of the truss above that of the truss in (b).

(d) A truss using the member shown is equally likely to buckle in either of the modes shown.

Member symmetrical about both the x and y axes

Member made stiffer to prevent buckling in the x-z plane when only midspan bracing is used

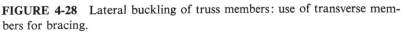

FIGURE 4-28 Lateral buckling of truss members: use of transverse members for bracing.

165

this type of buckling. From a design point of view, it is necessary to brace the truss as shown in Figure 4-28(c). This makes the effective length of the top chords the same in both directions. An alternative would be to change the cross-sectional geometry of the top chord members, making them more inherently resistive to buckling in the horizontal plane. This is diagrammatically illustrated in Figure 4-28(d), in which the member is made stiffer in the horizontal (x–z) plane. Using techniques discussed in detail in Section 7.4 it is possible to design member characteristics so that the potential for buckling in the vertical plane (about the weak axis of the member shown, but with an effective member length of a) is exactly identical to that for buckling in the horizontal plane (about the stronger axis of the member but with an effective member length of $2a$). Such a member is diagrammatically illustrated in Figure 4-28(d). Many member configurations could provide the necessary balance in stiffnesses.

Which of the two approaches discussed above is preferable for controlling the potential for member buckling in the horizontal plane depends on many other design considerations. Of primary concern would be the effect of the transverse beam spacing in the design of the roof decking. One spacing or the other might prove to be more desirable in this respect. A detailed look at the design of the decking is necessary to answer this question.

4.4.7 Lateral Buckling of Bar Assemblies

While usually there is the possibility for bracing the top chord in the horizontal plane, it is possible to have a truss that is essentially free-standing (Figure 4-29). In structures of this type, there is, in *addition* to the possibility of individual member buckling of the type discussed previously, the possibility that very large portions of the structure may buckle *laterally*. This type of buckling, which often makes the use of free-standing trusses undesirable, is illustrated in Figure 4-29. In the truss shown, the entire set of top chord members is in a state of compression, so the *entire* top of the truss can buckle as indicated. Preventing this phenomenon is no small design task. The whole

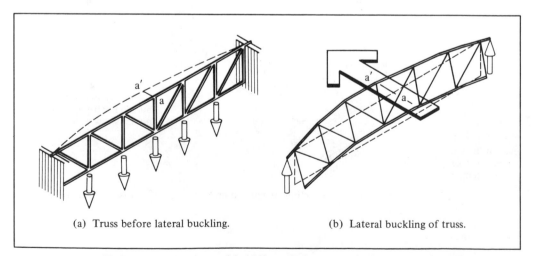

(a) Truss before lateral buckling. (b) Lateral buckling of truss.

FIGURE 4-29 Lateral buckling of planar truss without any bracing.

length of the top of the truss can be thought of as the effective length of a long composite compression member. The type of bracing provided when transverse beams are attached to the top chord, however, generally prevents this type of buckling.

When trusses must be free-standing and hence cannot be braced by transverse members, it is necessary to make the top chord of the structure sufficiently stiff in the transverse direction to resist lateral buckling. This can be done by increasing the lateral dimension of the top chord or by adopting a three-dimensional triangulization pattern (Figure 4-30).

The top plane of the three-dimensional truss shown in Figure 4-30(b) inherently provides a strong resistance to lateral buckling and is one of the primary reasons why such configurations are often used. A truss of the type shown in Figure 4-30(c) also inherently provides a measure of bracing against this phenomenon, owing to the

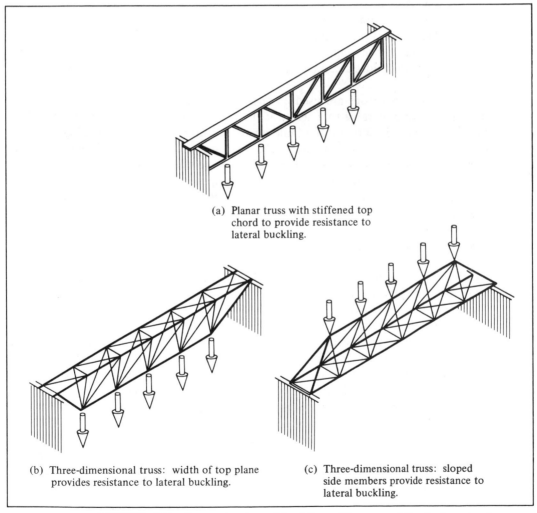

(a) Planar truss with stiffened top chord to provide resistance to lateral buckling.

(b) Three-dimensional truss: width of top plane provides resistance to lateral buckling.

(c) Three-dimensional truss: sloped side members provide resistance to lateral buckling.

FIGURE 4-30 Preventing lateral buckling of bar assemblies in free-standing trusses.

orientation of the diagonals. In these two cases the resistance to lateral buckling provided by three-dimensional structures is largely dependent on the spacing of members in the third dimension, with a larger spacing more desirable than a very small one. The determination of an optimum separation is a difficult problem beyond the scope of this book. The spacings illustrated are, however, quite reasonable.

4.4.8 Planar Versus Three-Dimensional Trusses

Resistance to lateral buckling in free-standing structures is only one of a number of issues which should be considered in making a decision to use either a planar or a three-dimensional structure. Which uses a greater amount of material to support a given load in space, or are planar and three-dimensional structures similar in this respect? Quantitative resolution of this very complex question is unfortunately beyond the scope of this book. It should be noted, however, that nothing a priori suggests that three-dimensional trusses necessarily use less material for equivalent loads and spans than planar trusses. In fact, it can be demonstrated that when trusses are used as one-way spanning elements, a planar truss usually requires less volume of truss material than does a comparable three-dimensional truss serving the same function.

Thus, when trusses are used as one-way spanning elements a series of planar trusses is often preferable to a series of three-dimensional trusses, particularly when they are used on the interior of a building and lateral bracing is intrinsically provided by the roof framing system or some other mechanism. Three-dimensional configurations often prove more efficient when the trusses are used in a free-standing way (e.g., on the exterior of a building with the roof suspended beneath the trusses). When trusses are used to form two-way systems, three-dimensional trusses are also often more advantageously used (see Chapter 10).

4.4.9 Funicular Trusses

Although virtually any pattern of triangulated elements that is stable can be made to work as a truss, it may well be that some overall truss shapes are preferable to others. This section explores one interesting class of trusses that exhibit uniquely simple load-carrying actions.

Consider the truss structures shown in Figure 4-31. Note that the overall shape of each truss is derived from a consideration of the shape that a cable would deform into if subject to the same loads. Since the shape that a flexible cable assumes under a loading is referred to as the funicular shape for that loading (see Section 1.5), trusses having comparable overall shapes are often referred to as *funicular trusses* for that loading. If these trusses were analyzed by any of the techniques previously described, a seemingly curious aspect would emerge. This is that in each case the only load-carrying members are those forming the external envelope of the truss. All interior members turn out to be zero-force members, no matter what specific internal triangularization pattern is used.

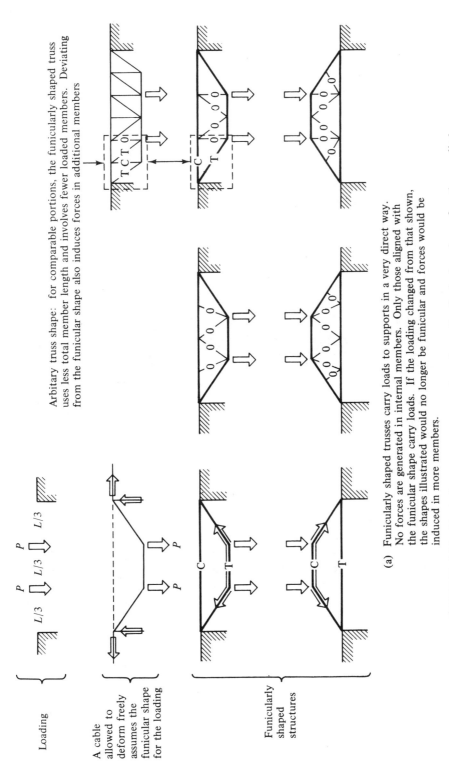

FIGURE 4-31 Funicular trusses: trusses based on the funicular shape for the applied loading demonstrate a uniquely simple load carrying action.

(a) Funicularly shaped trusses carry loads to supports in a very direct way. No forces are generated in internal members. Only those aligned with the funicular shape carry loads. If the loading changed from that shown, the shapes illustrated would no longer be funicular and forces would be induced in more members.

Arbitary truss shape: for comparable portions, the funicularly shaped truss uses less total member length and involves fewer loaded members. Deviating from the funicular shape also induces forces in additional members

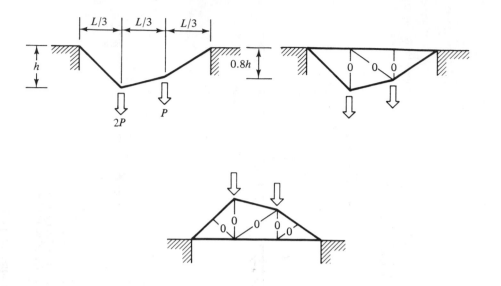

(b) There is only one general funicular shape for a given loading. When designing a funicularly shaped truss, care must be taken to assure that the slopes of members correspond exactly to the funicular shape (see Chapter 5).

FIGURE 4-31   (Cont.)

Obviously, what is happening in these trusses is that the specific members organized along the lines of the funicular shape for the loading involved have become the dominant, actually only, mechanism for transferring the external loads to the supports. Other members are zero-force members serving a bracing function at most. A remarkably simple load-carrying action is thus displayed by these trusses.

For the rather remarkable load-carrying actions described to occur, it is necessary that the basic shape of the structure correspond *exactly* to the funicular shape. Relative heights of the two interior points on the truss shown in Figure 4-31(b), for example, must be exactly determined. The relative heights are clearly a function of the loads and their locations. Chapter 5 describes in detail how to determine in a quantitative way the exact shape of a funicular structure for a specific loading condition. In general, the shaping leads to a configuration whose structural depth varies directly with the bending moment present. The external shear force present at a section is balanced by an equal internal resisting shear force provided by the vertical component of the force in the outer inclined member, with the consequence that interior members need not contribute any shear resistance, and thus become zero-force members. The external moment at the same section is balanced by an equal internal resisting moment provided by the horizontal components of the forces in the upper and lower members. Again, internal diagonal members assume no role.

Since different loading conditions lead to different overall configurations (as can be appreciated from the deflected string method of arriving at funicular shapes), a shape funicular for one loading will not be funicular for any other. This implies that forces are developed in interior members of trusses designed to be funicular for one loading when nondesign loads are present. The load-carrying action would again be more complex in such a case.

Comparing a truss not having an overall funicular shape for the loading involved with a funicularly shaped one (see Figure 4-31) reveals that deviating from the funicular shape not only serves to make the total member length involved longer but more members are subjected to primary forces (and would thus have to be designed as load-carrying rather than zero-force members). The consequence is that it can be expected that a strategy of using a funicular shape for the overall envelope of a truss may be more likely to lead to a lighter-weight or minimum-volume solution than a nonfunicular shape.

In a funicularly shaped truss, the choice of which internal triangularization pattern to use largely depends the role of the members as braces for compression members since internal members are zero-force elements. Minimum triangulation would not frequently brace compression members and relatively large members would be needed. Alternatively, a closely packed triangularization pattern could be adopted which would provide frequent bracing and allow for smaller compression members. The latter, of course, is at the expense of additional internal members. Trade-offs are obviously involved.

Figure 4-32(a) illustrates a truss that is funicularly shaped for a series of equal concentrated loads. If the structure were exactly shaped, each interior element would be a zero-force member. It is interesting to note that conceptually the zero-force members could simply be removed to form a nontriangulated configuration without changing the ability of the structure to carry design loads to the ground. The function of the lower chord, which obviously serves as a tie, could also be provided by abutments, as shown in Figure 4-32(b). The latter structure without diagonals is, however, of dubious practicality, since it is stable *only* under the exact loading pattern shown. If the loading pattern were changed in *any* way (e.g., if even one of the loads were removed or its magnitude altered), the configurations would no longer be stable and would collapse. This is because the required shear and moment resistance at the different locations is now no longer the same as that for which the structure was originally designed and the structure cannot intrinsically satisfy these new requirements as it did for the design loading. The same overall configuration with diagonals, however, is stable under not only the design loading but any other loading condition. A function of the interstitial members is to carry forces generated by other nonuniform loads the structure must occasionally bear. If the loads were suspended from the lower chord panel points instead of being applied to the upper chord, the vertical interior members would serve as suspension rods transferring the applied loads to the upper chord members.

While the structure shown in Figure 4-32(a) was conceived of as a truss, the transformation shown in Figure 4-32(b) perhaps fits the commonly perceived image of an *arch* better than that associated with the word "truss." The basic structural action, however, is the same. The stability question in the form without diagonals clearly points to the need to pay special attention to the issue of how various loading conditions affect the design of elements that are initially conceived of as "arches." This issue will be addressed in Chapter 5.

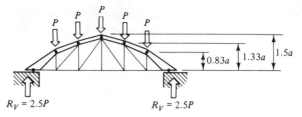

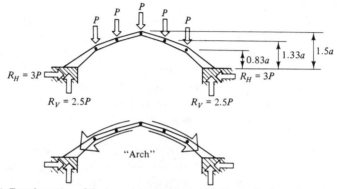

(a) Funicularly shaped truss for a series of identical and equally spaced concentrated loads. Interstitial members are zero-force members under the primary loading condition shown but carry forces when the loading changes. This is a common type of bridge structure.

(b) Transformation of the truss shown in (a) in which the non-load-carrying interstitial members are eliminated and the function of the lower chord (which acts as a tension tie) is replaced by buttresses. An "Arch" structure results. If the linear segments of the structure shown are pin-connected, the assembly is stable only under the loading condition shown.

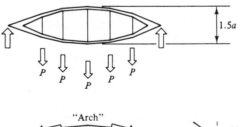

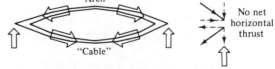

(c) A further transformation of the assemblies shown in (a) and (b) into a lens-shaped structure. The depth of the structure at any section is the same as before. One way of conceiving of the structural action of the assembly is to think of it as a combination "arch" and "cable" in which the horizontal thrusts of the arch are exactly balanced by those of the cable. Hence no net horizontal forces develop at the foundation, only vertical forces. The structure is stable only under the loading condition shown.

FIGURE 4-32 Funicular trusses: transformations of a funicular truss into related forms.

One final transformation of the truss shown in Figure 4-32(a) is of interest. The *lens*-shaped structure shown in Figure 4-32(c) has the same structural depth at all sections as the original structure. Its outer configuration is clearly part of the same family of shapes as that present in the original truss. An analysis of this truss would reveal that the interstitial diagonals are zero force members and thus only serve the function of stabilizing the assembly under variant loading conditions. The verticals transfer loads such that the upper and lower chord members are similarly loaded (a condition that must be met for the shape similarity to be correct).

Instead of conceiving of this lens-shaped structure as a special form of truss, it could equally well be conceived of as a combination "arch–cable" structure. In this conception the outward thrusts of the upper "arch" are identically balanced by the inward pulls of the "cable," with the consequence that there is no net lateral force present at the reaction, a result long acknowledged to be advantageous in foundation design.

The lens-shaped form was commonly used in many *lenticular* trusses built in the last century for use as bridges (see Figures 4-33 and 3-9). Additional diagonal bracing was typically used to allow the structures to carry nonuniform loading patterns.

FIGURE 4-33 A "lenticular" truss: the Smithfield Street Bridge across the Monogahela River in Pittsburgh, Pennsylvania (circa 1883). (Photograph courtesy of the American Society of Civil Engineers.)

4.4.10 Support Locations

Instead of manipulating with just the form of a truss alone, a good design technique is to simultaneously vary support locations. Quite often the location of the supports is not fixed, but is a design variable. From a structural viewpoint, the problem becomes one of finding out whether there is a particular location for the supports which causes a preferred behavior in the truss. As can be seen from looking at the magnitudes of the forces developed in the members, the support location shown in Figure 4-34(b) is preferable to that shown in Figure 4-34(a) for the loading condition indicated. The magnitude of the forces in many of the members are reduced, thus reducing the required member sizes. It can be expected, however, that pulling in the end supports to an even greater extent could become counterproductive. The subject of optimum support locations is a very interesting one and will be studied in detail in Chapter 6. In general, some amount of overhang on either end of a truss is desirable for a loading condition of the type shown.

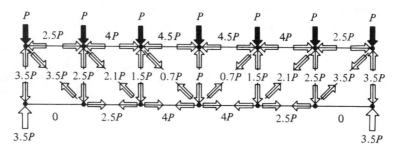

(a) Truss supported at ends.

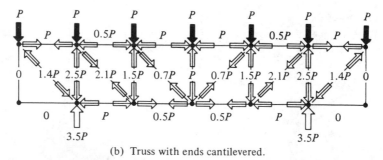

(b) Truss with ends cantilevered.

FIGURE 4-34 Effect of changing support locations: By pulling in the supports one panel point from each end, the magnitudes of the forces in many of the members are significantly reduced.

4.4.11 Heights of Trusses

One of the most frequently encountered questions during the design of a truss is that of how deep it should be or whether there is an optimal depth for a truss. It would seem that this should be a relatively easy design question to answer, but it is, on the contrary, one of the more difficult. As already noted in Section 4.3.4, the member forces, hence volumes, generated in a truss by an external load are invariably dependent on the dimensions, incluing the height (h) of the truss.

As can be appreciated, determining an optimum height which minimizes the total volume of a truss is no easy matter, but the process is conceptually straightforward. For each truss member, a volume expression based on the member forces written as a function of the variable height h is first determined and a total volume expression for the entire set of members obtained. This expression is then minimized with respect to the variable height.

In general, the optimization process will reveal that trusses which are relatively deep in relation to their span are most efficient and shallow trusses less so. Angles formed by the diagonals with respect to the horizontal are typically from 30° to 60°, with 45° are often a good choice for determining triangulization geometries.

Over the years, rules of thumb have evolved which help estimate what the depth of a truss should be. Trusses that carry relatively light loads and are closely spaced often have depths approximately $\frac{1}{20}$ of their span (e.g., roof load-transfer members). Secondary collector trusses which carry reactions produced by load-transfer members are usually deeper, since the loads carried are larger, and can be on the order of $\frac{1}{10}$ of their span. Primary collector trusses which support huge loads (e.g., a truss carrying the column loads from a multistory building over a clear span on the ground floor) are usually very deep and are often made equal to the depth of a story, say $\frac{1}{4}$ or $\frac{1}{5}$ of the span of the truss. Rules of thumb of this nature should not be taken as final truths. Some of the best and most efficient uses of trusses in building contexts have occurred when heights varied drastically from what the ratios above suggest. Still, when in doubt, they are a place to start. Chapter 15 discusses rule-of-thumb applications in greater detail.

4.4.12 Joint Rigidity

In all the previous discussions it has been assumed that the trusses studied were modeled according to the assumptions stated in the opening sections of this chapter. All joints between bars have been considered to be idealized pin connections. In many cases, actually making a connection may be neither possible nor desirable. If the actual joint conditions are such that the ends of the bars are not free to rotate, local bending moments may develop in the bars in addition to any axial loads present. If these bending moments were too large, the member would have to be designed so that it would still be safe under the action of the combined axial and bending stresses. The magnitudes of bending stresses typically induced by rigid joints are rarely more than 10 to 20% of the axial stresses normally developed. For preliminary design purposes these secondary bending stresses are often simply ignored, but at some later point in the design process they would be reconsidered. One positive effect of increasing joint rigidity is to increase the overall resistance of the truss to deflections. Making joints rigid is rarely an influential factor affecting the general shaping of the final truss.

4-1. Discuss in detail the historical development of quantitative methods of analyzing truss structures. Consult your library.

4-2. For the truss shown in Figure 4-35(a), qualitatively determine the nature of the force present in each of the members. Numerical values are not required.

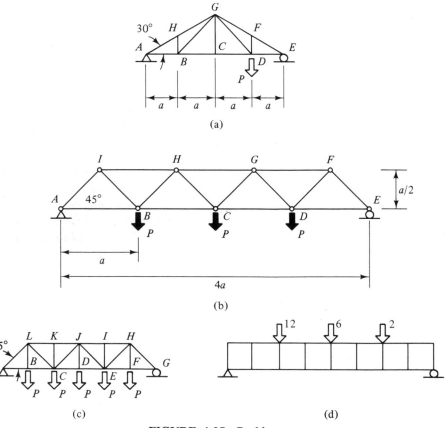

FIGURE 4-35 Problems.

4-3. For the truss shown in Figure 4-35(a), quantitatively determine the magnitudes of the forces present in all the truss members. A method-of-joints approach is suggested.

4-4. For the truss shown in Figure 4-35(b), quantitatively determine the magnitudes of the forces present in all the truss members. Use a method-of-joints approach.

4-5. For the truss shown in Figure 4-10, quantitatively determine the magnitudes of the forces present in all the truss members. A method-of-joints approach is suggested.

4-6. Determine the force in member GH in the truss shown in Figure 4-35(b) by using a method-of-sections approach.
Answer: $\overline{GH} = 4P$ (compression).

4-7. For the truss shown in Figure 4-9, quantitatively determine the magnitudes of the forces present in all the truss members. A method-of-joints approach is suggested.

4-8. Quantitatively determine the forces in members *JI*, *JE*, and *ED* in the truss shown in Figure 4-35(c) by a method-of-sections approach. How would the force in member *JI* be affected if the overall depth of the structure was doubled? How would the forces in members *JI* and *JE* be affected if the depth of the truss were continually decreased until it approached zero?

4-9. Consider the parallel chord truss shown in Figure 4-35(d). Obviously, the truss needs diagonal elements for stability. Add the diagonal elements in an arrangement such that *all* the diagonals are in a state of tension under the loading condition indicated.

4-10. Draw a diagram of the relative sizes of the individual members of the truss shown in Figure 4-35(c). Assume that cables are used for tension elements.

chapter 5

FUNICULAR STRUCTURES: CABLES AND ARCHES

5.1 INTRODUCTION

5.1.1 Funicular Structures

Few types of structures have so consistently appealed to the imagination of builders as either the arch or the hanging cable. These two apparently different types of structures share some fundamental characteristics that cause them to be more closely related than might initially appear, particularly in terms of their basic structural behavior.

A cable subjected to external loads will obviously deform in a way dependent on the magnitude and location of the external forces. The form acquired is often called the *funicular* shape of the cable (the term "funicular" is derived from the Latin word for "rope"). Only tension forces will be developed in the cable. Inverting the structural form obtained will yield a new structure that is exactly analogous to the cable structure except that compression rather than tension forces are developed. Theoretically, the shape found could be constructed of simply stacked elements that are nonrigidly connected (a "compression chain") and the resultant structure would be stable. Any slight variation in the nature of the applied loading, however, would mean that the structure would cease to be funicular in shape and bending would develop under the new loading, with the result that complete collapse could occur since the nonrigidly connected blocks cannot resist this bending. Since the forms of both the tensile and compressive structures that are derived in the manner described above are both related to the notion of a loaded hanging rope, they are collectively referred to as *funicular structures.*

5.1.2 Cables

Funicular structures have been used throughout history. Considering tension structures first, there are many early recorded examples existent. The flexible suspension bridge which was initially developed in China, India, and South America, for example, is of great antiquity. While most early bridges used rope, often made from bamboo, there is recorded evidence of bridges made using chains in China as early as the first century A.D. In addition to the varieties of tents that used ropes, cable structures also found application in some major buildings. A rope cable structure, for example, was used in about A.D. 70 to roof a Roman amphitheater (Figure 5-1). Still it was in connection with suspension bridges that the use of cables as theoretically understood structural elements was developed. This theoretical understanding is actually rather recent, since the suspension bridge remained relatively unknown in Europe (although a type of chain-suspended structure was built in the Swiss Alps in 1218), where most developments in structural theory were occurring, until the sixteenth century, when, in 1595, Fausto Veranzio published a drawing of a suspension bridge. It was not until 1741 that a permanent iron chain footbridge, located in Durham County, England, was finally built. This bridge was probably the first significant suspension bridge in Europe. It failed, however, to establish a precedent.

A major turning point in the evolution of the suspension bridge occurred in the early part of the nineteenth century in America, when James Findley developed a suspension bridge capable of carrying vehicular traffic. His initial bridge, built in 1801 over Jacobs Creek in Uniontown, Pennsylvania, used a flexible chain of wrought-iron links. Findley's real innovation was not the cable itself, however, but the intro-

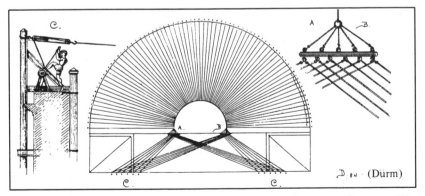

FIGURE 5-1 Cable roof structure over Roman Colosseum, circa A.D. 70. Rope cables anchored to masts spanned in a radial fashion across the open structure supported a movable sunshade that could be drawn across to cover the arena. The span of the structure was 620 ft (188 m) along the major axis and 513 ft (156 m) along the minor axis. (From Dürm.)

duction of a *stiffened bridge deck* in which the stiffening was achieved by longitudinal trusses made of wood. The use of a stiffened deck kept the supporting cable from changing shape, and consequently changing the shape of the road surface it supported, by distributing concentrated vehicular loads over a larger portion of the cable (Figure 5-2). With this innovation the modern suspension bridge was born. The work of Findley became known to others, possibly even to the great builder Thomas Telford in England, who designed the bridge over the Menai Strait in Wales (1818–1826). Louis Navier, the great French mathematician, discussed Findley's work in his classic book on suspension bridges, *Rapport et Mémoire sur les Ponts Suspendus*, published in 1823, in which he gave Findley credit for the introduction of the stiffened bridge deck and provided a way for other bridge builders to become aware of his work.

Other great suspension bridges were built in rapid succession, including the beautiful Clifton Bridge in England by Isombard Brunel, John Roebling's Brooklyn Bridge in 1883, and a host of other major bridges. Significant modern bridges include the bridge over the Messina Straits, with a middle span of over 5000 ft (1525 m), and the Verrazano-Narrows Bridge, with a middle span of 4260 ft (1300 m).

The application of cables to buildings other than tents developed more slowly because of the lesser need to span large distances and the intrinsic problems of using cables when they were applied. Although James Bogardus submitted a proposal for the Crystal Palace of the New York Exhibition of 1853 in which the roof of a circular cast-iron building, 700 ft (213 m) in diameter, was to be suspended from radiating chains anchored to a central tower, the pavilion structures of the Nijny-Novgorod

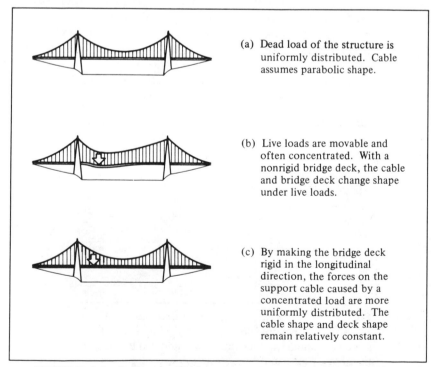

(a) Dead load of the structure is uniformly distributed. Cable assumes parabolic shape.

(b) Live loads are movable and often concentrated. With a nonrigid bridge deck, the cable and bridge deck change shape under live loads.

(c) By making the bridge deck rigid in the longitudinal direction, the forces on the support cable caused by a concentrated load are more uniformly distributed. The cable shape and deck shape remain relatively constant.

FIGURE 5-2 Suspension bridge structures: use of rigid bridge deck.

exhibition that were designed by V. Shookhov in 1896 marked the actual beginning of modern applications. Subsequent significant structures include the locomotive roundhouse pavillion at the Chicago World's Fair in 1933 and the Livestock Judging Pavillion built at Raleigh, North Carolina, in the early 1950s. Since then, a number of other significant cable-supported buildings have been constructed.

5.1.3 Arches

The counterpart of the cable, the arch, also has its roots in antiquity and examples are found in many early civilizations. Strictly speaking, many of the early arches, particularly the semicircular ones of the type typically used by the Romans, are not exactly funicular structures, since their shape deviates slightly from the nonuniform funicular curve (see Section 5.2) that would normally be associated with the loading usually existent. Because of the relatively large cross sections of the arches, however, the differences in behavior are not overly significant.

A characteristic of early curved arches is that they were constructed with wedge-shaped voussoirs built up side by side on temporary supports until the top or keystone piece was put in place. It was long recognized that the resultant structure was often made more stable by the superimposition of additional weight on its top, thus "firming up" the arch. The idea was that the greater the load, the higher were the compressive forces developed between blocks and the less likely were the tension cracks between blocks (which could develop from significant shape deviations or unusual loads) that could cause failure to develop. The initial theory of arches, largely formulated by French engineers in the eighteenth century, also held that any tension in a voussoir arch would be disastrous. Strictly speaking, however, a failure mechanism consisting of more than one crack must form before a masonry arch can collapse (see Figure 5-3). The compressive forces associated with the dead weight of a correctly shaped masonry arch typically dominate the possible tension forces that could result from live-load variations, with the consequence that such structures could withstand a reasonably diverse group of loading patterns without collapsing.

With the advent of steel and reinforced concrete, the inherent member rigidity possible with the new materials made the arch a more versatile form, since the material

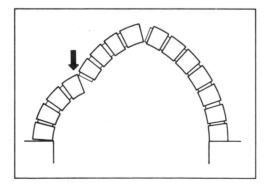

FIGURE 5-3 Arch collapse mechanism. Cracking must occur at several locations before the arch can collapse.

rigidity allowed the arch to maintain its basic shape and not collapse when subjected to unanticipated loads. A modern rigid arch is often shaped in response to a primary loading condition and carries loads in axial compression when this loading is actually present but is also designed to have sufficient bending resistance to carry load variations. The structure, however, cannot be funicular for the new loading. It is in this way that a rigid arch is fundamentally different from a flexible cable. A flexible cable must always change shape under changes in loading and thus essentially be momentless under all loadings, while a rigid arch is momentless only under one loading condition and is capable of carrying some bending due to load variations.

5.2 GENERAL PRINCIPLES

5.2.1 Funicular Shapes

Of fundamental importance in the study of arches and cables is a knowledge of what exact curve or series of straight-line segments defines the funicular shape for a given loading. The funicular shape is that naturally assumed by a freely deforming cable subjected to the loading. A cable of constant cross section carrying only its own dead weight will naturally deform into a *catenary* shape (see Figure 5-4). A cable carrying a load that is uniformly distributed along the horizontal projection of the cable, as is the primary loading in a suspension bridge supporting a horizontal bridge deck, will deform into a parabola. Cables carrying concentrated point loads will (ignoring the dead weight of the cable itself) deform into a series of straight-line segments.

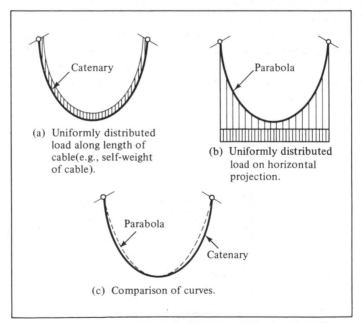

(a) Uniformly distributed load along length of cable(e.g., self-weight of cable).

(b) Uniformly distributed load on horizontal projection.

(c) Comparison of curves.

FIGURE 5-4 Catenary versus parabolic curves. The two curves are very similar.

Combinations of different loadings will produce combined forms with the largest load producing the dominating shape. Comparable arch shapes for these same loadings are simply inversions of the shapes described above.

As illustrated in Figure 5-5, there is a whole family of funicular shapes associated with a given loading, since only the shape and not the absolute dimensions (such as the height or depth of the funicular shape) is of consequence in defining a funicular response.

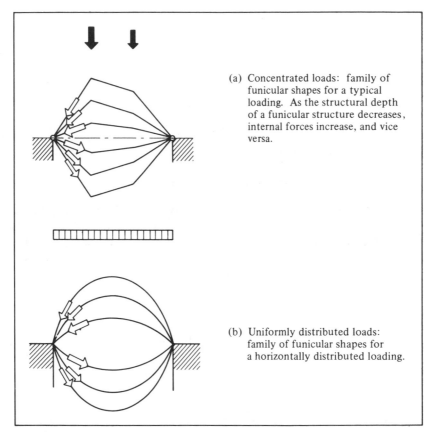

(a) Concentrated loads: family of funicular shapes for a typical loading. As the structural depth of a funicular structure decreases, internal forces increase, and vice versa.

(b) Uniformly distributed loads: family of funicular shapes for a horizontally distributed loading.

FIGURE 5-5 Typical funicular shapes for concentrated and uniformly distributed loadings. Funicular shapes are those obtained by loading a member incapable of resisting bending (e.g., a cable) and allowing it to deform freely. Structures having a funicular shape for a specified loading carry the loading by either axial tension or compression only.

The magnitude of the forces developed in the arch or cable, however, *are* dependent on the relative height or depth of the funicular shape in relation to its length as well as the magnitude and location of the applied loads (Figure 5-6). The greater the rise of an arch or the sag of a cable, the smaller are the internal forces developed in the structure, and vice versa. Reactive forces developed at the arch or cable ends also depend on these parameters. End reactions have both vertical and horizontal

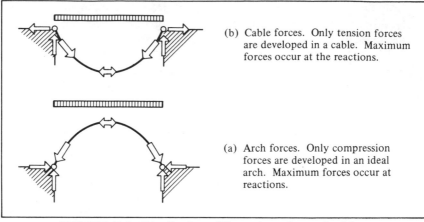

(b) Cable forces. Only tension forces are developed in a cable. Maximum forces occur at the reactions.

(a) Arch forces. Only compression forces are developed in an ideal arch. Maximum forces occur at reactions.

FIGURE 5-6 Distribution of forces in arches and cables.

components which must be resisted by the foundations or some other element such as a compressive strut or tie rod.

5.2.2 Special Design Considerations

While most of the discussions thus far have focused on the cable as a simple, singly curved or draped suspension element, cables are also used in a greater variety of ways than are arches. Cable structures are more correctly categorized into either suspension structures or cable-stayed structures (Figure 5-7). *Suspension structures* can be typically subclassified into the following: (1) single-curvature structures, in which roofs are made by placing cables parallel to one another and using a surface formed by beams or plates to span between cables; (2) double-curvature structures, in which a field of crossed cables of different and often reverse curvatures make up the primary roof surface; and (3) double-cable structures, in which interconnected double cables of different curvatures are used in the same vertical plane. *Cable-stayed structures* typically use vertical or sloping compression masts from which straight cables run to critical points or horizontally spanning members.

A critical problem in the design of any cable roof structure is the dynamic effect of wind, something that does not significantly affect an arch structure. Consider the simple roof structure supported by cables in Figure 5-8. As the wind blows over the top of the roof, a suction will be created. If the magnitude of the suction force due to the wind exceeds the dead weight of the roof structure itself, the roof surface will begin to rise. As it begins rising and dramatically changing shape, the forces on the roof begin to change, since the magnitude and distribution of wind forces on a body are dependent on the exact shape of the body. Since the wind forces change due to the changing shape of the roof, the flexible structure itself again changes shape in response to the new loading. The process is cyclical. The roof will not remain in any one shape, but will move, or flutter, as long as the wind is present. The primary ways that this clearly undesirable phenomenon can be prevented in a singly curved cable are to use a very heavy roof surface and rely on the dead load to keep fluttering from occurring, or to use some sort of system of crossed or staying cables (Figure 5-9).

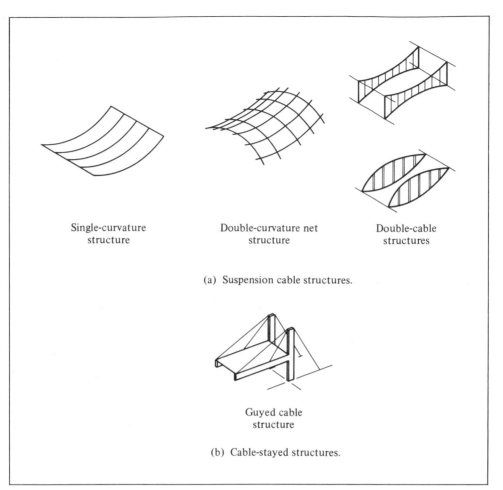

Single-curvature
structure

Double-curvature net
structure

Double-cable
structures

(a) Suspension cable structures.

Guyed cable
structure

(b) Cable-stayed structures.

FIGURE 5-7   Basic types of cable structures.

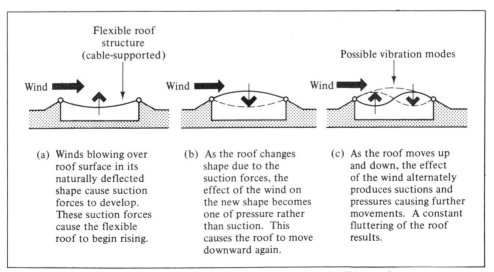

Flexible roof
structure
(cable-supported)

Possible vibration modes

Wind

Wind

Wind

(a) Winds blowing over roof surface in its naturally deflected shape cause suction forces to develop. These suction forces cause the flexible roof to begin rising.

(b) As the roof changes shape due to the suction forces, the effect of the wind on the new shape becomes one of pressure rather than suction. This causes the roof to move downward again.

(c) As the roof moves up and down, the effect of the wind alternately produces suctions and pressures causing further movements. A constant fluttering of the roof results.

FIGURE 5-8   Dynamic effects of wind on typical flexible roof structure.

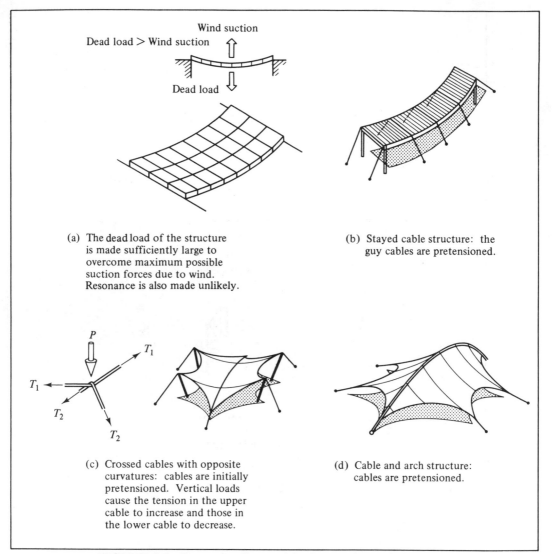

(a) The dead load of the structure is made sufficiently large to overcome maximum possible suction forces due to wind. Resonance is also made unlikely.

(b) Stayed cable structure: the guy cables are pretensioned.

(c) Crossed cables with opposite curvatures: cables are initially pretensioned. Vertical loads cause the tension in the upper cable to increase and those in the lower cable to decrease.

(d) Cable and arch structure: cables are pretensioned.

FIGURE 5-9 Preventing flutter of flexible roof structures due to dynamic effects of wind.

Double-curvature cable structures are often used to control the flutter problem described above. By using a net surface series of cables at right angles to one another and of reverse curvatures, a staying effect can be achieved which can either prevent or reduce the tendency of the roof to flutter. Double-surfaced cable also have characteristics that tend to prevent roof flutter. These two types of structures will be considered in more detail in Section 5.4.

5.3 ANALYSIS OF ARCHES AND CABLES

5.3.1 Three-Hinged Arches

As a way of getting into the analysis and design of arch and cable structures, it is useful to consider first a very special form of structure—the *three-hinged arch*—which may or may not be a funicular structure, depending on its exact shape. The three-hinged arch is a structural assembly consisting of two rigid sections that are connected to each other and ground foundations by hinged or pinned connections. Unless the two segments form a funicular shape for the loading involved, and they need not, the term three-hinged "arch" can be somewhat misleading. The term "arch"

FIGURE 5-10 *Galérie des Machines*, International Exhibition, Paris, 1889. Base of three-hinged arch. The structure spanned 364 ft (111 m).

is, however, still commonly used to describe such structures, whether they are exactly funicularly shaped or not.

Three-hinged arch structures were developed by French and German engineers in the mid-nineteenth century, partly as a way of overcoming calculation difficulties with older forms of fixed arches. The introduction of hinges at the crown and foundation connections of the structure allowed horizontal thrusts and internal forces to be precisely calculated and exact funicular shapes determined for each of the two rigid segments. Large structures could then be built with greater confidence. The famous *Galérie des Machines*, actually named *Palais des Machines*, of the Paris Exposition in 1889 was composed of a series of three-hinged arches, as were many other notable buildings (see Figure 5-10). With improved calculation techniques, their use fell out of favor because of unfavorable rigidity characteristics and construction difficulties (see Section 5.4.3).

Consider the three-hinged structure shown in Figure 5-11. It is possible to tell

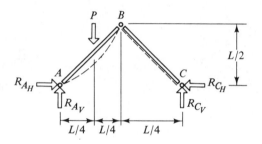

(a) Basic structure.

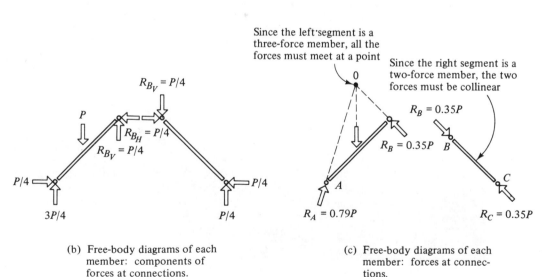

(b) Free-body diagrams of each member: components of forces at connections.

(c) Free-body diagrams of each member: forces at connections.

FIGURE 5-11 Three-hinged arch: connection forces.

by inspection (with reference to the deflected string analogy) that the structure is not funicularly shaped. At this point this is of no consequence. Later the shape will be redesigned into a funicular one.

Note that the nature of the connection at B is such that the end of member AB is free to rotate under the action of the load. The joint offers no potential restraint. This implies that member BC in no way participates in restraining the rotation of the end of AB. Ends of attached members can rotate freely and independently. The connection cannot transmit internal moments from one member to the other or provide any sort of rotational resistance (consequently, no internal moments are shown at connections in free-body diagrams). Hence, the sum of all the rotational effects produced about this point by the external and internal forces must be zero.

Since the external load acts directly on and transverse to the left member, the direction of the force transmitted through a pinned connection is not necessarily collinear with the member, as was the case in truss analysis. As the following example indicates, however, the magnitudes and directions of the forces at each of the connections can be found by considering the equilibrium of the whole structure first and then of each piece in turn.

EXAMPLE

Determine the internal forces present at each of the hinged connections of the structures shown in Figure 5-11.

Solution:

The structure is first decomposed into two parts, as illustrated, and equilibrium diagrams drawn for each. Forces at the top connection which are actually internal to the whole structure are shown equal and opposite, since they are reactive in nature.

Equilibrium in the vertical direction for whole structure: $\sum F_y = 0$ ↑+:

$$R_{A_v} + R_{C_v} - P = 0$$

Rotational equilibrium for whole structure about point A: $\sum M = 0$ ↻:

$$-P\left(\frac{L}{4}\right) + R_{C_v}(L) + R_{C_h}(0) + R_{A_h}(0) = 0 \qquad \text{hence } R_{c_v} = \frac{P}{4} \quad ↑$$

Knowing R_{C_v}, R_{A_v} can be found from $R_{A_v} + R_{C_v} - P = 0$. Hence, $R_{A_v} = \frac{3}{4}P$ ↑.

Equilibrium in the horizontal direction for the whole structure: $\sum F_x = 0$ →+:

$$R_{A_h} - R_{C_h} = 0 \qquad \text{or} \qquad R_{A_h} = R_{C_h}$$

It is not possible to find the magnitudes of R_{A_h} and R_{C_h} from considering the equilibrium of the whole structure. All that can be determined is that the forces are equal in magnitude and opposite in sense. To find these forces and those at the crown connection, B, it is necessary to consider the equilibrium of each of the two segments comprising the whole structure. The right segment will be considered first.

Equilibrium in the vertical direction for the right element: $\sum F_v = 0$ ↑+:

$$R_{C_v} - R_{B_v} = 0 \qquad \text{or} \qquad R_{C_v} = R_{B_v}$$

Since $R_{C_v} = P/4$, $R_{B_v} = P/4$ ↓.

Rotational equilibrium for the right element about B: $\sum M = 0$ ⤷:

$$+\left(\frac{P}{4}\right)\left(\frac{L}{2}\right) - R_{C_h}\left(\frac{L}{2}\right) = 0 \qquad \text{or} \qquad R_{C_h} = \frac{P}{4} \quad \leftarrow$$

Equilibrium in the horizontal direction for the right element: $\sum F_x = 0 \quad \rightarrow +:$

$$R_{B_h} - \left(\frac{P}{4}\right) = 0 \qquad \text{or} \qquad R_{B_h} = \frac{P}{4} \quad \rightarrow$$

The forces at the right support connection C, and the crown connection B is now known. The values found are components of the actual forces.

As Figure 5-11(c) illustrates, these components can be resolved into single forces. Since the right element is a two-force member, it follows that the two forces acting on the member must be collinear, equal in magnitude and opposite in sense. In order to find R_{A_h}, the only force as yet unknown, it is necessary to consider the equilibrium of the left segment of the structure.

Equilibrium in the vertical direction for the left element: $\sum F_y = 0 \quad \uparrow +:$

$$R_{A_v} + R_{B_v} - P = 0$$

Since R_{A_v} and R_{B_v} are already known, this is merely a check. Thus, $(3P/4) + (P/4) - P = 0$. Note that the force R_{B_v} as shown on this member is equal and opposite to R_{B_v} as it affects the right segment of the structure. This follows since it is a reactive force. The same is true for R_{B_h}.

Equilibrium in the horizontal direction for the left element: $\sum F_x = 0 \quad \rightarrow +:$

$$R_{A_h} - R_{B_h} = 0 \qquad \text{since } R_{B_h} = \left(\frac{P}{4}\right), \, R_{A_h} = \frac{P}{4} \quad \rightarrow$$

Both components of R_A are now known. As a check, rotational equilibrium of the left segment can also be considered.

Rotational equilibrium for the left element about B: $\sum M = 0$ ⤷:

$$-R_{A_v}\left(\frac{L}{2}\right) + P\left(\frac{L}{4}\right) + R_{A_h}\left(\frac{L}{2}\right) = 0$$

or

$$-\left(\frac{3P}{4}\right)\left(\frac{L}{2}\right) + P\left(\frac{L}{4}\right) + \left(\frac{P}{4}\right)\left(\frac{L}{2}\right) = 0 \qquad \text{Check!}$$

The components of the forces acting on the left element can be resolved into single forces, as illustrated in Figure 5-11(c). Since the left segment of the structure is a three-force member, it follows that the lines of action of all the forces must pass through a single point if the element is indeed in equilibrium. Inspection of Figure 5-11(c), which shows these lines of action, reveals that they meet at point 0.

It should be noted that the forces found at the base and crown hinges were influenced only by the location in space of the three hinges and the magnitude and line of action of the external load. *Nowhere* did the specific shape of the individual rigid segments from A to B and from B to C ever enter into or affect the calculations. As long as they are rigid and their end-point locations defined, any segmented shape (e.g., linear, arbitrarily curved, exact funicular shapes) could have been used and the magnitude of the forces developed at the hinged connections would not have been affected.

Although the connection forces found are independent of the shapes of the rigid segments, it is evident that the linear forms used in the structure shown in Figure 5-11 are not funicularly shaped, since the external load clearly produces a bowing or bending in the left member.

The independence of connection forces in three-hinged assemblies from the specific shape of the segmented pieces is an important concept, since it forms the basis for some useful design approaches. Consider the three loading conditions shown in Figures 5-12, 5-13, and 5-14, respectively. Assume that it is desired to determine appropriate structural shapes for the segmental pieces to be used for members *AB* and *BC* in each structure such that bending is minimized or eliminated.

Since each structure is statically determinate, it is possible to determine by the methods just described the connection forces for each structure *without* knowing their precise shapes (assuming that the dead loads of the structure are ignored). Rigid segments of unspecified shape are assumed. Vertical and horizontal forces determined through equilibrium calculations for each of the structures are shown in Figures 5-12(b), 5-13(b), and 5-14(b).

By analyzing the complete force system acting on each segment of each structure, it is possible to shape each segment so that bending is minimized. For the first structure shown in Figure 5-13, consider the right segment from *B* to *C*. As is evident, the forces acting on the segment are those at the connections only. The segment is thus a two-force member. These forces must be collinear, equal in magnitude, and opposite in sense for the segment to be in equilibrium. It is evident that *if* the objective is to shape the segment so that it is subject only to axial compression with no bending present, then the segment should be simply a straight member exactly aligned with the direction of the collinear forces and thus simply connect the two hinges. As illustrated in Figure 5-12(f), bending will develop in any *other* shape of structure used.

The same type of thinking can be applied to the left segment. This segment is obviously a three-force member. Consequently, all forces must meet at a point for the member to be in equilibrium. When the structure is lined up with the lines of action of the three forces, it is evident that the structure will be subject to axial compression only. Note that the reactive forces literally predetermine the direction of the structural shape at the support points if the member is to be axially loaded [see Figure 5-12(d)].

Figure 5-13 illustrates a similar analysis for a structure with a uniformly distributed load over the left half. Connection forces are again determined first. The right segment is shaped in a manner exactly the same as that for the previous structure. The left segment is more complex. Note that if the uniform load is considered to be replaced by a statically equivalent concentrated load, it is a three-force member in which the lines of action of all forces again meet at a point, as in the previous example. For the actual uniformly distributed load, however, the appropriate shape is not the same as previously. At the base connection it is evident that the structure must be aligned with the direction of the resultant reaction for the structure to be in com-

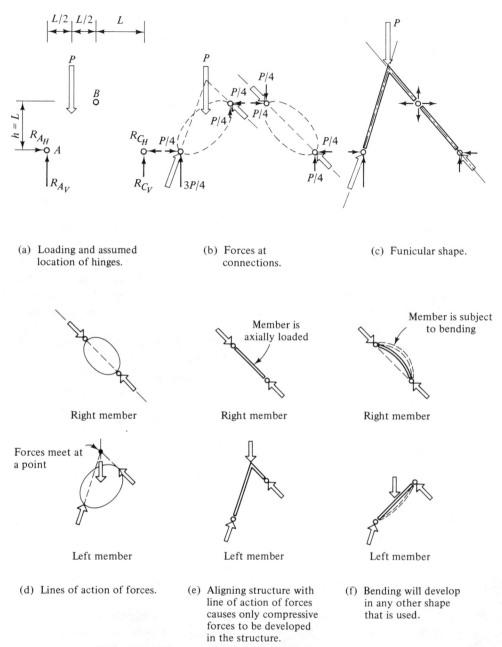

(a) Loading and assumed location of hinges.

(b) Forces at connections.

(c) Funicular shape.

Member is axially loaded

Member is subject to bending

Right member

Right member

Right member

Forces meet at a point

Left member

Left member

Left member

(d) Lines of action of forces.

(e) Aligning structure with line of action of forces causes only compressive forces to be developed in the structure.

(f) Bending will develop in any other shape that is used.

FIGURE 5-12 Shaping three-hinged arches for point loads.

pression. The same is true for the crown connection (point B). The directions of the reactions determine the alignment of the structure at each of these points. The real question is how is the structure shaped between these two points of known alignment. At this point and with this method of analysis, there is no way of determining the shape exactly. Intuitively, it would seem that the structure would form a continuous

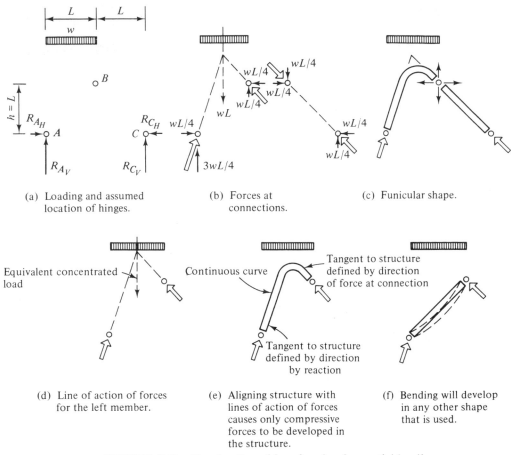

| (a) Loading and assumed location of hinges. | (b) Forces at connections. | (c) Funicular shape. |

| (d) Line of action of forces for the left member. | (e) Aligning structure with lines of action of forces causes only compressive forces to be developed in the structure. | (f) Bending will develop in any other shape that is used. |

FIGURE 5-13 Shaping three-hinged arches for partial loadings.

curve between the two known alignments, as shown in Figure 5-13(e). The basis for this assumption is the discussion in Section 1.5, in which it was noted that funicular structures have a shape that can be determined by considering the shape that a cable deforms to under the specified loading. A continuous load produces a continuous curve. The shape being found here is analogous and, indeed, is simply the inversion of the deformed cable shape under the same loading. The resultant structure is similar to the previous example, but is obviously curved rather than peaked.

The third example illustrated in Figure 5-14 is for the case when the loading is uniformly distributed across the whole length of the structure. Connection forces are again determined by equilibrium calculations. Note that the resultant force at the top connection (point B) is horizontal, since it has no vertical component. By isolating the right segment, it can be seen that the segment is again a three-force member if the uniformly distributed load is conceptually replaced by a statically equivalent concentrated load. As before, the forces meet at a point. The directions of the forces at the top and base connections again determine the alignment of the structure at these points if the structure is to be in axial compression. Assuming that

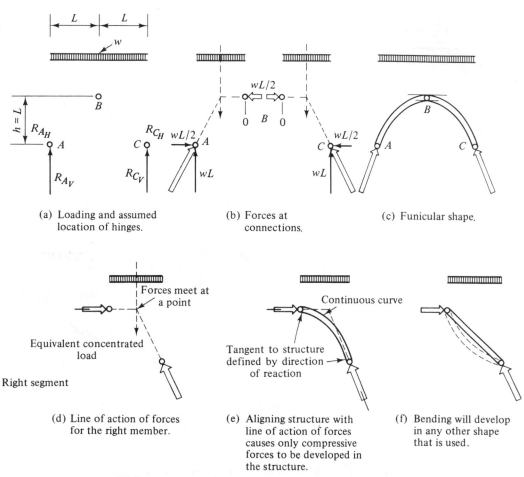

(a) Loading and assumed
location of hinges.

(b) Forces at
connections.

(c) Funicular shape.

(d) Line of action of forces
for the right member.

(e) Aligning structure with
line of action of forces
causes only compressive
forces to be developed in
the structure.

(f) Bending will develop
in any other shape
that is used.

FIGURE 5-14 Shaping three-hinged arches for full loadings.

the structure curves continuously and smoothly between these two alignments yields
the shape indicated in Figure 5-14(e). As before, it is not possible at this point to
exactly determine what this shape is, based on this type of analysis.

The inverted-string analogy, however, is again a quite useful tool in imagining
the appropriate shape for the member. The left segment is, of course, similar to the
right segment. The resultant shape is indicated in Figure 5-14(c). The shape is parabolic with the tangents to the curve defined at the supports by the direction of the
reactions and the tangent at the supports by the direction of the reactions and the
tangent at the crown connection (point B) horizontal. The shape is clearly that
associated with the common *arch*.

Each of the shapes of structures shown is in axial compression and is not subject
to any bending. Some interesting manipulations can be made with these shapes.
Although the shape of the structure was originally found by considering the top
connection, point B, to be located as shown in Figure 5-12, the actual location of
this pinned connection could be anywhere on the final shape found and neither the

shape of the structure nor the magnitudes and directions of the forces at the base connections would be affected. This could be demonstrated through equilibrium calculations. Interestingly enough, the whole structure could be imagined to have been made of similar pinned elements and the resultant configuration would theoretically be stable under the loading. This is obviously the same as the sphere analogy previously illustrated in Figure 2-4. In actuality, such imaginary structures are not appropriate for use because of their obvious instability under varying load conditions. Still, the approach is a highly useful conceptual tool for appreciating the implications of using this type of structure. A more realistic counterpart to the stacked spheres would be a structure made up of stacked block elements. As long as the block elements are stacked in a particular way (assuming they are weightless), they will form a stable structure for a particular loading condition no matter whether they are mortared together or not. The structure supporting the full continuous load, for example, could be made from wedge-shaped blocks to form a continuous curved structure stable under the loading. It is evident that if the loading were changed, the resultant structure would become unstable and would collapse immediately. All the shapes shown, by the way, are special forms of arches if an arch is defined as a structure that is subject to axial compressive forces only. The true value of thinking in terms of the block or sphere analogy is in appreciating *the unique relationship that must exist between load and structure when the criterion used to shape the structure is that bending must not be present.*

It is interesting to note that in the structures analyzed, the effect of specifying the location of the top hinge (point *B*) in space is that this establishes a location on the structure that cannot, by definition, provide any rotational resistance or resist moments in any way. The moment of the external force system around this point must be zero. Similarly, the structure cannot provide any internal resisting moments at this point. This means that any funicular structural response *must* pass through this point *B*, no matter where the point is located in space.

Following sections will continue to explore the problem of shaping structures. The following section will discuss ways of exactly defining the nature of the curve connecting the support points on a loaded structure known to act funicularly.

5.3.2 Cables and Funicular Arches: Concentrated Loads

This section focuses on the analysis of structures known to be funicularly shaped (e.g., cables). A cable can be conceived of as a continuous series of discrete elements connected to one another by hinged connections—a chain is an obvious image. Hence, each discrete element is free to rotate as it will under the action of a load. The connections are such that internal moments cannot be transmitted from one element to another. It then follows that the sum of all rotational effects produced about *any* such location by the external and internal forces must be zero. In a three-hinged arch having a nonfunicular shape, this observation about rotational equilibrium was true only at the three hinged connections. In a cable the observation is true for all locations. Use can be made of the observations described to determine the exact shapes of such members and the forces in them.

Another point that should be noted is that it has been stressed in Chapters 2 and 4 that the function of any structure can be defined as that of carrying the shears and moments generated by the effects of external loads. In a cable or funicular-shaped arch, the external shear at a section is balanced by an internal resisting shear force that is provided by the vertical component of the internal axial force developed in the cable or arch member. The external bending moment at the same section is balanced by an internal resisting moment that is provided by the couple formed by the horizontal component of the axial force in the funicular member and the horizontal force acting through the end support of the structure. This latter force may be either the horizontal thrust developed at the foundation or the internal force present in a tie-rod or compression strut used between end points of an arch or cable structure, respectively.

The following example illustrates how cable geometries and internal forces can be determined through application of the principles described.

EXAMPLE

Figure 5-15 illustrates a cable structure carrying two loads. Assume that the deformed structure has the maximum depth shown under the first load. The deformed structure has another critical depth under the second load. The latter depth is not precisely known beforehand, but it must be a value that is directly dependent on the maximum depth of the other point, which is a given.

Solution:

The first step in the analysis process is to determine the reactions at the ends of the member. Vertical reactions can be found in a straightforward manner by considering moment equilibrium of the whole structure about one or the other reaction. Horizontal reactions can be found by passing a section through the structure at a location of known structural depth and considering the equilibrium of either the right or left portion of the structure. Thus:

Vertical reactions:

$$\Sigma M_A = 0 \quad \text{↻:}$$

$$-15(10) - 30(5) + 45 R_{B_V} = 0$$

$$R_{B_V} = 6.7$$

$$\Sigma F_Y = 0 \quad +\uparrow:$$

$$-10 - 5 + R_{A_V} + R_{B_V} = 0$$

$$R_{A_V} = 8.3$$

Horizontal reactions:

Pass a vertical section through the structure immediately to the right of the location of the maximum depth and consider the equilibrium of the left portion of the structure:

$$\Sigma M_C = 0 \quad \text{↻:}$$

$$-15(8.3) + 10(R_{A_H}) = 0$$

$$R_{A_H} = 12.5$$

For the whole structure:

$$\Sigma F_X = 0 \quad \overset{+}{\rightarrow}:$$

$$-R_{A_H} + R_{B_H} = 0$$

$$R_{B_H} = 12.5$$

(a) Cable structure and loading. The structure deforms into a shape responsive to the relative magnitudes of the loads applied. The sag at only one point can be specified since the magnitudes of the relative deformations are interdependent.

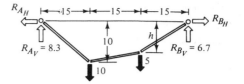

(b) The cable sag at points other than the sag specified can be calculated by considering the equilibrium of different pieces of the structure. For the structure shown, the first step is to determine the left horizontal reaction (R_{A_H}) by considering the equilibium of the left portion of the structure ($\Sigma M_C = 0$). The unknown sag at point D can then be calculated by considering the equilibrium of the right-hand portion by noting that $R_{A_H} = R_{B_H}$ and summing moments about point D ($\Sigma M_D = 0$), treating h_D as an unknown.

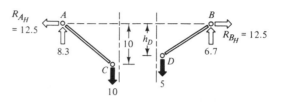

(c) Once the geometry of the structure is completely specified, internal forces in various segments of the cable can be determined through joint equilibrium considerations.

(d) An alternative analytical process is to first draw shear and moment diagrams for the basic loading condition and then use a method of sections approach to find vertical and horizontal components of cable forces.

(e) The internal shear and moment provided by the structure at any section must balance the external shear and moment present at that section. The internal shear is provided by the vertical component of the force in the cable. The internal moment is provided by the couple formed between the horizontal reaction and the horizontal component of the force in cable. Note that the horizontal component of the force in the cable is always equal to the horizontal thrust because of translatory equilibrium in the horizontal direction. The force in a cable segment is found by resolving vertical and horizontal components into a single resultant force.

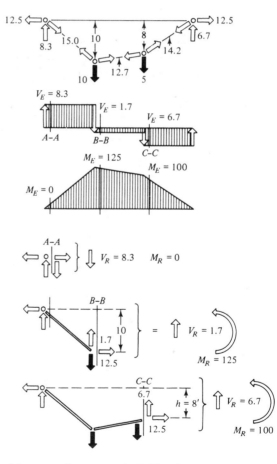

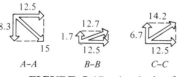

FIGURE 5-15 Analysis of a cable supporting concentrated loads.

197

Depth of structure at point D:

Pass a vertical section immediately to the left of point D and consider the equilibrium of the right portion of the structure. Let h_D equal the height of the structure at this point. Thus,

$$\Sigma M_D = 0 \;\; \text{⟲:}$$

$$-h_D(12.5) + 15(6.7) = 0$$

$$h_D = 8.3$$

Note that no other structural depth would satisfy rotational equilibrium considerations at this point.

Cable forces:

Forces in individual segments in the cable can now be found by using the method of joints. The results of this process are illustrated in Figure 5-15(c). It should be noted that the absolute magnitude of the forces in different segments vary. Cables are *not* constant-force structures.

    After finding vertical reactions as before, an alternative method of analysis for the same structure is to first draw shear and moment diagrams for the external loading condition and then use what is basically a *method-of-sections* approach.

Horizontal reactions:

The horizontal reactions can be found by observing that the structure must provide an internal moment resistance equal and opposite in sense to the external moment at every point. The internal moment is provided by a couple formed between the end horizontal reaction and the horizontal component of the cable force. Considering the structure at point C:

$$\underbrace{10R_{A_H}}_{\substack{\text{internal}\\\text{resisting}\\\text{moment}}} = \underbrace{125}_{\substack{\text{external applied}\\\text{moment (see moment}\\\text{diagram)}}} \quad \text{or} \quad R_{A_H} = 12.5$$

This is obviously essentially the same process as described before but is a conceptually different way of thinking about it.

Depth of structure at point D:

As before:

$$\underbrace{h_D(12.5)}_{\substack{\text{internal}\\\text{resisting}\\\text{moment}}} = \underbrace{100}_{\substack{\text{external}\\\text{moment}}} \quad \text{or} \quad h_D = 8.3$$

Cable forces:

Sections are passed through points immediately to the right of each critical point and equilibrium of the left portion considered [see Figure 5-11(e)]. Note that in each case the internal shear resistance that balances the external shear force is provided by the vertical component of the force in the cable. The vertical and horizontal forces are components of the actual forces in the cables and can be used to calculate these forces.

    The second method presented is extremely useful for conceptualizing how cable or arch structures provide a mechanism for carrying the shear and moment associated with the external load. It also shows particularly well how the reactions and internal forces depend on the height of the structure. Doubling the initial maximum depth would decrease horizontal thrusts, for example, by a factor of 2. Cable forces would thus be reduced. Still the new combination of cable height and force would still yield the same internal balancing shear and moments discussed above.

5.3.3 Cables and Funicular Arches:
Uniformly Distributed Loads

Cables or arches carrying uniformly distributed loads can be analyzed in much the same way as described for concentrated loads. Since the funicular shape is a constantly curving one, however, a variant of the method of sections is the exclusive analytical technique used.

EXAMPLE

Consider the cable shown in Figure 5-16. Assume the maximum cable sag to be h_{max}. The first step is to calculate vertical reactions which are formed through considering the equilibrium of the overall structure. Following steps are to calculate horizontal reactions, find the shape of funicular curve, and then determine the internal forces in the cable.

Solution:

Vertical reactions:

$$R_{Av} = R_{Bv} = \frac{wL}{2} \qquad \text{by inspection}$$

Horizontal reactions:

Pass a section through the structure at midspan (the location of known sag) and consider the equilibrium of the left portion of the structure:

$$\Sigma M_{L/2} = 0:$$

$$+R_{A_H} h_{max} - \left(\frac{wL}{2}\right)\left(\frac{L}{2}\right) + \left(\frac{wL}{2}\right)\left(\frac{L}{4}\right) = 0$$

or

$$R_{A_H} = \frac{wL^2}{8h_{max}}$$

Alternatively, using shear and moment diagrams directly:

$$\underbrace{R_{A_H} h_{max}}_{\substack{\text{internal} \\ \text{resisting} \\ \text{moment}}} = \underbrace{\frac{wL^2}{8}}_{\substack{\text{external} \\ \text{applied} \\ \text{moment}}} \qquad \therefore R_{A_H} = \frac{wL^2}{8h_{max}}$$

Shape of cable:

The shape of the cable can be found by considering the equilibrium of different sections of the structure. Consider a section of the structure defined by a distance x from the left connections. Let y be the depth of the cable at that point.

$$\Sigma M_x = 0:$$

$$\left(\frac{wL^2}{8h_{max}}\right)(y) - \left(\frac{wL}{2}\right)(x) + wx\left(\frac{x}{2}\right) = 0$$

$$y = \frac{4h_{max}(Lx - x^2)}{L^2}$$

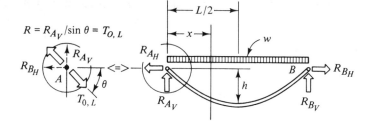

$R = R_{A_V}/\sin\theta = T_{0,L}$

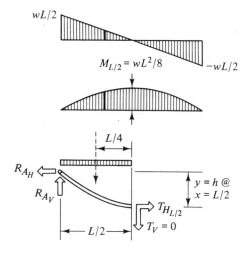

(a) Cable structure and loading.

$wL/2$

$M_{L/2} = wL^2/8$

$-wL/2$

(b) Shear and moment diagrams for loading condition.

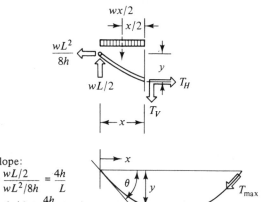

(c) The first step in the analysis process is to find the reactions. Vertical reactions can be found by passing a section through a point where the cable sag is known or specified and considering moment equilibrium about that point. Thus,

$$R_{A_H}(h) = wL^2/8$$

or $R_{A_H} = wL^2/8h$

(d) The equation of the deflected shape of the cable is found by considering the equilibrium of different sections of the cable. Thus,

$$(wL^2/8h)y - wLx/2 + wx^2/2 = 0$$

or $y = 4h(Lx - x^2)/L^2$

End slope:
$$\theta = \frac{wL/2}{wL^2/8h} = \frac{4h}{L}$$

or $\theta = dy/dx = \dfrac{4h}{L}$

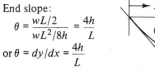

(e) The slope of the cable at any point can be evaluated by considering the deflected shape. At the ends of the cable, the slope is also given by the ratio of the vertical and horizontal components of the end reactions. The maximum cable force is merely the resultant of these two forces. The cable force at midspan is less and is numerically equal to the horizontal thrust: $T_{L/2} = wL^2/8h$.

FIGURE 5-16 Analysis of a cable supporting a horizontally distributed load.

Alternatively, equating the internal resisting moment to the external applied moment:

$$\left(\frac{wL^2}{8h_{max}}\right)y = \left(\frac{wL}{2}\right)x - (wx)\left(\frac{x}{2}\right)$$

$$y = \frac{4h_{max}(Lx - x^2)}{L^2}$$

The equation found is obviously that of a parabola. The absolute value of the slope at either end is given by

$$\left(\frac{dy}{dx}\right)_{x=0,x=L} = \theta_{x,L} = \frac{4h_{max}}{L}$$

This slope could have also been found by considering the ratio of the horizontal and vertical reactions:

$$\theta = \frac{wL/2}{wL^2/8h_{max}} = \frac{4h_{max}}{L}$$

Cable forces:

Since the slope of the cable is zero at midspan, where both the external and internal shear forces are also zero, the cable force can be seen to identically equal the horizontal reaction by considering the horizontal equilibrium of the section to the left or right of midspan. Thus:

$$T_{L/2} = \frac{wL^2}{8h_{max}}$$

The cable force in the cable at either end connection is found by simple joint equilibrium considerations:

$\Sigma F_x = 0$:

$$T_{0,L} \cos\theta = \frac{wL^2}{8h_{max}}$$

where θ is given by $4h_{max}/L$ and $\cos\theta$ by $1/\sqrt{1 + 16h_{max}^2/L^2}$. Thus,

$$T_{0,L} = \frac{wL^2}{8h_{max}}\sqrt{1 + 16h_{max}^2/L^2}$$

Alternatively,

$$T_{0,L} = \sqrt{(R_{A_H})^2 + (R_{A_v})^2}$$

$$= R_{A_H}\sqrt{1 + (R_{A_v}/R_{A_H})^2}$$

$$= \frac{wL^2}{8h_{max}}\sqrt{1 + 16h_{max}^2/L^2}$$

The force in the cable at the ends thus exceeds that at midspan.

EXAMPLE

A series of cables spaced at 15 ft (4.6 m) on center span 100 ft (30.5 m) between abutments and carry a live load of 50 lb/ft² (2394 N/m²). Each cable has a maximum sag of 20 ft (6.1 m). Assume that the dead load of the whole cable structure is 50 lb/ft² (2394 N/m²). Also, assume both the live load and dead load to be horizontally projected. Find the minimum required cable diameter assuming the allowable cable stress is 75,000 lb/in.² (517 N/mm²).

Solution:

Reactions:

$$\text{vertical} = R_{A_V} = \frac{wL}{2} = \frac{[(50 + 50) \text{ lb/f}^2](15 \text{ ft})(100 \text{ ft})}{2} = 75,000 \text{ lb}$$

$$= \frac{[(2394 + 2394) \text{ N/m}^2](4.6 \text{ m})(30.5 \text{ m})}{2} = 334 \text{ kN}$$

$$\text{horizontal} = R_{A_H} = \frac{wL^2}{8h} = \frac{[(50 + 50) \text{ lb/f}^2](15 \text{ ft})(100 \text{ ft})^2}{8(20 \text{ ft})} = 93,750 \text{ lb}$$

$$= \frac{[(2394 + 2394) \text{ N/m}^2](4.6 \text{ m})(30.5 \text{ m})^2}{8(6.1 \text{ m})} = 417 \text{ kN}$$

Maximum cable force (at ends):

$$T = \sqrt{R_{A_V}^2 + R_{A_H}^2} = (75,000^2 + 93,750^2)^{1/2} = 120,000 \text{ lb}$$

$$= (334^2 + 417^2)^{1/2} = 534 \text{ kN}$$

Alternatively,

$$T = R_{A_H}\sqrt{1 + 16h^2/L^2}$$

$$= 93,750 \text{ lb}\left[1 + 16\left(\frac{20^2}{100^2}\right)\right]^{1/2} = 120,000 \text{ lb}$$

$$= 417 \text{ kN}\left[1 + 16\left(\frac{6.1^2}{30.5^2}\right)\right]^{1/2} = 534 \text{ kN}$$

Cable diameter:

$$A = \frac{T}{f_{\text{allowable}}} = \frac{120,000}{75,000} = 1.60 \text{ in.}^2$$

$$= \frac{534,000 \text{ N}}{517 \text{ N/mm}^2} = 1033 \text{ mm}^2$$

Twisted strand cable would probably be used in which the total diameter is made up of a series of individual wire strands of smaller diameters. The gross area of the total diameter therefore does not equal the net area of the cross sections of individual strands. The resisting area of the whole cable is about two-thirds of its gross area. Thus, the gross area of the cable needed for the above example is $(\frac{3}{2})(1.60) = 2.4$ in.^{2} or $(\frac{3}{2})(1033) = 1550$ mm^2. This leads to a cable diameter of $1\frac{3}{4}$ in. or 45 mm.

5.3.4 General Funicular Equation

A more general approach can be taken to the analysis of cables than previously discussed, which is very useful in treating more complicated structures that are not symmetrically loaded or which have cable supports at different levels.

Consider the cable shown in Figure 5-17. As before, the first step in the analysis process is to find the reactions. If $\sum M_{B_E}$ is used to designate the sum of all the rotational moments about point B of all the external loads [e.g., $P_n(L - x_n) + P_{n-1}(L - x_{n-1}) + \ldots$], the total moment equilibrium expression for the entire cable about point B becomes:

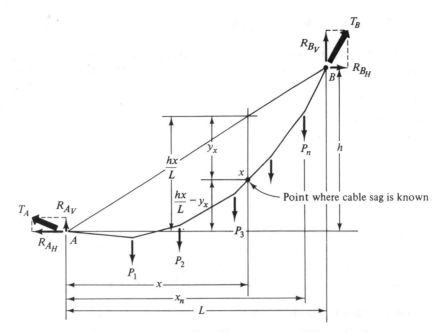

FIGURE 5-17 A cable with supports at different levels.

$$\sum M_B = 0:$$

$$-R_{A_H}(h) - R_{A_V}(L) + \sum M_{B_E} = 0$$

$$R_{A_V} = \frac{\sum M_{B_E}}{L} - \frac{R_{A_H}(h)}{L}$$

This expression cannot be directly solved because both R_{A_V} and R_{A_H} are unknown values. The next step is to pass a section through a point x on the cable *where the cable sag is known* and consider moment equilibrium of the left portion of the cable. Let $\sum M_{x_E}$ be used to designate the sum of the external moments of the loads on the left portion about this point. Noting that the moment arm of R_{A_H} at this point is given by $(hx/L - y_x)$, where y_x is the cable sag from a line connecting the two end points of the cable, the moment expression about x becomes:

$$\sum M_x = 0:$$

$$-R_{A_H}(hx/L - y_x) - R_{A_V}x + \sum M_{x_E} = 0$$

Substituting R_{A_V} from above:

$$R_{A_H}(y_x) = \left(\frac{x}{L}\right)\sum M_{B_E} - \sum M_{x_E}$$

If y_x is specified at a point, the R_{A_H} can be evaluated. Once R_{A_H} is known, it can be used to find unknown cable sags at points other than x. Cable forces can be found by equilibrium considerations in a manner similar to that previously described.

5.3.5 Cable or Arch Lengths

It is often very important to know the total length of a cable or arch given its span and sag or rise. This length can always be evaluated from considering the basic expression for the deformed shape of a cable. For a *uniformly loaded* cable with a horizontal chord (both supports on the same level), let L_{total} be the total length of the cable, L_h the span, and h the maximum sag. The total cable length can be shown to be approximately

$$L_{\text{total}} = L_h(1 + \tfrac{8}{3}h^2/L_h^2 - \tfrac{32}{5}h^4/L_h^4)$$

For the derivation for this and other important expressions defining the elongation of a cable, the reader is referred to other references.[1]

5.4 DESIGN OF ARCHES AND CABLES

This section treats the design of cables and arches by first looking at an issue common to both structures and then looking at each structure in turn.

5.4.1 Cable Sag or Arch Rise

One of the most important design variables in a cable or arch is the maximum depth of the structure. A review of Section 5.2 reveals that the forces in these structures, hence their size, are critically dependent on the amount of sag or rise relative to the span of the structure. The horizontal component of the force in a uniformly loaded level ended cable, for example, is given by $T_h = wL^2/8h_{\text{max}}$, where h_{max} is the maximum sag present. Obviously, doubling the maximum sag decreases T_h by a factor of 2. In general, as the sag h_{max} is made smaller, the cable force becomes larger, and vice versa. Clearly, it is not possible to have a zero sag cable, since this would imply indefinitely large cable forces and consequently indefinitely large cable cross sections.

Given this sensitivity to structural depth, finding the most appropriate sag or rise becomes an optimization problem. As h_{max} increases, the force in the cable decreases and thus the related cross-sectional area required also decreases. The cable length, however, simultaneously increases. Clearly, an indefinitely long cable requires an indefinitely large volume. Conversely, as h_{max} decreases, cable forces and required cross-sectional areas increase, but the cable length decreases. An optimum value of h_{max} must exist. This optimum value can be determined simply by developing an expression for the cable volume in terms of h_{max}, using the analytical expressions developed in Section 5.2, and minimizing the value of the expression with reference to the critical variable h_{max}. A depth equal to approximately one-third the span of the structure will be found to result in the minimum volume for a uniformly loaded structure.

The exact sag or rise chosen for use, however, is invariably dependent on the

---

[1]See, for example, C. H. Norris and J. B. Wilbur, *Elementary Structural Analysis*, 2nd ed., McGraw-Hill Book Company, New York, 1960; or J. Scalzi, W. Poddorny, and W. Teng, *Design Fundamentals of Cable Roof Structures*, U.S. Steel Corporation, Pittsburgh, Pa., 1969.

overall context in which the cable or arch is used. Most cable structures used in buildings have sag/span ratios between 1:8 and 1:10 rather than the 1:3 discussed above. Arch structures are usually deeper with rise/span ratios on the order of at least 1:5.

5.4.2 Design of Cable Structures

SUPPORTING ELEMENTS. In addition to the actual roof cables, other structural elements (e.g., masts, guy cables) are needed to make a building structure. The elements typically support the cable in space and provide a means for transferring its vertical and horizontal thrusts to the ground. The design of these elements is as crucial as is the design of the cable.

A basic design issue is whether to absorb the horizontal thrusts involved directly through the foundations or by using a supplementary horizontal compression strut. While designing foundations to absorb both vertical and horizontal thrusts is difficult, doing so is entirely feasible, depending on soil or other foundation conditions that are existent. The use of horizontal compression struts is much less frequent because of the long unbraced length of such members, which makes them highly susceptible to possible buckling since they are in compression. Required sizes for the struts would be very large and tend to offset any efficiencies gained by using a cable to span a long distance.

Figure 5-18 illustrates several types of common support elements found in buildings using cables to make a volume-forming enclosure. Figure 5-18(a) shows a cable supported by end piers. In this case the entire horizontal thrust of the cable must be resisted by the vertical pier acting like a cantilever beam. Owing to the very large horizontal thrusts, significant bending is induced in the pier and it must be made quite large (the sizing of members that are in bending is discussed in Chapter 6). The foundation of the pier must also be designed to resist the overturning moment. By and large, this approach is economically viable only in lightly loaded cables of relatively short span.

Figure 5-18(b) shows a cable supported by guyed masts. When the masts are vertical, the horizontal cable thrust is taken up completely by the guy cable, which transfers the force to the ground. The mast itself picks up only axial compression forces. No bending is present. The mast is designed as a column (see Chapter 7), which picks up the sum of the vertical force components in both the primary and guy cables. Since mast foundations need only carry vertical loads, they are fairly easy to design and construct. The foundation for the guy cable, however, is more complex, since it needs to withstand not only lateral forces but uplift forces as well. By and large, however, this use of guyed masts is a fairly efficient type of cable-support system.

A variant of the approach described above is shown in Figure 5-18(c), in which the masts are tilted. In this case part of the horizontal thrusts are picked up by the inclined masts acting in compression and the remainder by the attached guy cables. While this increases the axial force in the masts and their lengths as well, and thus increases their size, this tilting reduces the forces in the guy cables and their ground

(a) Pier supports: the vertical piers supporting the ends of the cables carry the vertical components of the cable reactions by axial compression and the horizontal component by bending. This system is good for cables of relatively short span only.

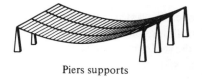

Piers supports

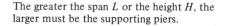

The greater the span L or the height H, the larger must be the supporting piers.

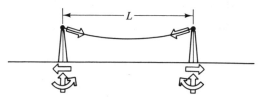

Basic force system. Note that the foundation must prevent the piers from overturning.

(b) Guyed mast supports: the horizontal components of the cable thrusts are absorbed by diagonal guy cables and transferred to the ground. The vertical masts act in axial compression only. This system is good for relatively long span cables.

Guyed mast supports

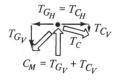

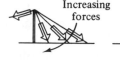

Increasing forces

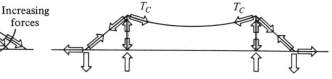

Free-body diagram of top of cable mast

As the guy cable angle changes as shown, the forces in the guy cable tend to increase. Mast forces also increase accordingly.

Basic force system. The masts carry axial forces. The guyed cable foundations must prevent guy uplift and sliding.

(c) Inclined guyed masts: inclining the masts causes them to pick up some of the horizontal cable thrusts, thus reducing the forces in the diagonal guy cables. This system is good for long-span cables.

Inclined guyed masts

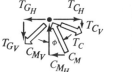

$$T_{G_H} + C_{M_H} = T_{C_H}$$
$$T_{G_V} + T_{C_V} = C_{M_V}$$
$$C_M = C_{M_V}/\cos \phi$$

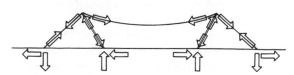

Free-body diagram of top of cable mast

Basic force system. The masts carry only downward forces. Guy cable foundations must prevent guy up-lift and sliding

FIGURE 5-18   Different types of cable support systems.

attachments. Since uplift forces in the guy cable ground attachments are reduced, foundations can be more easily made.

In all the above, the sag chosen for the cable is a significant variable since pier or mast lengths are directly related to this value for a given functional enclosed cable height above ground. The deeper the sag, the more substantial support members become. Sags smaller than the one-third depth/span value discussed in the last section are thus often used to achieve a more optimum overall solution. Sag/span ratios of around 1 : 8 to 1 : 10 are often used.

DESIGNING FOR WIND EFFECTS.   As mentioned earlier, the stability of a cable under a wind force is of fundamental interest because of the phenomenon of flutter. Although a treatment of vibration phenomena is beyond the scope of this book, a few general observations can be made.

A vibration in a structure that is induced by wind (or earthquake) effects is a reciprocating or oscillating motion which repeats itself after an interval of time. This interval of time is called the *period* of vibration. The *frequency* of vibration is numerically equal to the reciprocal of the period.

All suspension structures (and other structures as well) have a *natural frequency* of vibration when acted on by an externally applied force. When an external dynamic force acting on the structure comes within the natural frequency range, a state of vibration may be reached where the driving-force frequency and the body's natural frequency are in tune, a condition referred to as *resonance*. At resonance the structure undergoes violent vibration, resulting in structural damage (see Section 3.2.4).

The natural frequency of a suspended cable is given by $f_n = (N\pi/L)\sqrt{T/(w/g)}$, where L is the cable length, N any integer, w the applied load per unit length, T the cable tension, and g the acceleration due to gravity. The first three modes of vibration are shown in Figure 5-8. The cable could vibrate in these and other modes, depending on the frequency of the exciting force.

Resonance in a cable would occur when the external exciting force has a frequency exactly equal to the first natural frequency of the cable or any of its higher frequencies. Unfortunately, for many cable structures, the frequency of wind forces is often close to the natural frequency of the cable structure.

There are only several fundamental ways to combat flutter due to wind forces. One is to simply increase the dead load on the roof, thereby also increasing cable tensions and changing natural frequencies. Another is to provide anchoring guy cables to the cable at periodic points to tie the structure to the ground. A related method is to use some sort of crossed-cable or double-cable system. The latter method is extremely interesting, since it is possible to create an internally self-dampening system.

Consider the arrangements shown in Figure 5-19, in which there are upper and lower cables interconnected by interior elements. In each case both sets of cables are given an initial tension force. When the assembly shown in Figure 5-19(b) is put in place, it essentially acts much like a truss in that the top member tends to compress while the bottom member tends to elongate. Since both cables have already been pretensioned, these tendencies cause a reduction in the tension force in the top cable and an increase in the force on the bottom cable. The top cable must be designed and

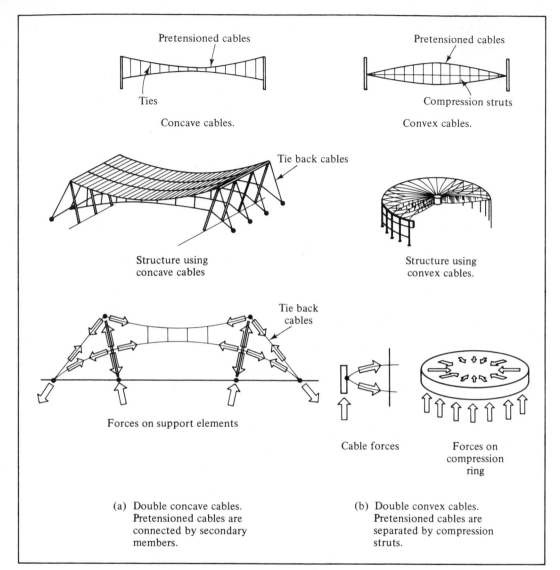

Pretensioned cables

Ties

Concave cables.

Pretensioned cables

Compression struts

Convex cables.

Tie back cables

Structure using
concave cables

Structure using
convex cables.

Tie back
cables

Forces on support elements

Cable forces

Forces on
compression
ring

(a) Double concave cables.
Pretensioned cables are
connected by secondary
members.

(b) Double convex cables.
Pretensioned cables are
separated by compression
struts.

FIGURE 5-19 Use of double-cable systems to prevent roof flutter due to wind effects. In both cases shown, the cable force in the upper member is slightly different from that in the lower member. Therefore, each member has a different natural frequency of vibration which causes the whole assembly to be a self-damping mechanism, since neither cable can freely vibrate in its natural mode because of the other cable.

prestressed so that under full loading the change in force in the top cable is not large enough to cause the cable to lose all its initial prestressing and go slack. The bottom cable must be sufficiently sized to carry both the initial pretensioning and the added tension that develops under loading. The structure shown in Figure 5-19(a) behaves differently under loading in that the pretensioned upper cable is subject to further tension while the pretensioning in the lower cable is reduced.

In both structures, the external loads and the pretensioning forces cause large horizontal forces to be developed on the supporting elements. A series of tie-back cables are used to take care of these forces in the structure shown in Figure 5-19(a). A large continuous compression ring, typically made out of steel, is used to take care of these forces in the structure shown in Figure 5-19(b) (since the ring takes up the horizontal forces the columns carry only vertical loads).

The dynamic behavior of the roof types described above is interesting. As previously mentioned, the natural frequency of vibration of a tight cable depends on the tension present in the cable. Since both cables are, as discussed above, at different tensions, it is evident that each has a slightly different natural frequency of vibration. The natural frequency of vibration of the lower cable will thus always tend to be slightly different from the upper cable. As an external exciting force tends to cause either cable to vibrate in one of its fundamental modes, the other cable will tend to damp out the vibrations in the cable, since it has a different natural frequency of vibration. The result is that any oscillations are damped out since neither cable can go into resonance because of the damping effect of the other.

It is possible, however, that the whole assembly (which must have its own natural frequency, which is distinct from those of either of its constituents) could go into resonance. The natural frequency of the whole system is related to the combination of individual cable frequencies and is higher than either. If this combined frequency can be made sufficiently high by careful design, the danger of there being a wind with a sufficiently high critical velocity to cause the whole system to resonate can be minimized to the point where no danger exists.

5.4.3 Design of Arch Structures

DESIGNING FOR LOAD VARIATIONS.   As mentioned earlier, one of the most significant aspects of the modern arch is that it can be designed to sustain some amount of variation in load without either changing shape or experiencing damage. Only arches designed and constructed using rigid materials, such as steel or reinforced concrete, have this capability.

The shape of an arch is usually initially determined as a response to its primary loading condition (e.g., parabolic for uniformly distributed loads). If a load of a different type comes to bear on the arch, bending is developed in arch members in addition to axial compressive stresses. For the arch to be able to carry this load variation it must be sufficiently large to handle both the bending and axial forces. Techniques for doing this are described in Chapter 6, which discusses members in bending. At this point, the aspects of primary importance are that designing a rigid member to carry bending is usually quite possible and that the size of the member is very sensitive to the amount of bending present. The higher the bending, the larger must be the member used. If bending is very large, the design is not feasible. The primary design issue faced, therefore, is to determine an arch shape in which a minimum of bending is present under any possible loading condition. Some bending, however, must invariably be present, since a single shape cannot be funicular for multiple loading conditions. Strictly speaking, therefore, it is not quite correct to refer to arches of this type as funicular structures since they are funicular for only one loading.

Consider the three-hinged arch shown in Figure 5-20(a) (which is actually shaped in response to an inwardly directed radial load) under the action of a concentrated load. Since the structure has three hinges, it can be analyzed as discussed in Section 5.3.1. The results of this analysis are shown in Figure 5-20(b). By looking at a free-body diagram of one of the two segments of the structure, it is possible to see that the base reactions and crown connection forces can be envisioned as external forces acting on the member. In this specific case the forces are collinear. As illustrated in Figure 5-20(c), this set of forces tends to cause bending to develop along the length of the member. The greatest bending would occur at the point labeled M. This is the location where, if the member were not designed to withstand the bending, cracks would first develop and the member would initially fail. The greater the magnitude and moment arm of the external forces about a point on the structure, the greater is the bending developed at that point, and vice versa.

Depending on the loads involved, the structure could be designed to withstand the bending developed. From a design point of view, however, it is preferable to attempt finding a shape that reduces or eliminates this bending altogether. For this particular example, finding such a shape is easy. If a linear element were simply placed in line with the line of action of the collinear applied forces, the member would be subjected only to axial forces. No bending would be present (the moment arm of the applied forces is zero). For the loading indicated, it is therefore structurally preferable to use the shape indicated in Figure 5-20(d) than the one in Figure 5-20(a). Member sizes required would be considerably smaller. Functional requirements may dictate the use of the original shape, but this is not a point of concern in this discussion.

It should be noted that the shape found above that carries the load without bending is nothing more than the funicular shape for the loading as envisioned by imagining the deflected shape of a cable and inverting it. It should also be noted that the amount of bending developed at a point on the original structure is directly proportional to the deviation of that point from the funicular line [see Figure 5-20(e)]. This is a generalizable result of usefulness in a design context. The reader should also refer back to the discussion in Section 5.3.1, in which essentially the same approach was taken to formulate the shape of the structure so that it carries load by axial compression.

Although it is not possible to devise a single shape that is funicular for *all* multiple loadings, the effects of bending can be minimized by careful design. Consider the situation shown in Figure 5-21, in which it is desired to design an arch structure having three hinges located in space as shown to carry any of the following loading conditions: (1) dead load only, (2) full dead load plus full live load, and (3) full dead load plus a live load over either the left or right segment of the structure. Figure 5-21(b)–(d) shows the funicular shapes for the different loading conditions. The middle hinge defines the height of the structure at midspan, since the funicular line must pass through the hinge (the hinge is a point at which the rotational and bending moments of the forces about that point are zero—the condition defining any point on a funicular curve). Superimposing the funicular shapes obtained yields the result shown in Figure 5-21(e). If the effects of the off-balanced live loads are small, the curves will be very close together. When the curves are close together, a structure can be sized so that the whole family of curves is contained within its cross section.

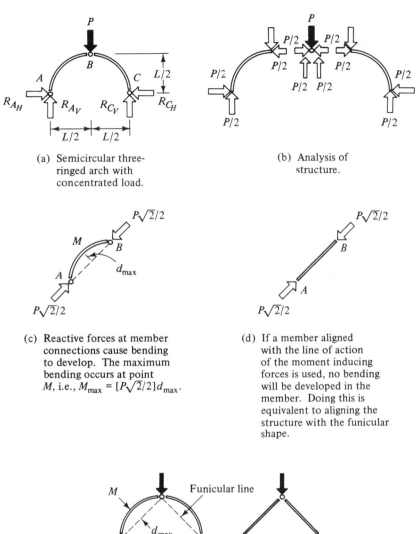

(a) Semicircular three-ringed arch with concentrated load.

(b) Analysis of structure.

(c) Reactive forces at member connections cause bending to develop. The maximum bending occurs at point M, i.e., $M_{max} = [P\sqrt{2}/2]d_{max}$.

(d) If a member aligned with the line of action of the moment inducing forces is used, no bending will be developed in the member. Doing this is equivalent to aligning the structure with the funicular shape.

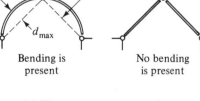

Bending is present

No bending is present

(e) The moment present in the structure is proportional to the deviation of the structure from the funicular or pressure line.

FIGURE 5-20 Bending is developed in arch structures not funicularly shaped for the applied load.

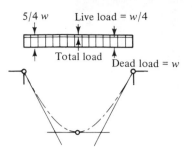

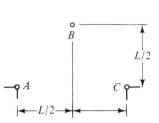

(a) Assume that the location of three hinges is given. A structure is to be designed to carry a full dead and live load and a full dead load and partial live load.

(b) Funicular shape for full-loading conditions.

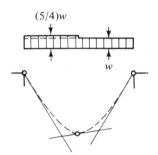

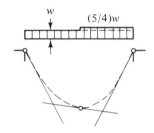

(c) Funicular shape for full dead load and partial live load (left segment carries live load).

(d) Funicular shape for full dead load and partial live load (right segment carries live load).

(e) Superimposed funicular shapes.

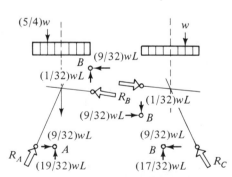

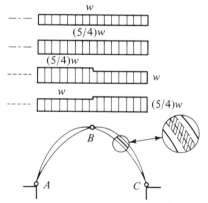

(f) A more rigorous analysis can be used to identify more accurately critical slopes for the structure. Analyses other than that shown are also needed.

(g) If a structure is designed to contain the full spread of funicular lines for all possible load variations, then blending is minimized. If the lines are contained within the middle third of the cross-section, then only compressive stresses are developed in the structure.

FIGURE 5-21 Designing to minimize the effects of load variations. The famous Salginatobel bridge designed by Robert Maillart and constructed between 1929 and 1930 reflects the same general shaping shown in (g) but is more sophisticated in structural concept. Several steel railroad bridges, such as one in Roxbury, Massachusetts, which were conceived by anonymous designers and built around the turn of the century, also had the same general shape.

If this is done, bending will be minimal in the structure under any of the possible load variations. If the funicular lines can all be contained within the middle of the cross section of the member used, the structure will still carry loads primarily by axial compression without any tension stresses being developed (see Section 7.2.1 for a discussion of why forces acting within the middle third of a cross section produce compressive stresses only). Figure 5-21(f) illustrates a more quantitative analysis of this same structure. A resultant structural form is shown in Figure 5-21(g).

SUPPORT ELEMENTS. As with cables, a basic issue in arch design is whether or not to absorb the horizontal thrusts by some interior element, in this case a tie-rod, or by the foundations. When it is functionally possible to do so, tie-rods are very frequently used (far more so than their counterpart struts in cable structures). Since the tie-rod is a tension element, it is a highly efficient way to take up the outward arch thrusts. The foundation then need only be designed to provide a vertical reaction and thus can be considerably less complex or substantial than one designed for both horizontal and vertical forces.

With regard to using vertical elements beneath arches, many of the same observations made in connection with designing support elements for cables are appropriate. Usually, there is less need, however, to support an arch on top of vertical elements because of the amount of head room naturally provided by the arch shape. When it is possible to place an arch directly on the ground, it should be done. Tie-rods that can be buried in the ground can then be used, which facilitates the foundation design. If arches must be used on top of vertical elements, the use of buttressing elements is generally preferable to attempting to design the verticals to act as vertical beams which carry the horizontal thrusts by bending.

The rise of most arches is made fairly deep, since there is no need to reduce the structural depth as was the case in cable design due to the natural headroom provided by the arch shape.

CHOICE OF END CONDITIONS. In the design of rigid arch structures the choice of member end conditions is important. There are three primary types of arches used that are normally described in terms of their end conditions: the three-hinged arch, the two-hinged arch, and the fixed-ended arch (see Figure 5-22). Most of the attention thus far has been placed on the three-hinged arch, since this structure is statically determinate. Reactions, connection faces, internal shears, and moments can consequently all be found by direct application of the basic equations of statics. The two-hinged arch and fixed-ended arch are more complicated to analyze in a rigorous way and are not treated in detail in this text.

The structural behavior of the three rigid arches that are comparable in every way (and carrying identical loads) except in the type of end condition used is not appreciably different *when* each is shaped as a funicular response to the applied loading. Internal compressive forces are similar. Significant differences arise, however, when other factors are considered. Important factors include the effects of possible support settlements, the effects of member expansion and contraction due to changing temperatures, and the relative amount of deflection induced by an applied load. Figure 5-22 illustrates how the different arches behave with respect to these factors.

In general, it is evident that different end conditions are preferable with respect

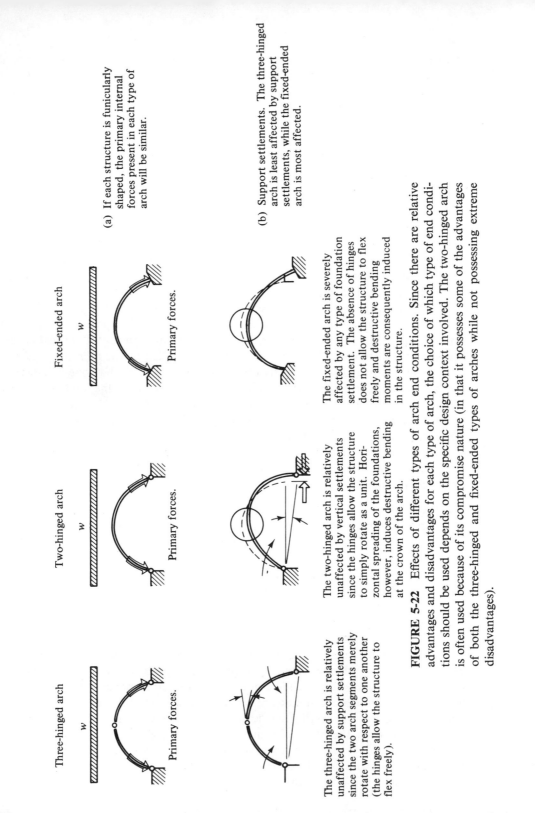

Three-hinged arch

Primary forces.

The three-hinged arch is relatively unaffected by support settlements since the two arch segments merely rotate with respect to one another (the hinges allow the structure to flex freely).

Two-hinged arch

Primary forces.

The two-hinged arch is relatively unaffected by vertical settlements since the hinges allow the structure to simply rotate as a unit. Horizontal spreading of the foundations, however, induces destructive bending at the crown of the arch.

Fixed-ended arch

Primary forces.

The fixed-ended arch is severely affected by any type of foundation settlement. The absence of hinges does not allow the structure to flex freely and destructive bending moments are consequently induced in the structure.

(a) If each structure is funicularly shaped, the primary internal forces present in each type of arch will be similar.

(b) Support settlements. The three-hinged arch is least affected by support settlements, while the fixed-ended arch is most affected.

FIGURE 5-22 Effects of different types of arch end conditions. Since there are relative advantages and disadvantages for each type of arch, the choice of which type of end conditions should be used depends on the specific design context involved. The two-hinged arch is often used because of its compromise nature (in that it possesses some of the advantages of both the three-hinged and fixed-ended types of arches while not possessing extreme disadvantages).

214

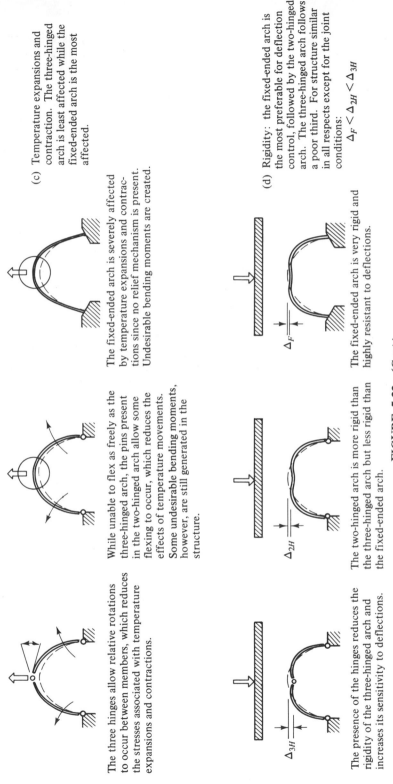

(c) Temperature expansions and contraction. The three-hinged arch is least affected while the fixed-ended arch is the most affected.

The three hinges allow relative rotations to occur between members, which reduces the stresses associated with temperature expansions and contractions.

While unable to flex as freely as the three-hinged arch, the pins present in the two-hinged arch allow some flexing to occur, which reduces the effects of temperature movements. Some undesirable bending moments, however, are still generated in the structure.

The fixed-ended arch is severely affected by temperature expansions and contractions since no relief mechanism is present. Undesirable bending moments are created.

(d) Rigidity: the fixed-ended arch is the most preferable for deflection control, followed by the two-hinged arch. The three-hinged arch follows a poor third. For structure similar in all respects except for the joint conditions:

$$\Delta_F < \Delta_{2H} < \Delta_{3H}$$

The presence of the hinges reduces the rigidity of the three-hinged arch and increases its sensitivity to deflections.

The two-hinged arch is more rigid than the three-hinged arch but less rigid than the fixed-ended arch.

The fixed-ended arch is very rigid and highly resistant to deflections.

FIGURE 5-22 (Cont.)

to different phenomena. The presence of hinges is very useful when supports settlements and thermal effects are considered, since they allow the structure to flex freely. Undesirable bending moments can be generated by similar phenomena in fixed-ended arches. The fixed-ended arch is far less likely to deflect excessively, however, under load than is the three-hinged or two-hinged arch and is thus preferable in this respect. Three-hinged arches tend to be flexible under load, to the point of being potentially problematical.

The choice of which end conditions should be used depends on the exact design conditions present and if one or the other is of dominant importance. The two-hinged arch is frequently used because it combines some of the advantages of the other two types of arches while not comparably sharing their disadvantages.

LATERAL BEHAVIOR OF ARCHES.   A primary design consideration is how to cope with the behavior of an arch in the *lateral* direction. It is obvious that a typical arch, which lies in one vertical plane, must be prevented from simply toppling over sideways. There are two mechanisms commonly used to prevent this. One is to use fixed-base connections. As was previously mentioned, however, there are certain disadvantages to using fixed connections which have to do more with the in-plane rather than out-of-plane behavior of an arch. For very large structures, the use of fixed-base connections for lateral stability also requires massive foundations to prevent overturning. Another more commonly used method for achieving lateral stability is by relying on members that are transversely placed to the arch. The approach diagrammed in Figure 5-23(c) is fairly commonly used. A pair of arches at either end of a com-

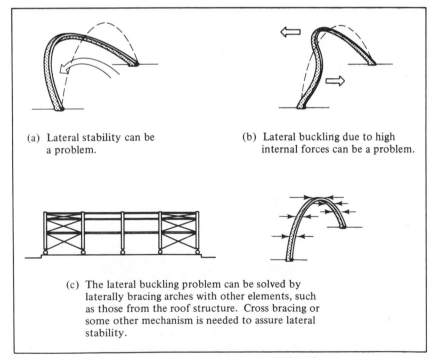

(a) Lateral stability can be a problem.

(b) Lateral buckling due to high internal forces can be a problem.

(c) The lateral buckling problem can be solved by laterally bracing arches with other elements, such as those from the roof structure. Cross bracing or some other mechanism is needed to assure lateral stability.

FIGURE 5-23 Arch design considerations.

plete structure are stabilized through the use of diagonal elements. Interior arches are stabilized by being connected to the end arches by connecting transverse members.

The second major problem with respect to the behavior of trusses in the lateral direction is that of lateral buckling. Since internal forces are often fairly low, arches designed with high-strength materials (e.g., steel) can be fairly slender elements. An out-of-plane buckling of the type illustrated in Figure 5-23(b) can consequently potentially occur. As with trusses, one solution to the lateral buckling problem is to increase the stiffness of the arch in the lateral direction by increasing its lateral dimensions. Another solution is to laterally brace the arch periodically along its length with transverse members (Figure 5-24). Fairly slender arches can then be used. It should be noted that the same system used to stabilize arches from overturning laterally also provides lateral bracing for arch members and prevents lateral buckling from occurring.

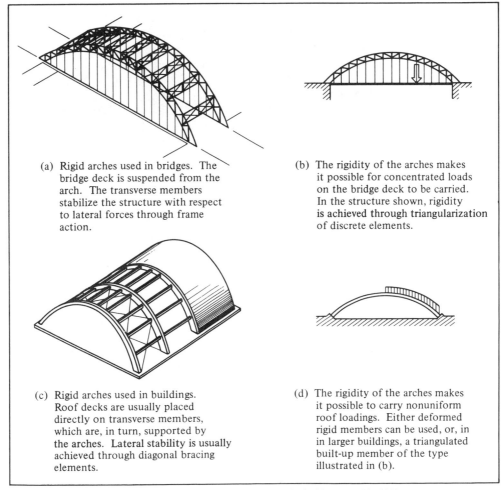

(a) Rigid arches used in bridges. The bridge deck is suspended from the arch. The transverse members stabilize the structure with respect to lateral forces through frame action.

(b) The rigidity of the arches makes it possible for concentrated loads on the bridge deck to be carried. In the structure shown, rigidity is achieved through triangularization of discrete elements.

(c) Rigid arches used in buildings. Roof decks are usually placed directly on transverse members, which are, in turn, supported by the arches. Lateral stability is usually achieved through diagonal bracing elements.

(d) The rigidity of the arches makes it possible to carry nonuniform roof loadings. Either deformed rigid members can be used, or, in larger buildings, a triangulated built-up member of the type illustrated in (b).

FIGURE 5-24 Use of rigid arches in bridges and buildings.

chapter 5 / FUNICULAR STRUCTURES: CABLES AND ARCHES   **217**

QUESTIONS

5-1. What is the maximum force developed in a cable that carries a single concentrated load of P at midspan and which spans a total distance of L? Assume that the cable has a maximum sag of $h = L/5$.

5-2. What is the maximum force developed in a cable carrying a uniform load of 500 lb/ft which spans 200 ft? Assume that the cable has a maximum midspan sag of 20 ft. Answer: 135,000 lb.

5-3. What is the maximum force developed in a funicularly shaped arch that carries a uniformly distributed load of 800 lb/ft and spans 120 ft? Where does this force occur?

5-4. What is the maximum force developed in a funicularly shaped arch that carries a uniformly distributed load of 10,000 N/m and spans 35 m? Assume that the rise of the arch is 8 m. What is the force in the arch at midspan? Answer: 191.4 kN.

5-5. Determine the exact shape of a cable that spans 100 ft, has a maximum sag of 10 ft, and supports three concentrated loads of 5000 lb apiece which are located at quarter-points along the span (i.e., at 25, 50, and 75 ft from the left support point). Determine the force present in each segment of the cable.

5-6. Draw shear and moment diagrams for the loading condition described in Question 5-5 and demonstrate how the cable structure supporting the loadings develops internal resisting shear forces and moments that equilibrate the externally applied shear forces and bending moments.

5-7. How would the maximum force developed in the cable described in Question 5-1 be affected if the left end support were elevated a distance of $h = L/10$ above the right support? Again assume that the cable sag is a maximum of $L/5$, but assume that this distance is measured with respect to a line connecting the two end supports.

5-8. Determine the funicular shape for a structure that carries a uniform load of w from $x = 0$ to $x = L/3$ and is unloaded elsewhere. First sketch the probable shape of the structure and then determine appropriate algebraic expressions that precisely define the shape.

5-9. Determine the funicular shape for a structure that carries a uniform load of w across its entire span and a concentrated load of P at midspan. Sketch the probable shape of the structure first and then determine appropriate algebraic expressions that precisely define the shape.

5-10. Is there any maximum to the distance that a cable can span? Discuss.

5-11. If your answer to Question 5-10 is essentially affirmative, find what the maximum span would be. Make *any* assumptions you need to but clearly specify them.

chapter 6

MEMBERS IN BENDING: SIMPLE BEAMS

6.1 INTRODUCTION

Few structural elements are as ubiquitous as the common beam. Beams have been used as devices for transferring vertical loads horizontally since buildings were first constructed. Post-and-beam systems in which a horizontal member is simply rested on two vertical supports have formed the basic construction approach for much of the architecture of both early and recent civilizations. This is at least partly attributable to the convenience and simplicity of the beam as an element of construction. This simplicity with respect to construction, however, belies the fact that internally beams have a far more complex load-carrying action than do many other structural members (e.g., trusses or cables). Indeed, little was known about the exact mechanisms by which beams carry loads until very recently, relative to the length of time such elements have been used. A sketch by Leonardo da Vinci indicates that he tried to describe beam action in a rational way.

Another product of the Renaissance, however, is more responsible not only for laying the groundwork for our current understanding of beam theory, but also for founding, in Hans Straub's words, ". . . an entirely new branch of science: the theory of the strength of materials." The man is Galileo Galilei. Galilei was the first to begin investigating the bending problem in a systematic way (see Figure 6-1). The term *bending problem* refers to the study of stresses and deformations generated in an ele-

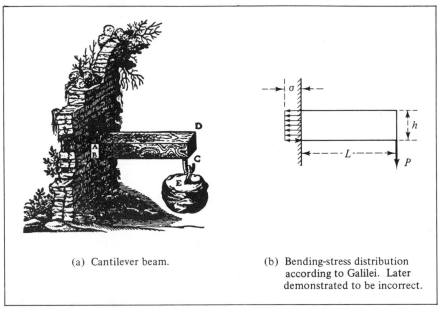

(a) Cantilever beam.

(b) Bending-stress distribution according to Galilei. Later demonstrated to be incorrect.

FIGURE 6-1 Early beam theory. (From Galileo Galilei, *Discorsi e dimostrazioni matematiche*, 1638.)

ment bowed by the action of forces (typically perpendicular to the axis of the element) so that the fibers on one face of the element are elongated and those on the opposite face contracted.

Although Galilei was not completely successful in solving the problem, his initial formulation laid the foundation for future investigators. The names of Hooke, Mariotte, Parent, Leibnitz, Navier, and Coulomb are all linked with this problem. Auguste Coulomb (1736–1806) is generally credited with the final solution of the bending problem. Investigators in the rapidly expanding field of *strength of materials* soon tackled other problems commonly associated with beams, such as the effects of torsion, until by now the field is amazingly well developed. As will be seen, many of the concepts historically developed in connection with the study of beams (such as the theory of bending) actually are applicable to other structural elements. This chapter, however, will focus exclusively on beams.

6.2 GENERAL PRINCIPLES

6.2.1 Beams in Buildings

Few buildings are constructed that do not make use of beams. In addition to the more evident beam assemblies that often serve as primary structural systems, many other common building components (such as parts of some stairways or large window mullions) are, from a structural viewpoint, beams.

When used to form the primary structural system in a building, beam elements are most typically used in a repetitive pattern forming way. Most often the pattern is comprised of a hierarchical arrangement of beams. Planar surface load-transfer members (e.g., decking or planks) usually have limited span capabilities and are therefore typically supported periodically by larger-span secondary members to form a two-level system. These members are, in turn, sometimes supported by collector beams to form a three-level system. Loads acting on the surface are first picked up by the surface members, transferred to the secondary members, which in turn transfer them to the collectors or supports. The amount of load carried by each member thus progressively increases. This increase in loading, coupled with an increase in length, typically leads to a progressive increase in member size or depth (Figure 6-2).

There can obviously be any number of levels in such a hierarchical arrangement. A three-level arrangement is typically the maximum used. One- and two-level arrangements are also quite common. For spanning any given space, it is obvious that, in general, a system of any level can be used. Member sizes for a given system could be determined on the basis of an analysis of the spans, loads, and materials involved. Simply because any arrangement could be made to be serviceable, however, does not imply that all possible arrangements are equally desirable based on considerations of efficiency, construction ease, costs, or other criteria. Making an informed decision on the basis of the criteria selected is the task of the structural designer. When a criterion is measurable (e.g., material economy), the selection procedure quite often consists of analyzing a postulated assembly, determining required member sizes and characteristics through a member design procedure, measuring the whole assembly with respect to the criterion involved, and comparing this measure with those obtained from similar evaluations of other structural assemblies serving the same function.

The process of sizing a specific beam of known span is relatively straightforward and consists of determining the loads acting on the member and then finding a set of related material properties, cross-sectional geometries, and dimensions such that the resulting member is serviceable under the specified loading conditions.

The actual stresses developed in a beam depend on the amount and distribution of material in the cross section of the member. Basically, the larger the beam, the smaller the stresses. The way this material is organized in space, however, is important (Figure 6-3). A thin rectangular member that is placed on its side cannot carry anywhere near the same amount of load as an equivalently sized member placed so that its maximum depth is in line with the applied loads. It is thus not just the area of the member that is important, but its distribution as well. Generally, the deeper a member is, the stronger it is vis-à-vis bending. A measure termed the *moment of inertia* (I) can be used to characterize the strength of a member in bending, as it depends on the amount of material present and the shape of the cross section used (see Section 6.3).

Important basic design variables include the magnitudes of the loads present, the distances between the loads, and the nature of the support conditions of the beam. Support conditions are particularly important. A member that has its ends restrained is much stiffer than one whose ends are free to rotate. A beam with fixed ends, for example, can carry a concentrated load at midspan twice that of a similarly sized beam with unrestrained ends. A beam that is *continuous* over several supports also offers some advantages as compared to a series of simply supported ones, although there are some trade-offs involved (see Chapter 8).

chapter 6 / MEMBERS IN BENDING: SIMPLE BEAMS   **221**

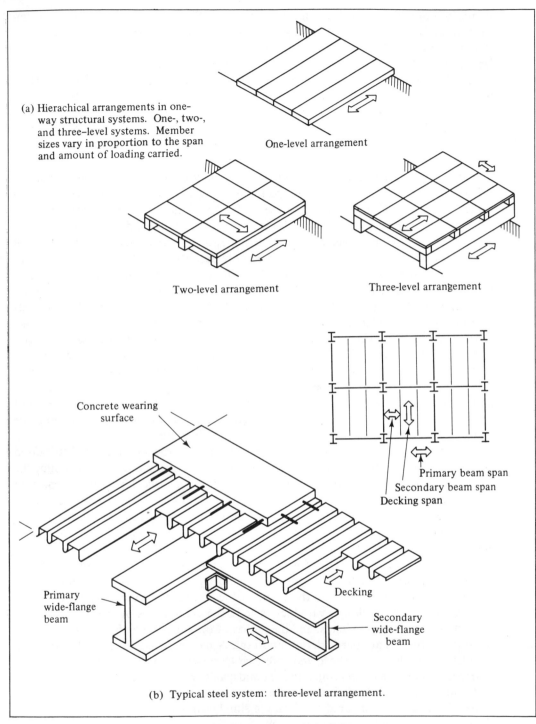

(a) Hierachical arrangements in one-way structural systems. One-, two-, and three-level systems. Member sizes vary in proportion to the span and amount of loading carried.

One-level arrangement

Two-level arrangement

Three-level arrangement

Concrete wearing surface

Primary beam span
Secondary beam span
Decking span

Primary wide-flange beam

Decking

Secondary wide-flange beam

(b) Typical steel system: three-level arrangement.

FIGURE 6-2 Beams in buildings.

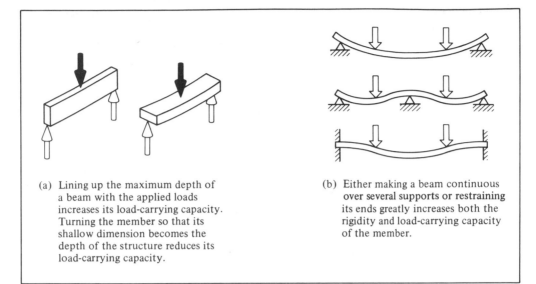

(a) Lining up the maximum depth of a beam with the applied loads increases its load-carrying capacity. Turning the member so that its shallow dimension becomes the depth of the structure reduces its load-carrying capacity.

(b) Either making a beam continuous over several supports or restraining its ends greatly increases both the rigidity and load-carrying capacity of the member.

FIGURE 6-3   Important factors in beam design.

6.2.2   Stresses and Deformations at a Cross Section

Consider the member illustrated in Figure 6-4. As is evident, at any one cross section of the beam, the action of the applied load is to produce a deformation pattern where some fibers of the beam are being elongated while others are shortened. In the beam shown, fibers in the upper portion of the beam are shortened and those in the lower portion elongated. When one looks closely at the geometry of the deformation of the beam at one cross section [see Figure 6-4(b)], it can be seen that the deformations vary in a linear or near linear way from a maximum elongation on one face to a maximum shortening at the other. There must exist a layer somewhere near the middle of the beam where the beam fibers are neither shortened nor elongated. This layer is of particular importance in the study of beams and is typically called the *neutral axis* of a beam. For a symmetrical cross section, as in a rectangular beam, it can be correctly expected that this plane is at the midheight of the beam. This need not be so if the section is not symmetrical. It will be shown later that the neutral axis corresponds with the *centroid* (see Appendix 2) of the cross section.

One of the primary reasons for studying the deformations in a beam is the fact that the magnitudes of the stresses induced in the beam by applied loads are related to the magnitudes of the deformations. If the beam is constructed of a material that is *linearly elastic*, the stresses produced by bending are directly proportional to the deformations present. Thus, in the beam under consideration, the stresses induced by the load are maximum at the outer fibers of the beam and decrease linearly to zero at the neutral axis. Tensile stresses are associated with elongations in beam fibers, and compressive stresses with contractions. Together these stresses are typically referred to as *bending stresses*. By looking at Figure 6-4, one can also see that the deformations present at a section, hence the bending stresses, are also related to the *curvature* $(1/R)$ of the beam at that point. As the radius (R) describing the curvature of the beam

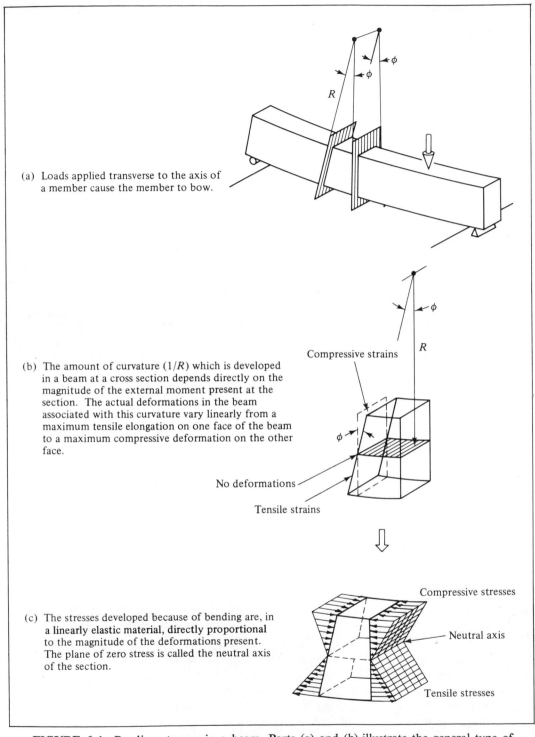

(a) Loads applied transverse to the axis of a member cause the member to bow.

(b) The amount of curvature $(1/R)$ which is developed in a beam at a cross section depends directly on the magnitude of the external moment present at the section. The actual deformations in the beam associated with this curvature vary linearly from a maximum tensile elongation on one face of the beam to a maximum compressive deformation on the other face.

Compressive strains

No deformations

Tensile strains

(c) The stresses developed because of bending are, in a linearly elastic material, directly proportional to the magnitude of the deformations present. The plane of zero stress is called the neutral axis of the section.

Compressive stresses

Neutral axis

Tensile stresses

FIGURE 6-4 Bending stresses in a beam. Parts (a) and (b) illustrate the general type of deformations produced by the external loading. Part (b) illustrates the deformations present at a cross section and (c) illustrates the bending-stress distribution at the same cross section.

(a) Loaded beam.

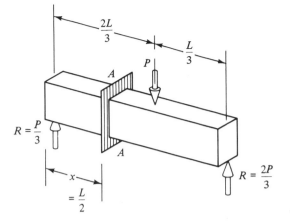

(b) The net effect of the external force system acting on a portion of the beam is to produce an upward shear force, V_E, and a rotational moment, M_E. An equilibrating set of shears and moments, V_R and M_R are developed in the beam.

$$\Sigma F_y \qquad V_E = P/3$$
$$\Sigma M_x \qquad M_E = (P/3)x$$
$$\text{at } x = L/2$$
$$M_E = PL/6$$

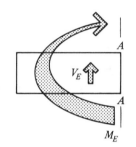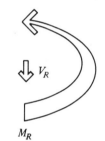

(c) Vertical shear stresses developed at the section and acting over the face of the section provide the internal resisting shear, V_R.

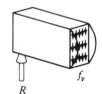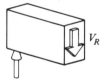

(d) Bending stresses developed at the section act over the face of the section to produce a couple which provides the internal resisting moment, M_R.

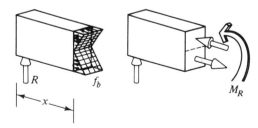

FIGURE 6-5 Basic load-carrying mechanisms in a beam.

becomes small, the curvature $(1/R)$ of the beam becomes greater, and greater deformations and stresses occur in the beam. When the radius (R) is large (i.e., the beam is flatter), the curvature is small and so are deformations and stresses.

Since bending stresses are maximum at the outer fibers of a beam, it follows that in designing beams it is generally best to concentrate material at these locations. As will be shown later in this chapter, deep beams with as much material as possible located away from the neutral axis are generally more efficient than other types of beams.

6.2.3 Basic Load-Carrying Mechanisms

The actual magnitudes of the deformations, bending stresses, and curvatures present at a section are dependent on the amount of bending produced at the section by the action of the external forces acting on the beam. The bending present can be found through studying the equilibrium of both the whole structure and elemental portions of it. Consider the illustration shown in Figure 6-5. With respect to the elemental portion of the structure to the left of section $A–A$, it is evident that the total effect of the set of external forces acting on this portion of the structure is to produce a net rotational moment of $PL/6$ and a net upward force of 0.33P. As already discussed in Section 2.4.3, these forces are referred to as the *external bending moment* and *shear force*, respectively, acting at the section. For equilibrium to obtain for the portion of the structure shown, a set of internal forces must be developed in the structure whose net effect is to produce a rotational moment equal in magnitude but opposite in sense to the external bending moment and a vertical force equal and opposite to the external shear force. These internal resisting shears and moments are diagrammed in Figure 6-5(b).

In trusses it was seen that these shears and moments are provided by forces in individual members (see Section 4.3). The internal resisting moment was provided by a couple ($\leftrightharpoons$) produced by member forces separated by a known distance. In beams, the internal moment resistance is still provided by a couple, but one whose forces are produced by tensile and compressive stress fields acting over portions of the face of the cross section. The effect of compressive stresses in the upper zone acting over the area can be equivalently described by a single resultant compressive force acting at a particular location. The same is true for the tensile stresses. The internal resisting couple, therefore, can be thought of as being provided by these two resultant forces separated by the moment arm indicated [see Figure 6-5(d)]. Unlike trusses, the moment arm is not a distance known a priori, since it depends on the locations of the resultant tension and compression forces, which in turn depend upon the nature of the way stresses are distributed and on the shape of the cross section on which they act. Techniques for finding what this moment arm should be are discussed later in this chapter.

The bending stresses just discussed are developed in response to the applied external moment and, in the examples discussed, act horizontally and consequently do nothing to balance the external shear forces. The external shear force is balanced by *shear stresses* which act on the face of the section [see Figure 6-5(c)]. The question of how these stresses are distributed is complex and will be covered later in this chapter, where it will be demonstrated that they are maximum at the neutral axis of the cross section and decrease nonlinearly toward the outer faces. The important point, however, is that the resultant effect of these stresses acting over the face of the cross section is to produce a resultant vertical shear force which equilibrates the applied external shear force.

Shear stresses of another type are also developed in beams. These stresses act horizontally rather than vertically. Their presence is most easily visualized by considering two beams, one made of a series of unconnected planes (similar to a deck of cards) and the other of a solid piece of material (see Figure 6-6). Under the action of a load, a slippage occurs between each of the planes in the former case as the whole assembly deforms. The structure is essentially behaving as a series of thin superimposed beams. The second structure behaves in a composite way and is much stiffer than the first and able to carry a much higher load. If the surfaces of the planes on the first structure were bonded together, the resulting structure would also behave as a composite whole. If this were done, stresses acting in the bond parallel to the adjacent surfaces would be developed, since the planes would still have a tendency to slide. These stresses are called *horizontal shearing stresses*. As long as the material can carry these stresses, adjacent planes in a beam remain in contact and the structure behaves as a composite whole. When a beam fails due to horizontal shear, slippage between planes occurs. Some materials, such as timber, are particularly weak with respect to stresses of this kind, and failures of this type are not uncommon [see Figure 6-6(d)]. It is interesting to note that the numerical magnitude of the horizontal shear stress at a point is exactly equal to the shear stress acting in the vertical direction at that same point. This phenomenon will be discussed more later in this chapter.

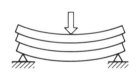 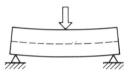 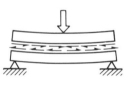

(a) Beam made of unbonded layers. Layers tend to slide horizontally with respect to one another under the action of a load.

(b) In a homogeneous beam there is a tendency for the same type of sliding to occur.

(c) Horizontal shear stresses are developed in homogeneous beam as a consequence of this tendency to slide.

 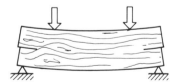

(d) Cracks developed in a timber beam as a result of excessively high horizontal shear stresses.

FIGURE 6-6 Horizontal shear stresses in a beam.

In addition to bending stresses and horizontal and vertical shear stresses, other considerations of importance in beam analysis include bearing stresses, torsional stresses, combined stresses, the shear center of a beam, principal stresses, and the deflection characteristics of a beam. These topics will be discussed later in this chapter.

6.2.4 Distribution of Forces in a Beam

The previous sections considered the nature of the stresses developed at a specific cross section in a beam. Of equivalent importance is knowing how these stresses vary along the length of the beam. Since the stresses at a cross section depend on the magnitude and sense of the external shears and moments at that cross section, the distribution of stresses along the length of the beam can be found by studying the distribution of shears and moments in a beam.

The basic principles for finding the distribution of shears and moments in a member have already been covered in Section 2.4.3. There it was noted that when the equilibrium of an elementary portion of a member is considered, the effect of the set of external forces (including reactions) is to produce a net rotational effect (the external moment, M_E) about the cut section considered and a net vertical translatory force (the external shear force, V_E) and that these quantities are balanced by an equilibrating set of shears and moments (V_R and M_R) developed internally in the

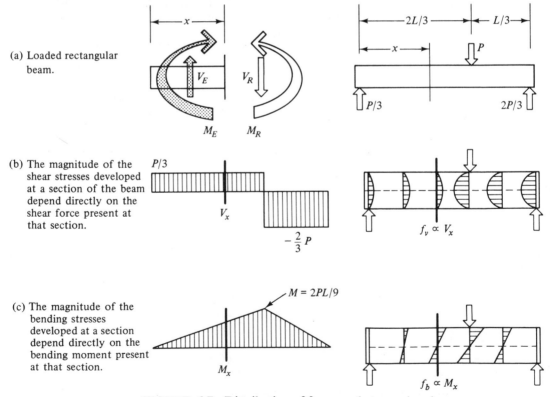

(a) Loaded rectangular beam.

(b) The magnitude of the shear stresses developed at a section of the beam depend directly on the shear force present at that section.

(c) The magnitude of the bending stresses developed at a section depend directly on the bending moment present at that section.

FIGURE 6-7 Distribution of forces and stresses in a beam.

structure (i.e., $V_R = V_E$ and $M_R = M_E$). Clearly, the magnitudes of these shears and moments depend on the extent of the elemental section considered and the set of forces acting on it. Thus, the distribution of these shears and moments can be found by considering in turn the equilibrium of different elemental portions of the structure and calculating the shear and moment sets for each elemental portion. As an aid in visualizing the distribution of these shears and moments, the values thus found can be plotted graphically to produce *shear and moment diagrams*. These diagrams were discussed previously in Section 2.4.3.

Figure 6-7 shows the general relationship between the shear and moment distribution in a structure and the shear and bending stresses developed. Figure 6-8 shows the relationship between the moment distribution present and the bending stresses developed in a beam having an inflection point.

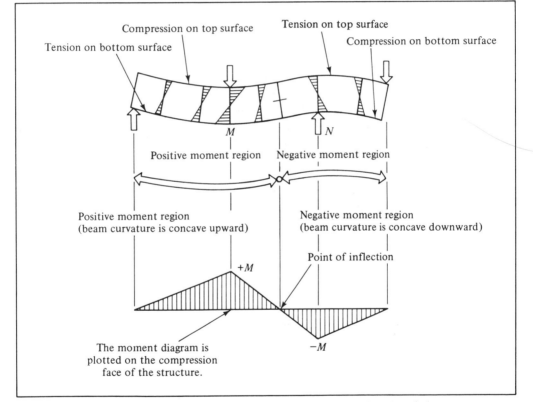

FIGURE 6-8 Positive and negative moment regions in a beam.

6.3 ANALYSIS OF BEAMS

This section briefly reviews topics normally addressed in detail in textbooks devoted to a study of the strength of materials. The following topics, which generally comprise the basic issues involved in the analysis of beams made of linearly elastic materials, are addressed: (1) bending stresses, (2) shearing stresses, (3) bearing stresses, (4) combined stresses, (5) torsional stresses, (6) shear centers, (7) principal stresses, and (8) deflections.

The objective of the coverage presented on this topic is to clarify the basic relationships between variables. On exactly what factors, for example, do the magnitude of the bending stresses at a point in a beam depend and how are these factors related? No attempt is made in this discussion to derive rigorously all the relationships discussed. Only a few of the more important relationships are developed in the Appendices. This is done primarily to illustrate the general spirit of the approach typically taken to derive other relationships. The reader is referred to any one of several traditional textbooks devoted exclusively to a study of the strength of materials, where all these topics are covered in great depth.[1]

6.3.1 Bending Stresses

INTRODUCTION. As noted in the discussion in Section 6.2 on the general behavior of beams, bending stresses are developed in a beam in response to the action of the external bending moment existing in the beam at that point.

Before looking at how to determine these bending stresses in a quantitative way, it is useful to review the general factors on which the bending stresses in a beam depend. Consider the beam illustrated in Figure 6-9. The stress distribution depicted is based on the assumption that stresses are linearly dependent on deformations, in other words, that the material used is linearly elastic. The magnitude of the bending stresses (f_y) present at a point is directly dependent on the magnitude of the external moment (M) present at the section. The magnitude of (f_y) must also directly depend on the distance (y), which defines the location of the point considered with respect to the neutral axis of the beam. It is also quite reasonable to expect that the stress (f_y) will be inversely dependent on some measure defining the size and shape of the beam itself. Increasing the size of the beam should decrease the stress level in the beam for a given moment. Let us temporarily simply designate this measure as I (typically called the *moment of inertia*) without as yet defining it further. There are no other parameters on which the bending stress can be dependent. The actual stresses developed, for example, are *not* influenced by the type of material used, in the same way that the forces in a truss member were not dependent on the material of which the member was made. Ascertaining whether a member is safe under a given loading does require consideration of the strength characteristics of the material, but this is a different problem than the one addressed here.

Since the bending stress (f_y) directly depends on the moment (M) and the locational parameter (y), and inversely the measure of the properties of the beam section (I), the variables generally relate in the following way:

$$f_y \Longrightarrow \left(M, y, \frac{1}{I} \right)$$

[1]See, for example, Ferdinand L. Singer, *Strength of Materials*, 2nd ed., Harper & Row, Publishers, New York, 1962.

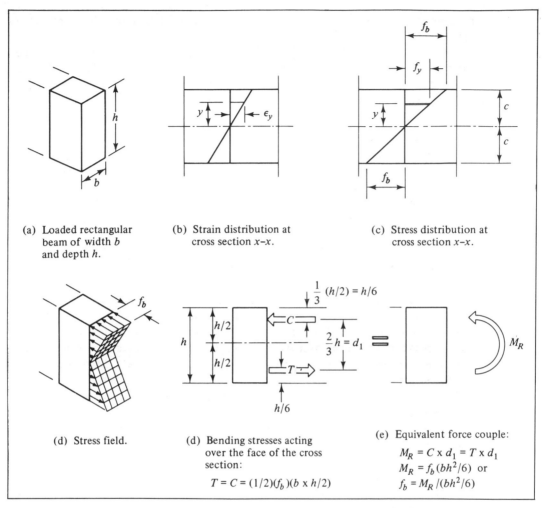

(a) Loaded rectangular beam of width b and depth h.

(b) Strain distribution at cross section x–x.

(c) Stress distribution at cross section x–x.

(d) Stress field.

(d) Bending stresses acting over the face of the cross section:

$$T = C = (1/2)(f_b)(b \times h/2)$$

(e) Equivalent force couple:

$$M_R = C \times d_1 = T \times d_1$$
$$M_R = f_b(bh^2/6) \text{ or}$$
$$f_b = M_R/(bh^2/6)$$

FIGURE 6-9 Bending stresses in a rectangular beam.

As

M increases	f_y increases
y increases	f_y increases
I increases	f_y decreases

Thus,

$$f_y \Longrightarrow \frac{My}{I}$$

It is interesting to note that if the units of the stress f_y are lb/in.², the units of M are in.-lb, and the units of y are in., then the units of I (the term characterizing the nature of the cross section) must be the rather unique unit of in.⁴, so that the expression My/I is dimensionally correct [i.e., lb/in.² = (in.-lb)(in.)/in.⁴]. Alternatively, in the SI system of units, the units of the stress f are (N/mm²), or MPa, the units of M are

N-mm, and the units of y are mm. The units of I must therefore be mm⁴. Thus, $N/mm^2 = (N\text{-}mm)(mm)/mm^4$.

There is nothing in the discussion above that will allow us to make any further statement about what the term I actually is. This must be done through equilibrium calculations. The following will illustrate the procedure for finding the bending stresses at any point in a rectangular beam and clarify what the term I actually represents.

RECTANGULAR BEAMS. The tensile and compressive components of the stress field present in a rectangular beam, or any other shaped beam for that matter, generate a force couple ($\leftrightarrows$) that provides an internal resisting moment (M_R) that equilibrates the rotational effect of the external moment (M_E) present at the section. An expression for this internal resisting moment can be found in terms of the unknown bending stresses developed at the section. The unknown internal stresses associated with the internal force couple can then be solved for by setting the expression equal to the known external moment present at a section ($M_R = M_E$). This procedure is demonstrated for the specific case of a rectangular beam in this section and for a general shape in Appendix A-4.

Consider the beam shown in Figure 6-9. The compressive stresses acting over the upper half of the beam are statically equivalent to a single force, C, acting as indicated in Figure 6-9(b). The magnitude of this force is simply the *average* compressive stress times the area over which it acts (i.e., force = average stress × area). Its location is determined by considering the way the stresses are distributed. For a triangular stress distribution, this location is at the third point from the base of the triangle, as illustrated in Figure 6-9(d). (See the discussion on the *centroids* of figures in Appendix 2.) The tension stress field can be similarly described in terms of an equivalent static force, T, acting as illustrated in Figure 6-9(d). The internal resisting moment in the beam is the couple formed by these two forces. By equating this resisting moment to the external moment, the stress at the outer fibers can be explicitly determined:

Equivalent forces:

compressive force = average stress × area:

$$C = \tfrac{1}{2}f_b \times \left(b \times \frac{h}{2}\right) = \frac{f_b bh}{4}$$

tension force = average stress × area:

$$T = \tfrac{1}{2}f_b \times \left(b \times \frac{h}{2}\right) = \frac{f_b bh}{4}$$

moment arm = distance between C and T:

Resultant forces T and C act at the centroids of their respective stress fields (see Figure 6-9). Thus:

$$\text{moment arm} = \left(\frac{2}{3} \times \frac{h}{2}\right) + \left(\frac{2}{3} \times \frac{h}{2}\right) = \frac{2h}{3} = d_1$$

internal resisting moment $= C \times d_1 = T \times d_1$

$$M_R = \left(\frac{f_b h}{4}\right) \times \left(\frac{2h}{3}\right) = f_b\left(\frac{bh^2}{6}\right)$$

The foregoing is an expression for the internal resisting moment at a section (which is equal to the external moment at the same section) in terms of the maximum extreme fiber stress, f_b, and the dimensions of the rectangular beam. Solving for f_b:

$$f_b = \frac{M}{bh^2/6}$$

A general expression for the bending stress, f_y, at a point defined by a distance y from the neutral axis, can also be found. By similar triangles, note that $f_y/y = f_b/c$ (see Figure 6-9). Thus, $f_b = f_y(c/y)$. Noting that $c = h/2$, the expression for f_b becomes

$$f_y \frac{1}{y} \frac{h}{2} = \frac{M}{bh^2/6}$$

or

$$f_y = \frac{My}{bh^3/12}$$

If the quantity $bh^3/12$ is defined as the moment of inertia (I) for a rectangular beam, we have

$$f_y = \frac{My}{I}$$

This is the same general expression discussed before. The value of I found above is valid only for rectangular beams. Note that it has the units of in.^{4} or mm^4.

An examination of the results found above indicate that the bending stresses developed in a beam are extremely sensitive to the depth of the beam. For a given applied moment, doubling the depth of a beam while holding its width constant would reduce bending stresses by a factor of 4. Alternatively, doubling the width of a beam while holding its depth constant would only reduce bending stresses by a factor of 2. These observations follow from manipulations with the expression $f = M/(bh^2/6)$.

EXAMPLE

A cantilever beam that is rectangular in cross section and of length L carries a concentrated load of P at its free end. Determine the maximum bending stress present in the member. Assume that the dimensions of the beam are as follows: $b = 4$ in. (101.6 mm), $h = 6$ in. (152.4 mm), $L = 8$ ft (2.44 m). Assume that $P = 500$ lb (2224 N). It is evident that the maximum bending stress occurs where the moment is a maximum and on the outer fibers of the beam at the same section.

Solution:

Maximum bending moment:

$$PL = 500 \text{ lb} \times 8 \text{ ft} = 4000 \text{ ft-lb} = 48,000 \text{ in.-lb}$$
$$= 2224 \text{ N} \times 2.44 \text{ m} = 5426.6 \text{ N-m} = 5.426 \times 10^6 \text{ N-mm}$$

Maximum bending stress $= f_b$ *at* $y = c$:

$$f_b = \frac{M}{bh^2/6} = \frac{48,000 \text{ in.-lbs}}{4 \text{ in.} \times (6 \text{ in.})^2/6} = 2000 \text{ lb/in.}^2$$

$$= \frac{5.426 \times 10^6 \text{ N-mm}}{101.6 \text{ mm} \times (152.4 \text{ mm})^2/6} = 13.79 \text{ N/mm}^2 = 13.79 \text{ MPa}$$

Alternatively, the maximum bending stress can be found by using the more general expression for bending stresses ($f_y = My/I$):

Moment of intertia:

$$I = \frac{bh^3}{12} \quad \text{(for a rectangular section)}$$

$$= \frac{4 \text{ in.} \times (6 \text{ in.})^3}{12} = 72 \text{ in.}^4$$

$$= \frac{101.6 \text{ mm} \times (152.4 \text{ mm})^3}{12} = 30.0 \times 10^6 \text{ mm}^4$$

Maximum bending stress $= f_b$ *at* $y = c$:

$$f_b = \frac{My}{I} = \frac{Mc}{I}$$

$$= \frac{(48,000 \text{ in.-lb})(6 \text{ in.}/2)}{72 \text{ in.}^4} = 2000 \text{ lb/in.}^2$$

$$= \frac{(5.426 \times 10^6 \text{ N-mm})(152.4 \text{ m}/2)}{30.0 \times 10^6 \text{ mm}^4} = 13.79 \text{ N/mm}^2$$

GENERAL THEORY. The basic method described above for finding the stresses in a beam is of general applicability. When the shape of the cross section is complex, however, finding statically equivalent forces and their locations for the stress fields present is very involved. To simplify calculations, the problem has been solved for in general terms and the results presented in the form $f_y = My/I$. In this formulation, a general expression for I appears which can be evaluated for any different shape of cross section. This general formulation is presented in detail in Appendix 4 to which the reader is referred. In general, the procedure used in deriving this expression is quite similar to that demonstrated for the rectangular beam, except that the shape of the member is not specified and operations are made with respect to an elemental area within the cross section. Thus, the moment resistance in a beam of any cross-sectional shape is generated by stresses acting over areas to produce forces, which in turn act through a moment arm to produce a resisting moment. In the case of an elementary area, dA, the force on the element is directly proportional to y [i.e., force $= f_y \, dA = f_b(y/c) \, dA$] and its moment is the force acting over the distance y [i.e., moment $= yf_y \, dA = (f_b/c)(y^2 \, dA)$]. Thus, the term $y^2 \, dA$ represents the *resistance to bending* associated with the elemental area dA and its location, defined by y, in the beam. The total moment resistance of the beam is the sum of the contributions of all elemental areas, or $M = \int_A (f_b/c)(y^2 \, dA)$. If I is defined as $\int_A y^2 \, dA$, then $M = f_b I/c$ or $f_b = Mc/I$, as before. The term $\int_A y^2\,dA$ represents the total resistance to bending associated with the sum of all elemental areas in the beam. This expression is commonly encountered as the *second moment* of an area in mathematics and is discussed in basic calculus textbooks, where the term is evaluated for different shapes. The expression is usually called the *moment of inertia* in the structural engineering field. Appendix A-3 contains an extensive discussion of this term, including exactly what it means, and illustrates its application to several typical beam shapes. The moment of inertia for a rectangular beam of width b and depth h, for example, is demonstrated to be $I = bh^3/12$ (as found before from equilibrium considerations).

The expression $f_b = Mc/I$ is applicable for finding the stresses at the extreme fiber of the beam, located by the distance c. For the stresses at any point located by the distance y, the expression becomes $f_y = My/I$. The distance y, like the distance c, is measured from the neutral axis of the member. It is thus necessary to know exactly where this axis lies. In Appendix 4 it is demonstrated that the neutral axis usually coincides with the *centroid* of the section of the beam. The centroid of a cross-sectional area, actually defined by $\int_A \bar{y}\,dA = 0$, can be visualized as that point at which the geometric figure defining the area would balance.

The location of the centroid of a figure also determines the location of the plane of zero deformations and bending stresses in the figure. In cross sections that are symmetrical about their horizontal axes, such as a rectangle, square, or circular shape, the centroid (and hence neutral axis) is invariably at the midheight of the section. Deformations and bending stresses vary linearly in the member and are proportional to the distance from the neutral axis of the member.

SYMMETRICAL SECTIONS. Several examples illustrating the application of $f_y = My/I$ are shown below. The first is for a simple rectangular shape. The example illustrates the procedure used in determining if a beam constructed of a given material carrying a certain load is adequately sized with respect to bending. This simply involves calculating the maximum critical bending stress actually developed in the beam (which occurs at the section of maximum moment along the length of the beam and at the fiber of the beam farthest removed from the neutral axis of the beam) and comparing this actual stress to a stress level predetermined as being acceptable for the material used (the *allowable* stress for the material). As discussed in Section 2.5, this allowable stress contains a specified factor of safety. Obviously, if the actual stress present at a point exceeds the allowable stress, the beam is considered *overstressed* at that point and is not considered acceptable.

EXAMPLE

A simply supported beam carries a concentrated load of P at midspan. Assume that the dimensions of the beam are as follows: $b = 5$ in. (127 mm), $h = 10$ in. (254 mm) and $L = 10$ ft (3.048 m). Also assume that $P = 4000$ lb (17.792 kN). If the allowable stress in bending for the specific type of timber used is $F_b = 1500$ lb/in.^{2} (10.34 N/mm^2 or 10.34 MPa), is the beam adequately sized with respect to bending?

Solution:

Maximum bending moment:

$$M = \frac{PL}{4} = \frac{(4000 \text{ lb})(10 \text{ ft})}{4} = 10,000 \text{ ft-lb} = 120,000 \text{ in.-lb}$$

$$= \frac{(17.792 \text{ kN})(3.048 \text{ m})}{4} = 13.56 \text{ kN-m} = 13.56 \times 10^6 \text{ N-mm}$$

Moment of inertia:

$$I = \frac{bh^3}{12} = \frac{(5 \text{ in.})(10 \text{ in.})^3}{12} = 416.7 \text{ in.}^4$$

$$= \frac{(127 \text{ mm})(254\text{mm})^3}{12} = 173.4 \times 10^6 \text{ mm}^4$$

Maximum bending stress $= f_b$ *at* $y = c$:

$$f_b = \frac{Mc}{I} = \frac{(120,000 \text{ in.-lb})(10 \text{ in.}/2)}{416.7 \text{ in.}^4} = 1440 \text{ lb/in.}^2$$

$$= \frac{(13.56 \times 10^6 \text{ N-mm})(254 \text{ mm}/2)}{(173.4 \times 10^6 \text{ mm}^4)} = 9.93 \text{ N/mm}^2$$

Note that the stress level calculated is independent of the type of material used in the beam (i.e., the same stress would be present no matter whether the beam was made of timber, steel, or plastic). If the material used has the allowable stress value originally noted, however, the beam is adequately sized with respect to bending since the actual stress developed is less than this allowable stress [i.e., $(f_b = 1440 \text{ lb/in.}^2) \leq (F_b = 1500 \text{ lb/in.}^2)$ or $(9.93 \text{ N/mm}^2) \leq (10.34 \text{ N/mm}^2)$].

NONSYMMETRICAL SECTIONS. The general principles discussed above are applicable to nonsymmetrical as well as symmetrical sections. The critical difference is that in a nonsymmetrical section, such as a T beam or triangular shape, the location of the centroid is no longer obvious and is usually never at the midheight of the section. Its location must be calculated along the lines discussed in Appendix 2. Calculating the moment of inertia I for a nonsymmetrical section is also more involved than for simpler sections. Instead of the simple expression $I = bh^3/12$, which is valid for rectangular beams only, either the basic formulation of $I = \int y^2 \, dA$ or the related *parallel-axis theorem* must be used to find moment of inertia values. Appendix 3 discusses such procedures for nonsymmetrical beams.

    In the T beam illustrated in Figure 6-10, the centroid is located near the top flange of the member. Deformations and bending stresses in the member still vary linearly in the member and are proportional to the distance from the neutral axis of the member. This implies that the stress levels at the top and bottom of the beam are no longer equal as they typically are in symmetrical sections. The stresses are greater at the bottom of the beam than they are at the top because of the larger y distance. A difference in stress levels between the top and bottom surfaces is characteristic of nonsymmetrical sections. This point has enormous design implications, as will be discussed in Section 6.4.

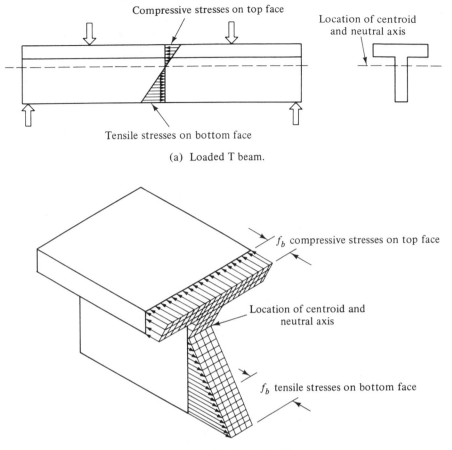

(a) Loaded T beam.

(b) Bending-stress distribution in T beam.

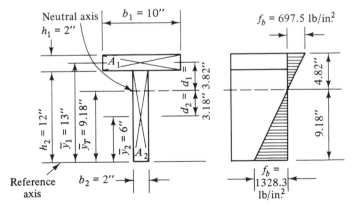

(c) T beam analyzed in example problem.

FIGURE 6-10 Bending stress distribution in a T beam. Since the cross section is non-symmetric, the centroid of the section is also nonsymmetrically located. The bending-stress distribution is still linear and passes through zero at the centroid (or neutral axis). This results in a difference in stress levels on the upper and lower faces of the beam.

EXAMPLE

Determine if the T beam illustrated in Figure 6-10 is adequately sized to carry an external bending moment of 120,000 in.-lb. Assume that the allowable stress in bending of the material used is $F_b = 1200$ lb/in.

Solution:

To find the actual bending stresses developed in the T beam, use can be made of $f_y = My/I$, as before. The real task in this problem is to evaluate I and y_{max}, a task more complicated than before because of the non-symmetrical nature of the cross section. The procedure is to first locate the centroid of the cross section and then to use the parallel-axis theorem to evaluate the moment of intertia about the centroidal axis. Appendix 3 explains the general process for doing this calculation and its underlying rationale in greater detail.

Location of centroid:

Consider the shape to be made up of two rectangular figures ($A_1 = b_1 \times h_1$ and $A_2 = b_2 \times h_2$), as illustrated in Figure 6-10(c). Assume a reference axis as illustrated. Let $\bar{y}_T$ represent the distance from the base of the figure to the centroid of the whole figure, and $\bar{y}_1$ and $\bar{y}_2$ be the distances to the centroids of the two rectangular figures from this same base. Thus,

$$\bar{y}_T = \frac{\sum\limits_i A_i \bar{y}_i}{\sum\limits_i A_i} = \frac{A_1 \bar{y}_1 + A_2 \bar{y}_2}{A_1 + A_2} = \frac{(2 \text{ in.} \times 10 \text{ in.})(13 \text{ in.}) + (2 \text{ in.} \times 12 \text{ in.})(6 \text{ in.})}{(2 \text{ in.} \times 10 \text{ in.}) + (2 \text{ in.} \times 12 \text{ in.})}$$

$$= 9.18 \text{ in.}$$

Moment of inertia:

Let $\bar{I}_T$ represent the moment of inertia of the whole figure about its centroidal axis. Let $\bar{I}_1$ and $\bar{I}_2$ represent the moments of inertias of the two rectangular figures about their own centroidal axes and d_1 and d_2 be the locations of these axes with respect to the centroidal axis of the whole figure. As explained in detail in Appendix 3, $\bar{I}_T$ can be found by using the parallel-axis theorem; thus,

$$\bar{I}_T = \sum_i (\bar{I}_i + A_i d_i^2)$$

$$= \left[\left(\frac{b_1 h_1^3}{12}\right) + (b_1 h_1)(d_1)^2\right] + \left[\left(\frac{b_2 h_2^3}{12}\right) + (b_2 h_2)(d_2)^2\right]$$

$$= \left[\frac{(10 \text{ in.})(2 \text{ in.})^3}{12} + (10 \text{ in.})(2 \text{ in.})(3.82)^2\right] + \left[\frac{(2 \text{ in.})(12 \text{ in.})^3}{12} + (2 \text{ in.})(12 \text{ in.})(3.18)^2\right]$$

$$= (6.7 + 291.9) \text{ in.}^4 + (288.0 + 242.7) \text{ in.}^4$$

$$= 829.3 \text{ in.}^4$$

Bending stresses:

The general type of bending stress distribution present is the T beam as illustrated in Figure 6-10(a) and (b). The stresses that occur at the top face of the beam are defined by $f_b = My/I$, where $y = 4.82$ in. (the distance from the top face to the centroidal axis of the figure). Thus,

$$f_b = \frac{My}{I} = \frac{(120,000 \text{ in.-lb})(4.82 \text{ in.})}{829.3 \text{ in.}^4}$$

$$= 697.5 \text{ lb/in.}$$

The stresses that occur at the bottom face of the beam are found similarly except that $y = 9.18$ in. Hence,

$$f_b = \frac{My}{I} = \frac{(120{,}000 \text{ in.-lb})(9.18 \text{ in.})}{829.3 \text{ in.}^4}$$

$$= 1328.3 \text{ lb/in.}^2$$

The stresses at the bottom face of the beam are clearly much larger than those at the top face of the beam. This is due to the nonsymmetrical nature of cross section. If the allowable stress of the material used in the beam is given by $F_b = 1200 \text{ lb/in.}^2$, then the beam is overstressed on its lower face (i.e., $1328.3 \text{ lb/in.}^2 > 1200 \text{ lb/in.}^2$). Note that stresses on the upper face are within the acceptable range. The beam is still considered inadequately sized, however, because of the overstress occurring on the lower face.

$$\bar{y}_T = 233.17 \text{ mm} \qquad \bar{I}_T = 345.15 \times 10^6 \text{ mm}^4 \qquad f_b = 4.81 \text{ N/mm}^2 \qquad f_b = 9.16 \text{ N/mm}^2$$

6.3.2 Shear Stresses

As noted in the opening section, equilibrium of an elemental portion of a beam with respect to vertical shear forces occurs through the development of vertical shearing stresses developed in the beam. The resultant force, $V_R = \int_A f_v \, dA$, equivalent to these stresses is equal in magnitude but opposite in sense to the external shear force, V_E. To gain an understanding of the magnitude and distribution of these stresses, it is actually more convenient to look first at the horizontal shear stresses also existing in the beam. In the example illustrated in Figure 6-11, the magnitude of these horizontal shear stresses can be found by considering the equilibrium, in the horizontal direction, of the upper left portion of the beam (remember that *any* elemental portion of a structure must be in equilibrium so any portion can be selected for analysis). With respect to equilibrium in the horizontal direction, it is obvious that the bending stresses acting on the right face of the element produce a horizontal force acting to the left. For equilibrium to obtain, the horizontal force due to bending must be balanced by other internal forces acting in the opposite direction, which implies the existence of horizontal shear stresses. The equilibrating force is provided by shear stresses acting in the horizontal direction over the horizontal face of the beam. Other horizontal planes in the beam also have shearing stresses but of varying magnitudes. In Figure 6-11(c), a plane near the top of the beam is shown. Again, the bending stresses produce a force in the horizontal direction which causes shearing stresses to develop on the horizontal plane. These forces and stresses, however, are smaller than those at the middle section of the beam, since the bending stresses act over a smaller area and thus produce a smaller horizontal force. Indeed, at the very top layer in the beam there can exist no forces or shear stresses at all. A close look at the magnitudes of these shearing stresses would indicate that they vary parabolically from a maximum at the neutral axis of a beam to zero at free surfaces. The horizontal stresses also vary along the length of the beam in most situations (in particular when the bending stress varies along the length of the beam). An exact expression, based on concepts similar to those discussed above, can be determined for the horizontal shearing stress in a beam. The horizontal shear stress at a layer located a distance y from the neutral axis can be

chapter 6 / MEMBERS IN BENDING: SIMPLE BEAMS   **239**

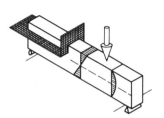

(a) Loaded beam.

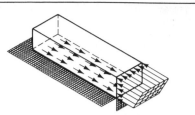

(b) Diagram of upper-left portion of beam showing bending stresses and horizontal shearing stresses. The bending stresses develop a resultant force acting to the left. An equilibrating force is developed by the horizontal shearing stresses acting over the horizontal plane indicated.

(c) Near the top of the beam, the horizontal force developed by the bending stresses is small since they act over a small area. Horizontal shearing stresses are also consequently small.

(d) At the neutral axis of the beam, the horizontal force developed by the bending stresses is the maximum possible. The horizontal shear stresses which equilibrate this force are consequently maximum.

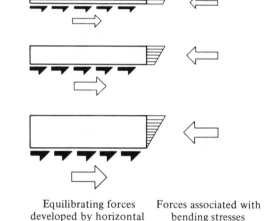

Equilibrating forces developed by horizontal shear stresses

Forces associated with bending stresses

FIGURE 6-11 Horizontal shear stresses in a beam. In a rectangular beam the horizontal shear stresses vary parabolically from a maximum at the neutral axis of the member to zero at the top and bottom faces. The distribution of stresses along the length of a horizontal plane depends directly on the variation in the external shear force along the same length. Where the external shear force is high, shear stresses will be high, and vice-versa.

shown to be given by the expression

$$f_v = \frac{VQ}{Ib}$$

where V is the vertical shear force at the vertical section considered, b is the width of the beam at the horizontal layer considered, I is the moment of inertia at the section, and Q is the first moment of the area (about the neutral axis) above the horizontal layer considered ($Q = \int y dA$). The derivation of this expression and applications of

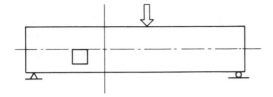

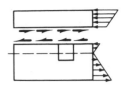

(a) Loaded beam.

(b) Stresses acting on a portion of the beam.

(c) Horizontal shear stresses acting on the upper and lower faces of an elemental area in the beam. The stresses produce a rotational moment.

(d) The rotational moment produced by the horizontal shear stresses is exactly balanced by a rotational moment associated with vertical shear stresses. The rotational moments, hence stresses, must be equal at the point.

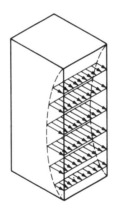

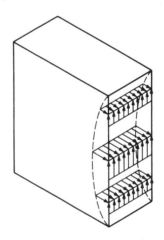

(e) Horizontal shear–stress distribution in a rectangular beam.

(f) Vertical shear–stress distribution in a rectangular beam.

FIGURE 6-12 The numerical magnitudes of the horizontal and vertical shear stresses at any point in a beam are always equal.

its use is contained in Appendix 5 to which reference is made. There it is also noted that this expression not only describes the horizontal shear stresses at a point, but the vertical shear stresses as well, since it can be demonstrated that the magnitudes of the horizontal and vertical shear stresses acting at a point are always equal (see Figure 6-12). Thus, the vertical shear stresses also vary parabolically with a maximum value occurring at the neutral axis.

The general shear-stress equation $f_v = VQ/Ib$ can be evaluated for the specific case of a rectangular beam (see the following example). In a rectangular beam the maximum shear stresses occur at the neutral axis of the member (at midheight) and are given by $f_v = \frac{3}{2}(V/bh) = \frac{3}{2}(V/A)$, where b and h are the dimensions of the cross section. Thus, the maximum shear stress in a rectangular beam is 1.5 times the value of and h are the average shear stress in a rectangular beam [see Figure 6-13(a)].

In beams constructed of thin-walled sections with a lot of material in flange areas, the shear stresses are more uniform because of the way the material is distributed throughout the cross section. The shear stress in such sections, such as a wide-flange beam, can be reasonably approximated by $f_v = V/A_{web}$ with relatively little error [see Figure 6-13(b)].

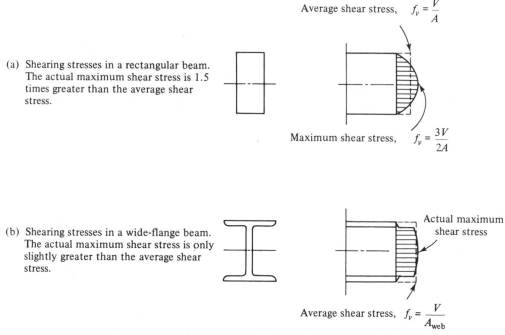

Average shear stress, $f_v = \dfrac{V}{A}$

(a) Shearing stresses in a rectangular beam. The actual maximum shear stress is 1.5 times greater than the average shear stress.

Maximum shear stress, $f_v = \dfrac{3V}{2A}$

(b) Shearing stresses in a wide-flange beam. The actual maximum shear stress is only slightly greater than the average shear stress.

Actual maximum shear stress

Average shear stress, $f_v = \dfrac{V}{A_{web}}$

FIGURE 6-13 Shearing-stress distributions in a rectangular beam and in a wide-flange beam.

EXAMPLE

Is the timber beam illustrated in Figure 6-14(a) adequately sized with respect to shear? Assume that the allowable shear stress for the material used is $F_v = 150$ lb/in. (1.03 N/mm²).

Solution:

The maximum shear stress occurs at the neutral axis of the beam at the cross section where the shear force, V, is a maximum. The maximum shear stress is given by

$$f_v = \frac{VQ}{Ib} = \frac{V(A'y')}{Ib} = \frac{V(bh/2)(h/4)}{(bh^3/12)b} = \frac{3V}{2bh} = \frac{3}{2}\frac{2000\ \text{lb}}{(5\ \text{in.})(10\ \text{in.})}$$

$$= 60\ \text{lb/in.}^2$$

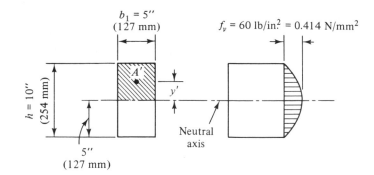

(a) Shear stresses in a rectangular beam.
Shear force: $V = 2000$ lb(8896 N).

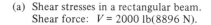

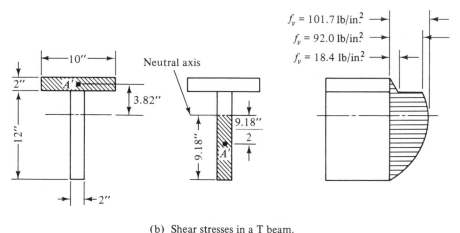

(b) Shear stresses in a T beam.
Shear force: $V = 2000$ lb.

FIGURE 6-14 Shearing stresses in beams.

This is the maximum vertical or horizontal shear stress that occurs in the beam. Since 60 lb/in. $\leq$ 150 lb/in., the beam is adequately sized with respect to shear. Alternatively,

$$f_v = \frac{3}{2}\frac{V}{bh} = \frac{3}{2}\frac{(8896 \text{ N})}{(127 \text{ mm})(254 \text{ mm})} = 0.414 \text{ N/mm}^2 < 1.03 \text{ N/mm}^2$$

EXAMPLE

For the same T-beam previously analyzed and again illustrated in Figure 6-14(b), determine the shear stress present at the interface between the top flange element and the web element if $V = 2000$ lbs. If glue were used to bond the elements together, what would be the stress on the glue? If nails spaced at 3.0 in. on center were used as connectors, what would be the shear force on each nail? What is the maximum shear stress present in the cross section?

Solution:

The expression $f_v = VQ/Ib$ will be used again, with $Q = A'y'$ evaluated for the area of the beam above the interface between the top flange and web element.

$$Q = A'y' = (b_1 \times h_1)(y') = (10 \text{ in.} \times 2 \text{ in.})(3.82 \text{ in.}) = 76.3 \text{ in.}^3$$

$$I = 829.3 \text{ in.} \quad \text{(see previous example)}$$

$$b = 2 \text{ in.}$$

$$V = 2000 \text{ lb}$$

$$f_v = \frac{VQ}{Ib} = (2000 \text{ lb})(76.3 \text{ in.}^3)/(829.3 \text{ in.}^4)(2 \text{ in.})$$

$$= 92.0 \text{ lb/in.}^2$$

This is the horizontal or vertical shear stress that occurs at the interface. This implies that if the two rectangular elements were glued together at this interface, the glue would have to be able to withstand this stress. If some other connector, such as nails, were used, the connector would have to be able to carry 2 in. $\times$ 92.0 lb/in., or 184 lb/in. If nails were placed 3.0 in. on center, each nail would have to carry 3 in. $\times$ 184 lb/in., or 552 lb of shear. This value could be compared to experimentally derived allowable forces on nails to determine if the nail spacing is adequate.

The maximum shear stress in the section occurs at the neutral axis. In evaluating this maximum stress, it is more convenient to consider the area below the neutral axis in evaluating $Q = A'y'$. Thus,

$$Q = A'y' = (2 \text{ in.} \times 9.18 \text{ in.})(9.18 \text{ in.}/2) = 84.3 \text{ in.}$$

$$f_v = \frac{VQ}{Ib} = \frac{(2000 \text{ lb})(84.3 \text{ in.})}{(829.3 \text{ in.}^4)(2 \text{ in.})}$$

$$= 101.7 \text{ lb/in.}^2$$

The shear-stress distribution for the whole cross section is illustrated in Figure 6-14(b). The discontinuity at the interface occurs because of the change from $b = 2$ in. to $b = 10$ in. Since shear stresses are inversely dependent on the width of the beam, actual shear stresses are significantly reduced in the wide top flange.

6.3.3 Bearing Stresses

The stresses developed at the point of contact between two loaded members are called *bearing stresses*. Such stresses, for example, are developed at the ends of a simply supported beam where it rests on end supports having certain dimensions. The magnitude of the stresses developed are dependent on the magnitude of the force transmitted through the point of contact and the surface area of the contact between the two elements. The smaller the contact area, the greater the bearing stresses. The stresses produce deformations in both elements at the point of contact. These deformations typically extend into the elements only a small distance.

The magnitude of the bearing stress at a point is simply equal to the load transmitted divided by the area of contact, or $f_{bg} = P/A$. This assumes that the bearing stresses are uniformly distributed over the contact area, an assumption that is not quite correct but reasonably so.

Many materials, such as timber, are particularly susceptible to bearing-stress failures. When a compressive load is transmitted, bearing-stress failures are usually not catastrophic and are typically manifested by an appearance of "crushing" in a material. The failure is generally localized but still should be avoided.

EXAMPLE

Assume that the timber beam illustrated in Figure 6-15 carries a uniformly distributed load of 500 lb/ft (7300 N/m) over a span of 16 ft (4.87 m). Also assume that the contact area between the beam and the column it rests on is 4 in. × 4 in. (101.6 mm × 101.6 mm). If the allowable stress in bearing of the timber in the beam is $F_{bg} = 400$ lb/in.^{2} (2.76 N/mm²), is the area of contact sufficiently large? Determine the actual bearing stress existing at the reaction. The reaction is 4000 lb (17792 N) and the contact area is 4 in. × 4 in. = 16 in.^{2} (10,322.6 mm²).

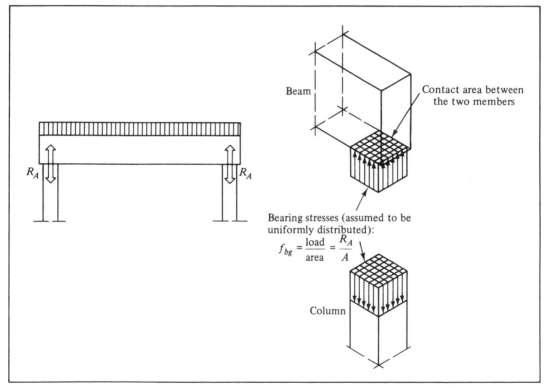

FIGURE 6-15 Bearing stresses at the end of a beam.

Solution:

$$\text{actual bearing stress} = f_{bg} = \frac{P}{A} = \frac{4000 \text{ lb}}{16 \text{ in.}^2}$$
$$= 250 \text{ lb/in.}^2$$
$$= \frac{(17792 \text{ N})}{(10{,}322.6 \text{ mm}^2)}$$
$$= 1.72 \text{ N/mm}^2$$

Since the actual bearing stress is less than the allowable bearing stress (250 lb/in.^{2} < 400 lb/in.^{2}) or (1.72 N/mm^2 < 2.76 N/mm^2), there is sufficient contact area between the two members.

6.3.4 Torsion

Torsion is a *twisting*. In many circumstances forces exist which cause a member to rotate about its longitudinal axis. Figure 6-16(a) illustrates a round member subjected to torsion. The analysis of the stresses produced by this type of force is not particularly complex but will not be covered here in detail. Figure 6-16(b) illustrates the type of stresses generated in the member by the applied torque. The force resultants of the stresses developed produce a couple that equilibrates the applied twisting moment. The stress at a point, τ, is dependent on the magnitude of the applied twisting moment, M_T; the location of the point, defined by the distance r from the centroid, and the

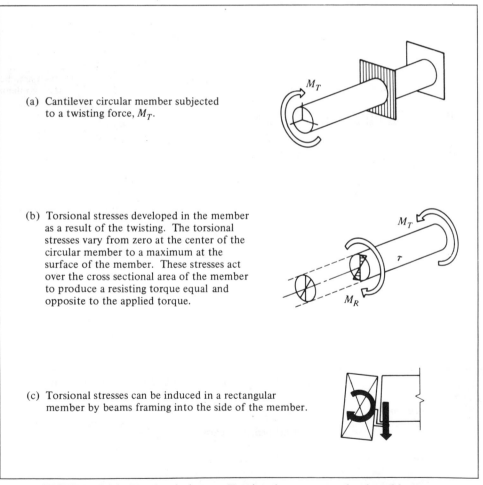

(a) Cantilever circular member subjected to a twisting force, M_T.

(b) Torsional stresses developed in the member as a result of the twisting. The torsional stresses vary from zero at the center of the circular member to a maximum at the surface of the member. These stresses act over the cross sectional area of the member to produce a resisting torque equal and opposite to the applied torque.

(c) Torsional stresses can be induced in a rectangular member by beams framing into the side of the member.

FIGURE 6-16 Torsion in beams. Torsional stresses are developed in members as a result of the application of a twisting force to the member.

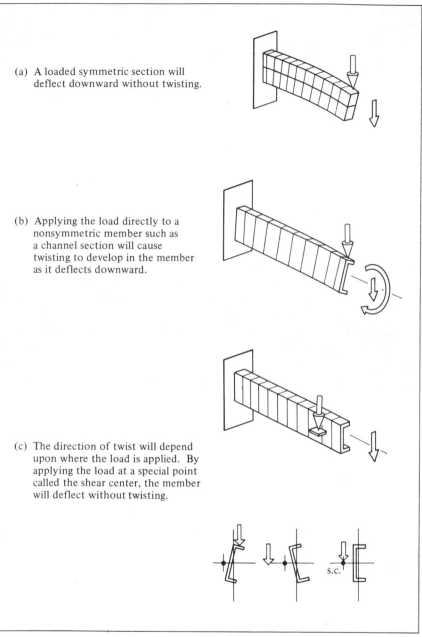

(a) A loaded symmetric section will deflect downward without twisting.

(b) Applying the load directly to a nonsymmetric member such as a channel section will cause twisting to develop in the member as it deflects downward.

(c) The direction of twist will depend upon where the load is applied. By applying the load at a special point called the shear center, the member will deflect without twisting.

FIGURE 6-17 Shear center in beams. In nonsymmetric members, applying a load directly to the member will cause the member to twist. By applying the load at the "shear center" of the beam, the member will deflect downward without twisting. The shear center of many nonsymmetric members often lies outside the member.

properties of the cross section, typically designated by the symbol J. The relation between these parameters is given by $\tau = Mr/J$. This expression is clearly analogous to the expression for bending stresses ($f_y = Mc/I$). J is analogous to I and is again given by $\int_A r^2\, dA$, except that now polar coordinates are used and J becomes the *polar moment of inertia*.

In instances where the section considered is not circular, the analysis of torsional stresses becomes fairly complex. Figure 6-16(c) illustrates a rectangular beam subjected to torsional stresses. Twisting is induced by the beam framing rigidly into the beam. The torsional stresses in such a beam are again dependent on the twisting moment present, M_T, the location of the point considered, and the properties of the rectangular section, and are given by an expression of the form $\tau = \alpha\, M_T/b^2 d$, where α depends on the relative proportions of b and d. Other shapes of arbitrary cross section can be analyzed but are beyond the scope of this book.[2]

6.3.5 Shear Center

Of importance in the use of members that are not symmetrical about the vertical axes and typically composed of thin-walled sections (e.g., a "channel section") is the phenomenon of twisting under the action of transverse loads. A symmetrical section, such as a rectangular beam, used as illustrated in Figures 6-17(a), simply deflects downward when loaded. A channel section carrying the same load would twist as indicated in Figure 6-17(b) because the line of action of the load does not pass through what is called the *shear center* of the member. This twisting leads to the development of torsional stresses in the member. The twisting itself can be predicted by a close investigation of the action of the stresses present in the section and can be described in terms of *shear flows*, but this topic is outside the scope of this book.[3] The presence of this twisting does not exclude the use of such members as beam elements, since a section can still be designed to safely carry external forces, but it does complicate their use as well as detract from their efficiency. If it were desired to use such a member to carry transverse forces without twisting, it is possible to do so by locating the load so that it passes through the *shear center* of the beam. In Figure 6-17(c), application of the load as indicated causes the twisting in the directions shown. There must, therefore, exist some point to the right of the section where the load could be applied and cause no twisting. This is the shear center. Other references describe how to locate this point.[4] Note that in the example illustrated, the shear center lies outside the section, thus complicating the use of this location as a point of load application.

6.3.6 Combined Stresses

In many situations a structural element is subjected not just to purely bending forces or purely axial forces, but to a combination of the two. Such cases often arise in the context of designing columns or in the use of prestressed beams. This section will

[2]For a further discussion of torsion, see, for example, Stephen Crandall and Norman Dahl, *An Introduction to the Mechanics of Solids*, McGraw-Hill Book Company, New York, 1959.

[3]See, for example, Crandall and Dahl.

[4]See, for example, Crandall and Dahl.

consider briefly how axial and bending stresses interact. The analysis is based on the fundamental principle of superposition.

Consider the beam shown in Figure 6-18(a) subjected to both axial forces acting through the centroid of the element and bending forces. The stress distribution associated with the axial forces is of uniform intensity and of magnitude $f_a = P/A$. The stress distribution associated with bending is given by $f_y = My/I$. The stress at any point on the cross section of a beam is merely the combination of these two stresses added or subtracted according to whether or not the stresses acting at the point are similar. Thus, either of the stress distributions shown in Figure 6-18(b) or (c) is possible, depending on the relative magnitudes of the axial and bending stresses.

By increasing the magnitude of a compressive axial force, any tensile stresses associated with bending can be completely overshadowed and the entire cross section put in a state of compression. As will be discussed in Section 6.4.7, this principle forms the rationale for the use of prestressing in concrete elements.

In members of this type the neutral axis, or plane of zero stress, no longer corresponds with the centroidal axis of the section.

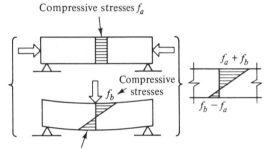

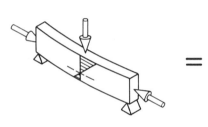

(a) Beam carrying both axial forces and bending moments.

(b) Using the principle of superposition, the stresses associated with each loading condition can be separately determined. The final stress distribution is a combination of these stresses. Similar stresses are additive while unlike stresses tend to counteract each other

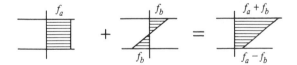

(c) In this example, the compressive stresses caused by the axial forces are sufficiently high to overshadow tensile stresses associated with bending. The entire cross section is thus in compression.

FIGURE 6-18 Combined stresses in beams. The final stress distribution in a beam carrying both axial and bending forces is simply a combination of the stresses associated with the axial forces and those associated with the bending moments.

6.3.7 Principal Stresses

One of the most interesting aspects of beam analysis is the way bending and shearing stresses interact. In Section 6.3.6, combined stresses resulting from the interaction of bending and axial stresses were considered. There the different stresses could be added algebraically, since they acted in the same direction. Stresses acting in different directions cannot simply be added algebraically, but their resultant interaction still can be found in much the same way that a resultant force can be found to represent the combined action of several different vector forces acting at a point. In a beam, the result of the interaction between bending and shear stresses is to produce a set of resultant tensile and compressive stresses, typically called *principal stresses*, that act in different directions from either the bending or shear stresses individually.

In a cantilever beam, the stresses acting on several typical elements are illustrated in Figure 6-19. A set of equivlent principal tension and compressive stresses could be found for each of the elements. Note that for an element at the neutral axis of the beam, where bending stresses are zero, only shear stresses exist. As diagrammed in Figure 6-19, these stresses can be resolved into equivalent principal tensile and compressive stresses acting at 45° angles to the neutral axis. At the extreme surfaces of the beam, an element carries only bending stresses, since shear stresses are zero. Thus, the principal stresses in tension become aligned with the bending stresses in tension and have the directions indicated in Figure 6-19. For an intermediate element subject to both shear and bending stresses, the principal stresses have an inclination depending on the relative magnitudes of the shear and bending stresses. By considering elements at other sections in the beam, stress trajectories can be drawn as illustrated. It is important to note that the lines are not lines of constant stress, but are lines of principal stress *direction*. Stress intensity, therefore, can vary depending upon the relative magnitudes and distributions of shear and bending stresses in the beam.

Figure 6-20 illustrates the stress trajectories present in a simply supported beam. Calculation of the magnitudes of principal stresses is a straightforward procedure but beyond the scope of this book. In general, it can be shown that maximum and minimum principal tensile and compressive stresses occur on planes of zero shearing stress and maximum shearing stresses occur on planes at 45° angles to the planes of principal tensile and compressive stresses.

6.3.8 Deflections

In the use of beams in a building context, few considerations are of the same importance as the amount the beam deflects under load. This section will briefly introduce the basic concepts involved in understanding the deflection characteristics of beams.

Consider the beam indicated in Figure 6-21(a). Also illustrated in highly exaggerated form is the deflected shape of the beam. The deflection Δ at a particular point in the beam depends on the load w, the length L of the beam, inversely on the stiffness of the beam as it depends on the amount and way material is distributed in the cross section (as characterized by the moment of inertia, I), and inversely on the stiffness of the beam as it depends on the stress-deformation characteristics of the material used in the beam (as characterized by the modulus of elasticity, E). Thus, Δ depends on w, L, $1/I$ and $1/E$. Note that as

w increases	Δ increases
L increases	Δ increases
I increases	Δ decreases
E increases	Δ decreases

The exact way that these variables quantitatively relate is discussed in Appendi-

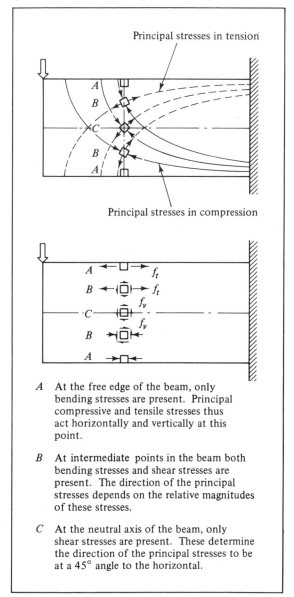

Principal stresses in tension

Principal stresses in compression

A At the free edge of the beam, only
bending stresses are present. Principal
compressive and tensile stresses thus
act horizontally and vertically at this
point.

B At intermediate points in the beam both
bending stresses and shear stresses are
present. The direction of the principal
stresses depends on the relative magnitudes
of these stresses.

C At the neutral axis of the beam, only
shear stresses are present. These determine
the direction of the principal stresses to be
at a 45° angle to the horizontal.

FIGURE 6-19 Principal stresses in a cantilever beam. Principal stresses
result from the interaction of bending stresses and shearing stresses. The
lines shown are often called stress trajectories and depict the direction of the
principal stresses in the member. They are not lines of constant stress.

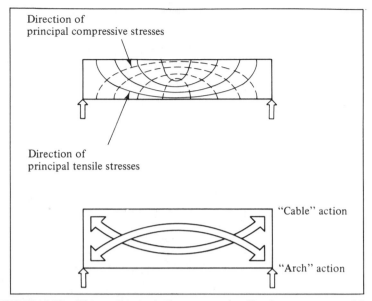

FIGURE 6-20 Lines of principal stresses: implications on general load-carrying mechanisms present in beams.

ces 6, 7, and 8, where deflection expressions for beams with several different types of loading and boundary conditions are developed.

For a simply supported and uniformly loaded beam, the maximum deflection can be demonstrated to be given by $\Delta = 5wL^4/384EI$. Evidently, deflections are *highly* sensitive to the length of the beam. It is also interesting to note that for a beam that is exactly similar in all respects except that the ends of the member are fixed rather than simply supported, the deflection is given by: $\Delta = wL^4/384EI$. Thus, fixing the ends of a simply supported beam carrying a uniformly distributed load reduces the midspan deflection by a factor of 5. Note that the form of both expressions is the same, but there is a difference in the modifying constant—which obviously reflects the different boundary conditions. The increased rigidity associated with fixed-ended beams, or also continuous beams, is one of the primary reasons why such members are used extensively in practice.

Figure 6-21 indicates the deflection expressions for other types of beams. Deflections are again considered in Section 6.4.1.

EXAMPLE

A simply supported timber beam that is 20 ft (6.1 m) long carries a uniformly distributed load of 200 lb/ft (2920 N/m). The beam has a rectangular cross section with dimensions $b = 8$ in. (203.2 mm) and $d = 16$ in. (406.4 mm). The modulus of elasticity of the timber is $E = 1.6 \times 10^6$ lb/in.^{2} (11,032 N/mm^2). What is the maximum deflection at midspan?

Solution:

Moment of inertia:

$$I = \frac{bd^3}{12} = \frac{(8 \text{ in.})(16 \text{ in.})^3}{12} = 2730 \text{ in.}^4$$
$$= \frac{(203.2 \text{ mm})(406.4 \text{ mm})^3}{12} = 1136.6 \times 10^6 \text{ mm}^4$$

Deflection:

$$\Delta = \frac{5wL^4}{384EI}$$

$$= \frac{5(200 \text{ lb/ft/12 lb/in.})(20 \text{ ft} \times 12 \text{ in.})^4}{384(1.6 \times 10^6 \text{ lb/in.}^2)(2730 \text{ in.}^4)} = 0.165 \text{ in.}$$

$$= \frac{5(2.920 \text{ N/mm})(6100 \text{ mm})^4}{384(11,032 \text{ N/mm}^2)(1136.6 \times 10^6 \text{ mm}^4)} = 4.2 \text{ mm}$$

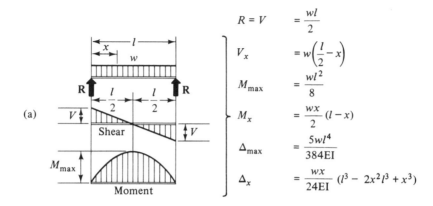

(a)

$R = V \quad = \dfrac{wl}{2}$

$V_x \quad = w\left(\dfrac{l}{2} - x\right)$

$M_{max} \quad = \dfrac{wl^2}{8}$

$M_x \quad = \dfrac{wx}{2}(l - x)$

$\Delta_{max} \quad = \dfrac{5wl^4}{384EI}$

$\Delta_x \quad = \dfrac{wx}{24EI}(l^3 - 2x^2l^3 + x^3)$

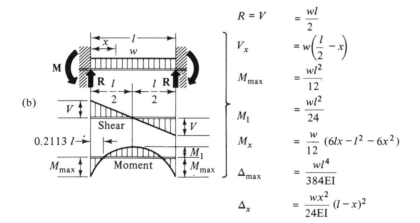

(b)

$R = V \quad = \dfrac{wl}{2}$

$V_x \quad = w\left(\dfrac{l}{2} - x\right)$

$M_{max} \quad = \dfrac{wl^2}{12}$

$M_1 \quad = \dfrac{wl^2}{24}$

$M_x \quad = \dfrac{w}{12}(6lx - l^2 - 6x^2)$

$\Delta_{max} \quad = \dfrac{wl^4}{384EI}$

$\Delta_x \quad = \dfrac{wx^2}{24EI}(l - x)^2$

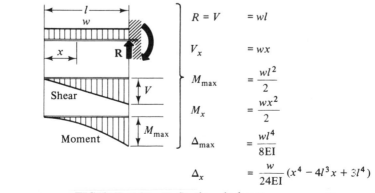

(c)

$R = V \quad = wl$

$V_x \quad = wx$

$M_{max} \quad = \dfrac{wl^2}{2}$

$M_x \quad = \dfrac{wx^2}{2}$

$\Delta_{max} \quad = \dfrac{wl^4}{8EI}$

$\Delta_x \quad = \dfrac{w}{24EI}(x^4 - 4l^3x + 3l^4)$

FIGURE 6-21 Deflections in beams.

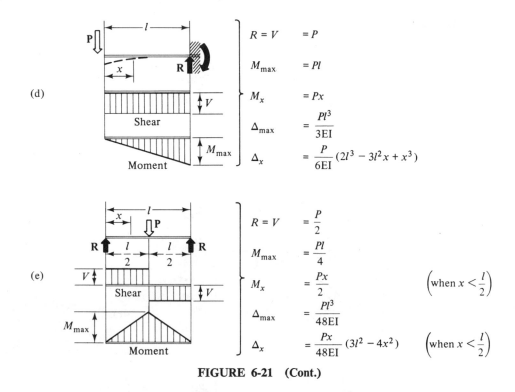

(d)

$$R = V \quad = P$$

$$M_{max} \quad = Pl$$

$$M_x \quad = Px$$

$$\Delta_{max} \quad = \frac{Pl^3}{3EI}$$

$$\Delta_x \quad = \frac{P}{6EI}(2l^3 - 3l^2 x + x^3)$$

(e)

$$R = V \quad = \frac{P}{2}$$

$$M_{max} \quad = \frac{Pl}{4}$$

$$M_x \quad = \frac{Px}{2} \qquad \left(\text{when } x < \frac{l}{2}\right)$$

$$\Delta_{max} \quad = \frac{Pl^3}{48EI}$$

$$\Delta_x \quad = \frac{Px}{48EI}(3l^2 - 4x^2) \qquad \left(\text{when } x < \frac{l}{2}\right)$$

FIGURE 6-21 (Cont.)

6.3.9 Plastic Behavior in Beams

The analyses presented thus far have been based on the assumption that the beam material is *linearly elastic* (i.e., that stresses are proportional to deformations). As discussed in Section 2-5, however, many materials such as steel are linearly elastic only up to a certain point, when the material begins to deform massively under a relatively constant stress level. Only after significant deformation has occurred does the material actually rupture. These are *plastic deformations*. A simplified stress–strain curve depicting this behavior is shown in Figure 6-22. This behavior is of crucial importance in understanding how beams actually fail under load.

Consider the rectangular beam shown in Figure 6-23. When external loads on the beam are low, the material in the beam is in the elastic range and bending stresses are linearly distributed across the cross section. As loads increase, the bending stresses increase until the material at the outer fibers reaches a point, f_y, where it begins to *yield* and enter the plastic range. As indicated in Figure 6-23(d), this corresponds to a resisting moment in the beam of $M_y = F_y(bh^2/6)$. Continuing to increase loads causes increased deformations in the beam fibers (a linear variation in strain still occurs, since this is a function of the gross geometry of deformations in bending in the entire beam). There is not, however, a corresponding increase in the stress level in the material in the region where beam fibers are deformed into the plastic region. At this stage, some fibers near the neutral axis are still below f_y. In the instance shown in Figure 6-23(e), the beam is still capable of carrying load (the outer fibers have yielded but not ruptured). Increasing the external load further causes increased deformations,

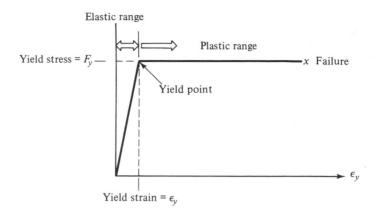

FIGURE 6-22 Idealized stress–strain curve for ductile material. At a certain stress level, F_y, the material begins to undergo increased deformations without any additional increase in stress level. Actual failure in the material does not occur until relatively large deformations occur.

until all the fibers in the cross section begin to yield. The load that finally causes the fibers nearest the neutral axis to yield is the maximum one the beam can carry. This load corresponds to a maximum resisting moment of $M_p = F_y(bh^2/4)$. Up to this point, the beam has always been able to provide an increased moment resistance capacity to balance the applied moment associated with the external load. After this point, however, the beam has *no* further capacity to resist the external moment and continuing to apply loads would simply cause additional deformations until rupture and collapse finally occur. A *plastic hinge* is said to develop when all fibers are fully yielded at a cross section.

For a rectangular beam, the ratio between the moment associated with the formation of a plastic hinge and that associated with the initial yielding is $M_p/M_y = F_y(bh^2/4)/F_y(bh^2/6) = 1.5$. The term $bh^2/4$ is often called the *plastic section modulus, Z*. In the statically determinate beam illustrated in Figure 5.23(a), the 1.5 ratio implies that the load, P_f, required to cause the beam to actually fail is 1.5 times the load, P_y, which would cause initial yielding to occur in the member. Other shapes have different ratios of this type. This ratio is often called the *shape factor* for a beam, since it is identical to the ratio of the plastic section modulus (Z) to the elastic section modulus (S) (i.e., shape factor = Z/S). For a wide-flange beam, shape factors vary, but average around 1.14. Other factors are 1.7 for a round bar and 2.0 for a diamond.

In statically indeterminate structures, the formation of a single plastic hinge need not lead directly to beam collapse. A number of hinges must form until a collapse mechanism is created. Consider the fixed-ended beam shown in Figure 6-24 and the moments that are developed in the structure. As the load, w, is increased, yielding in the beam first occurs at the ends of the beam where moments are maximum. As load intensity is increased, plastic hinges begin forming at these points. The moment at the center of the structure is also increased. Note that the formation of plastic hinges at the ends of the beam do not cause collapse to occur in the beam, since the structure still has moment-carrying capacity at midspan. Further beam loading would eventually cause a plastic hinge to develop at midspan. A collapse mechanism would then exist

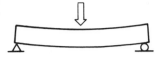
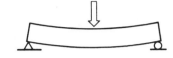
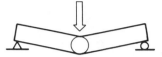

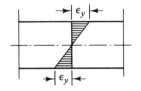

(a) Strain distribution when outer fibers just begin to yield.

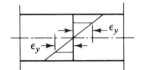

(b) Strain distribution when outer fibers have yielded. Strain distribution is still linear.

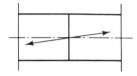

(c) Strain distribution when all fibers in cross-section have yielded.

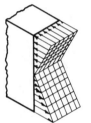

(d) Stress distribution when outer fibers just begin to yield.

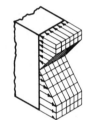

(e) Stress distribution when the outer, but not the inner, fibers of the beam have yielded. The beam still has a measure of reserve strength at this section.

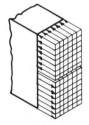

(f) Stress distribution when all fibers in the cross-section have yielded. The beam can support no more moment at this section.

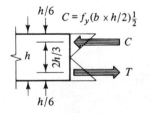

Elastic resisting moment = $F_y(bh^2/6)$

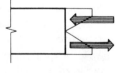

Elastic/plastic resisting moment

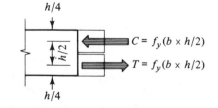

Plastic resisting moment = $C(h/2)$
$= F_y(bh^2/4) = M_P$

FIGURE 6-23 Plastic behavior in beams. As the material in a beam begins to yield under load, the stress distribution present begins to change. The beam can continue to carry moment until all the fibers at a cross section have yielded.

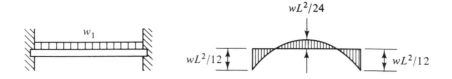

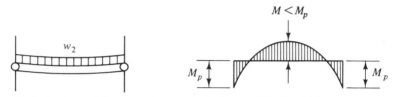

(a) Fixed-ended beam. Elastic moment distribution.

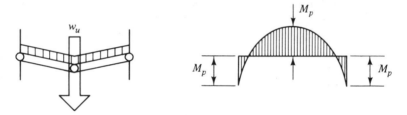

(b) As loads on the beam are increased, the beam will begin yielding at the regions of highest moment (at the ends in this case). Plastic hinges begin forming at these points. The beam begins to act like a simply supported beam with a fixed moment resistance at each end. Increases in loads cause a moment redistribution to occur (the moment at midspan begins to increase). Although hinges have formed at two points, the beam can still carry additional load.

(c) Further increases in load will eventually cause the formation of a third plastic hinge at midspan. The beam will collapse when this hinge forms.

FIGURE 6-24 Formation of plastic hinges in a fixed-ended beam. Plastic hinges initially develop where moments are the highest. The beam does not fail, however, until a sufficient number of hinges have formed to cause a collapse mechanism to develop in the beam. Beams of this type have a large measure of reserve strength.

and the beam could carry no additional load (total failure is imminent). The failure load, w_u, of the beam is greater than would be implied by a consideration of just the shape factor of the beam alone. It is shown elsewhere that for a restrained rectangular beam the load required to cause failure, w_u, is double that of the load required to initiate yielding, w_y.

6.3.10 Reinforced Concrete Beams

The previous sections have considered the behavior of loaded beams made of linearly elastic materials where the stresses are proportional to deformations. With many materials, however, the assumption of linear elasticity is not realistic. Reinforced concrete is such a material. Concrete itself is not capable of carrying anything other than nominal deformations, hence stresses, in tension before cracking occurs. As discussed in Chapter 2, neither is it exactly elastic in compression either. Relationships such as $f_y = My/I$ are based on the *assumption* that the material is elastic and capable of carrying tension stresses. A modified theory, therefore, is needed.

The theory used to analyze reinforced concrete beams is, like $f_y = My/I$, based on the fundamental principle that the internal forces developed in a beam serve to maintain the equilibrium of portions of the beam. Consider the beam illustrated in Figure 6-25. As loads induce bending in the member, a deformation pattern of the type illustrated develops, indicating that the region below the neutral axis is in tension. Since concrete cannot take tension, reinforcing steel is put into this region and serves the function of carrying the tensile force equivalent to the tensile stresses. Note that the concrete is assumed to be cracked in this tensile region. These cracks actually occur and generally extend upward until they terminate at the edge of the compressive region. An internal resisting moment is generated to balance the external moment by the couple formed by the tension force in the reinforcing steel and the resultant compressive force in the upper region of the beam. Hence, in a plain reinforced concrete beam only a portion of the concrete is actually participating in carrying the load. When the magnitude of the external moment present is known, it is possible to determine what internal forces must exist and what stress levels in the concrete and reinforcing steel correspond to these stress levels. This topic, however, is generally outside the scope of this book, particularly since the exact behavior of reinforced concrete is fairly complex due to the nonlinear stress deformation characteristics of concrete. Calculations are sometimes made on an idealized model based on the assumption that concrete in compression behaves elastically (*working stress analysis*). This approach has proved unrealistic, since actual stresses under service loads are more similar to those illustrated in Figure 6-25(b). An alternative model based on the experimentally determined behavior of reinforced concrete beams at failure loads is generally preferred. This model, illustrated in Figure 6-25(d), is called *ultimate strength analysis*.

As would be expected, the amount of tension steel required depends directly on the magnitude of the moment present at the section considered. The location of the neutral axis depends largely on the ratio of the amount of steel used relative to the size of the cross section. The final strength of a beam can be governed either by failure in the concrete or failure in the steel. If a large amount of steel is used, failure of the beam would occur by crushing of the brittle concrete. This type of failure is potentially highly dangerous, since total collapse occurs instantaneously with little prior warning. As the amount of steel is reduced, a point is reached where failure in either the steel or concrete is equally likely. This is a *balanced beam*. As the amount of steel is further reduced, failure would occur by yielding of the steel first, followed by crushing of the concrete afterward. Collapse occurs with the crushing of the concrete. The yielding of the steel is invariably accompanied by large deformations and hence deflections, thus giving prior warning of impending collapse. For this reason, the amount of steel in beams is limited to a "underreinforced" condition, so that this type of behavior occurs rather than the more dangerous type associated with "overreinforced" beams, where crushing in the concrete occurs before yielding of the steel.

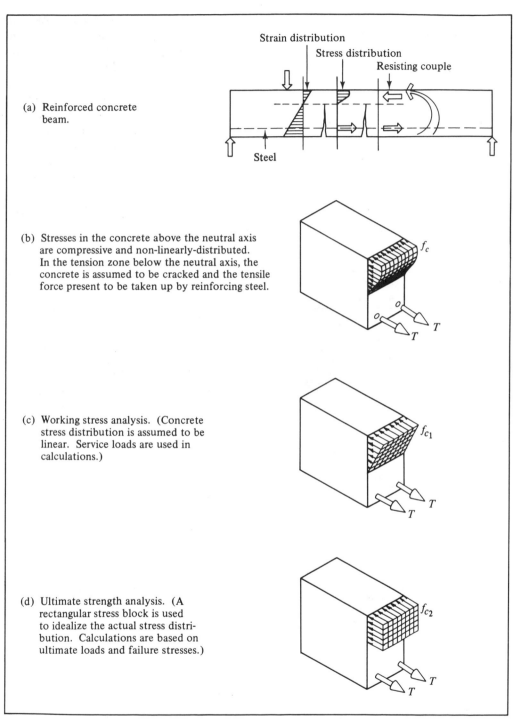

(a) Reinforced concrete beam.

Strain distribution

Stress distribution

Resisting couple

Steel

(b) Stresses in the concrete above the neutral axis are compressive and non-linearly-distributed. In the tension zone below the neutral axis, the concrete is assumed to be cracked and the tensile force present to be taken up by reinforcing steel.

f_c

T

(c) Working stress analysis. (Concrete stress distribution is assumed to be linear. Service loads are used in calculations.)

f_{c_1}

T

(d) Ultimate strength analysis. (A rectangular stress block is used to idealize the actual stress distribution. Calculations are based on ultimate loads and failure stresses.)

f_{c_2}

T

FIGURE 6-25 Reinforced concrete beams.

With respect to the shear resistance of a reinforced concrete beam, plain concrete can carry some shear stresses before cracking. The real difficulty is typically not the shear stresses, but the principal stresses in tension that result from the interaction of shear and bending stresses. The lines of principal tensile stresses in a uniformly loaded beam are as previously illustrated in Figure 6-20. The result is that cracks are often generated in members of the type illustrated in Figure 6-26(a). These cracks are referred to as *diagonal tension* cracks and are potentially quite dangerous, since they can lead to rapid failure of the whole beam. To prevent this, steel reinforcement is used to intersect such cracks. The preferable way to organize diagonal tension reinforcement would be to follow the patterns suggested by the stress trajectories or to approximate them. As a matter of construction convenience, vertical steel reinforcement is often used, although it is less efficient than other reinforcement patterns. The amount of steel in this reinforcement and how the steel is distributed generally depends on the magnitude of the shear force present.

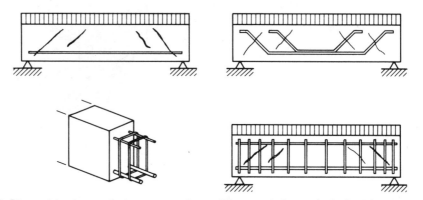

(a) Diagonal tension cracks in a concrete beam. These result from principal tension stresses, which act at 45° to the horizontal. Reinforcing steel must be placed to intersect these cracks. Using diagonal steel (by bending up reinforcing bars) is one way of doing this. Using vertical steel "stirrups" is another way. Care should be taken to provide a sufficient number of stirrups to intersect any possible 45° diagonal tension crack. The amount and placement of stirrup steel depends on the magnitude of the shear force present at a section.

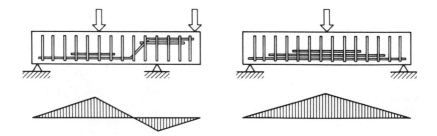

(b) Typical steel placement in reinforced concrete beams. The amount of longitudinal steel required at a section depends on the magnitude of the bending moment present at the section. The amount of vertical steel required depends on the magnitude of the shear force present at the section.

FIGURE 6-26 Reinforced concrete beams.

6.4 DESIGN OF BEAMS

6.4.1 Selecting Beam Properties

STRENGTH CONTROL. Conceptually, the design of beams is a relatively straightforward process. If the loading condition is known, the distribution of internal forces (e.g., shear and moments) that are generated in a member can be readily calculated. The problem then becomes one of finding section properties for a member such that the actual stresses generated in the beam at any point are limited to a predetermined level dependent on the strength properties of the material used. Various combinations of material and section properties can yield adequately sized beams. Not only should bending stresses be controlled, but other stresses as well (e.g., shear, bearing).

The process of designing beams is facilitated by the series of well-defined relationships between forces, section properties, stresses, and deformations discussed in the previous sections. The examples discussed below illustrate the general process used for determining beam characteristics at one section where the shears and moments are known. A material is first selected and *allowable stresses* in bending and shear are defined. Required section properties are next found on the basis of making the actual bending stress level in the beam equal to, or less than, the allowable stress level in bending. This process is based on the use of the bending stress relationship $f_y = M_y/I$. From a knowledge of how stresses are distributed in a beam, we know that the maximum stress occurs when y is maximum, (or $y_{max} = c$). Thus, in order to make the actual bending stress at $y_{max} = c$ equal to the allowable stress, we simply substitute F_b (the allowable stress for the material) for f_y, c for y_{max}, and solve for the other terms in the relationship $f_y = My/I$. Hence, for *any* beam shape,

$$\left(\frac{I}{c}\right)_{req'd} = \frac{M}{F_b}$$

The I/c measure is termed the section modulus, S, of a beam. The design problem is then to find an S or I/c value for a beam equal to, or greater than, the value of M/F_b.

For a *rectangular* beam (b = width, d = depth),

$$S = \frac{I}{c} = \frac{bd^3/12}{d/2} = \frac{bd^2}{6}$$

Thus,

$$\left(\frac{bd^2}{6}\right)_{req'd} = \frac{M}{F_b} \quad \text{or} \quad bd^2 = \frac{6M}{F_b}$$

Any rectangular beam with a combination of b and d dimensions which yields $bd^2 = 6M/F_b$ will be acceptable with respect to bending.

Just because a beam is adequately sized with respect to bending does not necessarily mean that it is acceptable with respect to other considerations, such as shear stresses or deflections. The process is usually to determine a beam size on the basis of one consideration, such as bending, and check it for other phenomena. It may be necessary to redesign the beam based on shearing, stress, or deflection control if the trial beam proves inadequate in this respect.

DEFLECTION CONTROL. Controlling deflections is always a major problem in beam design. It is also very difficult to even establish what constitutes an allowable deflection in a beam. Some objective criteria can be established. If a beam deflects such that it interferes with or impairs the functioning of another building element, for example, the allowable deflection could be based on acceptable tolerances for other systems. In many cases, however, the problem is more subjective. Many beams are, for example, perfectly safe from a strength viewpoint but are said to visually "sag" too much. Likewise, when people walk across floors and the floors feel bouncy or springy, the supporting beams are said to deflect too much. With respect to the last point it is true that attempts are made to control floor bounciness by limiting beam deflections. This is, however, a misconception in that people do not feel deflections, they rather sense the accelerations associated with deflections. This is an important point in that there may be other ways to control accelerations than by artificially limiting deflections.

The problem of what determines excessive deflections is tough. Empirical guidelines are often used. A common empirical criterion used to control both the visual sag and bounciness problems is that the deflection of a floor should not exceed $\frac{1}{360}$ of its span (the criterion is usually expressed as $\Delta_{\text{allowable}} = L/360$, where L is the span of the member). Deflection limitations for roofs are similarly expressed but are less restrictive. A common roof deflection criterion is $L/240$. If a member deflects more than these guidelines, it is usually considered not acceptable and a member of increased stiffness (i.e., increased I) must be used no matter how low the stress level in the member might be. Evidently, the size of such a member can be found by equating the appropriate deflection expression with the maximum allowable deflection and solving for the required stiffness (e.g., $\Delta_{\text{allowable}} = L/360 = 5wL^4/384EI$ or $I_{\text{required}} = 5(360)wL^4/384EL$ for a uniformly loaded beam).

The criteria noted above, $L/360$ and $L/240$ for floors and roofs, respectively, are felt by many designers to be conservative in view of the origins of these criteria. The criterion of $L/360$ has its roots in historical building traditions in which the rule was used largely as a measure to prevent plaster affixed to the underside of a floor (or on the ceiling of the room beneath) from cracking. The criterion has since been widely used for many other applications. There is no doubt that floors designed to this limitation are usually perceived by occupants as comfortable and not excessively saggy. It is interesting to note that our perceptions of what constitutes an acceptable level of both visual sag and floor bounciness are probably derived from our cultural conditioning in accepting prior experiences as a measure of correctness. These experiences are in turn based largely on an antiquated plaster-cracking criterion.

EXAMPLE

Determine the required size of one of the rectangular beams illustrated in Figure 6-27. Assume the following: live load = 50 lb/ft^2, dead load = 15 lb/ft^2 (includes weight of floor deck, flooring, and estimate of beam weight), allowable stress in bending = F_b = 1200 lb/in.^{2}, allowable shear stress = F_v = 150 lb/in.^{2}, allowable bearing stress, F_{bg} = 400 lb/in.^{2} and allowable deflection, $\Delta = L/360$. Also assume that $E = 1.6 \times 10^6$ lbs/in.^{2}, $L = 16.0$ ft, and $a = 16$ in.

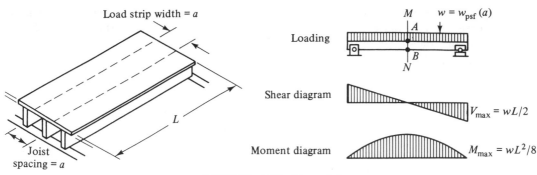

FIGURE 6-27 Floor joist system.

Solution:

Loads:

$$\text{load/ft} = \underbrace{(50 \text{ lb/ft}^2 + 15 \text{ lb/ft}^2)}_{\text{distributed load}} \underbrace{(16 \text{ in.}/12 \text{ in.}/\text{ft})}_{\substack{\text{width of floor deck} \\ \text{carried by one beam}}}$$

$$= w = 86.5 \text{ lb/ft}$$

Bending stresses:

$$\left(\frac{I}{c}\right)_{\text{req'd}} = S_{\text{req'd}} = \frac{M}{F_b}$$

$$\frac{bd^2}{6} = \frac{M}{F_b}$$

$$M = \frac{wL^2}{8} = \frac{(86.5 \text{ lb/ft})(16 \text{ ft})^2}{8} = 2768 \text{ ft-lb} = 33{,}216 \text{ in.-lb}$$

$$\frac{bd^2}{6} = \frac{M}{F_b} = \frac{33{,}216 \text{ in.-lb}}{1200 \text{ lb/in.}^2}$$

$$bd^2 = 166.1 \text{ in.}^3$$

Any beam with a value of $bd^2 = 166.1$ in.^{3} will be adequately sized with respect to bending. *Assume* that $b = 1.5$ in. Thus,

$$(1.5 \text{ in.})d^2 = 166.1 \text{ in.}^3 \qquad \text{or} \qquad d = 10.5 \text{ in.}$$

A beam 1.5 in. × 10.5 in. would be adequately sized with respect to bending. The nearest stock timber size is a 2 × 12 (actual dimensions are $1\frac{1}{2}$ in. × $11\frac{1}{2}$ in.). This beam would be slightly oversized but is reasonably close and will be used. The next step is to check this beam for other types of stresses and deflections.

Shear stresses:

The maximum shear force occurs at either end of the beam:

$$V = \frac{wL}{2} = \frac{(86.5 \text{ lb/ft})(16 \text{ ft})}{2} = 692 \text{ lb}$$

$$f_v = \frac{3}{2}\frac{V}{bd} = \frac{3}{2}\left(\frac{692 \text{ lb}}{1.5 \text{ in.} \times 11.5 \text{ in.}}\right) = 65.9 \text{ lb/in.}^2$$

This is the maximum shearing stress in the beam. It occurs at the neutral axis of the member at the reactions. Since the actual shear stress is less than the allowable shear stress (65.9 < 150), the beam is adequately sized with respect to shear stresses.

Bearing stresses:

The force at the ends of the beam is 692 lb. Assume that the area in bearing is 1.5 in. × 2 in.

$$f_{bg} = \frac{P}{A} = \frac{692 \text{ lb}}{1.5 \text{ in.} \times 2 \text{ in.}} = 231 \text{ lb/in.}^2$$

The actual bearing stress is less than the allowable bearing stress (231 < 400), so the beam is not overstressed in bearing at the reactions.

Deflections:

$$\Delta = \frac{5wL^4}{384EI}$$

$$= \frac{5[(86.5 \text{ lb/ft})/(12 \text{ in./ft})](16 \text{ ft} \times 12 \text{ in./ft})^4}{384(1.6 \times 10^6 \text{ lb/in.}^2)[1.5 \text{ in.} \times (11.5 \text{ in.})^3/12]}$$

$$= 0.42 \text{ in.}$$

The actual deflection at midspan is 0.42 in. The allowable deflection is $L/360$ or (16 ft × 12 in./ft)/360 = 0.53 in. Since the actual deflection is less than the allowable deflection, the floor system is sufficiently rigid under load.

The same general procedure is followed when SI units are used. For example:

$$\frac{\text{load}}{\text{unit length}} = 86.5 \text{ lb/ft} = 1262.31 \text{ N/m}$$

$$\text{bending moment} = 2768 \text{ ft-lb} = 3753.4 \text{ N-m}$$

$$\text{allowable stress in bending} = F_b = 1200 \text{ lb/in}^2 = 8.27 \text{ N/mm}^2$$

$$\text{beam size} = S_{\text{req'd}} = \frac{M}{F_b} = \frac{3753.4 \times 10^3 \text{ N-mm}}{8.27 \text{ N/mm}^2}$$

$$\frac{bd^2}{6} = S_{\text{req'd}} = 453,857 \text{ mm}^3$$

If b = 38.1 mm (1.5 in.), then d = 267.34 mm (10.5 in.). Shear stresses, bearing stresses, and deflections would be checked similarly.

EFFICIENCY OF CONSTANT-DEPTH RECTANGULAR BEAMS. A couple of points are of interest in connection with the previous example. One is that rectangular beams of the type illustrated are not particularly efficient. In the rectilinear beam illustrated in Figure 6-27, there are only *two* spots on the entire beam (points A and B at section $M–N$), where the beam fibers are stressed to the maximum extent possible. At all other points in the beam the material is understressed. Consequently, the maximum potential of the material is not being fully utilized. The reason for this lies in the nature of bending at a cross section where deformations and bending stresses vary from zero at the neutral axis to a maximum at the extreme fibers. Thus, the stresses doing the most to generate an internal resisting moment to equilibrate the external moment are those near the outer fibers. Those near the neutral axis are of little consequence in this respect. This implies that from a design viewpoint more efficient use of beam material could be made by moving the beam material away from the neutral axis and toward the extremities of the beam.

A second reason for why the beam material in the example illustrated in Figure 6-27 is underutilized is that the size of the section was determined in response to the maximum external bending moment present in the beams (at section *M–N*). At other points in the beam, the external bending moment is less (see the moment diagram). Thus, using a section designed in response to the maximum moment will result in inefficiencies at other points. From a theoretical viewpoint, it should be possible to vary the size of the beam in response to the internal forces present. A beam with variable dimensions along its length would result (see Section 6.4.5).

The following sections begin to explore ways of designing beam properties to make more efficient use of material with respect to issues of the type discussed above.

6.4.2 Cross-sectional Shapes

As noted in the previous section, the moment of inertia (*I*) and the section modulus (*S*) are of primary importance in beam design. A common design objective is to provide the required *I* or *S* for a beam carrying a given loading with a cross-sectional configuration that has the smallest possible area. The total volume of material required to support the load in space would then be reduced. Alternatively, the objective could be stated in terms of taking a given cross-sectional area and organizing this material so that the maximum *I* or *S* value is obtained, thus allowing the beam to support the maximum possible external moment (hence loading).

The basic principle for maximizing the moment of inertia obtainable from a given area is contained in the definition of the moment of inertia, which is $I = \int y^2 \, dA$. The contribution of a given element of area (*dA*) to the total moment of inertia of a section depends on the square of the distance of this elemental area from the neutral axis of the section. The sensitivity to the square of the distance is important. This would lead one to expect that for a given amount of material, the best way to organize it in space is to remove it as far as practically possible from the neutral axis of the section (i.e., make the section deep with most of the material at the extremeties). Consequently beams with high depth/width ratios are usually more efficient than ones with shallower proportions.

A highly economic cross section also based on the preceding principle is illustrated in Figure 6-28(a), where a thin web member is used to connect two widely separated thick flanges. The question arises of just how thin the web can be. It will be recalled that maximum shearing stresses always occur at the neutral axis of a beam. Thus, it is necessary that the web be of a certain minimum thickness to carry these shear stresses safely. In addition, there are lines of principal stresses which cross the beam in the middle. The principal compressive stresses could cause the web to buckle locally if it were too thin. Beams of the general proportions indicated in Figure 6-28(a) result after those factors are taken into account. Sections of this type made of steel, called *wide-flange beams*, are commonly used in building construction where bending stresses are typically a more important consideration than shearing stresses. When shear forces are high, it can be expected that beams with thicker webs will be used.

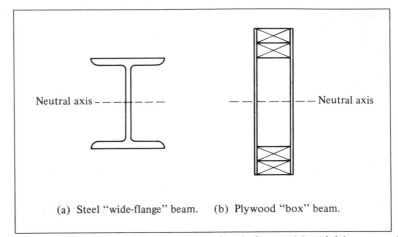

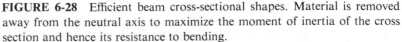

(a) Steel "wide-flange" beam. (b) Plywood "box" beam.

FIGURE 6-28 Efficient beam cross-sectional shapes. Material is removed away from the neutral axis to maximize the moment of inertia of the cross section and hence its resistance to bending.

The discussion thus far has focused on ways of increasing the moment of inertia (I) of a section. It should be recalled, however, that in design of a beam, the measure that is actually of primary importance is the I/y_{max} value of a section. In cases where the section is symmetric about the neutral axis, increasing the I value automatically increases the section modulus, S. It is interesting to note that in nonsymmetric sections such as T beams, the design must be based on the condition that the maximum bending stress at any point on the beam is limited to the allowable stresses. At a section this point is defined by y_{max} and occurs on one face of the section. The other face, with a value of y less than y_{max}, is therefore understressed. Hence, the use of such a section is not advantageous in beams made of homogeneous materials. This is not the sole measure for evaluating this shape, however, since T beams do have some advantages, which will be discussed in the next section and in Section 6.4.4.

EXAMPLE

A simply supported steel wide-flange beam that is 25 ft long supports a uniformly distributed load of 600 lb/ft. Assume that the allowable stress in bending is $F_b = 24,000$ lb/in.^{2}. Determine the required size of the wide flange based on bending stress considerations.

Solution:

Maximum bending moment:

$$M = \frac{wL^2}{8} = \frac{(600 \text{ lb/ft})(25 \text{ ft})^2}{8} = 46,875 \text{ ft-lb}$$

Required section modulus:

$$S = \frac{M}{F_b} = \frac{(46,875 \text{ ft-lb} \times 12 \text{ in./ft})}{24,000 \text{ lb/in.}^2}$$

$$= 23.4 \text{ in.}^3$$

Beam size:

By referring to Appendix A-13, it can be seen that either of the following will work:

$$\text{W } 12 \times 22 \qquad S = 25.3 \text{ in.}^3$$
$$\text{W } 8 \times 28 \qquad S = 24.3 \text{ in.}^3$$

The beam having the least weight is the most economical. The W 12 × 22 would therefore be selected for use. Shear stresses, deflections, and bearing stresses, however, would also have to be checked.

6.4.3 Lateral Buckling of Beams

In the previous section it was implied that making a beam as deep as possible was generally advantageous, since the moment of inertia I and section modulus S was usually maximized when this was done. There are, of course, limits to how deep a beam can be made. These limits are often imposed by how a beam is used in the context of the building. There are also other reasons. Consider the thin, deep beam illustrated in Figure 6-29. Application of a load may cause *lateral buckling* in the beam and failure will occur before the strength of the section can be utilized. The phenomenon of lateral buckling in beams is similiar to that found in trusses. An instability in the lateral direction occurs because of the compressive forces developed in the upper region of the beam coupled with insufficient rigidity of the beam in this direction. In all of the examples discussed thus far it was assumed that this type of failure does not occur. Depending upon the proportions of the beam cross section, lateral buckling can occur at relatively low stress levels.

Prevention of lateral buckling can be provided in two primary ways: (1) by using transverse bracing, or (2) by making the beam stiff in the lateral direction. When a beam is used to support a roof deck or a secondary framing system, transverse bracing is automatically provided by these elements. If a beam is used in a situation where this type of bracing is not possible, the beam can be made sufficiently stiff in the lateral direction by increasing the transverse dimension of the top of the beam. In a

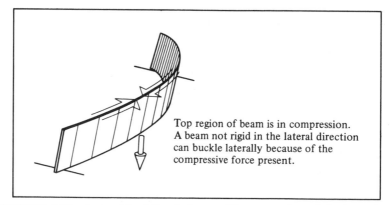

Top region of beam is in compression.
A beam not rigid in the lateral direction
can buckle laterally because of the
compressive force present.

FIGURE 6-29 Lateral buckling in beams. The top region of the beam is in compression. A beam not rigid in the lateral direction can buckle laterally because of the compressive force present.

rectangular beam, the basic proportions of the cross section can be controlled to accomplish the same end. The exact determination of these dimensions is beyond the scope of this book, but Table 6-1 illustrates when lateral bracing is not required for timber beams and recommends types of bracing when it is.

Table 6-1. LATERAL BRACING REQUIRED FOR TIMBER BEAMS

Beam depth/width ratio	Type of lateral bracing required	Example
2 to 1	None.	
3 to 1	The ends of the beam should be held in position.	Concrete blocks / Nailed
5 to 1	Hold compression edge in line (continuously).	
6 to 1	Diagonal bridging should be used.	
7 to 1	Both edges of the beam should be held in line.	

EXAMPLE

A cantilever beam 10 ft in length supports a concentrated load of 1000 lb at its free end. The maximum moment developed in the beam is therefore $M = PL = (10 \text{ ft} \times 12 \text{ in./ft})(1000 \text{ lb}) = 120{,}000$ in.-lb. Assume that the allowable stress in bending is $F_b = 1200$ lb/in.^{2} Determine the required member size.

Solution:

$$f = \frac{Mc}{I} = \frac{M}{S} \quad \text{or} \quad S_{\text{req'd}} = \frac{M}{F_b}$$

$$\frac{bd^2}{6} = \frac{M}{F_b} = \frac{(120{,}000 \text{ in.-lb})}{1200 \text{ lb/in.}^2} = 100$$

$$bd^2 = 600$$

Any beam with a bd^2 value of 600 will be satisfactory with respect to bending. The following table indicates acceptable beams.

Trial beam size (in.)		Cross-sectional area (in.²)	Depth/width ratio	Lateral bracing
b	d			
20	5.5	110	0.27 : 1	None req'd
10	7.7	77	0.77 : 1	None req'd
5	11.0	55	2.2 : 1	None req'd
3.5	13.1	46	3.7 : 1	Bracing req'd[a]
2	17.2	34.4	8.6 : 1	Bracing req'd[a]

[a]See Table 6-1.

As is evident from the previous example, there are often many different beam cross sections that could potentially be used in a common situation. Beams that are relatively wide and shallow typically require much more material to support a given load than do beams that are relatively thin and deep. The choice of which type to use is up to the designer. There are invariably design trade-offs involved. If relatively efficient thin deep beams are used, the designer *must* assure that lateral bracing of the appropriate type is provided. Providing such bracing usually requires additional material and is also an added cost item. If there is no way to provide the necessary lateral bracing, or if the designer chooses not to provide it, the proportions of the beam *must* be selected so that the member itself provides sufficient inherent resistance to lateral buckling.

Note that if a beam is used in the common situation of supporting floor decking, an inspection of the bracing requirements indicates that using beams that are relatively shallow (e.g., 2:1 or 3:1 depth/width ratios) is inefficient since the decking inherently provides lateral bracing and it is *not* necessary to have a beam capable of doing so independently. Consequently, much thinner and deeper beams are used in such instances (beams with depth/width ratios of between 5:1 and 7:1 are typically found in common house construction). So much for the exposed beams using 2:1 or 3:1 depth/width ratios that are typically found in brand-new suburban chalets.

6.4.4 Varying Materials in a Cross Section

With respect to bending, the general principle of moving as much as possible of the material in a beam toward its outer fibers (as is done in a wide-flange shape) effectively reduces some of the inefficiencies in beams which stem from the nature of the bending-stress distribution. This is not the only way of designing a beam to improve its efficiency. Instead of changing the way a specific type of material is distributed in a cross section, it is possible to vary the type of material used at different locations in a cross section. This is often easy to accomplish in certain situations. Consider the laminated timber beam illustrated in Figure 6-30(a). A laminated beam is one made up of a series of thin layers of wood bonded together to form a composite whole. The technique is

Material with allowable stress in bending: f_{b_1}

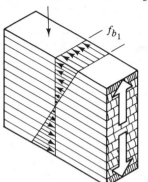

(a) Laminated timber beam materials
with higher allowable stresses in
bending are used at the extremeties
of the beam where bending stresses
are the highest.

Material with high
allowable stress

Material with lower
allowable stress

Material with high
allowable stress

(b) T beam. Top flange and web
are made from different grades
of timber having different
allowable stresses which match
the actual bending stresses present.

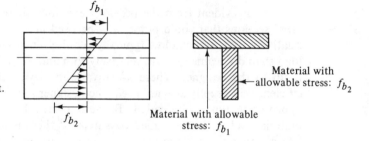

Material with
allowable stress: f_{b_2}

Material with allowable
stress: f_{b_1}

(c) Reinforced concrete T beam.
The shape of the section causes
the stresses on the top flange to
be lower than stresses in the web
of the member. Concrete is used
at the top and steel below.

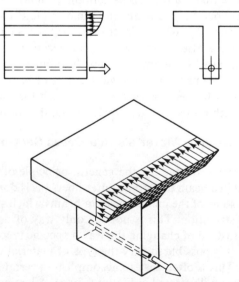

FIGURE 6-30 Varying the materials in beams to match the bending stresses
present.

often used for making large timber beams. In a case like this, the design of the section is still based on the process discussed in Section 6.4.1 [i.e., the required section modulus S of the beam is given by $S_{req'd} = M/F_b$, where F_b is the allowable stress in bending of the beam material *at the outer fiber of the beam*].

Timber comes in different grades having different allowable stresses. Grades with high allowable stresses are typically made from high-quality, hence costly, wood. Lower-stress grades are cheaper. This is the basis for achieving an economic section. In the typical bending-stress distribution illustrated in Figure 6-30(a), it is evident that it is not necessary to use wood layers made of the same-quality wood throughout the beam. Layers using wood having high allowable stresses could be used at the extremities of the beam (away from the neutral axis). Successive layers toward the neutral axis could be made of less expensive wood having lower allowable stresses. This technique of putting the stronger material on the outer faces of a beam and weaker material toward the middle can result in economic sections.

This same principle can apply to making more effective use of T-beams. It was noted previously that an inefficiency resulted from the nature of the stress distribution wherein one face is typically more highly stressed than the other. Material properties could be matched accordingly, as is illustrated in Figure 6-30(b). The same principle also explains the desirability of a T-beam constructed of reinforced concrete. The use of a T configuration naturally causes the stresses in the top flange to be relatively low (since the centroid of the shape is high and near the flange), and stresses in the bottom leg of the T to be high. This is a naturally advantageous situation for reinforced concrete. It is easy to construct such a shape from concrete. Concrete can carry compressive stresses, but not ones that are exorbitantly high. This is no problem, since the compressive stresses the flange must carry are fairly low as a result of the ways stresses are distributed. Steel is, of course, used to carry the tensile forces in the section. Since a relatively small amount of steel can carry high tensile forces, the actual amounts of steel required are usually fairly small. The thin leg of the T is usually quite adequate to house the steel. The location of the neutral axis higher in the section nearer the flange also allows for effective use to be made of the steel because of the high moment arm involved.

6.4.5 Shaping a Beam Along Its Length

The previous sections have generally considered ways of making a beam more efficient at a single cross section. This section considers the shaping of a beam along its axis as a way of improving the overall efficiency of a beam. The intention behind varying the shape of a member along its axis is to provide a better fit between the characteristics of a beam and the shears and moments which typically vary along the length of a beam.

The subject of shaping beams in response to the shears and moments present has a long history in the field of structures. In the early 1800s investigators were already determining appropriate responses for cantilever beams. It is interesting to review this same problem. Consider the cantilever illustrated in Figure 6-31(a), which carries a concentrated load at its end. Let us attempt to find an appropriate beam configuration based on the criterion that the bending stresses developed on the top and bottom

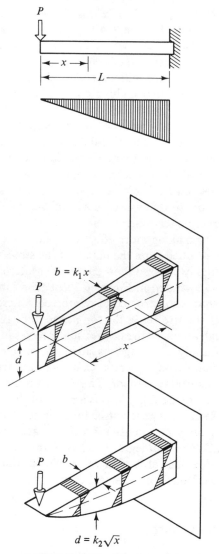

(a) Cantilever beam with concentrated load.
$M_x = P_x$.

(b) Beam shaped along its axis so that constant bending stresses are developed on its upper and lower faces. In this beam, the depth of the beam was assumed to be constant and the width b allowed to vary.

(c) Beam with constant bending stresses. In this case, the width b of the beam was assumed to be constant and the depth d of the beam allowed to vary.

FIGURE 6-31 Shaping a beam in response to the bending moment distribution present.

surface of the beam should be constant along the length of the beam. This is done by setting up an expression for the dimensions of the beam as a function of the external moment present in the beam. An expression for the moment in the beam with respect to an arbitrary reference point is also needed. Assume that a rectangular cross section is used.

Beam dimensions as a function of M:

$$S = \frac{bd^2}{6} = \frac{M}{F_b} \qquad \text{or} \qquad bd^2 = \frac{6M}{F_b}$$

External moment as a function of x:

$$M = Px$$

Hence,

$$bd^2 = \frac{6Px}{F_b} = x\left(\frac{6P}{F_b}\right)$$

If the beam depth d is assumed constant, then the width b varies directly with x, since the quantity $k_1 = 6P/d^2F_b$ is a constant. Consequently, a beam shaped as illustrated in Figure 6-31(b) would result. If the width b is assumed constant, the depth d obviously depends on $\sqrt{x}$ [i.e., $d = \sqrt{x}(\sqrt{6P/F_b b}) = k_2\sqrt{x}$]. Consequently, a beam shaped as illustrated in Figure 6-31(c) would result. If it is assumed that the beam section should always be square ($b = d$), then $d^3 = x(6P/F_b)$ or d varies as $\sqrt[3]{x}$.

For comparison purposes, it is useful to repeat this exercise with a cantilever carrying a uniformly distributed load. Again assume that a rectangular cross section is used.

Beam dimensions as a function of M:

$$S = \frac{bd^2}{6} = \frac{M}{F_b} \quad \text{or} \quad bd^2 = \frac{6M}{F_b}$$

External moment as a function of x:

$$M = \frac{wx^2}{2}$$

Hence,

$$bd^2 = x^2\left(\frac{6w}{2F_b}\right)$$

If the depth d is assumed constant, then the width b varies with x^2 since $k_3 = 6w/2F_b d^2$ is a constant. If b is assumed constant, then d varies directly with x:

$$d = x\sqrt{\frac{6w}{2F_b b}} = k_4 x$$

and the shape illustrated in Figure 6-32(c) results.

It should be emphasized that the shapes derived above are not necessarily intended to represent practical structural responses to the loadings indicated. Only bending stresses have been considered at the exclusion of other considerations (e.g., shear stresses, deflections) that might influence the final shapes found for the loadings. They are very much idealized responses to the loadings shown and their value derives largely from this. They are particularly useful as a tool in visualizing whether an actual structural response is appropriate for a given situation. The appropriateness of a structural response for a cantilever beam of the type illustrated in Figure 6-32(d), for example, can be immediately questioned by reference to the idealized responses.

The shapes found above are not necessarily generalizable to other beams carry-

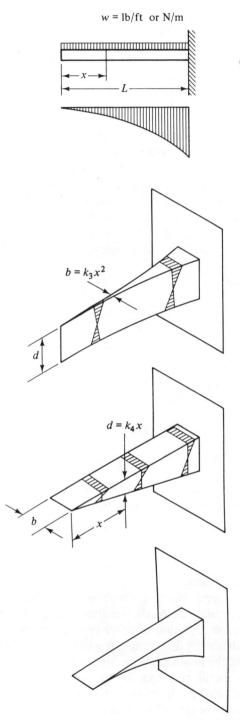

$w = \text{lb/ft} \ \text{or} \ \text{N/m}$

(a) Cantilever beam with uniformly distributed load. $M_x = wx^2/2$.

$b = k_3 x^2$

d

(b) Beam shaped along its axis so that constant bending stresses are developed on its upper and lower faces. In this beam, the depth of the beam was assumed to be constant at d and the width b allowed to vary.

$d = k_4 x$

b

x

(c) Beam with constant bending stresses. In this case, the width b of the beam was assumed constant and the depth d of the beam allowed to vary.

(d) Inappropriate structural response. Simply because a beam has the same shape as the moment diagram for the loading does not mean it is a preferred design.

FIGURE 6-32 Shaping a beam in response to the bending moment distribution present.

ing similar loads but supported differently. The moment equations for different types of beams differ, so it can be expected that shapes would also. The student is encouraged to explore types of beams other than the cantilevers illustrated and determine appropriate shapes.

Cross sections other than rectangular can also be used as the basis for determining idealized responses. Figure 6-33(b) illustrates a steel wide-flange beam designed in response to the moment present assuming that the beam should remain of constant depth and that the width of the flange is the only variable. Varying the width of the flange causes the section modulus S to change. This variation could be coupled with

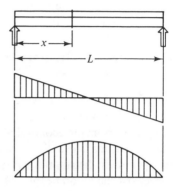

(a) Uniform loading condition.

(b) Wide-flange shape in which the width of the top flange has been varied to match the required section modulus of the beam with the amount of bending moment present at a section.

(c) Built-up member. The section modulus of the beam has been matched with the moment present by varying the thickness of the flange (in this case, "cover plates" are used to accomplish this).

FIGURE 6-33 Varying the sectional properties of beams in response to the bending moment distribution.

the moment distribution present in the beam to obtain a configuration where the bending stress level in the flanges remains constant.

In a wide-flange shape, the beam depth and flange width could be assumed as constant and the flange thickness allowed to vary. Figure 6-33(c) illustrates a beam of this type where, as a matter of construction simplicity, the variation in flange thickness is accomplished by use of horizontal layers of thin steel plates bonded (welded or bolted) together. Again the properties of this beam at each section can be coupled with the amount of moment present to create a situation of essentially constant bending stresses in the beam flanges. Stresses cannot be held purely constant, since discrete changes occur in the flange thickness due to the layering used. Often standard structural steel wide-flange shapes have additional steel plates (commonly called *cover plates*) added at regions of high moment.

6.4.6 Varying Support Locations and Boundary Conditions

The discussion thus far has dwelt on ways of improving the efficiency of a beam without altering how the structure is supported. Manipulating support conditions can also lead to major economies in the use of materials. The immediate intent of such manipulations is usually to reduce the magnitudes of the design moments present or to alter their distribution. A classic way of reducing design moments it to use cantilever overhangs on beams. The effect of cantilevering one end of a simply supported beam with uniform loads is to cause a reduction in the positive moment present while a negative moment develops at the base of the cantilever over the support. The greater the cantilever, the higher the negative moment becomes and the lower the positive moment becomes. The cantilever can be extended until the negative moment even exceeds the positive moment.

Of interest here is that there must exist a certain overhang such that the numerical value of the positive moment is exactly equal to that of the negative moment. It can be expected that these moments are considerably smaller than either the case of those with no overhang or those with a large overhang. As such, they would represent minimum possible design moments.

Figure 6-34(b) illustrates a simple beam with a variable length cantilevers on both ends. If no cantilevers were present ($x = 0$), it is evident that the critical design moment would be $M = 0.125\,wL^2$ (the maximum positive moment at midspan). If the ends were cantilevered to a certain specific extent, the structure could be designed for a moment of $M = 0.021\,wL^2$ at midspan and $M = -0.021\,wL^2$ at either support. The moment can be reduced no farther than this. Reducing the design moment in this way leads to considerable economies when member sizes are determined. The exact extent of the overhang on either end which results in the positive and negative moments being equal can be found either by a trial-and-error process or by setting up one equation in terms of the variable x for the negative moment at a support and another for the positive moment at midspan and equating them. Figure 6-34(b) illustrates the results of such an analysis. The optimum overhang is approximately one-fifth of the overall span.

The same approach can be taken when the location of only one instead of both

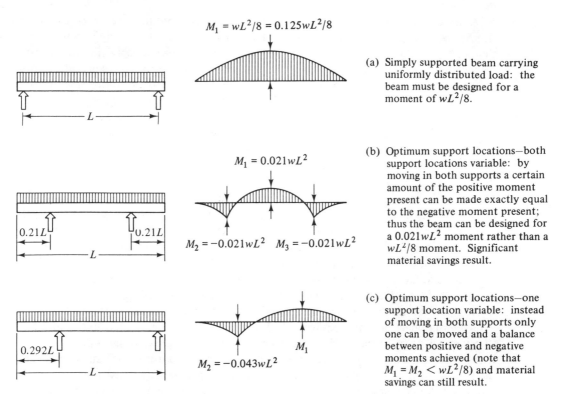

$$M_1 = wL^2/8 = 0.125wL^2/8$$

(a) Simply supported beam carrying uniformly distributed load: the beam must be designed for a moment of $wL^2/8$.

$$M_1 = 0.021wL^2$$

$$M_2 = -0.021wL^2 \quad M_3 = -0.021wL^2$$

(b) Optimum support locations—both support locations variable: by moving in both supports a certain amount of the positive moment present can be made exactly equal to the negative moment present; thus the beam can be designed for a 0.021wL^2 moment rather than a $wL^2/8$ moment. Significant material savings result.

$$M_2 = -0.043wL^2$$

$$M_1$$

(c) Optimum support locations—one support location variable: instead of moving in both supports only one can be moved and a balance between positive and negative moments achieved (note that $M_1 = M_2 < wL^2/8$) and material savings can still result.

FIGURE 6-34 Locating supports to minimize design moments in beams.

supports can be varied. The optimum overhang is about one-third of the span. Considerations of the type above are one way of answering the question of how far is it possible to cantilever a beam. Rules like the above, however, should be tempered with other considerations. In many cases, for example, the amount of deflection at the end of the cantilever will control how far one can extend the member rather than using criteria based on moment considerations. The type of construction is also important. The measures are not applicable to prestressed or post-tensioned reinforced concrete members, for example, without special considerations. By and large the one-fifth and one-third measures given above are best applied to relatively short or moderate span members made of homogeneous materials.

6.4.7 Prestressing and Post-tensioning

A way of making beams more efficient that is particularly suitable for concrete structures is through the use of prestressing or post-tensioning. These are techniques for permanently loading a beam in a controlled way with the intent of building up stresses in the member opposite to those developed by the external service loads. The basic concepts underlying this approach have already been covered in Section 6.3.6, which dealt with the combined effects of axial and bending loads.

From a design viewpoint, good use can be made of the way stresses interact. Consider the bending stresses developed in the beam previously illustrated in Figure

6-18. By applying an axial force to the beam it is possible to develop compressive stresses of a magnitude sufficient to dominate over the tensile bending stresses and thus put the entire cross section in compression. This is particularly important with respect to concrete, which cannot withstand tension stresses and cracks when subjected to them. By putting the entire cross section in compression, more effective use can be made of this material. It is important to note that the compressive stresses from bending and the applied axial force are additive. Thus, the advantages of doing this to a beam made of a material (e.g., steel) inherently capable of withstanding tension stresses is dubious, since the increased level of the combined compressive stresses would govern the amount of load the member could carry. Note that when the axial load is eccentrically applied, the types of stresses induced are particularly useful for offsetting the bending stresses associated with the service loads. There are a number of other issues involved, however, which tend to make the analysis and design of such members far more complex than implied in the discussion above.

Consider the first of the two most common ways of applying an axial force to a member, that of *prestressing* (see Figure 6-35). In this process, which is typically done in an off-site factory, high-strength steel wires or tendons are stretched between two piers so that a predetermined tensile force is developed. Concrete is then cast in formwork placed around these wires and cured. Wires are then cut. The tension force in the wire then in effect becomes equivalent to a compressive force applied to the member (the tension stress in the wires is transferred to the concrete through bond stresses). If the wire is eccentrically placed, the effect of cutting the wires is to induce stresses of the type illustrated in Figure 6-35(c) and to cause the member to bow upward. When upward bowing commences, the dead load of the concrete member itself begins inducing stresses of the type illustrated. Neither of the stress patterns shown in Figure 6-35(d) occurs independently and a combined stress pattern of the type shown to the right of the figure is present after wire cutting. Note that the exact amount of prestress force must be very carefully controlled. Indeed, during this stage of manufacture is when failure often occurs. Turning the beam upside down or on its side would lead to a rather dramatic failure, since dead-load stresses would no longer offset tensile stresses developed by the eccentric load, but would accentuate them. When the member is carefully put in place (in this case by supporting it from its ends) the live load can then be applied, which in general results in a final stress distribution of the type illustrated in Figure 6-35(e).

The second most common method of applying a normal force to a member is that of *post-tensioning*. In this method, which is typically done on-site, a tube, conduit, or equivalent containing unstressed steel tendons is set in place and concrete cast around the tube. After curing, the tendons are clamped on one end and jacked against the concrete on the other end until the required force is developed. The tendons are then anchored on the jacking end and the jacks removed. The tendons are then usually bonded in place by injecting grout into the tube. The net effect is to do the same thing that the prestressing technique accomplishes (see Figure 6-36).

In both of the approaches discussed above, a host of problems exist. An important one is the effect of "creep" in concrete (deformation with time in a constant-stress situation), which has the effect of causing a loss in prestress or post-tension

(a) High-strength steel strands are
pretensioned between abutments
by using jacks.

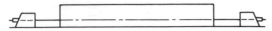

(b) Concrete is cast around the pretensioned
strands and allowed to cure.

(c) After the concrete has cured, the strands are
cut. Since the strands are in the lower part
of the beam, cutting of the cables is equivalent
to putting a compressive force in the beam at
this level. The beam bows upwards.

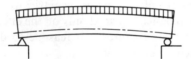

Dead load $\quad$ P/S $\qquad$ Total

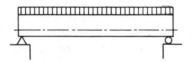

(d) Beam in place. An upward camber is
present. Dead load and prestress force
stresses are present.

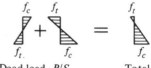

Dead load $\quad$ P/S $\qquad$ Total

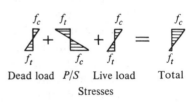

Dead load $\quad$ P/S $\quad$ Live load $\qquad$ Total
Stresses

(e) Beam under live load. The live load causes
the beam to deflect downwards. The beam
thus shows no sag under load. Dead load,
live load, and prestress force stresses are
present.

FIGURE 6-35 Prestressed concrete members.

force. These forces must be calculated very carefully so that such losses with time are
taken into account. Another consideration is that the stresses due to an eccentric force
of the type illustrated are constant along the length of the beam, while the stresses
associated with live and dead loads vary along the length of the beam. Indeed, with a
constant eccentric force, critical prints in the beam are often near the ends and are
due to the prestress force, since the offsetting effects of stresses due to the live and
dead loads are not present. One advantage of post-tensioning over prestressing is that

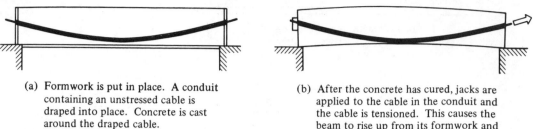

(a) Formwork is put in place. A conduit containing an unstressed cable is draped into place. Concrete is cast around the draped cable.

(b) After the concrete has cured, jacks are applied to the cable in the conduit and the cable is tensioned. This causes the beam to rise up from its formwork and break the bond present. The formwork is removed and the cable force locked in. Grout is often inserted into the conduit to cause a bond to develop between the cable and the member.

FIGURE 6-36   Post-tensioned concrete members.

the cable can be draped fairly easily so that the eccentricity of the cable force is a variable which can be adjusted to account for this phenomenon. Deforming prestressing steel in this way is more difficult.

In situations where the beam undergoes a reverse curvature, special care must be taken with prestressed or post-tensioned members. In small members (e.g., concrete planks) this is often handled by not placing wires eccentrically but putting them at the centroid of the member. Thus, reliance is placed only on the effect of the uniform stresses produced by the wires. In large structures it is feasible to drape a post-tensioned cable to reflect the anticipated reverse curvature.

A final point is of importance in the ultimate load-carrying capacity of prestressed or post-tensioned members. Usually, the primary criterion in designing such beams is their behavior under service or working loads. As loads increase beyond design loads, the beam begins behaving much like a standard reinforced concrete beam (i.e., cracks develop, etc.). The ultimate strength of a prestressed or post-tensioned beam is not overly superior to that of a similarly proportioned plain reinforced concrete beam. The primary value of the tensioning operation is thus to improve the performance of the beam at design or service loads. In particular, a very significant reason for prestressing or post-tensioning is to create a larger effective section for resisting deflections. In plain reinforced concrete beams cracks develop in the tension region and propagate until they reach the compressive zone. This cracked section makes no significant contribution to the stiffness of the section (the magnitude of I would be based only on the extent of the compressive zone and on the amount of steel present). In a prestressed or post-tensioned beam, no tensile stresses are developed and hence no cracks develop. Thus, the entire cross section of a beam contributes to the stiffness of the beam (i.e., the I would be based on the gross dimensions of the section). Thus, for two beams of similar gross dimensions, a prestressed or post-tensioned one is better able to resist deflections than is its plain reinforced counterpart. In addition, the upward camber produced by eccentric prestressing forces is clearly useful in controlling deflections. All members should be carefully designed, however, since prestressed members can sometimes exhibit an undesirable "springiness." All in all, however, such members are highly versatile and useful.

QUESTIONS

6-1. Discuss in detail the historical development of quantitative methods of analyzing beams. Consult your library.

6-2. Determine the value of the moment of inertia, I, and the section modulus, S, for each of the following cross-sectional shapes:
(a) A 4 in. $\times$ 10 in. rectangle.
(b) A 10 in. $\times$ 20 in. rectangle.
(c) A 10 in. $\times$ 20 in. rectangle with a symetrically located interior rectangular hole of 8 in. $\times$ 12 in.
Consider the moment of inertia about the centroidal strong axis of each section only.
Answer: (a) 333 in.^{4}, 66.6 in.^{3}; (b) 6,666 in.^{4} 666 in.^{3}; (c) 5514 in.^{4}, 551.4 in.^{3}.

6-3. Determine the value of the moment of inertia about the centroidal axis of a triangular cross section having a base dimension of 5 in. and a height of 15 in.
Answer: 468.75 in.^{4}.

6-4. Floor joists having cross-sectional dimensions of $1\frac{1}{2}$ in. $\times$ $9\frac{1}{2}$ in. are simply supported, span 12 ft, and carry a floor load of 50 lb/ft^2. Compute the center-to-center spacing between joists to develop a maximum bending stress of 1200 lb/in.^{2}. Compute what safe floor load could be carried if the centerline spacing was 16 in. Neglect dead loads.

6-5. Consider a cantilever beam that is 10 ft long and which supports a concentrated load of 833 lb at its free end. Assume that the beam is 2 in. $\times$ 10 in. in cross section and that lateral bracing is present. Draw a diagram of the beam showing the bending-stress distribution present at the base of the cantilever and at every 2.5-ft increment toward the free end of the beam (a total of five locations). Indicate numerical values for the maximum bending stresses at each section. Do a similar exercise for shearing stresses. What generally happens to the magnitudes of the bending and shearing stresses found when the width of the beam is doubled and its depth held constant? What happens when the width of the beam is held constant and its depth doubled?

6-6. Repeat Question 6-6, except assume that the beam is simply supported at either end and carries a uniformly distributed load of 200 lb/ft.

6-7. A simply supported beam 12 ft long carries a uniformly distributed load of 100 lb/ft. Assume that the beam is $1\frac{1}{2}$ in. $\times$ $9\frac{1}{2}$ in. in cross section and is laterally braced. Also assume that the beam is made of timber which has an allowable stress in bending of 1200 lb/in.^{2} and in shear of 150 lb/in.^{2}. Is the beam safe with respect to bending and shear stress considerations? What is the maximum deflection of the beam? Assume that $E = 1.6 \times 10^6$ lb/in.^{2}. Is this deflection acceptable?
Answer: $(f_b = 956) < (F_b = 1200)$, $\therefore$ safe in bending; $(f_v = 63.1) < (F_v = 150)$, $\therefore$ safe in shear; and $(0.29) < (L/360 = 0.4)$, $\therefore$ deflections are okay.

6-8. A steel beam is to be used to span 30 ft and support a uniformly distributed load of 400 lb/ft. Assume that the allowable stress in bending is $F_b = 24,000$ lb/in.^{2}. Determine the most efficient wide-flange shape to be used based on a bending stress analysis. Use one of the shapes listed in Appendix A-13. Ignore dead loads.

6-9. Assume that a laminated timber beam having cross-sectional dimensions of 8 in. $\times$ 20 in. is available. Based on bending-stress considerations only, how far could this beam span if it carried a uniformly distributed load of 250 lb/ft and were simply supported at either end? How far could it span if it carried the same load but had fixed ends? How far could it span if it carried the same load but were cantilevered? Assume that the allowable stress in bending is $F_b = 2400$ lb/in.^{2} and that the beams are all adequately laterally braced. Ignore dead loads.

Answer: 58.4 ft if simply supported.

6-10. Assume that a series of laminated timber beams are to be used at 5 ft on center to span 25 ft and carry a uniformly distributed floor load of 60 lb/ft^2 (assume that this figure represents all live and dead loads present). Completely design an appropriate proto-typical beam. Assume that the allowable stress in bending is $F_b = 2200$ lb/in.^{2}, in shear $F_v = 400$ lb/in.^{2}, in bearing $F_{bg} = 400$ lb/in.^{2}, and that $E = 1.8 \times 10^6$ lb/in.^{2}. Make any assumptions that you need to (e.g., about lateral bracing), but clearly state them and/or illustrate with sketches.

6-11. With respect to bending-moment considerations only, draw a sketch of the shape variation present in a beam carrying two equal concentrated loads located at third points in the structure such that a constant bending-stress level is maintained on the top and bottom surfaces of the member. Draw one sketch assuming the width of the beam is held constant and the depth varies, and another assuming that the depth of the beam is held constant and the width allowed to vary.

chapter 7

MEMBERS IN COMPRESSION: COLUMNS

7.1 INTRODUCTION

Along with load-bearing walls, columns are the most common of all vertical support elements. Even a load-bearing wall, however, can be considered as a column extended in one plane. Strictly speaking, columns need not be only vertical. Rather, they are rigid linear elements that can be inclined in any direction, but to which loads are applied only at member ends. They are not normally subject to bending directly induced by loads acting transverse to their axes.

Columns can be usefully categorized in terms of their length. *Short columns* tend to fail by crushing (a strength failure). *Long columns* tend to fail by buckling, which is an instability failure rather than a strength failure. Material rupture occurs only after the member has buckled. Long columns are members in which the length of the element is relatively great as compared to its least lateral dimension. The potential for buckling tends to limit the load-carrying capacity of long members. The buckling phenomenon itself was recognized quite early as being of unique interest. A number of investigators attempted analytical solutions for predicting exactly what load would cause a slender member to buckle. The problem was finally solved by Leonard Euler (1707–1783). Euler was a mathematician born in Switzerland and related through training and association to the celebrated Bernoulli family, long recognized for their contributions to mathematics.

Euler correctly analyzed the action of a long, slender member under an axial load while in St. Petersburg, Russia, in 1759. The form that he gave to the solution is still in use today. It is one of the few early contributions to the structural engineering field that has survived in a virtually unchanged form.

7.2 GENERAL PRINCIPLES

The ultimate load-carrying capacity of a compression members depends on the relative length and cross-sectional dimensional characteristics of the member (with the least cross-sectional dimension playing a critical role), as well as the properties of the materials used. Columns are usually classified as either short or long, depending on the relative type of failure that might occur under excessive load.

7.2.1 Short Columns

Members in which the least cross-sectional dimension present is appreciable relative to the length of the member are short columns. The load-carrying capacity of a short column is *independent* of the length of the member and, when excessively loaded, the short column typically fails by crushing. Consequently, its ultimate load-carrying capacity depends primarily on the strength of the material used and the amount of it present in the cross section. The problem of what exactly defines a short column will be addressed in Section 7.3.2 of this chapter. Many commonly encountered elements in buildings (e.g., masonry piers) are actually short columns.

7.2.2 Long Columns

As a compressive member becomes longer and longer, the relative proportions of the member change to the extent that it can be described as a slender element, or long column. The behavior of a slender element under compressive load differs dramatically from that of a short column.

Consider the long compression member illustrated in Figure 7-1 and its behavior under increasing loads. When the applied load is small, the member maintains its linear shape and continues to do so as the load is increased. As a particular load level is reached, the member will quite suddenly become unstable and deform into the shape illustrated. This is the *buckling* phenomenon. When the member has buckled, it is no longer capable of carrying any additional load. Added loads will immediately cause added deformation, which will eventually cause the member to snap. This snapping, however, is regarded as a secondary failure, since the maximum load-carrying capacity is that associated with initial buckling. A structure in a buckled mode is *not* serviceable.

The phenomenon of buckling is a curious one. It is a mode of failure that is due to an instability of the member induced through the action of the load. Failures due to instability need not initially involve any kind of material rupture or distress. Indeed, internal force levels can be quite low and buckling can still occur if the member is quite long. The buckling phenomenon is associated with the *stiffness* of the member. A member with low stiffness will buckle before one with high stiffness. Increasing member length causes a reduction in stiffness.

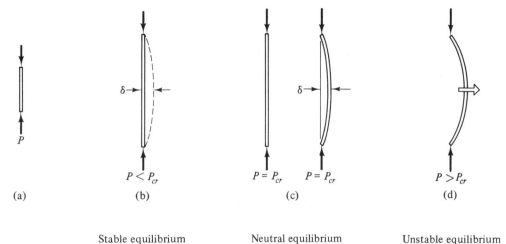

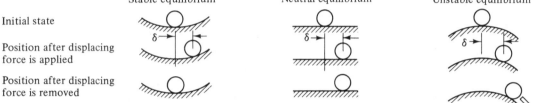

Representation of stable, neutral, and unstable equilibrium.

(a) Short column: the member fails by a crushing action.

(b) Long column (load less than the buckling load): the member is in a state of stable equilibrium. If the member is displaced slightly at mid-height it will spring back into its original configuration.

(c) Long column (load = buckling load): when the load on the column reaches the critical buckling load, the member changes to a state of neutral equilibrium. If the member is displaced from its original linear configuration, it will remain exactly in the new configuration and not spring back. The buckling load is the maximum load the column can carry.

(d) Long column (load greater than the buckling load): if the load on the column is increased beyond the critical buckling load, the member changes to a state of unstable equilibrium. The member will continue to deform under a constant load level until it finally snaps.

FIGURE 7-1 Behavior of columns under load.

When a member initially becomes unstable, as does a column exactly at the buckling load, the member does not and cannot generate internal forces that tend to restore the structure to its original linear configuration. Such restoring forces are developed below the buckling load. Figure 7-1 illustrates in a diagrammatic way systems that are in stable, unstable, and neutral equilibrium. A column exactly at the buckling load is similar to the system in neutral equilibrium. The system does not have restoring characteristics which tend to reestablish the original configuration of the element once it is displaced.

There are numerous factors that influence the buckling load, denoted P_{cr}, of a long compressive member. *Length,* of course, is a critical factor. In general, the load-carrying capacity of a column varies *inversely* as the square of its length. In addition to the length of the member, the other factors that influence the load necessary to cause a member to buckle are associated with the stiffness characteristics of the member that result from the inherent properties of the specific material used and those that result from the amount of material used in a cross section and the way the material is distributed.

The *stiffness* of the member is strongly influenced by the amount and distribution of the material present. In the thin rectangular member illustrated in Figure 7-2(a), it can be easily experimentally demonstrated, or theoretically proven, that the member will always buckle in the direction indicated. The member is less rigid about this axis than the other. A symmetrical member (e.g., a round or a square one) would not have a characteristic direction of buckle. By considering the thin rectangular member with respect to its two primary axes, it can be seen that the member has much less of an ability to resist bending about one axis than the other. The member will tend to buckle *about* the weak axis (associated with the lesser ability to resist bending). The same member, however, can be sufficiently stiff about the other axis to resist buckling associated with that axis. The load-carrying capacity of a compressive member is thus not dependent on just the amount of material present in the cross section (as is the case in tension members), but on its distribution as well. A useful measure in this connection is the *moment of inertia, I,* which combines the amount of material present with the way it is distributed into a single stiffness characteristic. Moments of inertia are discussed in Chapter 6 and Appendix 3.

Another factor of extreme importance in influencing the amount of load a member can carry is the nature of the *end conditions* of the member [see Figure 7-2(b)]. If the ends of a column are free to rotate, the member is capable of carrying far less load than if the ends are restrained. Restraining the ends of a member increases its stiffness and thus its ability to resist buckling. *Bracing* the element in some direct way also increases stiffness. These important influences are explored in detail in Section 7.3.2.

Material properties also influence the load-carrying capacity of a member. Some materials are inherently stiffer than others (and deform less under a given load). A measure useful in this connection is the *modulus of elasticity, E.* The higher the modulus of elasticity for a material, the stiffer it is and the more resistant a column made of this material is to buckling, and vice versa. Section 2.5.3 discusses the concept of the modulus of elasticity of a material in detail.

7.2.3 Summary

In summary form, Figure 7-2(c) illustrates the general relationship between the load-carrying capacity and length of a member in direct compression. As illustrated, the phenomenon of buckling generally causes a reduction in the load-carrying capabilities of compression members. The shape of the buckling curve illustrates that the load-carrying capacity of a long compressive member varies inversely with the square of the length of the member. The maximum load a column can carry, however, is invari-

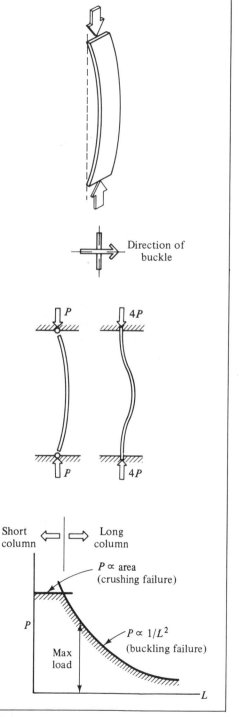

(a) Long members that are stronger about one axis than the other always buckle about the weak axis.

Direction of buckle

(b) End conditions are very important in affecting how much load is required to buckle a member. Assuming that the members above are identical except with respect to end conditions, the member with fixed-ends can carry four times the load that one with pinned ends can carry.

(c) General relationships between member length and buckling load. Short members tend to crush while longer ones tend to buckle. The longer the member, the lower its load-carrying capacity.

FIGURE 7-2 Long columns: The shape of the cross section, materials used, end conditions and the member length all significantly influence the load-carrying capacity of a column.

ably associated with *crushing* (in the short-column range) rather than buckling. The crushing load is given by $P_y = AF_y$, where A is the cross-sectional area of the column and F_y the yield or crushing stress of the material. Long, slender columns invariably fail at loads less than the crushing load. It follows that the actual stress level present when buckling occurs (the *critical* buckling stress) is invariably less than the crushing or yield stress (i.e., $f_{cr} = P_{cr}/A \leq f_{\text{yield}}$). Long columns can fail by buckling at stress levels considerably lower than that associated with yielding in the material.

It should always be remembered that the actual axial stress level present in any symmetrically loaded compressive member, long or short, is always given by $f = P/A$. The stress level at which failure or buckling occurs, however, depends on whether the member is long or short.

7.3 ANALYSIS OF COMPRESSION MEMBERS

7.3.1 Short Columns

AXIAL LOADS. Compression members that fail primarily by crushing or direct stresses, and whose ultimate load-carrying capacity is consequently independent of member lengths, are relatively easy to analyse. When the load is applied to the centroid of the cross section of the loaded element, uniform compressive stresses of a magnitude $f = P/A$ are developed. Failure occurs when the actual direct stress exceeds the crushing stress of the material (i.e., $f_a \geq F_y$).

ECCENTRIC LOADS. When loads are eccentrically applied (i.e., not applied at the centroid of the cross section), the resultant stress distribution will not be uniform. The effect of eccentric loads is to produce bending stresses in the member, which in turn interact with direct compressive stresses. If the load is very eccentrically applied, tensile rather than compressive stresses can even be developed at a cross section.

Consider the member shown in Figure 7-3 which is subjected to an eccentric load P acting at a distance e from the centroidal axis of the member. Stresses produced by this load can be found by resolving the eccentric load into a statically equivalent axial force producing only uniform stresses f_a and a couple (moment) which produces only bending stresses f_b. This resolution is illustrated in Figure 7-3. Final stresses are merely the combination of the two stress distributions.

$$\text{uniform stresses} = f_a = \frac{P}{A}$$

$$\text{bending stresses} = f_b = \frac{Mc}{I}$$

Since

$$M = Pe$$

$$f_b = \frac{(Pe)c}{I}$$

$$\text{combined stresses} = f_{\text{actual}} = f_a + f_b = \pm \frac{P}{A} \pm \frac{(Pe)c}{I}$$

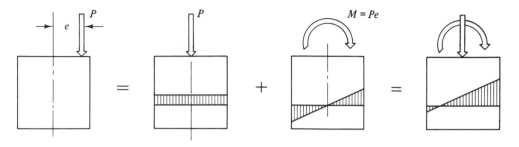

(a) Stress distributions under eccentric loads.

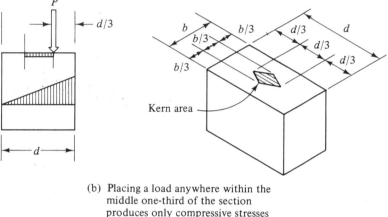

(b) Placing a load anywhere within the middle one-third of the section produces only compressive stresses in the cross section. Placing a load outside of the middle one-third leads to the development of tensile stresses on one face of the member.

FIGURE 7-3 Eccentric loads.

An inspection of the stress distributions illustrated in Figure 7-3 reveals that the magnitude of the bending stresses is proportional to the eccentricity e of the load. In a situation of the type indicated, the vertical load P can produce tensile stresses on one face of the element if the eccentricity is large (i.e., bending stresses f_b dominate over axial or normal stresses). When $e = 0$, clearly only compressive stresses f_a exist. There must exist a limit to the eccentricity of the load if the intention is to have only compressive stresses in the member. This point can be found by simply equating the resultant stress to zero ($f_a + f_b = 0$) and solving for the eccentricity; or $P/A = (Pe)c/I$ and $e = I/Ac$. For a rectangular cross section (see Figure 7-3),

$$\frac{P}{bd} = \frac{Pe(d/2)}{bd^3/12} \quad \text{and} \quad e = \frac{d}{6}$$

Thus, if the load were placed within this maximum value, the stresses produced would all be compressive. Placing the load exactly at this point ($e = d/6$) produces

chapter 7 / MEMBERS IN COMPRESSION: COLUMNS   **289** zero stresses on the opposite face. Exceeding this eccentricity will cause tensile stresses to develop on the opposite face. This location is obviously of some importance and is referred to as the *Kern point*. Note that for a load which can vary in either direction, Kern points exist on either side of the centroidal axis. These locations are at third points on the face of the element. This gives rise to the *middle third rule*, often referred to in the design of masonry structures (particularly in a historical context), where the design intent is to keep the load inside the middle third to prevent tensile stresses, which masonry cannot withstand, from developing. If the third dimension is considered, a *Kern area* can be found, as indicated in Figure 7-3.

COMPOSITE SECTIONS.   In many situations several different materials are bonded together lengthwise to form a compressive member. A reinforced concrete column is an obvious example. In cases of this type, the compressive load is shared among the materials present. In a reinforced concrete column, both the concrete and steel will deform exactly the same amount under load since they are bonded together. Stress levels in the steel and concrete must consequently be such as to produce deformations that are compatible. Since the steel is stiffer than the concrete, higher stress levels must be present in the steel than in the concrete to produce equivalent deformations (see Section 2.6). The consequence is that the steel in a reinforced concrete column carries a far larger proportion of the total load in the column than a simple consideration of relative areas would indicate. More advanced texts discuss how to quantitatively determine stress distributions in composite members.[1]

Reinforced concrete columns are particularly sensitive to creep effects (see Section 2.6). Creep effects may cause the relative stress levels present to change with time (with the relative stress level in the concrete decreasing and that in the steel increasing). This important phenomenon is discussed in more detail elsewhere.[2]

7.3.2 Long Columns

The analyses discussed in the previous sections were all strength-oriented. This section introduces methods of analyzing the behavior of long columns subject to buckling.

EULER BUCKLING.   As previously mentioned, Leonard Euler was the first investigator to formulate an expression for the critical buckling load of a column. The critical buckling load for a pin-ended column, termed the *Euler buckling load*, is

$$P_{cr} = \frac{\pi^2 EI}{L^2}$$

where E = modulus of elasticity
$\quad I$ = moment of inertia
$\quad L$ = length of column between pinned ends
$\quad \pi$ = constant pi, = 3.1416

The derivation for this expression is contained in Appendix 12. The expression clearly shows that the load-carrying capacity of a column depends inversely on the

---

[1]F. L. Singer, *Strength of Materials*, 2nd ed., Harper & Row, New York, 1962.
[2]P. M. Ferguson, *Reinforced Concrete Fundamentals*, 3rd ed., John Wiley & Sons, Inc., New York, 1973.

square of the member's length, directly on the value of the modulus of elasticity of the material used, and directly on the value of the moment of inertia of the cross section. The moment of inertia of concern is the minimum one about any axis for the cross section if the member is not braced.

The Euler buckling expression predicts that when a column becomes indefinitely long, the load required to cause the member to buckle begins to approach zero (see Figure 7-4). Conversely, when the length of a column begins to approach zero, the load required to cause the member to buckle becomes indefinitely large. What actually happens, of course, is that as the member becomes short, the failure mode changes into that of crushing. Consequently, the Euler expression is not valid for short members, since it predicts impossibly high values. The crushing strength becomes a cutoff for the applicability of the Euler expression.

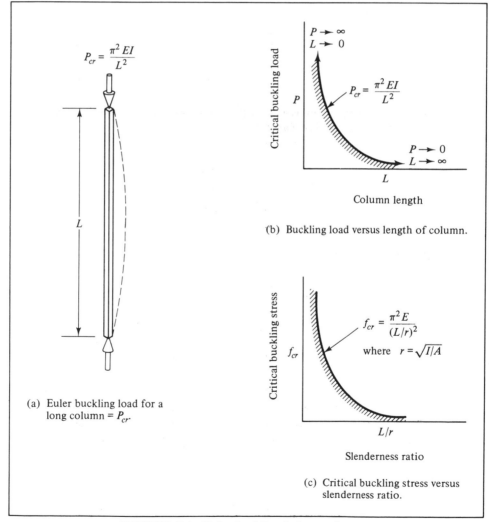

FIGURE 7-4 Euler buckling in long columns.

The dependency of the buckling load on the inverse of the square of the length of the column is important. Doubling the length of a column reduces its load-carrying capacity by a factor of 4. Thus, if $P_1 = \pi^2 EI/L_1^2$, and $L_2 = 2L_1$, then $P_2 = \pi^2 EI/(L_2)^2$ or $\pi^2 EI/(2L_1)^2 = P_1/4$. Similarly, halving its length increases its load-carrying capacity by a factor of 4 [i.e., $P_2 = \pi^2 EI/(\frac{1}{2}L_1)^2 = 4P_1$]. The buckling load of a column is thus extremely sensitive to changes in the length of the member.

EXAMPLE

Determine the critical buckling load for a 2 in. × 2 in. (50.8 mm × 50.8 mm) steel column that is 180 in. (4572 mm) long and pin-ended. Assume that $E = 29.6 \times 10^6$ lb/in.² (0.204×10^6 N/mm²).

Solution:

Moment of inertia, I_x or I_y:

$$I = \frac{bh^3}{12} = \frac{2 \text{ in. } (2 \text{ in.})^3}{12} = 1.33 \text{ in.}^4$$

$$= \frac{50.8 \text{ mm}(50.8 \text{ mm})^3}{12} = 555 \times 10^3 \text{ mm}^4$$

Since $I_x = I_y$, the column is equally likely to buckle about either axis, or about a diagonal since that I value also is equal to I_x or I_y.

Critical buckling load:

$$P_{cr} = \frac{\pi^2 EI}{L^2} = \frac{\pi^2 (29.6 \times 10^6 \text{ lb/in.}^2)(1.33 \text{ in.}^4)}{(180 \text{ in.})^2} = 12{,}000 \text{ lb}$$

$$= \frac{\pi^2 (204{,}000 \text{ N/mm}^2)(555{,}000 \text{ mm}^4)}{(4572 \text{ mm})^2} = 53{,}458 \text{ N}$$

The actual stress level corresponding to this critical buckling load is

$$f = \frac{P}{A} = \frac{12{,}000 \text{ lb}}{2 \text{ in. } \times 2 \text{ in.}} = 3000 \text{ lb/in.}^2$$

$$= \frac{53{,}458 \text{ N}}{50.8 \text{ mm} \times 50.8 \text{ mm}} = 20.7 \text{ N/mm}^2$$

Thus, the column buckles at a relatively low actual stress level.

EXAMPLE

For what length will the same square column previously analyzed begin to crush rather than buckle (i.e., what is the transition length between short- and long-column behavior for this specific member). Assume the yield stress of the steel to be $F_y = 36{,}000$ lb/in.² (248 N/mm²).

Solution:

For failure by crushing:

$$P_{max} = F_y A = (36{,}000 \text{ lb/in.}^2)(2 \text{ in. } \times 2 \text{ in.}) = 144{,}000 \text{ lb}$$

$$= (248 \text{ N/mm}^2)(50.8 \text{ mm} \times 50.8 \text{ mm}) = 640 \text{ kN}$$

Buckling length for $P_{\max}$:

$$P_{cr} = \frac{\pi^2 EI}{L^2}$$

$$= P_{\max}$$

$$144,000 \text{ lb} = \frac{\pi^2 (29.6 \times 10^6 \text{ lb/in.}^2)(1.33 \text{ in.}^4)}{L^2} \qquad \therefore L = 52 \text{ in.}$$

$$= 640,000 \text{ N} = \frac{\pi^2 (204,000 \text{ N/mm}^2)(555,000 \text{ mm}^4)}{L^2} \qquad \therefore L = 1320 \text{ mm}$$

Consequently,

For $L < 52$ in. (1320 mm)	the member will crush
For $L > 52$ in. (1320 mm)	the member will buckle
For $L = 52$ in. (1320 mm)	either failure mode is likely

In actuality, such an exactly defined transition point does not really exist. Rather, a more general transition occurs.

When an unbraced member is nonsymmetrical, it is necessary to take into account the different moments of inertia of the member (Figure 7-5). Members of this type will generally buckle in the direction of their least dimension, or, more precisely, about their weaker axis. A typical rectangular column has two primary moments of inertia, I_x and I_y. Associated with each is a load that will cause the member to buckle about each respective axis, P_{cr_x} and P_{cr_y}. The load that actually causes the whole mem-

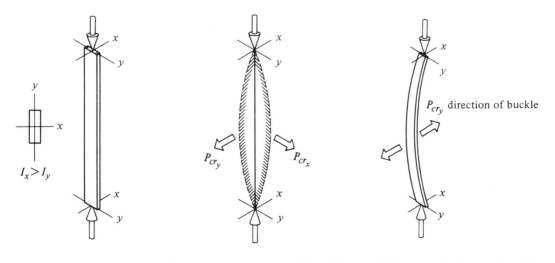

(a) The moment of inertia about one axis is greater than that about the other.

(b) The member can potentially fail by buckling about either axis. The load required to cause it to buckle about the stronger axis, however, exceeds the load which will cause buckling about the weaker axis.

(c) Consequently the member will buckle at $P_{cr_y} = \pi^2 EI_y / L^2$ in the direction shown.

FIGURE 7-5 Buckling of asymmetric cross sections.

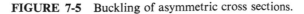

ber to buckle is the smaller of these two values:

$$P_{cr_z} = \frac{\pi^2 E I_x}{L_x^2} \quad \text{and} \quad P_{cr_y} = \frac{\pi^2 E I_y}{L_y^2}$$

EXAMPLE

Determine the critical buckling load for a column with the same cross-sectional area as the column previously analyzed but which is rectangular: $b = 1$ in., $d = 4$ in. (25.4 mm $\times$ 101.6 mm). Assume that $L = 180$ in. (4572 mm). $E = 29.6 \times 10^6$ lb/in.^{2} (204,000 N/mm^2), as before.

Solution:

Moment of inertia:

$$I_x = \frac{bd^3}{12} = \frac{(1)(4^3)}{12} = 5.33 \text{ in.}^4$$

$$= \frac{(25.4 \text{ mm})(101.6 \text{ mm})^3}{12} = 2.20 \times 10^6 \text{ mm}^4$$

$$I_y = \frac{db^3}{12} = \frac{(4)(1^3)}{12} = 0.33 \text{ in.}^4$$

$$= \frac{(101.6 \text{ mm})(25.4 \text{ mm})^3}{12} = 0.139 \times 10^6 \text{ mm}^4$$

Critical buckling loads:

Load causing buckling about the x axis:

$$P_{cr_z} = \frac{\pi^2 E I_x}{L_x^2}$$

$$= \frac{\pi^2 (29.6 \times 10^6)(5.33)}{(180)^2} = 48,060 \text{ lb}$$

$$= \frac{\pi^2 (204,000 \text{ N/mm}^2)(2.20 \times 10^6 \text{ mm}^4)}{(4572 \text{ mm})^2} = 213.8 \text{ kN}$$

Load causing buckling about the y axis:

$$P_{cr_y} = \frac{\pi^2 E I_y}{L_y^2}$$

$$= \frac{\pi^2 (29.6 \times 10^6)(0.33)}{(180)^2} = 2975.5 \text{ lb}$$

$$= \frac{\pi^2 (204,000 \text{ N/mm}^2)(0.139 \times 10^6 \text{ mm}^4)}{(4572 \text{ mm})^2} = 13.3 \text{ kN}$$

Since the load required to cause the member to buckle about the weaker y axis is much less than the load that is associated with buckling about the stronger x axis, the critical buckling load for the entire column is 2975.5 lb (13.3 kN). The member will buckle in the direction of the least dimension. As compared to the member having an equivalent area but square in shape, the load-carrying capacity of the rectangular member analyzed is greatly reduced.

The Euler expression is often rewritten in a slightly different form that is more useful as a design tool. The critical buckling load for a column can be converted into a *critical buckling stress* f_{cr} simply by dividing both sides of the Euler expression by the area A of the column. This $f_{cr} = P/A = \pi^2 EI/AL^2$. This expression contains two measures related to the dimensional properties of the column, I and A, that can be combined into a single measure. This single measure is called the *radius of gyration r* and is *defined* by $r = \sqrt{I/A}$. Note that $I = Ar^2$. Simply accepting this as a definition, the expression for the critical buckling stress of a column can be rewritten as

$$f_{cr} = \frac{P}{A} = \frac{\pi^2 E}{(L/r)^2}$$

The term L/r is called the *slenderness ratio* of the column. The critical buckling stress depends inversely on the square of the slenderness ratio. The higher the slenderness ratio, the lower is the critical stress that will cause buckling, and vice versa. The slenderness ratio (L/r) is an important way of thinking about columns, since it is the single measurable parameter on which the buckling of a column depends.

The radius of gyration, r, can be interpreted in the following way. The moment of inertia of the cross-sectional area of the column is equal to the product of the area times the square of the distance r by definition (i.e., $I = Ar^2$). Consequently, the radius of gyration of this area with respect to an axis is a distance such that if the total area were conceived of as concentrated at this point, its moment of inertia about the axis would be the same as that of the original distributed area about that axis. The higher the radius of gyration of a section, the more resistant the section is to buckling (although the true measure for resistance to buckling is the L/r ratio). Radius-of-gyration values for different sections are often tabulated in much the same way that moments of inertia are.

ACTUAL VERSUS IDEAL COLUMN STRENGTHS. When columns are actually tested, there is usually a difference found between actual buckling loads and theoretical predictions. This is particularly true for columns of lengths near the transition between short- and long-column behavior. Reasons for this include such factors as minor accidental eccentricity in the column loading, initial crookedness in the member, initial stresses present before loading due to the fabrication process, nonhomogeneity in the material, and others. Bending stresses not accounted for in the Euler expression often exist. The result is that buckling loads are often slightly lower than predicted, particularly near the transition zone between short and long columns, where failure is often partly elastic and partly inelastic (crushing). For this reason a number of additional theoretical and empirical expressions have been developed to take into account this phenomenon. Expressions are available, for example, for more accurately predicting behavior in the *intermediate* range (see Figure 7-6). A full exploration of these expressions, however, is beyond the scope of this text.

ECCENTRIC LOADS AND MOMENTS. Many columns are subjected not only to axial loads but to large bending moments as well (see Chapter 9). For short columns, ways of taking such moments into account have already been discussed in Section 7.3.1. The moment is simply expressed as $M = Pe$ and combined stresses are considered. For long columns, the Euler expression does not take into account such additional considerations. Again expressions are available for situations of this type, but are beyond the scope of this text.

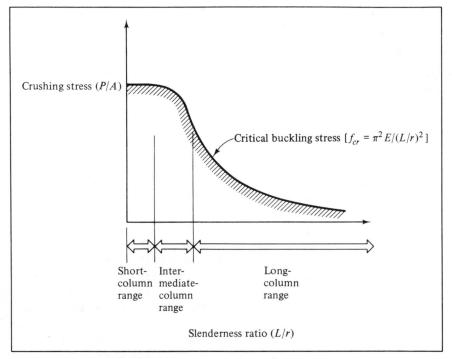

Crushing stress (P/A)

Critical buckling stress $[f_{cr} = \pi^2 E/(L/r)^2]$

Short-column range

Inter-mediate-column range

Long-column range

Slenderness ratio (L/r)

FIGURE 7-6 Actual versus ideal columns.

Despite the limitations discussed above, the Euler expression is still extremely useful as a tool for conceptually understanding the behavior of columns.

END CONDITIONS.   The discussion in the previous section focused on columns having pin-ended connections in which the ends of the members were free to rotate (but not translate) in any direction at their ends. This condition allows the member to deform in the manner illustrated in Figure 7-7(a). Obviously, other end conditions were possible. Restraining the ends of a column from a free-rotation condition generally increases the load-carrying capacity of a column. Allowing translation as well as rotations at the ends of a column generally reduces its load-carrying capacity.

This section considers the relative load-carrying capacity of four columns that are identical in all respects except for their end conditions. The conditions are illustrated in Figure 7-7 and generally represent theoretical extremes, since end conditions in actual practice are often combinations of these primary conditions. Column A represents the standard pin-ended column already discussed. Column B has both ends fixed (i.e., no rotations can occur). Column C has one end fixed and the other pinned. Column D has one end fixed and the other completely free to both rotate and translate (and is thus similar to a flag pole).

The theoretical buckling load for each of these columns can be computed in a way similar to that presented in Appendix 12 for the pin-ended column. Different boundary conditions would be used. Another way of finding the load-carrying capacity of these columns is to consider the effects of the end conditions on the deformed shape of the columns. Consider the deformed shape of the fixed-ended member

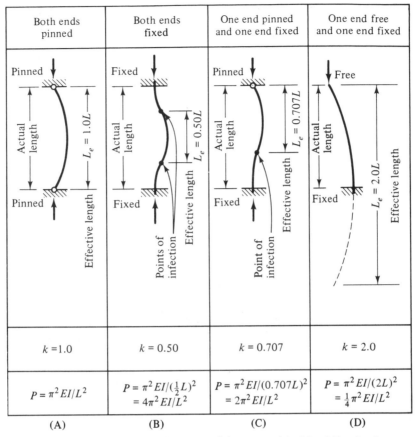

Both ends pinned	Both ends fixed	One end pinned and one end fixed	One end free and one end fixed
$k = 1.0$	$k = 0.50$	$k = 0.707$	$k = 2.0$
$P = \pi^2 EI/L^2$	$P = \pi^2 EI/(\frac{1}{2}L)^2$ $= 4\pi^2 EI/L^2$	$P = \pi^2 EI/(0.707L)^2$ $= 2\pi^2 EI/L^2$	$P = \pi^2 EI/(2L)^2$ $= \frac{1}{4}\pi^2 EI/L^2$
(A)	(B)	(C)	(D)

FIGURE 7-7 Effects of end conditions on critical buckling loads.

(column B) illustrated in an exaggerated way in Figure 7-7(b). The shape of the curve can be simply sketched with a high degree of accuracy by noting that a fixed-end condition of the type illustrated causes the tangent to the member to remain vertical at each end. Under the buckling load, the member would begin bowing as indicated, with the curvature beginning immediately outside the connection. In order for the curve to be continuous, as it must be, the resultant curve must be similar to that illustrated. Of critical importance is that the member must undergo a change in curvature. Evidently, points of inflection must exist at two locations where the sense of the curvature changes. The location of these points can be estimated fairly accurately. In this case they are $L_1/4$ from each end, where L_1 is the actual length of the column. Note that the shape of the column between those two points of inflection is similar to that of a pin-ended column. This is sensible, since both pin-ended connections and points of inflection are analogous. It follows that this portion of the column behaves as if it were a pin-ended column of a length equal to the distance between the points of inflection, in this case one-half the actual length of the column, or $L_1/2$. This distance is called the *effective length* (L_e) of the fixed-ended column. The buckling load of the whole column is controlled by when this portion of the column buckles. The buckling load of this portion is given by $P_B = \pi^2 EI/L_e^2$, where L_e is the distance between the points of inflection. Consequently, the buckling load for the column in terms of its original length is $P_B = \pi^2 EI/(L_1/2)^2 = 4\pi^2 EI/L_1$. Since the initial buckling load for a pin-ended column of the same actual length is $P_A = \pi^2 EI/L_1^2$, the effect of fixing both ends of the column is to increase its load-carrying capacity by a factor of 4. This is a substantial increase that is equivalent to that caused by halving the length of the column.

This process can be repeated for other end conditions. Deformed shapes are sketched and locations of inflection points are determined. Effective lengths are next determined. Critical buckling loads are then given by $P = \pi^2 EI/L_e^2$. The effective length of the column with one end pinned and the other fixed is $L_C = (1/\sqrt{2})L_1$. Its buckling load is consequently $P_C = \pi^2 EI/[1/\sqrt{2}\,L]^2 = 2\pi^2 EI/L_1^2$, or twice that of a column with pins on both ends. For the flagpole column, the deformed shape of the actual column can be seen to be one-half the shape analogous to that of a pin-ended column [see Figure 7-7(d)]. Its effective length is thus $2L_1$. Consequently, $P_D = \pi^2 EI/(2L_1)^2 = \frac{1}{4}\pi^2 EI/L_1^2$, or one-fourth of that for a pin-ended column of the same length.

The concept of effective length is thus very useful in analyzing columns with different end conditions, since it provides a shortcut for making predictions about their load-carrying capacities. The numerical value that modifies the actual length is often called the *k* factor of the column.

Quite often a column has different end conditions with respect to one axis than another (e.g., it may be pin-ended with respect to one axis and fixed with respect to the other). Consequently, care must be taken to couple the correct effective length with the appropriate moment of inertia or radius of gyration. Thus,

$$P_{cr_z} = \frac{\pi^2 EI_x}{L_{e_z}^2} \qquad \text{and} \qquad P_{cr_y} = \frac{\pi^2 EI_y}{L_{e_y}^2}$$

$$f_{cr_z} = \frac{\pi^2 E}{(L_{e_z}/r_x)^2} \qquad \text{and} \qquad f_{cr_y} = \frac{\pi^2 E}{(L_{e_y}/r_y)^2}$$

As in one of the examples previously discussed, the critical buckling load for the entire column is governed by the smaller of P_{cr_z} or P_{cr_y}.

The effects of imperfect end conditions in field applications are often taken roughly into account by simply modifying the effective length of a member. If a column is intended to have fixed ends but the actual field connection is such that only partial restraint is obtained and some rotations occur, for example, it is evident that the effective length of the column lies between $0.5L$ and $1.0L$. The more nearly full restraint is obtained, the closer is the effective length to $0.5L$. The more it rotates, the closer the effective length is to $1.0L$. Values such as $0.6L$ to $0.7L$ are often used.

In the case of frames, only partial restraint is often obtained due to the rotation of end joints (see Chapter 9). The result of this end rotation is to increase the effective lengths of members. When the end rotation is coupled with translation of the joint in space, the effective member length can be longer than the physical length of the column and can constitute a serious design problem.

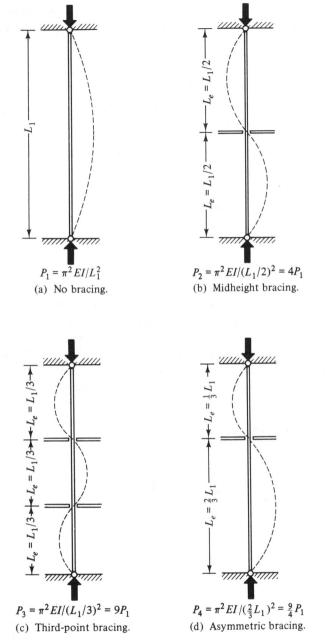

$P_1 = \pi^2 EI/L_1^2$

(a) No bracing.

$P_2 = \pi^2 EI/(L_1/2)^2 = 4P_1$

(b) Midheight bracing.

$P_3 = \pi^2 EI/(L_1/3)^2 = 9P_1$

(c) Third-point bracing.

$P_4 = \pi^2 EI/(\tfrac{2}{3}L_1)^2 = \tfrac{9}{4}P_1$

(d) Asymmetric bracing.

FIGURE 7-8 Effects of lateral bracing on column buckling. Bracing a column changes its buckling mode and consequently its effective length. The more a column is braced, the shorter its effective length becomes and the greater the load that is required to cause buckling. If bracing is used, it is usually more effective when placed symmetrically.

BRACING. To reduce column lengths and increase their load-carrying capacities, columns are frequently braced at one or more points along their length (see Figure 7-8). The bracing can actually be part of the structural framework for the rest of the building, which also serves other functions.

Figure 7-8(b) illustrates a pin-ended column braced at its midheight. When the member buckles, it will deform into an S shape as illustrated. Since the shape of the member between the bracing and the point of inflection is analogous to that of a pin-ended column, the effective length of the column is equal to this distance. The length of the column shown has thus been effectively halved, which increases its load-carrying capacity by a factor of 4: $P = \pi^2 EI/L_e^2 = \pi^2 EI/(L_1/2)^2 = 4\pi^2 EI/L_1^2$.

If the bracing were placed two-thirds of the way up from the bottom, the member would again deform into a modified S shape. Its effective length would then be $\frac{2}{3}L$, or $\frac{1}{3}L$. It is evident that the longer unbraced portion would buckle prior to the shorter portion. Consequently, the critical load for the column would be

$$P = \frac{\pi^2 EI}{L_e^2} = \frac{\pi^2 EI}{(\frac{2}{3}L_1)^2} = (\tfrac{9}{4})\left(\frac{\pi^2 EI}{L_1^2}\right)$$

Clearly, bracing the member at this point is not as effective in increasing the load-carrying capacity of the column as is midheight bracing.

Note that a column can be braced about one axis but not the other. The column will have a tendency to buckle in the direction associated with the highest slenderness ratio. The symmetrical member shown in Figure 7-9 will buckle about the unbraced x axis that corresponds to a column length of L_1 rather than in the S mode associated with buckling about the y axis and an effective length of $L_1/2$.

7.4 DESIGN OF COMPRESSION MEMBERS

The design of columns involves several aspects. In its simplest terms the notion of design might involve only the sizing of the cross section of a column to carry a known load. More generally, the design problem also encompasses decisions about what type of end conditions to use, and where to brace the member. Usually, member sizing is done only after decisions of this type have been made.

7.4.1 Design Objectives

A general column design objective is to support the design loading by using the smallest amount of material, or, alternatively, by supporting the greatest amount of load with a given amount of material. Whenever the buckling phenomenon enters in, it is evident by looking at the buckling curves discussed in the previous section that the full strength potential of the material used in the compressive member is not being exploited to the maximum degree possible.

As was previously noted, the primary factor of importance in connection with buckling is the *slenderness ratio* (L/r) of a column. A low slenderness ratio means that the column is not prone to buckling. High ratios are not desirable. The most efficient use of material can be achieved by either minimizing the column length or maximizing

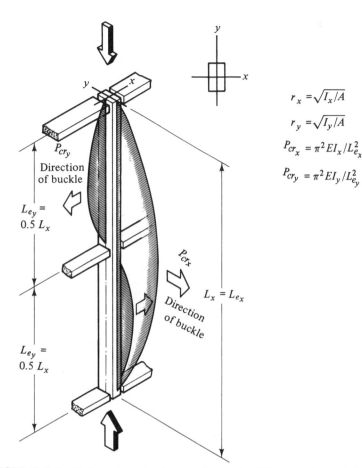

$$r_x = \sqrt{I_x/A}$$

$$r_y = \sqrt{I_y/A}$$

$$P_{cr_x} = \pi^2 EI_x/L_{e_x}^2$$

$$P_{cr_y} = \pi^2 EI_y/L_{e_y}^2$$

FIGURE 7-9 Column bracing in one plane only. When a column is braced in only one plane, it can potentially buckle in two modes. The column will buckle in the mode associated with the higher slenderness ratio (L/r).

the value of the radius of gyration for a given amount of material. Either or both will reduce the slenderness ratio of a member, and hence increase its load-carrying capabilities for a given amount of material. Most design techniques hinge around one or the other of these operations.

7.4.2 End Conditions

A review of Section 7.3 indicates that, by increasing the end restraints on a column (e.g., by changing from pinned to fixed connections), the load necessary to cause a member to buckle is greatly increased. It follows that if the intent is to support a specified load, a member with restrained ends will generally require less material volume to support the load than will one with unrestrained ends. Reducing the effective length of the member allows more efficient use to be made of material. Consequently, restrained ends are far preferable with respect to column design than are ends free to rotate. The end conditions associated with the flagpole column illustrated in Figure 7-7(d) are least desirable.

The choice of whether to use restrained ends, however, should be tempered by other considerations. Fixed joints are invariably more difficult and costly to make than pinned joints (this is particularly true when materials such as timber are used). If the joint were at the foundation, the foundation itself would have to be designed to provide restraint to rotation rather than simply receive axial forces. For relatively small columns, this usually does not require significant additional material or effort, but may do so as column sizes and loads increase. In general, most trade-offs still favor end fixity.

7.4.3 Bracing

One of the most common tools that the designer has available to increase the efficiency of compressive elements is that of bracing. The addition of bracing decreases the effective length of columns. The addition of bracing must be done with care, however, since quite often, if not done properly, no benefit is gained and even a loss is incurred due to the added material and effort expenditures in the bracing. Consider the columns shown in Figure 7-10. In these cases the addition of bracing does *not* increase the

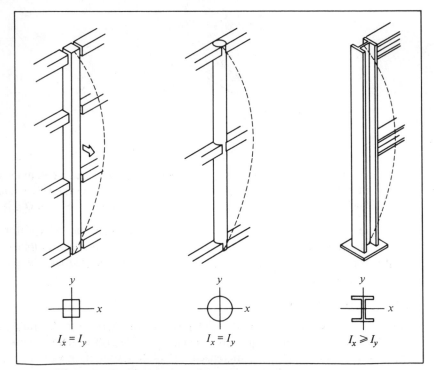

FIGURE 7-10 Ineffective use of bracing. Columns always buckle in the mode associated with the highest slenderness ratio (L/r). The columns shown will buckle as illustrated. The corresponding buckling loads relate to an unbraced or unsupported column length of L and are the same as if the columns were not braced at all. The bracing patterns shown are ineffective in increasing the load-carrying capability of the columns.

load-carrying capacity of any of the columns. As a result of either their symmetrical cross sections or relative I_x and I_y values, all would buckle in a mode associated with their actual rather than braced lengths simply because of their higher slenderness ratios with respect to this direction. Buckling loads would be the same as if there were no bracing present at all. Evidently, if a symmetrical member is to be effectively braced, the bracing must be such as to reduce the slenderness ratios in all directions (by providing bracing in more than one plane).

A very effective type of bracing is illustrated in Figure 7-11(a). Here a symmetrical member is frequently braced in all directions at several points along its length. The ideal number of brace points is quite easy to determine. As previously discussed, for a member of a certain size and material, there exists a certain length that marks the transition between short- and long-column behavior. By putting in a sufficient number of brace points, it is possible to completely eliminate the possibility of buckling anywhere in the member. Consequently, the member could be sized for direct stresses only (i.e., $A = P/F_{\text{allowable}}$) and no premium (in terms of additional area) would be paid to prevent the possibility of buckling. The minimum number of brace points thus corresponds to making column lengths between brace points exactly equal to the transition length between short- and long-column behavior. Putting in more brace points than this does not accomplish anything more in terms of reducing required areas, because the column size is already governed by direct stresses.

In many situations it is not possible to use bracing in more than one plane. Columns are often used in walls, for example, where the wall can serve as lateral bracing in one plane but where no bracing can be provided in other planes for functional reasons. In situations of this type, nonsymmetrical members can often be usefully employed wherein the strong axis of the member is so organized to be associated with the possible out-of-braced-plane buckling mode and the weak axis with in-plane modes. When this is done, the properties of the column can be exactly related to the number and kind of brace points to achieve a high level of efficiency. This is done by creating a set of conditions such that the critical load associated with buckling about one axis is exactly equal to that associated with buckling about the other axis (i.e., $P_{cr_x} = P_{cr_y}$). Thus,

$$P_{cr_x} = P_{cr_y}$$

$$\frac{\pi^2 E I_x}{L_{e_x}^2} = \frac{\pi^2 E I_y}{L_{e_y}^2}$$

$$\frac{I_x}{I_y} = \frac{L_{e_x}^2}{L_{e_y}^2}$$

When the proportion shown immediately above involving moment of inertia and effective lengths obtains, the member is equally likely to buckle in either direction. Evidently, either the member configuration (I_x and I_y) can be varied, or effective member lengths (L_{e_x} and L_{e_y}) by bracing, to achieve this proportion; or both may be varied simultaneously.

The general relationship $I_x/I_y = L_{e_x}^2/L_{e_y}^2$ can also be used to determine the appropriate number of brace points for a given column in which I_x/I_y is fixed.

Instead of I_x/I_y ratios, use is often made of r_x/r_y ratios in actual practice. The

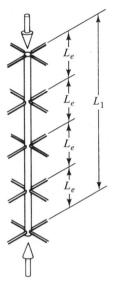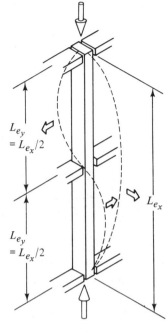

(a) The load-carrying capacity of a symmetrical column can be increased to the maximum extent possible by putting in a sufficient number of brace points such that the effective length of the column is reduced to the point where short-column behavior obtains. Consequently, no portion of the column at all would be susceptible to buckling. The maximum load would be associated with crushing rather than buckling.

(b) When the plane of the bracing possible is fixed, the load-carrying capacity of a column of a given cross sectional area can be increased by varying the relative proportions of the cross section with the objective of making the column equally likely to buckle in any of its possible modes (i.e., making the slenderness ratios of the column similar in all directions). This occurs when the following ratios obtain:

$$I_x/I_y = (L_{e_x}/L_{e_y})^2$$
or
$$r_x/r_y = L_{e_x}/L_{e_y}$$

These ratios can be achieved by careful proportioning of the column cross section. When they are achieved,

$$P_{cr_x} = P_{cr_y}.$$

FIGURE 7-11 Effective use of bracing.

concepts involved are similar except that now $f_{cr_x} = f_{cr_y}$ or $\pi^2 E/(L_x/r_x)^2 = \pi^2 E/(L_y/r_y)^2$. Hence, $r_x/r_y = L_x/L_y$.

7.4.4 Columns in a Building Context

The previous discussion has focused on ways of improving the load-carrying capacity of a column or reducing size requirements to carry a given load by manipulating the bracing. It was tacitly assumed that there was no penalty involved in using as many brace points as necessary. On the other hand, the brace points themselves must be built. This, in turn, requires material volume. While bracing members can be quite small since the prevention of buckling requires little force, an expenditure of construction effort is still required to put them in place. If column design is placed in the context of a building, it is evident that trade-offs occur in which the objective of optimizing the complete column and bracing system rather than the column alone makes the most sense. Quite often it may be preferable to use fewer brace points and a slightly larger column.

Figure 7-12 illustrates a column in a wall of a simple industrial building. The wall is made by attaching vertical siding to horizontal elements, which in turn frame into the column. Clearly either a greater or fewer number of horizontal elements (girts) could be used from those illustrated. Using more horizontal elements implies that the span of the siding, which functions as a vertical beam carrying wind loading, is reduced (and thus a lighter gage siding could be used). Similarly, the columns would be braced at more points on one plane. This, in turn, would influence the choice of the properties of the column itself. More horizontal elements are involved, however, and more construction difficulty. Using fewer horizontal elements increases the loads on the girts (hence their sizes), decreases the number of bracing points and increases the span of the siding, but tends to increase construction ease. Typically, the spacing of the horizontal elements is based on the optimum span for the most economic form of siding available. This, in turn, establishes where the column is braced. Column properties would be selected on the basis of these bracing points (i.e., an L_{e_x}/L_{e_y} ratio would be found and from this an r_x/r_y ratio). A column with this r_x/r_y ratio would then be sized to carry the axial load involved. This example, it should be noted, implies that it is preferable to optimize the siding system rather than the column itself. This is often, but not necessarily always so. The larger the loads on the column, and the longer it is, the more important it is to pay attention to optimizing the column itself. There are invariably trade-offs involved and each situation must be looked at individually.

7.4.5 Column Cross Sections

Determining the required cross-sectional shape for a column intended to carry a given load is a task that is conceptually straightforward. In most cases the objective is to find a cross section that provides the necessary r_x or r_y values by using the smallest amount of material in cross section as possible. The problem is similar to that of designing beams, where the objective is to find a section that provides the greatest moment of inertia for the smallest amount of material. In column design, the task is

Ineffective orientation of column. Its weak axis corresponds to the maximum unsupported length of the column.

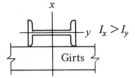

$I_x > I_y$

Effective orientation of column. Its strong axis corresponds to the maximum unsupported length of the column.

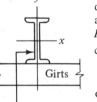

Since $L_{e_y} = L_c/3$ and $L_{e_x} = L_c$, it is desirable to use a column stronger about the x than the y axis so that $P_{C_y} = P_{c_x}$ can be obtained. For this condition

$$r_x/r_y = L_{e_x}/L_{e_y}$$

or $r_x = 3r_y$ for the case shown. A column having these proportions would be selected.

Typical interior column

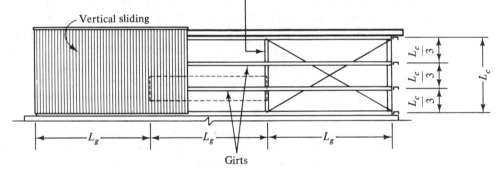

Girts

The horizontal girts serve two functions: (1) To support the vertical siding which picks up wind loads. Consequently, the girts serve as beams having a span of L_g. The siding itself functions in beam fashion with a span of $L_c/3$. (2) To brace the column in one plane, thus reducing its effective length about the y axis to $L_c/3$. The effective length about the x axis remains L_c.

FIGURE 7-12 A small industrial building.

more complicated because of the need to consider radii of gyration about different axes. In general, however, an efficient column section is one that removes material as far from the centroidal axis of the column as is feasible. This, in turn, increases the moment of inertia I for the section for a given amount of material A. Consequently, the radius of gyration r is increased since $r = \sqrt{I/A}$. Increasing the r decreases the slenderness ratio L/r for a column, thus making it less susceptible to buckling. The techniques for optimizing moments of inertia discussed in Chapter 6 are consequently also relevant in column design.

7.4.6 Column Sizes

Figure 7-13 illustrates the critical buckling-stress curve derived from the Euler expression. From a design viewpoint, it is evident that a member must be sized such that the actual axial stress in the member due to the applied axial load (i.e., $f = P/A$) never reaches either the critical stress associated with buckling *or* the stress associated with crushing. By using a factor of safety, a new allowable stress curve of the type illustrated in Figure 7-13 can be established. The factor of safety used depends on the material used. One difficulty in using curves of the type illustrated is that, in the long-column range, the stress that can be allowed to exist in a column depends upon the L/r ratio of the column itself. If the size of the column is not known to begin with, the L/r ratio is not known since r is not known. Hence, the stress that is actually allowable cannot be specified a priori. Consequently, an iterative procedure is typically used to size columns. A trial size is selected and the allowable stress for the corresponding L/r ratio determined. The actual column stress as determined by $f = P/A$ is then compared to the allowable stress as determined from the modified buckling curve. On the basis of the closeness of actual to allowable stresses, the member is either accepted or a new size estimated.

As noted above, the factor of safety used is typically dependent on the material used. These factors are often directly incorporated into an allowable stress expression appropriate for that material alone. The reader is referred elsewhere for more detailed information regarding the design of columns of different materials.[3]

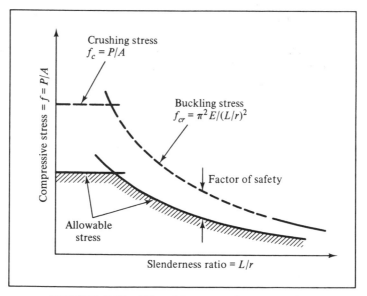

FIGURE 7-13 Allowable stresses in compression.

[3]W. Scofield, *Modern Timber Engineering*, 3rd ed., Southern Pine Association, New Orleans, La., 1963; P. M. Ferguson, *Reinforced Concrete Fundamentals*, 3rd ed., John Wiley & Sons, Inc., New York, 1973; L. Beedle et al., *Structural Steel Design*, The Ronald Press Co., New York, 1964.

QUESTIONS

7-1. An unbraced pin-ended square steel column has cross-sectional dimensions of 1.5 in. × 1.5 in. and is 20 ft long. What is the critical buckling load for this column? Assume that $E_s = 29.6 \times 10^6$ lb/in.^{2}.
Answer: 2139 lb.

7-2. An unbraced pin-ended steel column has rectangular cross-sectional dimensions of 3 in. × 2 in. and is 25 ft long. What is the critical buckling load for this column? Assume that $E = 29.6 \times 10^6$ lb/in.^{2}.

7-3. A pin-ended cylindrical steel column has a diameter of 20 mm and a length of 5 m. What is the critical buckling load for this column? Assume that $E_s = 204{,}000$ N/mm^2.
Answer: 632.5 N.

7-4. Assume that a pin-ended column of length L has a square cross section of dimensions $d_1 \times d_1$ and has a critical buckling load of P_1. What is the relative increase in load-carrying capacity if the cross-sectional dimensions of the column are doubled?

7-5. An unbraced steel column of rectangular cross section 1.5 in. × 2 in. and pinned at each end is subjected to an axial force. Assume that $F_y = 36{,}000$ lb/in.^{2} and $E = 29.6 \times 10^6$ lb/in.^{2}. Find the transition point between short- and long-column behavior.

7-6. Compare the relative load-carrying capacity of a square column and a round column of equal lengths, similar end conditions, and identical cross-sectional areas. Which can carry the greater load? What is the ratio of the critical buckling loads found? Explain your answer in qualitative terms.

7-7. A pin-ended column of length L is to be braced about one axis at quarter points (i.e., there are three equally spaced bracing points). There is no bracing about the other axis. If a rectangular section of $b \times d$ dimensions is to be used, what is the most appropriate b/d ratio for maximizing the load-carrying capacity of the section? Assume that the dimension d corresponds to the strong axis of the section.

7-8. Using the same bracing and end conditions noted in Problem 7-7, determine the most appropriate I_x/I_y ratio for a column having any cross-sectional shape. Also determine the most appropriate r_x/r_y ratio. Assume that I_x and r_x correspond to the strong axis of the section.

chapter 8

CONTINUOUS STRUCTURES: BEAMS

8.1 INTRODUCTION

This chapter explores the analysis and design of beams that are continuous over several supports or are in some other way statically indeterminate due to the support conditions present (Figure 8-1). Mathematically, a statically indeterminate structure is one in which reactions, shears, and moments cannot be determined directly by applying the basic equations of statics ($\sum F_y = 0$, $\sum F_x = 0$, and $\sum M_0 = 0$). Typically, there are more unknowns than there are equations available for solution. Another way of viewing this class of structures which is of more intrinsic importance from a design viewpoint is that in indeterminate structures the values of all reactions, shears, and moments are *dependent* on the physical characteristics of the cross section and the specific material used in the structure (and any variation with length), as well as the span and loading. This contrasts with statically determinate structures in which reactions, shears, and moments can be found through direct application of the basic equations of statics and, consequently, are *independent* of the shape of the cross section and the materials used (as well as any variations along the length of the member). Members having more than two points of support or with multiple fixed ends are typically indeterminate structures.

    As will be explored later in more detail, indeterminate structures quite often have advantages over determinate ones in terms of increased stiffness and reduced magni-

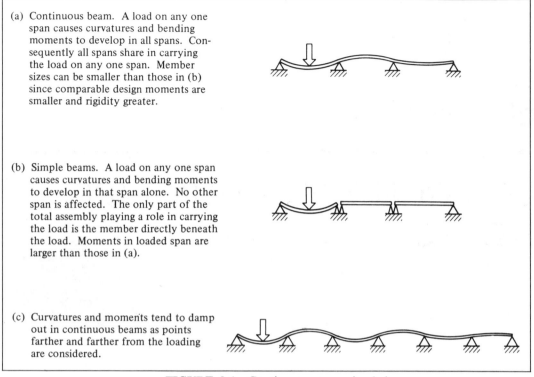

(a) Continuous beam. A load on any one span causes curvatures and bending moments to develop in all spans. Consequently all spans share in carrying the load on any one span. Member sizes can be smaller than those in (b) since comparable design moments are smaller and rigidity greater.

(b) Simple beams. A load on any one span causes curvatures and bending moments to develop in that span alone. No other span is affected. The only part of the total assembly playing a role in carrying the load is the member directly beneath the load. Moments in loaded span are larger than those in (a).

(c) Curvatures and moments tend to damp out in continuous beams as points farther and farther from the loading are considered.

FIGURE 8-1 Continuous versus simple beams.

tudes of the moments developed in them in response to loads. Consequently, member sizes can often be smaller.

8.2 GENERAL PRINCIPLES

Two types of indeterminate structures are illustrated in Figure 8-2. The first [Figure 8-2(a)] is a continuous beam over three simple supports. Reactive forces are developed at each support. It is evident that the magnitudes of these reactive forces cannot be found through the direct application of the basic equations of statics, since there are more unknown forces (R_A, R_B, and R_C) than there are independent equations ($\sum F_y = 0$ and $\sum M_0 = 0$) that can be used to solve for these unknown reactions. The same is generally true for the fixed-ended beam illustrated in Figure 8-2(e). In this case there are four unknown reactions (R_A, R_B, M_A^F, and M_B^F) and still only two independent equations of statics ($\sum F_y = 0$ and $\sum M = 0$) available for use. Both of these beams are consequently termed *statically indeterminate*. Obviously, there is an added layer of difficulty involved in analyzing structures of this type.

Despite added analytical difficulties, however, indeterminate beams are frequently used in actual practice. One primary reason for this is that such structures generally tend to be more rigid under a given loading and span situation than do comparable determinate structures. Another reason for their extensive use is that the internal moments generated in the structure by the external loads are often smaller in magnitude than in comparable determinate structures. Consequently, member sizes can often be reduced. More efficient use is made of smaller amounts of material. Disadvantages in structures of this type include their sensitivity to support settlements and thermal effects. Support settlements, for example, can cause undesirable bending moments to develop in continuous beams over several supports while not necessarily affecting a comparable series of simply supported beams.

8.2.1 Rigidity

The improved rigidity of beams having restrained ends or resting on several supports can be demonstrated through deflection studies. Using techniques discussed in Appendix 8, for example, it is possible to determine that the midspan deflection for a simply supported beam carrying a concentrated load at midspan is given by $\Delta = PL^3/48EI$ [see Figure 8-2(g)]. If the ends of the same beam were restrained by changing the nature of the support conditions to create "fixed ends," the deflection at midspan would be given by $\Delta = PL^3/192EI$ [see Figure 8-2(f)]. Consequently, fixing the ends of the beam reduces maximum deflections by a factor of 4. This is an enormous difference. A careful look at the deflected shapes of the two members reveals the role of end fixity in increasing the rigidity of the structure under loadings [see Figure 8-2(h) and (i)]. Note that the deflection pattern associated with a fixed-ended beam can be obtained from that of a simply supported beam by applying a rotational moment at either end of the simply supported beam. The end of the beam can then be rotated until the tangent of the beam is exactly horizontal. Rotating the ends of the beam would, of course, cause the center of the beam to rise upward, thus causing the absolute midspan deflection of the member to decrease. The value of the moment corresponding to that needed to rotate the tangent of the beam to the horizontal corresponds to that naturally developed as a reaction, in this case, a fixed-ended moment, when the end of the member is initially restrained.

Similar observation could be made concerning the differences in rigidity between a beam resting on multiple supports as compared to a series of simply supported beams. The continuous structure is generally more rigid than its simply supported counterparts. In a design context this means that the continuous beam is often favored.

8.2.2 Force Distributions

The second reason why continuous or fixed-ended beams are frequently used in preference to simple beams is that the magnitudes of the internal shears and moments generated in the structure by the external force system are often smaller than those in simply supported beams. One way of visualizing why this is so is to think in terms of *effective* rather than actual beam lengths. With reference to the fixed-ended beam shown in Figure 8-2(e), for example, it is evident that the deflected shape of the beam between the two points of inflection (where the curvature of the beam changes sense) is quite similar to that of a simply supported beam of the type illustrated in Figure 8-2(h). The portion of the fixed-ended beam that is between the two points of inflection, (a)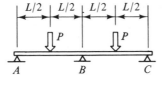

(e)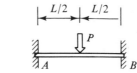

(b)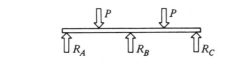

(f)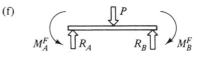

(g)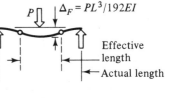

(c)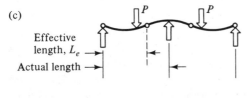

Effective
length, L_e
Actual length

(d)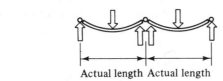

Actual length Actual length

(h)

(i)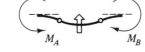

(a) Continuous beam.

(b) The beam is statically indeterminate, since there are three unknown reactions and only two independent equations of statics ($\Sigma F = 0$ and $\Sigma M = 0$).

(c) Deflected shape of beam. There are two points of inflection (where the beam undergoes a reversal in curvature) in this beam. These are also points of zero moment. The distance between points zero moment is often called the "effective length" of the beam to create a direct comparison with a simply supported beam. The shorter the effective length, the smaller are the bending moments present. Consequently, the moments in the continuous beam are smaller than those in two comparable simply supported beams since its effective span length is shorter than its actual span length.

(d) In simply supported statically determinate members, the effective length of a member is equal to its actual length.

(e) Fixed-ended beam.

(f) The beam is statically indeterminate, since there are four unknown reactions.

(g) Two points of inflection are developed in the beam. This makes its effective length much shorter than its actual length. Consequently the moments developed in the beam are smaller than those of a comparable simply supported beam. A fixed-ended beam is also far more rigid than a comparable simply supported beam.

(h) Deflections in a comparable simply supported beam are greater by a factor of four for this loading condition.

(i) The deflected shape of a simply supported beam can be transformed into that of a fixed-ended beam by applying rotational moments at each end. Doing this causes the midspan deflection of the member to decrease. The magnitudes of the rotational moments sufficient to twist the end tangents of the beam to a horizontal position correspond to the moments naturally developed as reactions in a fixed-ended beam.

FIGURE 8-2 Statically indeterminate structures.

which are also points of zero moment, can consequently be thought of as simply supported. The distance between the two points of inflection is termed the *effective length* of the fixed-ended member. The magnitude of the bending moment developed in this segment of the beam depends on this length as well as the value of the loading. The shorter the effective length, the smaller is the magnitude of the bending moment present, and vice versa. In a simply supported beam, the effective length of the beam is identical to its actual length. In a fixed-ended beam of identical actual length, the effective length is considerably less than the actual length. Thus, the midspan moment is smaller in the fixed-ended beam than in a comparable simply supported beam. It is evident, however, that bending moments are present in parts of the fixed-ended beam, at its ends, that are absent in the simply supported beam. This can be seen by looking at the deflected shapes of the two members and recalling that moments and beam curvatures are related. Fixing the ends of the beam changes not only the magnitude of the moments present in the structure but their distribution as well.

Thinking in terms of effective lengths is useful in analyzing any type of indeterminate structure. By sketching the probable deflected shape of a structure, the location of points of inflection can be estimated. Effective beam lengths can then be estimated. In the beam over three supports, illustrated in Figure 8-2(c), for example, it is evident that points of inflection are developed as shown. The effective length of each span is less than the actual distance between support points. Consequently, moments in their middle portion of each span would be expected to be less than those present in two comparable simply supported beams whose effective lengths are identical to their actual lengths.

This way of looking at indeterminate structures is obviously similar to that described for columns in Chapter 7. It is important to note, however, that the points of inflection are induced by different types of loading. If compressive loads were applied to each end of the member over three supports illustrated in Figure 8-2(a), the member would certainly not buckle into a shape having two points of inflection as illustrated in Figure 8-2(c) but would rather assume the shape of an *S*.

The following sections will explore approximate methods for determining reactions, shears, and moments in indeterminate beams. The methods are based on the concepts discussed above. More rigorous methods of analysis capable of yielding exact results are presented in Appendices 9, 10, and 11. Also to be explored are the effects on indeterminate structures of support settlements and variable loading conditions, two highly important considerations in connection with the use of indeterminate beams.

8.3 ANALYSIS OF INDETERMINATE BEAMS

8.3.1 Approximate Methods of Analysis

This section discusses a simplified method for determining the approximate values of reactions, shears, and moments in indeterminate structures. The method is based on the concept of effective member lengths discussed in the previous section. The method involves sketching the deflected shape of the structure, noting the location of points of inflection, and then using the condition that bending moments are known to be zero at points of inflection as a device for reducing the number of unknown reactions present. The accuracy of the method depends upon the accuracy of the original sketch of the deflected shape of the structure.

Consider the fixed-ended member illustrated in Figure 8-3(a). Through a careful sketch of the deflected shape, points of inflection can be estimated to be at the locations indicated. Since the internal moment is known to be zero at these points, the structure can be decomposed as illustrated in Figure 8-3(c). Each of these three pieces can now be treated as a statically determinate beam. The reactions for the center portion can be determined as indicated. These reactions then become loads on the cantilever elements shown. Vertical reactions and moments can then be found for these elements. Moment diagrams can also be drawn for each piece analyzed. The final moment diagram for the whole structure is a composite of individual diagrams.

The maximum bending moment developed in the structure, $M = wL^2/12$, occurs not at midspan but at the fixed ends. The midspan moment, $M = wL^2/24$, is only one-half of this value. The structure must be designed to carry these moments. A comparable simply supported beam has a midspan moment of $M = wL^2/8$. Consequently, fixing the ends of the member greatly reduces the maximum design moment as compared to the simply supported case. A substantial savings in materials is thus possible compared to the simply supported case because of these reduced design moments. It is interesting to observe that the sum of the absolute values of the positive and negative moments ($wL^2/12 + wL^2/24$) is equal to $wL^2/8$ (the same moment present at midspan in a simply supported beam carrying a uniformly distributed load). The moments the structure must be designed to carry at the ends and middle, however, are still those illustrated in Figure 8-3(d) ($wL^2/12$ and $wL^2/24$, respectively), and it is these moments that the beam would have to be sized to carry.

Figure 8-4 illustrates a similar analysis made for a beam with one end pinned and the other fixed. The approximate shape of the structure is sketched and the location of the point of inflection estimated. The structure is then decomposed into two determinate pieces. These pieces are analyzed and reactions, shears, and moments determined. The shear and moment diagrams for the whole structure are found by combining individual diagrams. In this case the point of inflection was estimated to be 0.3L from the right support. This yielded a moment at the support of $-0.15wL^2$. A more rigorous analysis made using the techniques described in Appendix 10 indicates the correct location of the point of inflection to be 0.25L and a moment of $-0.125wL^2$. Thus, the estimated moment is not exactly correct. It is, however, certainly close enough for most preliminary design purposes.

Figure 8-5 illustrates an analysis for a continuous beam over three supports. By carefully sketching the deflected shape of the structure under the loading shown, the member can be seen to have two points of inflection. These points are symmetrically located approximately 0.25L from the center support. The structure can then be decomposed for analytical purposes in the manner indicated in Figure 8-5(c). Each piece can then be treated as determinate beam and appropriate reactions, shears, and moments found. The shear and moment diagrams for the actual structure are found by combining individual diagrams. The maximum reaction in this type of continuous

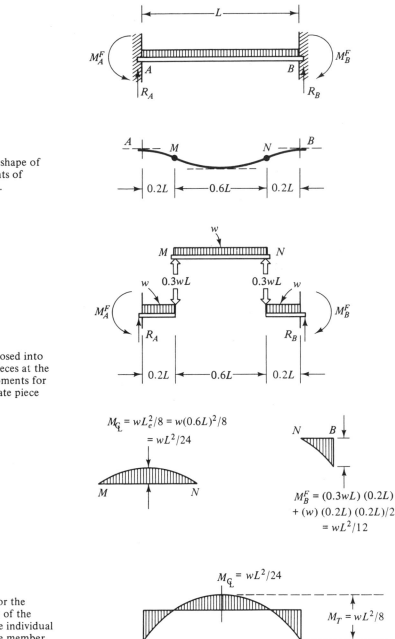

(a) Beam.

(b) Sketch of the deflected shape of the structure. Two points of inflection are developed.

(c) The structure is decomposed into statically determinate pieces at the points of inflection. Moments for each statically determinate piece are calculated.

$$M_{\text{C}} = wL_e^2/8 = w(0.6L)^2/8$$
$$= wL^2/24$$

$$M_B^F = (0.3wL)(0.2L)$$
$$+ (w)(0.2L)(0.2L)/2$$
$$= wL^2/12$$

(d) The moment diagram for the structure is a composite of the moment diagrams of the individual determinate pieces. The member must be designed to carry $wL^2/12$ and $wL^2/24$.

$$M_{\text{C}} = wL^2/24$$

$$M_T = wL^2/8$$

$$M_A^F = -wL^2/12$$

$$M_B^F = -wL^2/12$$

FIGURE 8-3 Approximate analysis of a fixed-ended beam.

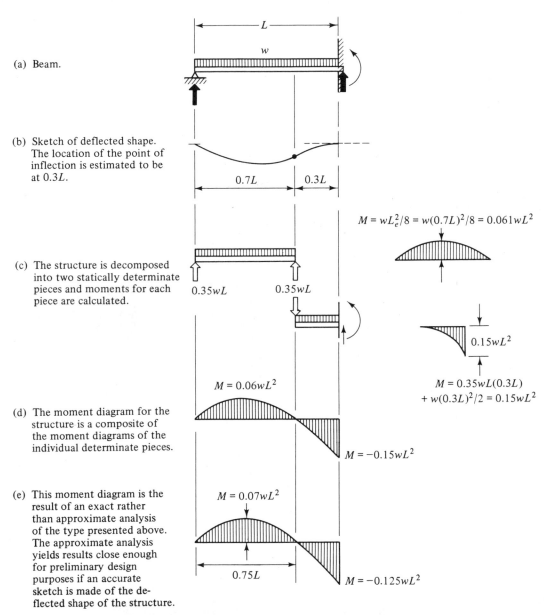

(a) Beam.

(b) Sketch of deflected shape. The location of the point of inflection is estimated to be at $0.3L$.

$$M = wL_e^2/8 = w(0.7L)^2/8 = 0.061wL^2$$

(c) The structure is decomposed into two statically determinate pieces and moments for each piece are calculated.

$0.35wL$ $0.35wL$

$$M = 0.35wL(0.3L) + w(0.3L)^2/2 = 0.15wL^2$$

$0.15wL^2$

(d) The moment diagram for the structure is a composite of the moment diagrams of the individual determinate pieces.

$M = 0.06wL^2$

$M = -0.15wL^2$

(e) This moment diagram is the result of an exact rather than approximate analysis of the type presented above. The approximate analysis yields results close enough for preliminary design purposes if an accurate sketch is made of the deflected shape of the structure.

$M = 0.07wL^2$

$0.75L$

$M = -0.125wL^2$

FIGURE 8-4 Approximate analysis of a propped cantilever beam.

beam occurs at the center support. This is also where the maximum bending moment occurs.

The technique of sketching deflected shapes, determining points of inflection, and then decomposing the structure into elemental pieces can be applied to beams having virtually any support condition. It is also a useful tool for learning how to qualitatively sketch the approximate shape of moment diagrams in complex structures. From the deflected shape, the locations of points of inflection can be estimated. Given

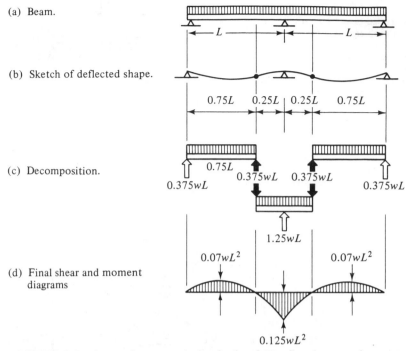

(a) Beam.

(b) Sketch of deflected shape.

(c) Decomposition.

(d) Final shear and moment diagrams

FIGURE 8-5 Approximate analysis of a beam continuous over three supports.

the loading, one can then infer the general shape of the moment diagram (i.e., uniformly distributed loads produce curved moment diagrams and concentrated loads diagrams composed of linear line segments) and it is known that the moment diagram must pass through zero at the points of inflection. With practice it is possible to sketch the moment diagrams of complex beams and from the sketches infer approximate values of moments.

8.3.2 Effect of Variations in Member Stiffness

In the opening of this chapter it was noted that one interpretation of why reactions, shears, and moments could not be found directly through application of the basic equations of statics was that these values are dependent on the exact geometrical cross section of the member used, its material characteristics and any variation in either along the length of the structure. The analyses in the previous section were based on the assumption that the E and I of the member studied were constant along its length.

To study the effects of a member not having a constant E and I, it is useful to consider first two extreme cases of the fixed-ended beam analyzed in the previous section. The extreme cases involve assuming dramatic differences in the cross-sectional properties at different points in the beam. With respect to Figure 8-6(a), assume that at the ends of the member, either E or I (or both), begin to be very small or approach zero. Both E and I, it will be recalled, are measures of the stiffness of the member. The modulus of elasticity, E, is a measure of stiffness as related to material. The smaller the value of E, the more flexible is the material (e.g., steel has a very high E

(a) Fixed-ended beam in which E or I approaches zero at each end of the beam.

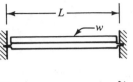

(b) The reduced E or I at the ends reduces the capacity of the beam to resist rotations (induced by the loading) at these points. Consequently this causes points of inflection to be developed at these locations rather than further out toward the center as in a member with a constant E and I along its length.

(c) The changed location of the inflection points causes a change in the moment diagram. In this case the moment diagram is quite similar to that of a simply supported beam. A maximum moment of $wL^2/8$ is present.

$$M_T = wL^2/8$$

(d) Fixed-ended beam in which E or I approaches zero at the center of the beam.

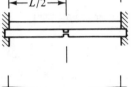

(e) The reduced E or I at the center causes a point of inflection to develop at this point.

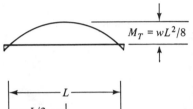

(f) The member begins behaving quite like two cantilever beams, each of length $L/2$. The total moment diagram for the structure reflects this. Again a maximum moment of $wL^2/8$ is present.

$$M = w(L/2)^2/2 = wL^2/8$$

(g) A fixed-ended beam with "haunches" or increased depths at each end.

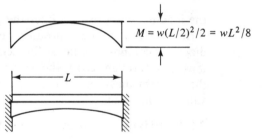

(h) Since the beam is stiffer at either end than in the middle, the points of inflection tend to develop more toward the middle of the beam than they do in beams having a constant cross-section.

$$>0.2L \qquad >0.2L$$

(i) The movement of the points of inflection inward affects the moment diagram. Compared to a beam with a constant cross-section, more moment is developed at the fixed-ends and less at mid-span. More moment was, so to speak, attracted to the ends by increasing member stiffnesses at these points. Note that the sum of the positive and negative moments remains $wL^2/8$. The total moment present has thus not been changed, only its distribution.

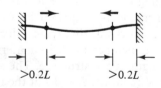

$$M_{\mathbb{C}} < wL^2/24$$

$$M_T = wL^2/8$$

$$M_B > wL^2/12$$

FIGURE 8-6 Effects of variations in member stiffness.

and rubber has a very low E). The moment of inertia, I, is a measure of the amount and distribution of material at a cross section. The higher the I value, the stiffer is the section. The overall stiffness of a member at a cross section (i.e., its ability to resist deformations or rotations induced by the external load) depends on both parameters. If either or both are low at a cross section, the stiffness of the member is low at that cross section.

When the whole member is loaded, the beam is more likely to undergo internal rotations at cross sections of lower stiffness than elsewhere. The capacity to resist rotation is smaller. If E and I actually approach zero at certain locations, the structure begins behaving (with respect to how it rotates) as if a pin connection were present. With respect to the fixed-ended beam, instead of points of inflection occuring at $0.2L$ from each end as they naturally did for a constant E and I member, the points of inflection occur at the locations of the drastically reduced E and I values. The deflected shape of the structure is influenced accordingly. By making the same type of analysis as was done for the beam with uniform cross section, the moment diagram illustrated in Figure 8-6(c) is obtained (this is obviously the same diagram as for a simply supported member). Alternatively, consider the behavior of the structure when it is assumed that the I or E value of the member becomes very small or approaches zero at the midpoint of the structure. Again, the capacity of the structure to resist rotation is reduced at this point and the structure begins behaving as if a pin connection were present at this location. The resulting deflected shape and associated moment diagram are illustrated in Figure 8-6(e) and (f), respectively. Again note that the total moment present is $wL^2/8$.

Instead of extreme cases, consider the behavior of the structure if the ends are made stiffer than the central portion of the beam (often referred to as *haunching* the member). In sketching the deflected shape of the member, it is evident that by making the beam stiffer at the ends, the member is better able to resist rotation at these locations. Consequently, the net effect is to cause the points of inflection to move more toward the center of the beam relative to their location in a beam with uniform cross section throughout. The exact location of these points can only be estimated at this stage. The net effect, however, of these points moving inward is to increase the magnitude of the moment present at the ends of the member and to reduce that present at the central portion of the member.

The examples above generally illustrate why the internal moments in an indeterminate structure are dependent on the properties of the cross section and their variation along the length of the member. In general, it is interesting to note that the net effect of increasing the size of the member at a particular location (relative to other parts of the structure) was to increase the magnitude of the moment present at that point. Conversely, reducing the capacity of a member at a particular point resulted in a reduction in the moment present at the point. Moment was, so to speak, attracted by the stiffer portions of the member. This is a point of fundamental interest from a design viewpoint.

In all the variations discussed above, the sum of the positive and negative moments was always $wL^2/8$. The appearance of a constant total is of interest. The effect of end fixity and variations in E and I was to alter the way this total moment was distributed and carried by the beam, but the total moment associated with a simply supported beam remained present. The change in distribution, however, was advantageous since the actual moments the member should be sized to carry, of course, are the positive and negative moments themselves (each less than $wL^2/8$), which make up the total of $wL^2/8$.

8.3.3 Effects of Support Settlements

Continuous or fixed-ended beams are quite sensitive to support settlements. Support settlements can occur for a variety of reasons with the most common being consolidation of the soil beneath a support. The larger the load on the soil, the more likely is consolidation to occur. Rarely is the amount of settlement exactly the same beneath all support points on a structure. If this were so, the whole structure would simply translate vertically downward and consequently cause no distress in the structure. The more usual case of differential settlement, however, can potentially cause undesirable bending moments to develop in continuous to fixed-ended structures.

Consider the fixed-ended beam illustrated in Figure 8-7. The beam is not loaded with any external force. Assume that a differential settlement of an amount Δ occurs. When one support simply translates with respect to the other, the fixed-ended nature of the support still restrains the end of the beam from rotating. The net effect is that curvatures are developed in the beam. Associated with these curvatures are internal bending moments. Clearly, the greater the differential settlement, the greater are the internal bending moments developed in the structure. This behavior is in obvious contrast to that of a simply supported beam. As differential settlement occurs under a simply supported beam, the member simply follows along. Since the ends of the member are unrestrained, the whole member simply rotates when the settlement occurs. The differential settlement does not cause curvature, hence bending moments, to develop in simply supported members. For this reason, simple support conditions are often preferable to rigid ones when problems with support settlement are anticipated.

Settlements also affect beams that are continuous over several supports. Consider the beam illustrated in Figure 8-7(c). If differential settlements of the type illustrated occurred, curvature and associated bending moments could potentially be induced in the beam. If settlements occur when the beam is fully loaded, the support settlement causes a change in the moment distribution present in the structure. The general effect of the center support settling relative to the two end supports, for example, is to cause the points of inflection to move inward toward the center support. This, in turn, increases the effective lengths of the end spans, thus causing an increase in the positive moment present in the member. Clearly, if the member is not sufficiently sized to carry this increased moment, the member could potentially fail or become seriously overstressed.

8.3.4 Effect of Partial-Loading Conditions

One of the more interesting aspects of the behavior of indeterminate structures under load is illustrated in Figure 8-8(a)–(c). A typical indeterminate beam over four supports is shown carrying three different sets of loads. Figure 8-8(c) shows the structure under full-loading conditions, in which all three spans of the structure are similarly

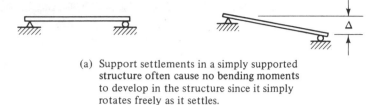

(a) Support settlements in a simply supported structure often cause no bending moments to develop in the structure since it simply rotates freely as it settles.

(b) Support settlements in a fixed-ended structure cause undesirable bending moments to be developed.

$M_{A_2} > M_{A_1}$
$M_{B_2} < M_{B_1}$
$M_{C_2} > M_{C_1}$

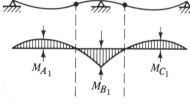

M_{A_1}

M_{C_1}

M_{B_1}

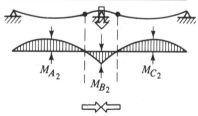

M_{A_2}

M_{C_2}

M_{B_2}

(c) Support settlements in continuous beams can cause increased bending moments to develop at different locations in the structure over those normally associated with the loading. Downward settlement of middle support causes points of inflection to move inward, thus increasing the positive moment present. The reaction at the beam ends also increase (that at the middle support decreases).

FIGURE 8-7   Effects of support settlements.

321

Continuous beam—three equal spans— one end span unloaded

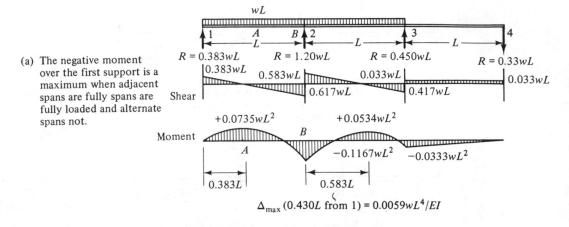

(a) The negative moment over the first support is a maximum when adjacent spans are fully spans are fully loaded and alternate spans not.

$R = 0.383wL$ $R = 1.20wL$ $R = 0.450wL$ $R = 0.33wL$

Shear

$0.383wL$ $0.583wL$ $0.033wL$ $0.033wL$

$0.617wL$ $0.417wL$

Moment

$+0.0735wL^2$ $+0.0534wL^2$

B

$-0.1167wL^2$ $-0.0333wL^2$

A

$0.383L$ $0.583L$

Δ_{max} (0.430L from 1) = 0.0059wL^4/EI$

Continuous beam—three equal spans—end spans loaded

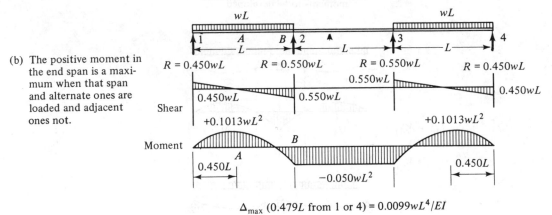

(b) The positive moment in the end span is a maximum when that span and alternate ones are loaded and adjacent ones not.

$R = 0.450wL$ $R = 0.550wL$ $R = 0.550wL$ $R = 0.450wL$

$0.550wL$

Shear

$0.450wL$ $0.550wL$ $0.450wL$

Moment

$+0.1013wL^2$ $+0.1013wL^2$

B

A

$0.450L$ $0.450L$

$-0.050wL^2$

Δ_{max} (0.479L from 1 or 4) = 0.0099wL^4/EI$

Continuous beam—three equal spans—all spans loaded

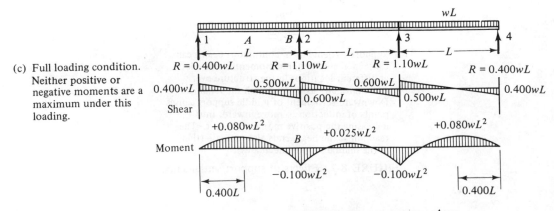

(c) Full loading condition. Neither positive or negative moments are a maximum under this loading.

$R = 0.400wL$ $R = 1.10wL$ $R = 1.10wL$ $R = 0.400wL$

$0.500wL$ $0.600wL$

Shear $0.400wL$ $0.600wL$ $0.500wL$ $0.400wL$

Moment

$+0.080wL^2$ $+0.025wL^2$ $+0.080wL^2$

B

$-0.100wL^2$ $-0.100wL^2$

$0.400L$ $0.400L$

Δ_{max} (0.466L from 1 or 4) = 0.0069wL^4/EI$

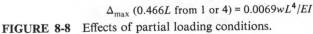

FIGURE 8-8 Effects of partial loading conditions.

loaded. Figure 8-8(a) and (b) illustrates two different partial-loading conditions in which one span is not loaded at all. These different conditions reflect different load patterns that might exist on the structure at one time or another. These figures also show the moments found from a rigorous analysis made using one of the methods discussed in Appendix 10.

When different locations on the structure are considered (e.g., *A* and *B*), it is evident that each loading condition produces a different moment at each of the respective locations. An inspection of the moments associated with the three loadings shown reveals some seemingly curious results. The maximum midspan positive moment at *A* does *not* occur when the structure is fully loaded but rather under the partial-loading condition indicated in Figure 8-8(b). Neither does the maximum negative moment at support *B* occur when the structure is fully loaded but rather under the partial-loading condition indicated in Figure 8-8(a). In no case does the maximum moment at a point result from a full-loading condition. A partial-loading condition always produces the critical design moment. Similarly, no one partial-loading condition simultaneously produces the critical maximum moment at all locations.

Obviously, what is happening in the situations mentioned above is that the curvature of the member, hence the bending moment present, at a location is affected by loads placed *anywhere* on the structure. With respect to point *A*, for example, it is evident that a load on span 1–2 would cause a positive curvature and bending moment to develop. If span 3–4 is also loaded, the resultant effect is to further increase the curvature and, consequently, the positive moment at *A* in the first span. A more critical moment at *A* is thus produced than when the first span alone is loaded [see Figure 8-9(a)]. Note that loading the middle span, however, tends to cause the left span to rise upward. The effect of the rise is to decrease the curvature and bending moment at point *A*. This loading condition thus produces a less critical moment than the partial-loading condition previously considered. Similar arguments could be made about other locations. .

Loadings that produce maximum moments of the type discussed above are called *critical loading conditions*. There are several formal techniques for establishing what loading conditions are critical on a structure. Going into these techniques in detail, however, is beyond the scope of this text.

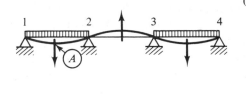

(a) Loadings on span 1–2 and span 3–4 tend to increase the curvature at *A*. Increasing the curvature means that the moment in the beam is also increased. Therefore, loadings on both 1–2 and 3–4 contribute to the development of positive moment at *A*.

(b) Loading span 2–3 would cause a decrease in positive curvature at *A* thus tending to decrease the positive moment there. Therefore, this span should not be loaded to develop maximum moment at *A*.

FIGURE 8-9 Critical loading conditions in continuous beams.

8.4 DESIGN OF INDETERMINATE BEAMS

8.4.1 Introduction

The process of designing a continuous beam is not unlike that of designing a simple beam. Once the value of the maximum moment that can be present under any loading is known, the process of determining the required member size at that point is straightforward. The process for doing this is identical to that already discussed in Section 6.4 in connection with simple beams. All the techniques for member sizing discussed there are appropriate for continuous beams as well. Principles concerning how best to optimally distribute material at a cross section are similarly applicable. This section, therefore, addresses only those general issues that are unique to the design of continuous beams.

8.4.2 Design Moments

Of particular importance in the design of continuous beams is to assure that the member is sized to account for the moments associated with all possible loading conditions. The discussion in Section 8.3.6 is particularly relevant here. There it was noted that the maximum moment that could be developed at any particular point in the structure rarely, if at all, resulted when the structure was fully loaded but typically occurred when the structure was only partially loaded. It should be noted, however, that both maximum positive and maximum negative moments conditions occur with full load on the span under consideration. The question of the effects of partial loads releates primarily to loading on adjacent spans. From a design viewpoint, it is therefore necessary to consider all the possible variants of the loading that might exist on a structure and to determine the moments produced by each of these loadings at all points in the structure. Some loadings will produce higher moments at certain locations than others. The size of the structure at any specific point would be based on the critical loading, which produces the maximum possible moment at that point. Other points would be based on maximum moments associated with other loads critical for those points. Consequently, under any one loading condition the structure would be somewhat oversized everywhere except where the critical moment associated with that loading condition exists.

In complex situations, determining appropriate design moments for the different critical points in a structure can obviously be tedious. For commonly encountered situations, appropriate design moments have been tabulated. For two or more approximately equal spans (the larger of the two adjacent spans not exceeding the smaller by more than 20%) carrying uniformly distributed loads in which the live load does not exceed the dead load by more than a factor of three, the shears and moments listed in Table 8-1 are reasonable for rough design purposes. More rigorous analyses must be made for unusual situations.

While of obvious importance, the following discussions will not further consider the effects of partial loadings in detail and the actual need to design a structure for a variety of loading conditions. Rather, the discussion will focus on shaping the structure in response to its primary design loading condition. Quite often the effects of partial loadings in structures used in buildings are taken into account after the structure is initially shaped for its primary loading.

Table 8-1. Typical Design Moments

Positive moment	
End spans	
If discontinuous end is unrestrained	$\frac{1}{11}wL^2$
If discontinuous end is integral with the support	$\frac{1}{14}wL^2$
Interior spans	$\frac{1}{16}wL^2$
Negative moment at exterior face of first interior support	
Two spans	$\frac{1}{9}wL^2$
More than two spans	$\frac{1}{10}wL^2$
Negative moment at other faces of interior supports	$\frac{1}{11}wL^2$
Negative moment at face of all supports for (a) slabs with spans not exceeding 10 ft, and (b) beams and girders where ratio of sum of column stiffnesses to beam stiffness exceeds eight at each end of the span	$\frac{1}{12}wL^2$
Negative moment at interior faces of exterior supports for members built integrally with their supports	
Where the support is a spandrel beam or girder	$\frac{1}{24}wL^2$
Where the support is a column	$\frac{1}{16}wL^2$
Shear in end members at first interior support	$1.15\frac{wL}{2}$
Shear at all other supports	$\frac{wL}{2}$

8.4.3 Shaping Continuous Beams

It is evident that the size of the cross section at a point of a continuous beam depends on the magnitude of the moment present at that point. There is a wide variation in the way moments are distributed in a continuous structure. Obviously, the structure could be sized for the absolute maximum moment present anywhere in the structure and the same size simply used throughout. This is often done in building structures as a matter of construction convenience.

Alternatively, the size of the structure could be varied along its length in response to the moments actually present in the structure. Techniques for doing this have already been generally discussed in connection with simple beams (see Section 6.4.5). Shaping members is fairly frequently done in connection with highway bridge design when the possible material savings overshadow the added construction difficulties.

The issues involved in shaping a beam along its length are relatively straightforward with one major exception. Consider the continuous beam illustrated in Figure 8-10. If the member depth was designed to be dependent on the magnitude of the moment present at a point, and *no* deviations were allowed to this relation, a structure having roughly the configuration illustrated in Figure 8-10(d) would result. Clearly, where there are points of inflection, there is no moment in the structure, and its depth consequently can approach zero (if shear forces are ignored). Actually doing this, of course, is absurd. The resultant structure is not a configuration that is stable under any other than the exact loading condition illustrated. Any slight deviation in the

(a) Loading.

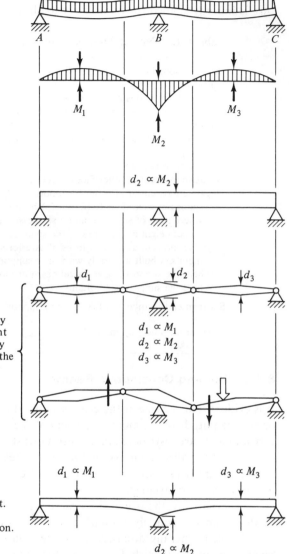

(b) Moment diagram:
$M_2 > (M_1, M_3)$.

M_1

M_2

M_3

(c) Structure with constant depth. The depth of the structure, d_2, reflects the maximum moment present. Structure is oversized elsewhere.

$d_2 \propto M_2$

(d) The depth of the structure at any point reflects the moment present at that point. While theoretically stable under the design loading, the structure is unstable under other loads.

d_1 d_2 d_3

$d_1 \propto M_1$
$d_2 \propto M_2$
$d_3 \propto M_3$

(e) Partly shaping the structure in response to the moments present. The resultant configuration is stable under any loading condition.

$d_1 \propto M_1$ $d_3 \propto M_3$

$d_2 \propto M_2$

FIGURE 8-10 Alternative strategies for designing a continuous structure.

loading (which is, of course, bound to occur) would cause immediate collapse. A reasonable alternative, however, is to design the structure to reflect maximum positive and negative moment values and simply ignore the points of inflection. When this is done, and the structure also designed for the shear forces present, a configuration of the general type illustrated in Figure 8-10(e) results. This configuration is frequently used as a practical structure, particularly in steel highway bridges.

Even if the effects of partial loadings are considered, the shape illustrated in

Figure 8-10(e) is probably still appropriate unless the live load/dead load ratio is very high.

The *exact* shape of a member as it varies from points of maximum positive and negative moments depends on the choice of structure used. The exact optimum shape for a truss structure for a given moment variation differs from the optimum shape for a solid rectangular beam (see Chapters 4 and 6). Other beam cross sections (e.g., wide-flange shapes) would result in different variations of depth with moment. For this reason, the shapes sketched showing possible variations of depth with moment are strictly diagrammatic only.

8.4.4 Use of Construction Joints

Because of construction difficulties, it is often difficult to make a long continuous member out of one piece and it is desired to introduce construction joints. Figure 8-11 illustrates the use of construction joints in a beam continuous over three spans but constructed of several pieces. To simplify the making of the joint and thus the connection between discrete pieces making up the total span, the joints are placed at or near points of inflection. Consequently, the joints need carry no moment and are sometimes designed as simple pinned connections. This type of connection is much easier to design and fabricate than rigid joints that can carry moment.

In the beam illustrated there are four points of inflection. Pinned construction joints, however, cannot be placed at each point of inflection, since this could result in a structure that could be unstable if the loading changed. This is obviously the same issue encountered in varying the depth of the beam in response to the moment present discussed in the previous section. Pinned construction joints can only be used when the resulting structural configuration is stable. For the beam illustrated in Figure 8-11(a), there are two basic ways of doing this. The first is by putting joints in the end spans, leaving a center span with two cantilever ends. The structure would be built by putting this center span in place first and then adding the end pieces. Alternatively, the joints could be placed in the middle span, leaving two stable beams with single cantilevers. The structure would be built by first putting in place the two end members and then adding the center piece. In either case the structure is converted into an assembly of statically determinates structures which function together in a way that *reflects* the behavior of a continuous member.

The advantage of using pinned construction joints occurs when the behavior of a continuous member can be exactly reflected. Pinned construction joints are most effectively used when the design loading, in this case a uniformly distributed load over the full structure, is expected to remain the primary loading. If loading conditions change from primary design loadings, these structures would still behave as assemblies of determinate structures, but ones that would *not* reflect the behavior of a continuous member and thus would not possess the implied advantages. For this reason, rigid rather than pinned construction joints are often used.

The general technique illustrated in Figure 8-11(e) is often coupled with a shaping of the structure in response to the extreme positive and negative moments present [Figure 8-11(f)]. Many large bridge structures reflect this general approach. In struc-

(a) Loading.

(b) Moment diagram.

(c) Use of contruction joints (typically
pinned connections) at each point
of inflection. The resultant structure
is unstable when partially loaded.
Rigid construction joints capable
of carrying some moment could
be used to achieve stability.
Otherwise, one of the options
shown below must be adopted.

(d) Use of construction joints in end span
only. The resultant configuration is
stable under any loading. The joints
facilitate construction. The center span
would be put in place followed by the
end spans.

(e) Use of construction in middle
span. Again the resultant structure
is stable under any loading. The end
spans would be put in place first
followed by the middle span.

(f) Use of construction joints in middle
span coupled with shaping of structure
to reflect moments present.

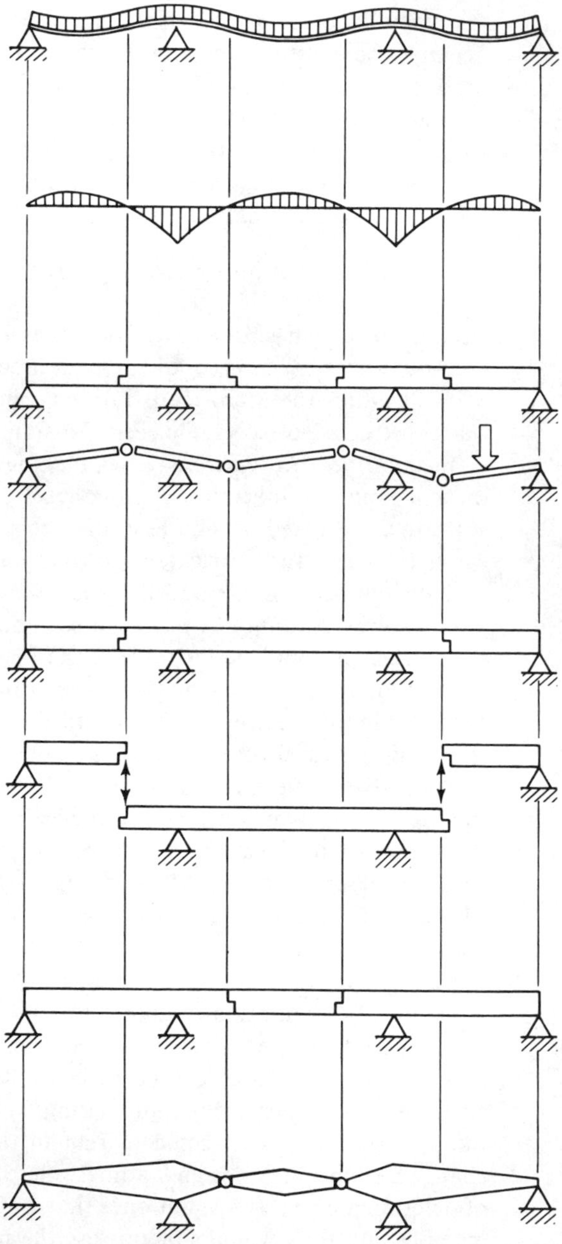

FIGURE 8-11 Use of construction joints in continuous members. Construction joints often facilitate construction. By creating a condition of zero moment by design at points of inflection, the behavior of a continuous member
is modeled by a series of statically determinate members.

tures of this type the dead load typically is the primary design load and far exceeds in magnitude the variable live load.

8.4.5 Controlling Moment Distributions

The moments developed in a continuous and fixed-ended member can be affected to a significant degree by decisions made by the designer. This can occur in several ways. One way is through simply paying careful attention to the spans and loadings involved. Often, by not using identical span lengths, the moment distribution can be affected in an advantageous way. As discussed in Section 6.4, the use of cantilevers on the ends of beams is often desirable, particularly as a device for reducing positive moments on end spans. Deflections are also reduced by using end cantilevers.

Another way of controlling moment distributions in an advantageous way is through the use of construction joints. In the preceding section the use of such joints at the locations of point of inflection was discussed. Obviously, pinned construction joints could be put anywhere and automatically cause a point of zero bending moment to develop at that point in the structure rather than where they would naturally occur. By locating such pin connections carefully, it is possible to reduce design moments.

Consider the fixed-ended beam previously analyzed (Figure 8-12). The moments naturally developed are $wL^2/12$ at the beam ends and $wL^2/24$ at the beam midspan. Points of inflection occur naturally at the $0.21L$ from each end. While the maximum design moment of $wL^2/12$ is considerably less than the $wL^2/8$ associated with a comparable simply supported span, it is possible to reduce the design moment to an even smaller value by inserting pin connections at points nearer the ends than where the inflection points naturally develop. Doing this would increase the effective span of the midsection (increasing the positive moment) and decrease the effective length of the end cantilever portions (decreasing the negative moment). Indeed, the maximum positive moment can be made exactly equal to the maximum negative moment by placing the pin connections at $0.15L$. As before, the total moment (the absolute sum of maximum positive and negative moments) remains $wL^2/8$. By forcing the positive and negative moments to become equal, design moments are reduced to $wL^2/16$ for each. This maximum design moment is considerably less than that associated with the unaltered fixed-ended member. Member sizes are reduced.

Insertion of pins at locations other than inflections points does, however, have other consequences. One is that there is no longer a smooth curve between regions of positive and negative moments as occurs at a natural point of inflection or when a pin is located exactly at such a point. Rather, a discontinuity is developed due to different end rotations of the midspan and cantilever portions of the member. If the magnitude of this discontinuity is large, it could potentially cause problems if the beam is used in a building context. If a roof area was directly over the beam, for example, the sharp discontinuity could cause cracking, hence leaking, in the roof. This is typically not a problem when such discontinuities are not present and the reversal in curvature is smooth.

Another example of locating pins to control moments is illustrated in Figure 8-12(c). By carefully locating pins, positive and negative moments can be made approximately equal, thus minimizing the maximum design moment present. Again,

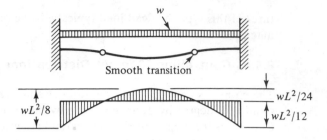

(a) Fixed-ended beam. Deflected shape and moments developed naturally.

Smooth transition

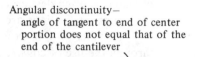

$wL^2/8$

$wL^2/24$

$wL^2/12$

Angular discontinuity—
angle of tangent to end of center
portion does not equal that of the
end of the cantilever

Construction joint (pin connection)

(b) By putting in pinned connections (points of known zero moment) at locations other than where the points would naturally develop, the maximum design moment in the structure can be reduced from $wL^2/12$ to $wL^2/16$. Material savings can therefore be accomplished.

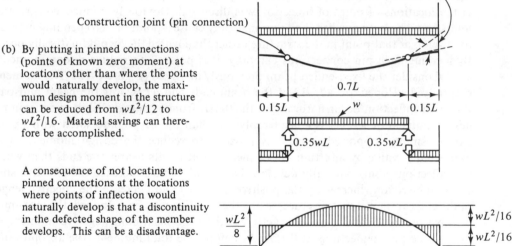

$0.7L$

$0.15L$ w $0.15L$

$0.35wL$ $0.35wL$

A consequence of not locating the pinned connections at the locations where points of inflection would naturally develop is that a discontinuity in the defected shape of the member develops. This can be a disadvantage.

$\dfrac{wL^2}{8}$

$wL^2/16$

$wL^2/16$

(c) Moments in structures continuous over several supports can also be controlled by carefully locating pinned connections. Again discontinuities in the deflected shape of the member would develop at the pins. By carefully locating construction joints (or pinned connections) and varying span lengths and overhangs, maximum positive and negative moments throughout the member can be made approximately equal.

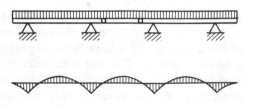

FIGURE 8-12 Controlling moment distributions by controlling the location of construction joints (pinned connections).

discontinuities in curvatures occur at pin locations when they are moved from the natural points of inflection.

8.4.6 Continuous Beams Made of Reinforced Concrete

Reinforced concrete is a particularly suitable medium with which to construct continuous beams. Continuity is achieved by the way the reinforcing is arranged.

Figure 8-13 illustrates the reinforcing pattern for a two-span continuous beam. Obviously, reinforcing steel is put into regions where tension stresses would normally develop. The amount of steel used at a particular location depends on the magnitude of the moment present.

Continuous beams can also be post-tensioned or prestressed. Figure 8-13(d) illustrates how a post-tensioning cable would be draped to be effective for the type of moments present. Draping a cable in this fashion is not easy nor is it abnormally difficult. Post-tensioning is frequently done. It is, however, difficult to prestress a continuous beam. The nature of the prestressing process is such that getting reverse curvatures in the prestressing strands is difficult. Since prestressing is most normally done in a factory circumstance (although not absolutely necessary), there is the added difficulty of transportation (i.e., a beam to be used over three supports would have to be transported with similar support conditions to keep moments in undesirable locations from developing). Prestressed beams are more normally used in simple support conditions.

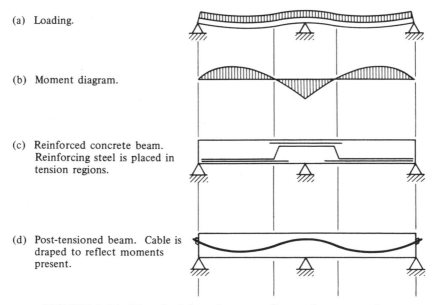

(a) Loading.

(b) Moment diagram.

(c) Reinforced concrete beam. Reinforcing steel is placed in tension regions.

(d) Post-tensioned beam. Cable is draped to reflect moments present.

FIGURE 8-13 Use of reinforced concrete for continuous members.

QUESTIONS

8-1. Investigate several multiple-span steel girder highway bridges in your local area, preferably including one that uses shaped elements. Make careful sketches of the shapes and the locations of any construction joints of the type discussed in Section 8.4.4. Discuss your findings.

8-2. A beam that is pinned on one end and fixed at the other end carries a concentrated load of P at midspan ($x = L/2$). Using an approximate method of analysis, estimate the reactions, shears, and moments present in the structure.

8-3. Using the expressions suggested in Section 8.4.2, determine the design moments for a three-span beam that is continuous over four supports (ends are integral with column supports). Determine the critical design positive and negative moments for each span. Assume that the structure carries a uniformly distributed load of 300 lb/ft and that each span is 30 ft.

8-4. Consider the two-span continuous beam illustrated in Figure 8-5. How would you expect the moment diagram to change if the right support began settling vertically downward with respect to the other two supports?

chapter 9

CONTINUOUS STRUCTURES: RIGID FRAMES

9.1 INTRODUCTION

A rigid frame structure is one made up of linear elements, typically beams and columns, that are connected to one another at their ends with joints that do not allow any relative rotations to occur between the ends of the attached members. Members are essentially continuous through the joints. As with continuous beams, rigid frame structures are statically indeterminate.

Many rigid frame structures resemble in appearance simpler post-and-beam systems but are radically different in structural behavior, owing to the joint rigidity. The joint rigidity can be sufficient to enable a framed structure to carry significant lateral loads, something a simpler post-and-beam system cannot do without additional bracing elements. Variants of rigid frame structures have been in use for a long time. A common table, for example, typically derives its stability from the rigid joints that are used to connect the legs to the table top. The traditional knee-braced timber structure can also be thought of as a type of rigid frame structure. Still the rigid frame as a widely used structural device in major buildings is a relatively recent phenomenon. The development of the steel rigid frame in cities such as Chicago during the latter part of the nineteenth and early part of the twentieth centuries is a major event in the history of structures. The related movement made possible by the introduction of the rigid frame of separating and differentiating enclosure surfaces from the supporting structural skeleton also marks a major turning point in the history of architecture. This movement was a marked departure from traditional building practices which made extensive use of dual functioning elements, such as the exterior load-bearing wall, that served simultaneously as both structure and enclosure.

9.2 GENERAL PRINCIPLES

A useful way to understand how a simple framed structure works is to compare and contrast its behavior under load with a post-and-beam structure that is identical in all respects except that members in the post-and-beam structure are not rigidly connected as they are in the framed structure (see Figure 9-1).

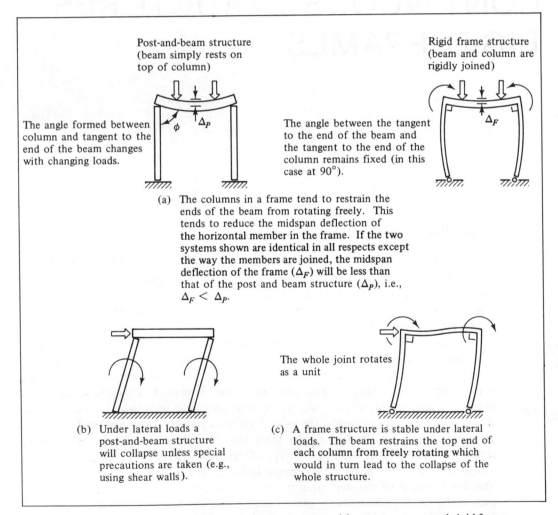

Post-and-beam structure
(beam simply rests on
top of column)

Rigid frame structure
(beam and column are
rigidly joined)

The angle formed between column and tangent to the end of the beam changes with changing loads.

The angle between the tangent to the end of the beam and the tangent to the end of the column remains fixed (in this case at 90°).

(a) The columns in a frame tend to restrain the ends of the beam from rotating freely. This tends to reduce the midspan deflection of the horizontal member in the frame. If the two systems shown are identical in all respects except the way the members are joined, the midspan deflection of the frame (Δ_F) will be less than that of the post and beam structure (Δ_P), i.e., $\Delta_F < \Delta_P$.

The whole joint rotates as a unit

(b) Under lateral loads a post-and-beam structure will collapse unless special precautions are taken (e.g., using shear walls).

(c) A frame structure is stable under lateral loads. The beam restrains the top end of each column from freely rotating which would in turn lead to the collapse of the whole structure.

FIGURE 9-1 Differences between post-and-beam structures and rigid frame structures.

9.2.1 Vertical Loads

First with respect to vertical loads, a load on a post-and-beam structure is picked up by the horizontal beam and is transferred by bending to the columns, which in turn carry the load to the ground. The beam is typically simply supported and merely rests on top of the columns. Consequently, as the vertical load causes bending to develop in the beam, the ends of the beam rotate on the tops of the columns. The angle formed between the beam and a column thus changes slightly with increasing loads on the beam. The columns do not in any way restrain the ends of the beam from rotating. No moment is transferred to the columns since the beam rotates freely. The columns carry only axial forces.

When a rigid frame structure is subjected to a vertical load, the load is again picked up by the beam and eventually transferred through the columns to the ground. The load again tends to cause the ends of the beam to rotate. In the frame, however, the column tops and beam ends are rigidly connected. Free rotation at the end of the beam cannot occur. The joint is such that the column tends to prevent or restrain the beam end from rotating. This has several important consequences. One is that the beam now behaves more like a fixed-ended beam than a simply supported one. Coupled with this are many of the advantages of fixed-ended members discussed in Chapter 8 (e.g., increased rigidity, decreased deflections, and decreased internal bending moments). On the other hand, the fact that the column top is offering restraint to rotation means that it must be picking up bending moments in addition to axial forces, thus complicating its design.

The rigid joint does not actually provide full end fixity for either the beam or the column. As the load tends to cause the end of the beam to rotate, the connected top of the column rotates as well. Consequently, there is a tendency for the whole joint between the column and the beam to rotate as a unit. While the whole joint rotates, however, the joint rigidity causes the members to retain their initial angular relationship to one another (e.g., if the members are initially at 90° to one another, they will remain so). The amount of the rotation that occurs depends on the *relative stiffnesses* of the beam and the column. The stiffer the column relative to the beam, the less total rotation occurs. *Some* rotation, however, invariably occurs. Thus, the end condition of the beam actually lies somewhere between a fully fixed-end condition (in which the joint offers full restraint and no rotations occur at all) and a pin connection (in which the member is completely free to rotate). The same is true for the column. Each element enjoys some, but not all, of the advantages of full fixity.

From a member-design viewpoint, the behavior described above generally means that beams in a rigid frame system carrying vertical loads can be designed to be somewhat smaller than those in a comparable post-and-beam system (since moments are reduced), while the columns may need to be somewhat larger than their post-and-beam counterparts (since they pick up both axial loads and moments, whereas the columns in a post-and-beam system pick up only axial forces). Relative column sizes may be further affected when buckling is considered since the column in the framed structure has some end restraint, while the column in a post-and-beam system has none.

Another way that frames are uniquely different from comparable post-and-beam

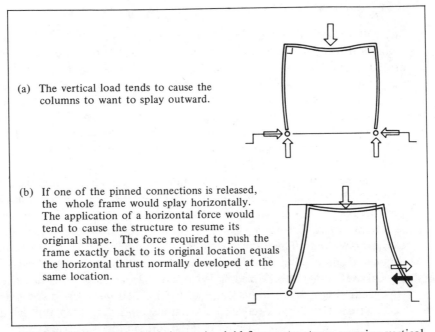

(a) The vertical load tends to cause the columns to want to splay outward.

(b) If one of the pinned connections is released, the whole frame would splay horizontally. The application of a horizontal force would tend to cause the structure to resume its original shape. The force required to push the frame exactly back to its original location equals the horizontal thrust normally developed at the same location.

FIGURE 9-2 Horizontal thrusts in rigid frame structures carrying vertical loads.

structures in carrying vertical loads that is intimately related to the presence of the moments in the columns discussed above is that frames typically develop horizontal as well as vertical reactions at the ground supports. The presence of these forces can most easily be visualized by imagining the deflected shape the structure would assume if the bases of the columns were not pin-connected to the ground foundation and allowed to freely translate horizontally (see Figure 9-2). The columns would naturally tend to splay outward. The application of horizontal forces that are inwardly directed at the column bases would cause the bases of the columns to tend to resume their original location. The amount of horizontal force that would cause the columns to be forced back into their exact original location equals the amount of horizontal thrust that the frame naturally exerts on the foundation when the column bases are normally attached to the foundation. As will be discussed later in this chapter, the magnitudes of the moments in the columns and those of the horizontal thrusts are directly related.

The foundations for a frame must be designed to carry the horizontal thrusts generated by the vertical loads. No such horizontal thrusts develop in post-and-beam structures carrying vertical loads. Foundations can consequently be simpler in post-and-beam systems than in frames.

9.2.2 Horizontal Loads

While the differences in behavior under vertical loads between frame structures and post-and-beam structures are pronounced, the differences with respect to horizontal or lateral loads are enormous. A post-and-beam structure is largely incapable of resisting significant lateral forces. What lateral force resistance such a structure does have stems largely from the large dead weight of the structure when stone or masonry is used or from the presence of some other element, such as an enclosure wall, that provides a bracing function. Column ends in timber structures might be sunk into the ground and thus provide a measure of lateral resistance by virtue of the end fixity achieved. By and large, however, most post-and-beam structures are not suited to carrying high lateral forces of the type associated with earthquakes and cyclones.

A rigid frame structure, on the other hand, is well capable of carrying lateral forces if designed properly. By virtue of the presence of a rigid connection, the beams restrain the columns from freely rotating in a way that would lead to the total collapse of the structure. The joints, however, do rotate (to a limited extent) as whole units. The stiffness of the beam contributes to the lateral-load-carrying resistance of a frame. It also serves to transfer part of the lateral load from one column to the other.

The action of a lateral load on a frame produces bending, shear, and axial forces in all members. Bending moments induced by wind loads are often the highest near the rigid joints. Consequently, members are either made larger or specially reinforced at joints when lateral forces are high.

Rigid frames are applicable to both large and small buildings. Many high-rise

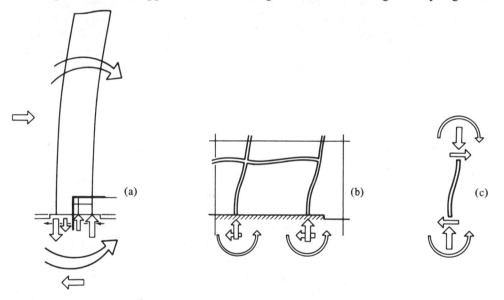

(a) The wind force acting against the face of the building creates an overturning moment and horizontal translatory force. The reactions developed at the base of the building provide a resisting moment and force that balance the overturning and sliding effect of the wind.

(b) Lower story columns: exaggerated diagram of deformations. The laterally acting wind force causes the columns to rack to the right. Curvatures and bending moments are induced in columns and beams.

(c) Free-body diagram of column: a typical column is subjected to axial forces, shears and bending moments. Beams are also subjected to similar forces.

FIGURE 9-3 Effects of lateral loads on a multistory frame.

buildings use rigid frames to carry both vertical and lateral loads. The higher the building, however, the larger become the forces and moments developed in individual members. The lower columns in a high-rise structure in particular are subjected to very large moments and axial forces because of the huge lateral loads involved (see Figure 9-3). A point is often reached where it becomes unfeasible to design members for these forces and moments and other bracing systems (e.g., diagonals or shear walls) are introduced to help carry the lateral loads and thus reduce forces and moments in the frames. Even in a lower building, supplementary bracing systems are typically used whenever possible, simply because carrying lateral loads by frame action alone is relatively inefficient.

9.3 ANALYSIS OF RIGID FRAMES

9.3.1 Approximate Methods of Analysis

GENERAL APPROACHES.   This section discusses simplified methods for determining the approximate values of reactions, shears, and moments in a rigid frame structure. As with continuous beams, rigid frame structures are statically indeterminate and these reactions, shears, and moments cannot be determined directly through the application of the basic equations of statics ($\sum F_x = 0$, $\sum F_y = 0$, and $\sum M_0 = 0$) alone. There are typically more unknowns than equations. This implies that reactions, shears, and moments are dependent on the precise characteristics of the structure itself, including the relative stiffnesses of beams and columns, as well as the overall geometry of the whole structure and its loading condition.

The analytical methods discussed below are based on many simplifying assumptions and yield approximate results only. These approximate methods, however, yield results that are extremely useful during preliminary design stages to determine an initial set of member sizes and properties. These estimates are then used in more exact analyses made during design development stages, since some member-size estimates must be available before an exact analysis can take place.

The number of assumptions that must be made in order to analyze a frame in a simple fashion is crucial, since the more that are made, the less exact will be the results obtained. Usual practice is to make a sufficient number of assumptions, each of which supplies an independent equation or relation, to allow the frame to be analyzed directly by application of the basic equations of statics alone. An effective method of analysis is based on a procedure that assumes the locations of points of zero internal moment (points of inflection). These locations are initially estimated by looking at the deflected shape of the structure. This is the same basic technique described in Chapter 8 in connection with the approximate analysis of beams.

SINGLE BAY FRAMES: LATERAL LOADS.   Two simple rigid frames subjected to lateral loads will be analyzed by approximate methods in this section. The first to be considered is a single bay frame with rigid connections between the beam and the columns and pin connections at the column bases. The second is identical except that it has fixed rather than pinned connections at the column bases. In both cases it is assumed that the lateral load carried by the structure can be characterized as a point force acting at the top joint. The exact type and location of the lateral forces carried by a frame in a real building depends on what type of force is being considered (wind or earthquake). An analysis should be made for the building considered to determine exactly how to model the load. A simple point load will be assumed in the following examples.

Figure 9-4 shows the reactions for the first frame. There are four unknown reactions (R_{A_H}, R_{A_V}, R_{D_H}, R_{D_V}) and only three equations of statics ($\sum F_x = 0$, $\sum F_y = 0$, and $\sum M_0 = 0$) available for solution. Consequently, the frame is statically indeterminate to one degree. In this particular frame it is still possible to determine the vertical reactions, R_{A_V} and R_{D_V}, by simply summing the moments of the external and reactive forces around either of the pin connections (locations of known zero moment resistance). Thus,

For the whole structure: $\sum M_A = 0$, $\sum F_y = 0$, and $\sum F_x = 0$:

$$-Ph + R_{A_V}(0) + R_{A_H}(0) + R_{D_V}(L) + R_{D_H}(0) = 0 \qquad R_{D_V} = \frac{Ph}{L} \uparrow$$

$$-R_{A_V} + R_{D_V} = 0 \qquad -R_{A_V} + \frac{Ph}{L} = 0 \qquad R_{A_V} = \frac{Ph}{L} \downarrow$$

$$+P - R_{A_H} - R_{D_H} = 0 \qquad R_{A_H} + R_{D_H} = P$$

It is evident that the vertical reactive forces can be found only because of the particular condition that the two unknown horizontal reactions pass through the moment center selected. It is not possible to determine any further information by direct application of the equations of statics. It can be surmised that the horizontal reactions, R_{A_H} and R_{D_H}, are equal but this is an assumption only.

In order to get a complete solution to the problem, use can be made of the fact that a point of inflection develops in the beam when the frame carries a horizontal load. By sketching the probable deflected shape of the structure, the location of this point of inflection can be accurately estimated. In this case the point of inflection is estimated to be at midspan [see Figure 9-4(a)]. The knowledge that at points of inflection the internal moment in the structure is zero can be used to provide another independent equation which will lead to a complete solution. For analytical puposes, stipulating that the internal moment is zero at this point is equivalent to inserting a pin connection at that location. The resultant structure can thus be modeled as a statically determinate three-hinged assembly of the type discussed earlier (see Section 5.3.1). By decomposing the structure at the point of inflection into two separate assemblies, the horizontal reactions can be determined by considering the moment equilibrium for each piece [see Figure 9-4(b)]. Thus:

For the left assembly, $\sum M_N = 0$:

$$P(0) + R_{A_V}\left(\frac{L}{2}\right) - R_{A_H}(h) = 0$$

$$\left(\frac{Ph}{L}\right)\left(\frac{L}{2}\right) = R_{A_H}(h) \qquad R_{A_H} = \frac{P}{2} \leftarrow$$

By considering the horizontal force equilibrium of the whole structure, the remaining unknown reaction, R_{A_H}, can be found. Thus:

Point of inflection "N"

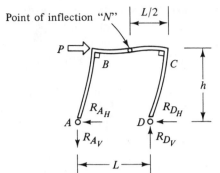

(a) Deflected shape of frame.

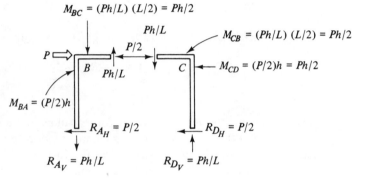

$M_{BC} = (Ph/L)(L/2) = Ph/2$

Ph/L

$M_{CB} = (Ph/L)(L/2) = Ph/2$

$P/2$

$M_{CD} = (P/2)h = Ph/2$

Ph/L

$M_{BA} = (P/2)h$

$R_{A_H} = P/2$

$R_{D_H} = P/2$

$R_{A_V} = Ph/L$

$R_{D_V} = Ph/L$

(b) Free-body diagrams for parts of frame separated at points of inflection (points of known zero moment). Since the structure is separated at points of zero moment, no internal moments need be shown or considered.

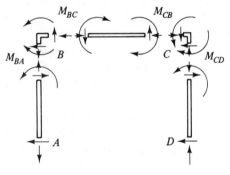

(c) Free-body diagram of individual beam, column, and joint elements. Since the structure is not separated at points of zero moment, the free-body diagrams must show internal moments.

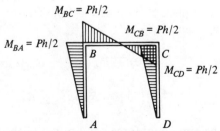

(d) Final moment diagram. Shear and axial forces are present in the members as well as the moment indicated.

FIGURE 9-4 Simplified analysis of a single bay rigid frame carrying a lateral load.

For the whole structure $\sum F_x = 0$:

$$R_{A_H} + R_{D_H} = P \qquad \frac{P}{2} + R_{D_H} = P \qquad R_{D_H} = \frac{P}{2} \quad \leftarrow$$

All reactions are now known ($R_{A_H} = P/2$, $R_{Av} = Ph/L$, $R_{D_H} = P/2$, $R_{Dv} = Ph/L$). Note that the assumption that the point of inflection in the beam was located at midspan is equivalent to assuming that the horizontal reactions are equal. Once the reactions are known, the shears V, moments M, and axial forces N in the structure can be found by considering each element in turn [see free-body diagrams in Figure 9-4(b)]. The typical designation system used is as follows:

M_{xy} = moment in member x–y at the end of the member that frames into joint x

Shears and axial forces are similarly denoted. For the frame considered:
Left column:

$$M_{BA} = \left(\frac{P}{2}\right)h = \frac{Ph}{2}$$

$$V_{BA} = \frac{P}{2}$$

$$N_{BA} = \frac{Ph}{L} \qquad \text{(tension)}$$

Right column:

$$M_{CD} = \left(\frac{P}{2}\right)h = \frac{Ph}{2}$$

$$V_{CD} = \frac{P}{2}$$

$$N_{CD} = \frac{Ph}{L} \qquad \text{(compression)}$$

Beam:

$$M_{BC} = \left(\frac{Ph}{L}\right)\left(\frac{L}{2}\right) = \frac{Ph}{2}$$

$$M_{CB} = \left(\frac{Ph}{L}\right)\left(\frac{L}{2}\right) = \frac{Ph}{2}$$

$$V_{CB} = V_{BC} = \frac{Ph}{L}$$

$$N_{CB} = N_{BC} = \frac{P}{2} \qquad \text{(compression)}$$

Moments are thus typically found by simply multiplying a force times a distance. A column, for instance, is treated quite like a vertical cantilever beam. The beam moments found above can also be determined by a slightly different procedure making use of another set of free-body diagrams. The free-body diagrams shown in Figure

9-4(c) indicate how the same structure can be broken down into individual beam, column, and joint elements. The concept of isolating joints in a frame, and considering their equilibrium, is similar to that used in the method of joints in truss analysis. The primary distinction is that in trusses the members were pinned together at joints, hence moments could not be developed and only translatory equilibrium needed to be considered. In the case of a frame, the member connections are rigid, hence internal moments are developed at member ends which must be reflected in the free-body diagrams of individual joints. Because of these moments, both translational and rotational equilibrium need to be considered for any individual joint. Using this type of free-body diagram, moments can be found as shown below. Considering rotational equilibrium only:

Joint Equilibrium Approach. As before, a moment is developed at the top of column *B–A* due to the horizontal reaction:

Column *B–A*:

$$M_{BA} = \left(\frac{P}{2}\right)h$$

There is an equal and opposite moment acting on joint *B*. For rotational equilibrium for the joint to be maintained, a balancing moment must be developed in *B-C*. The beam provides this moment restraint:

Joint *B*:

$$-M_{BA} + M_{BC} = 0$$

$$M_{BC} = \frac{Ph}{2}$$

Similar observations can be made about column *C–D* and joint *D*.

Column *C–D*:

$$M_{CD} = \left(\frac{P}{2}\right)h$$

Joint *C*:

$$-M_{CD} + M_{CB} = 0$$

$$M_{CB} = \frac{Ph}{2}$$

The end moments found for the beam are clearly the same as those previously found. Figure 9-4(c) shows not only how the various beam, column, and joint elements are in moment equilibrium but shows how they are in equilibrium with respect to shear and axial forces as well.

Moment diagrams can now be drawn for each of the beams and columns in the frame. By inspecting the type of forces producing the moments, it can be seen that all members have linearly varying moment diagrams with the maximum moments previously found occurring at the joints and the minimums occurring at either the pins or point of inflection.

In order to plot moment diagrams, a sign convention other than the one used for horizontal members (with positive moment being defined as compression on the top surface and tension on the bottom) must be used since the notion of top and bottom surfaces is meaningless in a vertical member. Common practice for plotting moment diagrams for vertical members is to look at the member from the right and employ the usual convention (this is the same as turning the member 90° in a clockwise direction). An equivalent convention that is perhaps more generally useful since it is applicable to any member having any orientation is simply to plot the moment diagram on the compression face of the member and to not be concerned with whether the moment is called a positive one or a negative one. An inspection of moment diagrams previously drawn reveals that this practice has been consistantly followed. Figure 9-4 illustrates the moment diagram for the frame considered. This completes the analysis for this frame.

A more complex type of frame is illustrated in Figure 9-5(a). This frame is identical to the one previously studied except that the column bases are fixed rather than pinned. Figure 9-5(a) shows the reactive forces and moments developed at the foundations. There are six unknown quantities (R_{A_H}, R_{A_V}, M_A^F, R_{D_H}, R_{D_V}, and M_D^F) and only three equations of statics available for use. Hence, the frame is statically indeterminate to the third degree. Consequently, three assumptions must be made if a static analysis is to be used. As before, a sketch of the probably deflected shape of the structure is made. As is evident, there are three points of inflection developed. One is in the midspan of the beam as before. Two others are near the midheights of the columns. They are not exactly at midheight since the top joints rotate slightly. If the joints simply translated horizontally without rotation, the points of inflection would have to be identically at midheight (by a symmetry argument). The slight rotation of the upper joints causes the point of inflection to rise somewhat to the location indicated in Figure 9-5(a).

Fixing the location of these three points of inflection makes a static analysis possible. The frame can be decomposed or separated at these points of zero moment in a manner similar to that previously described and each piece treated in turn (i.e., axial forces and shears can be found based on the knowledge that the net rotational moments around these points for any piece must be zero). This analysis is illustrated in Figure 9-5(b).

Note that if the frame is decomposed into two sections at the inflection points in the column, the upper part is quite analogous to the frame previously discussed (with pin-counnected bases) except that column heights differ. The analysis technique is the same. This structure can then be imagined as simply resting on two vertical cantilever elements. The horizontal thrusts associated with the upper part produce moments in the lower elements. A final moment diagram is shown in Figure 9-5(d).

Note that the final moments in individual members are lower than in the pinned frame previously analyzed.

SINGLE BAY FRAMES: VERTICAL LOADS. The general procedure for making an approximate analysis of a frame carrying vertical loads is much the same as the one described in the previous section for lateral loads. Consider the rigid frame shown in Figure 9-6(a), which has pinned connections at the column bases. The first step in the analysis is to sketch the probable deflected shape of the structure and to locate points of inflection [see Figure 9-6(a)].

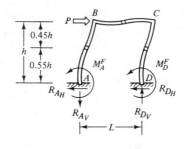

(a) Deflected shape of frame.

$M_{BC} = (0.45Ph/L)(L/2)$ $M_{CB} = (0.45Ph/L)(L/2)$
$= 0.225Ph$ $= 0.225Ph$

(b) Free-body diagram for parts
of structure separated at points
of inflection (points of known
zero moment).

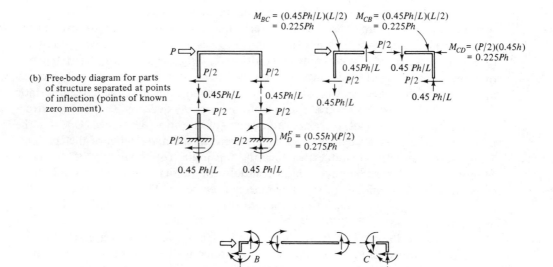

$M_{CD} = (P/2)(0.45h)$
$= 0.225Ph$

$M_D^F = (0.55h)(P/2)$
$= 0.275Ph$

(c) Free-body of individual beam column,
and joint elements. Each element must
be in a state of translatory and rotational
equilibrium.

(d) Final moment diagram. Shear and axial
forces are present in the members as well
as the moments indicated.

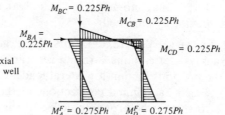

$M_{BC} = 0.225Ph$
$M_{CB} = 0.225Ph$
$M_{BA} = 0.225Ph$
$M_{CD} = 0.225Ph$
$M_A^F = 0.275Ph$ $M_D^F = 0.275Ph$

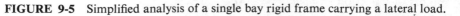

FIGURE 9-5 Simplified analysis of a single bay rigid frame carrying a lateral load.

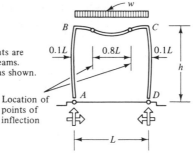

(a) Frame and loading. Inflection points are
developed near both ends of the beams.
Their locations are assumed to be as shown.

Location of
points of
inflection

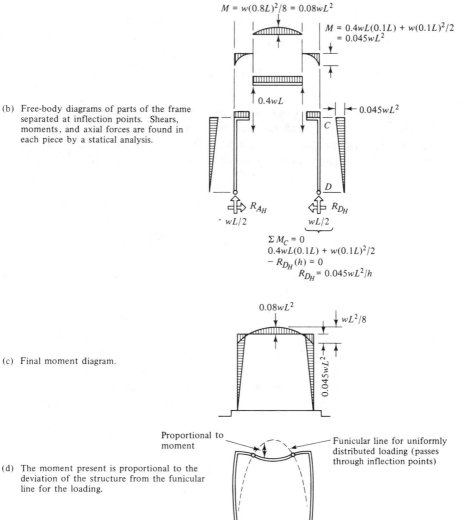

$M = w(0.8L)^2/8 = 0.08wL^2$

$M = 0.4wL(0.1L) + w(0.1L)^2/2$
$= 0.045wL^2$

(b) Free-body diagrams of parts of the frame
separated at inflection points. Shears,
moments, and axial forces are found in
each piece by a statical analysis.

$0.4wL$

$0.045wL^2$

C

D

R_{A_H}

R_{D_H}

$wL/2$

$wL/2$

$\Sigma M_C = 0$
$0.4wL(0.1L) + w(0.1L)^2/2$
$- R_{D_H}(h) = 0$
$R_{D_H} = 0.045wL^2/h$

$0.08wL^2$

$wL^2/8$

(c) Final moment diagram.

$0.045wL^2$

Proportional to
moment

Funicular line for uniformly
distributed loading (passes
through inflection points)

(d) The moment present is proportional to the
deviation of the structure from the funicular
line for the loading.

FIGURE 9-6 Simplified analysis of a single bay rigid frame carrying vertical
loads.

It is somewhat more difficult to locate the inflection points associated with vertical loads than the ones associated with lateral loads. If the joints provided full fixity (and no end rotations occurred at all at beam ends), then the points of inflection would be at $0.21L$ from either end (see Section 8.3). Since some rotations do occur, but the joint does not freely rotate, the actual end conditions lie somewhere between pinned connections and fully fixed connections; the inflection points in the beam thus lie somewhere between $0L$ and $0.21L$ from the joints. For beams and columns of normal stiffnesses, the inflection points are normally found about $0.1L$ from either end. As will be discussed more fully in Section 9.3.2, the precise location of the inflection points is very sensitive to the relative stiffnesses of the beams and columns.

If the inflection points are assumed to be at $0.1L$ from either end, the structure can be decomposed into three statically determinate elements and analyzed as illustrated in Figure 9-6(b). The final moment diagram obtained is shown in Figure 9-6(c).

As is evident, the vertical loads produce moments in both the beam and in the columns. Maximum moments in the beam normally occur at midspan but critical moments are present at end joints as well. Maximum column moments invariably occur at member ends.

MULTISTORY FRAMES: LATERAL LOADS.   There are several methods used to analyze multistory frames subjected to lateral forces. A time-honored approach called the *cantilever method* was first introduced in 1908. It involves making many of the same types of assumptions previously made, plus others. Generally, it assumes the following:

1. There is a point of inflection at the midspan of each beam in a complex frame;
2. there is a point of inflection at the midheight of each column; and
3. the magnitude of the axial force present in each column of a story is proportional to the horizontal distance of that column from the centroid of all the columns of the story under consideration.

These assumptions are diagrammatically illustrated in Figure 9-7. These illustrations also provide an insight into the behavior of a multistory frame under load. Moments are generated in all members. The magnitude of these moments depend upon the magnitude of the resultant shear force V_L (the sum of the lateral loads above the floor considered). This shear force is balanced by resisting shear forces (V_{c_1}, $V_{c_2}, \ldots, V_{c_n}$) developed in the columns at the same level ($V_L = \sum_i V_{c_i}$). The moment in any column is the shear force in the column multiplied by its moment arm (one-half the column height). Moments in beams are generated to balance moments at column ends. Since the total shear force V_L is greater at lower floors than upper floors, bending moments in beams and columns are greater at lower rather than upper floors.

The magnitudes of the axial forces in the columns depend on the magnitude of the overturning moment M_L associated with the lateral loads above the section considered. Consequently, axial forces are greater at the base of the building where the overturning moment is the greatest; they tend to reduce in upper story columns.

Other more exact methods of analysis that are computer oriented are now more

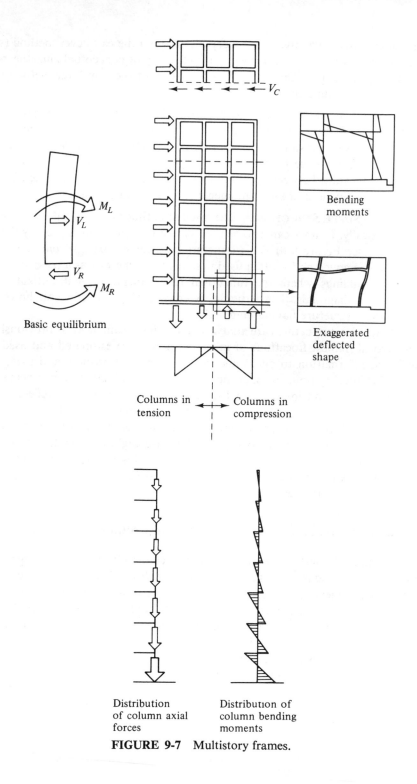

V_C

M_L

V_L

V_R

M_R

Basic equilibrium

Bending
moments

Exaggerated
deflected
shape

Columns in
tension

Columns in
compression

Distribution
of column axial
forces

Distribution of
column bending
moments

FIGURE 9-7 Multistory frames.

347

commonly used to analyze multistory frames than the cantilever method (see Appendix 11). Nonetheless it still provides a useful way of conceptualizing the behavior of multistory structures. Details of the application of the cantilever method to complex frames can be found in other texts.[1]

MULTISTORY FRAMES: VERTICAL LOADS. The magnitudes of forces and moments in multistory frames that are due to vertical loads can be estimated in the same way as was illustrated for single bay rigid frames. Inflection points can be assumed at $0.1L$ from either end of the beam. This assumption has the effect of creating a statically determinate beam between the two inflection points that is supported by short cantilevers. Positive and negative moments can be found by statics.

THE VIERENDEEL STRUCTURE. The discussion thus far has dealt with frames that are used vertically. Frames can also be used horizontally as is illustrated by the *Vierendeel structure* (see Figure 9-8). This structure resembles a parallel chord truss with the diagonals removed and is often used in much the same way as a truss. The structure is used in buildings where the functional requirements present mitigate against the presence of diagonals. The structure is considerably less efficient, however, than a comparable structure having diagonals.

Vierendeel structures are analyzed in much the same way as previously described for vertical frames. Locations of inflection points are estimated and used to provide sufficient information to determine internal shears, moments, and axial forces.

The internal moments in elements tend to be the highest in members near the end of the structure and lower toward the middle. This distribution reflects the fact that the overall shear forces associated with the total loading which cause local bending moments and shears in specific members are highest toward the ends and decrease toward the middle. Axial forces in members are highest in middle top and bottom chord elements and decrease toward the ends. This distribution reflects that of the overall moment associated with the total loading. This pattern is similar to that present in parallel chord trusses.

9.3.2 Importance of Relative Beam and Column Stiffness

In any statically indeterminate structure, including a frame, the magnitudes of the internal forces and moments are, in the final analysis, dependent on the relative properties of the members themselves. None of the approximate analyses discussed thus far have reflected the importance of this fact. Implicit in the analyses, however, have been assumptions about the relative characteristics of the members used. Normal or typical stiffnesses were considered.

The importance of different member properties can be seen by looking at Figure 9-9. If it is initially assumed that one column is stiffer than the other (i.e., one has a higher relative I value), the stiffer column will end up taking a greater share of the horizontal load than the more flexible one. An assumption that horizontal reactions are equal is thus not tenable. Higher moments would also be developed in the stiffer

---

[1]See, for example, J. McCormac, *Structural Analysis*, 1st ed., International Textbook Company, Scranton, Pennsylvania, 1960.

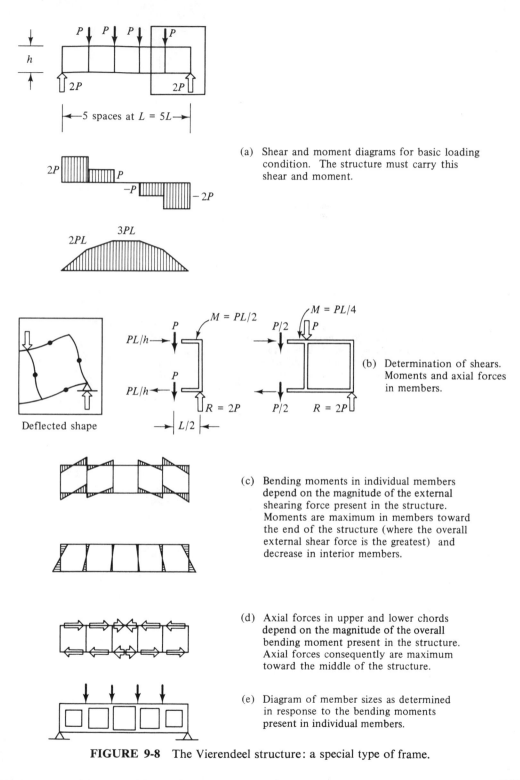

(a) Shear and moment diagrams for basic loading condition. The structure must carry this shear and moment.

(b) Determination of shears. Moments and axial forces in members.

Deflected shape

(c) Bending moments in individual members depend on the magnitude of the external shearing force present in the structure. Moments are maximum in members toward the end of the structure (where the overall external shear force is the greatest) and decrease in interior members.

(d) Axial forces in upper and lower chords depend on the magnitude of the overall bending moment present in the structure. Axial forces consequently are maximum toward the middle of the structure.

(e) Diagram of member sizes as determined in response to the bending moments present in individual members.

FIGURE 9-8 The Vierendeel structure: a special type of frame.

349

(a) Structure having columns of different stiffnesses. Assume $I_A = 3I_B$.

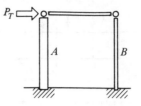

(b) Despite the different stiffnesses, each member must deflect the same amount ($\Delta_A = \Delta_B$).

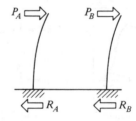

(c) A larger force must be applied to the stiffer column to cause it to deflect the same amount as the more flexible column. Consequently, it picks up a greater portion of the applied load ($P_A > P_B$).

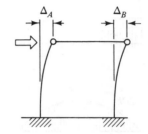

(d) The stiffer column picks up a greater portion of the applied load. Higher moments are also generated in the member.

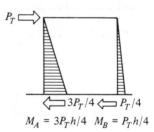

$$M_A = 3P_Th/4 \quad M_B = P_Th/4$$

FIGURE 9-9 Effects of different column stiffnesses on forces and moments in a rigid structure.

member than in the more flexible one as a consequence of its taking a greater portion of the load.

Different relative stiffnesses between beams and columns also affects moments due to vertical loads. As Figure 9-10 illustrates, the location of the inflection points are affected by the relative stiffness of the beams and columns. The stiffer the column relative to the beam, the more it restrains the end of the beam from rotating. The consequence is that higher moments are developed in the column when it is very stiff relative to the beam than when the column is more flexible. Negative moments in the beam

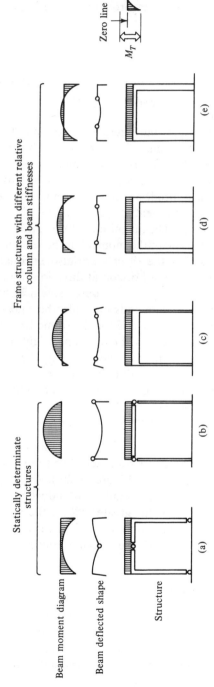

Statically determinate structures

Frame structures with different relative column and beam stiffnesses

Beam moment diagram

Beam deflected shape

Structure

(a)   (b)   (c)   (d)   (e)

M_T Zero line $\frac{M_-}{}$ M_+

(f)

(a) Three-hinged arch structure. Large negative moments are developed in the beam.

(b) Post-and-beam structure. Large positive moments are developed in the beam. The columns do not restrain the rotations of the beam ends at all.

(c) Frame with very flexible columns and an extremely stiff beam. The flexible columns do not provide significant restraint to the ends of the beam. The beam behaves similarly to a pin-ended one.

(d) Frame with normal beam and column stiffness. The columns offer partial restraint to the ends of the beams. Some rotations occur.

(e) Frame with very stiff columns and a flexible beam. The columns offer almost full restraint to the ends of the more flexible beam.

(f) The sum of the positive and negative beam moments remains the same in each case. The location of the base line and hence of the design moments varies.

FIGURE 9-10 Effects of different relative beam and column stiffness on internal forces and moments in a rigid frame structure. The moments generated in the frames by the vertical load vary with different locations of the inflection points. The stiffer the section of a frame, the greater is the moment developed at the section.

351

are also increased while positive moments are decreased. When the beam is stiff relative to the column, converse phenomena occur.

Exact methods of analysis are available which take into account these effects. These analyses must be based on an assumed set of beam and column stiffnesses. The approximate analyses previously discussed are useful in providing these initial estimates.

9.3.3 Sidesway

A phenomenon of particular interest in frames carrying vertical loads is that of *sidesway*. If a frame is not symmetrically shaped and loaded, the structure will sway (translate horizontally) to one side or the other.

The reason for sidesway can be seen by inspecting Figure 9-11. Assume that the columns are very stiff and completely restrain the ends of the beam, which can then be modeled as a fixed-ended beam. Corresponding fixed-ended moments are shown in Figure 9-11(b). Since the joints must be in rotational equilibrium, this means that the moments at the column tops are the same magnitude as those in the adjacent beam. For equilibrium of the column, the presence of a moment at the top implies the necessity of a horizontal thrust at the base of the frame. The magnitude of this thrust at the base of a column is directly related to the magnitude of the moment at the column top. Since the moments are unequal, this means that the thrusts are also unequal. By looking at the overall equilibrium of the frame in the horizontal direction, however, it can be seen that the thrusts *must* be equal so that $\sum F_x = 0$ obtains, which in turn means that the moments at the column tops and beam ends must also be equal. The only way that such an equality can be naturally obtained is for the frame to sway to the left. As it sways to the left, there will be a tendency for the upper right joint to open up slightly, thus reducing the moment present (and the thrust of the column base), and for the upper left joint to close up slightly, thus increasing the moment present (and the thrust at the column base). Just enough sway will occur so that moment and horizontal thrust equality will be achieved.

9.3.4 Support Settlements

As was the case with continuous beams, rigid frames are quite sensitive to differential support settlements. Consider the rigid frame shown in Figure 9-12(a). Any type of differential support movement, either vertically or horizontally, will induce moments in the frame. The greater the differential settlements, the greater will be the moments induced. These moments can lead to failures in the design of the frame if not anticipated. For this reason, special care must be taken with the design of foundations for rigid frame structures to minimize this risk.

9.3.5 Effects of Partial Loading Conditions

As was the case with continuous beams, the maximum moments developed in a frame often do not occur when the frame is fully loaded but only partially loaded. This, of course, complicates the analysis process enormously. The first problem is simply that of predicting which type of loading pattern produces the most critical moments. Only

(a) A rigid frame which is not
symmetrically loaded.

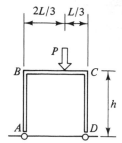

(b) Frame moments assuming that
no joint rotation at all occurs.

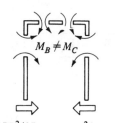

$$M_B^F = (2/27)PL^2 \qquad M_C^F = (4/27)PL^2$$

(c) The assumption that no joint
rotation at all occurs leads to the
result that horizontal frame reactions
at the foundation are unequal.

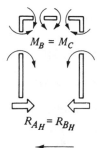

$M_B \neq M_C$

$$(R_{A_H} = 2PL^2/27h) \neq (4PL^2/27h = R_{B_H})$$

(d) From a consideration of the overall
equilibrium of the frame in the
horizontal direction, it is evident
that the result found in (c) is not
plausible. Horizontal reactions
must be equal. This also implies
that moments at the column tops
must also be equal.

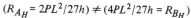

$M_B = M_C$

$R_{A_H} = R_{B_H}$

(e) The only way that the frame can naturally
adjust itself so that moments at column
tops are equal (so that horizontal reactions
are equal) is to sway to the left. This has a
tendency of closing up the upper left joint
(and increasing the moments present) and
opening up the upper right joint (and
decreasing the moments present). The frame
will sway until the moments present become
equal.

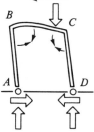

FIGURE 9-11 Sidesway in rigid frames. The absence of complete symmetry
in a frame or its loading leads to a horizontal sway in the structure.

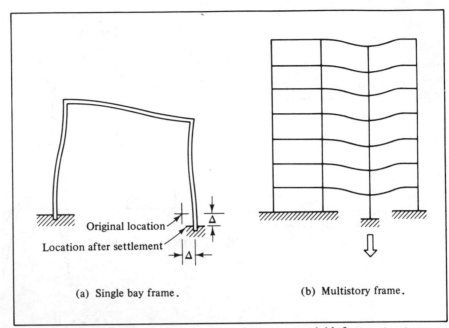

(a) Single bay frame.

(b) Multistory frame.

FIGURE 9-12 Effects of support settlements on rigid frame structures. Curvatures, hence bending moments, are induced in a frame because of differential joint settlement.

after this is done can an analysis take place. Usually, some variant of a checkerboard loading pattern over the whole frame produces maximum positive or negative moments at the location considered. There are several techniques for establishing exactly which loading conditions are most critical (e.g., the Miller-Breslau technique). Going into the problem of partial loadings in detail, however, is beyond the scope of this text.

9.4 DESIGN OF RIGID FRAMES

9.4.1 Introduction

Designing a frame structure can be an involved process. One of the first questions to be asked is simply whether or not it makes sense to use a frame at all. They are not, for example, particularly efficient types of structures to be used in situations where lateral loading conditions are high. Simply because frames can be designed to carry lateral loads does not mean they should be given preference over other approaches, such as structures using shear walls or diagonal bracing, that are better suited for carrying this type of loading. Frames make sense when the functional requirements of a building do not easily allow other solutions to be used.

When the functional requirements of a building warrant the use of frames, the general geometry and dimensions of the frame to be designed are often fixed in the building context and the design problem often becomes a more limited one of developing strategies for selection of connection types, selecting materials, and sizing members.

The first section will discuss member sizing. The process of designing a member in a frame is like that of designing any member in bending. The crucial step is that of determining the critical or maximum internal moments, shears, and critical forces in the member. Then the member can be sized using techniques discussed in Chapter 6.

9.4.2 Design Moments

Frames used in buildings normally carry both vertical and lateral loads, although occasionally some sort of supplementary structure may be used to help carry the lateral loads. For situations where the frame considered carries both types of loads, it is evident that members must be sized to provide adequate capacity to resist the aggregate moments generated by the loadings when they act on the structure simultaneously.

The previous sections on analysis have discussed how to determine moments from either vertical or lateral loads. To determine design moments, it is necessary to superimpose moments generated by both types of loadings. Figure 9-13 illustrates this process for a simple rigid frame. In some instances the moments from the vertical and lateral loads are additive and in other cases they tend to negate each other. Critical moments occur when moments are additive. It should be remembered that lateral loads can typically come from either side; therefore, the subtractive cases are rarely of importance and all members are designed for additive effects.

In cases where the lateral load is very high relative to the vertical load, moments from the lateral forces will be dominant and maximum design moments in either beams or columns will usually occur at joints. When lateral loads are less dominant, the critical design moments in a beam occur at midspan or even at some other point. In columns critical moments almost invariably occur at joints.

The discussion above has not addressed the complicating problem of partial loadings potentially creating higher moments than primary or full loadings. An analysis could be made for each relevant loading, but this is often too exacting a task for preliminary design purposes and is not often initially done. In this connection the discussion in Section 8.4.2 on design moments in continuous beams is relevant. The approximate moments given in Table 8-1, which at least partly reflect the effects of partial loadings, are often used for frames as well as other continuous beams for preliminary design purposes.

While of obvious importance, further consideration of the effects of partial loadings is beyond the scope of this text and the discussion will focus on designing frames in response to their primary loading conditions only. Quite often the effects of partial loadings are considered only after the structure is initially designed for its primary loading.

Once the critical maximum moments have been determined, and other internal shears and axial forces as well, then member sizing can take place. Two options are present for sizing a member. The first is to identify the maximum internal forces and moments present at any spot along the length of the member, size a member for these values, and then use a constant member size throughout. This would mean that the member would be oversized at every location but the critical one. The second general option is to shape the member in response to the variation in critical moments and

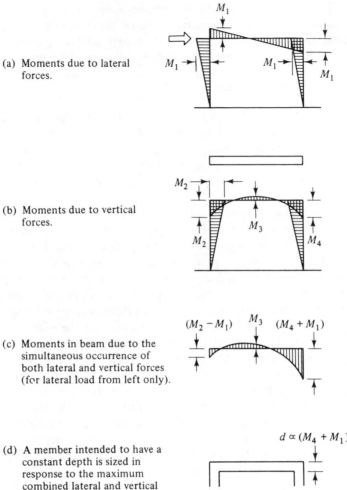

(a) Moments due to lateral forces.

(b) Moments due to vertical forces.

(c) Moments in beam due to the simultaneous occurrence of both lateral and vertical forces (for lateral load from left only).

(d) A member intended to have a constant depth is sized in response to the maximum combined lateral and vertical load moment present.

FIGURE 9-13 Critical design moments in a single bay rigid frame.

forces that are present. These are the same options faced in the design of continuous beams (see Chapter 8). The first option of using constant-size members is an inefficient use of material in contrast to the second, but often preferred because of construction considerations.

Member sizing for constant-depth members is discussed in Chapter 6. The implication of the second option of shaping members is discussed in the next section.

9.4.3 Shaping of Frames

SINGLE BAY STRUCTURES. As discussed before in connection with the design of beams, members can be shaped in response to the way internal moments and forces are distributed within the member. While doing so is not always practical from a construction viewpoint, it is interesting to look at problems involved in the shaping of members as a way of better understanding the general issues present in frame design. This discussion is in many ways simply a variant of the one on design moments presented in the previous section.

Consider the single bay frame shown in Figure 9-14. This is the same frame previously analyzed in Section 9.3.1. If the member depth is designed to be dependent only on the magnitude of the moment present at a point (and thus temporarily ignoring other internal forces that are present), and no deviations were allowed to this relation, configurations of the type shown in Figure 9-14 would result for each type of loading condition considered. Since the type of moments generated by lateral loads are drastically different in distribution than those for vertical loads, resultant structural responses are also very different.

It is important to follow through the implications of having a frame designed in response to one type of loading subjected to the other type (since both types would

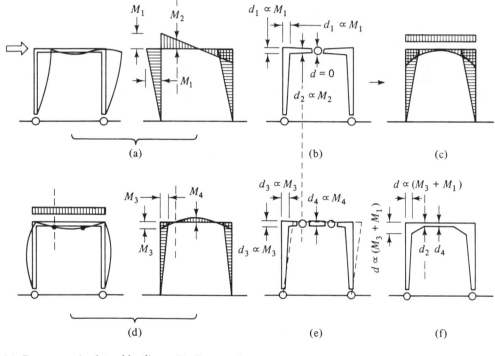

(a)　　　　　　　　(b)　　　　　　　(c)

(d)　　　　　　　　(e)　　　　　　　(f)

(a) Frame carrying lateral loading. (b) Structural response to moments induced to lateral loads.

(c) Moments in the frame shown in (b) due to vertical load. These moments are higher than those in the frame shown in (d).

(d) Frame carrying vertical loading. (e) Structural response to moments induced by vertical loading. Frame shown is an unstable four-hinged structure and cannot be used.

(f) Structural response to both lateral and vertical load requirements. The structure is shaped to meet critical requirements from both loading conditions but is optimum for neither.

FIGURE 9-14 Shaping frame members in response to internal moments.

occur in a real building structure). The structure designed in response to lateral loads can carry vertical loads as well. It is interesting to note that this structure is basically of the same type as the three-hinged arches discussed in Chapter 5. When vertical loads are applied to this structure, moments of the type illustrated in Figure 9-14(c) result. The final structure could be designed for the combined effects of moments associated with both lateral and vertical loads. The only question present is one of whether or not the general strategy outlined above is a reasonable one leading to an efficient frame. By looking at the magnitudes of the moments due to the vertical loads in the three-hinged arch type of structure in comparison with those generated in the original two-hinged frame structure [see Figure 9-14(d)], the answer can be seen to be no. The insertion of a pin in the beam (dictated by the response to lateral loads) causes an unfavorable moment distribution to develop in the beam in that the moments generated are considerably higher than those present in the two-hinged frames. The consequence is that member sizes are probably larger than need be.

If the strategy of using the response to lateral loads as the initial generator of the final structure does not lead to efficient structure, the alternative strategy of using a converse approach should be studied. As illustrated, however, the approach of using the response to vertical loads as the initial generator can be dismissed immediately because the structure (a four-hinged assembly) is clearly an unstable one.

An option that makes some sense is to shape critical parts of the structure in response to the positive and negative moment values present from either loading condition. A general configuration of the type shown in Figure 9-14(f) results. The configuration is optimum for neither the lateral-load condition nor for the vertical-load condition but works for the simultaneous presence of both loading types and fairly well for each individual type.

The frame shown in Figure 9-14(f) is characteristic of the compromise nature of the design of many frames, whether such a drastic shaping approach is taken or not. Often many frames are simply given additional local strengthening at the joints, which reflects the fact that moments are usually higher there than elsewhere. The general point to be remembered is that it is doubtful if there is any such thing as an optimum frame design when multiple loading conditions need to be considered.

MULTISTORY FRAMES. When multibay structures are considered, the same issues discussed in connection with single bay structures arise (Figure 9-15). Again, resultant structures are characterized by a compromise approach that works for the presence of both loading conditions but not optimally for either individual condition.

9.4.4 End Conditions

A primary design consideration is the selection of member end conditions. The primary issue faced is whether to use rigid or fixed joints exclusively or to introduce some hinged connections.

Figures 9-16 (a)–(d) illustrate four single bay structures identical in all respects except the type of member connections used. Moment diagrams for lateral loads which were determined by methods discussed in the section on analysis are also shown. Clearly, equivalent lateral loads produce different moments in the structure. Maxi-

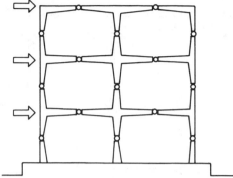

Moments in a
typical bay

(a) Structural response to lateral loads. This
diagram is not meant to show what is
necessarily a practical construction
approach but rather an extreme design
approach based on the criterion that
structural depths should be proportional
to moments present.

Possible practical building
elements partly based on
the design approach shown
above

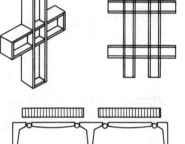

Moments in a
typical bay

(b) Structural response to vertical loads. This
is clearly an impractical and even dangerous
solution, since the shaping of members in
response to moments results in an unstable
structure.

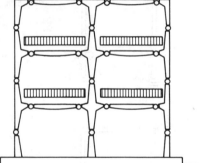

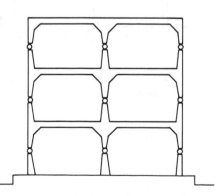

(c) Structural response to both lateral and
vertical load requirements. The structure
is shaped to meet critical requirements
from both loading conditions but is not
necessarily optimum for either.

FIGURE 9-15 Shaping frame members in response to internal moments in
a multistory structure.

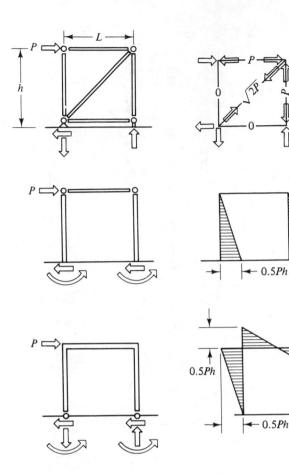

(a) Truss system. No moments at all are present in this hinged assembly. Member sizes are thus relatively small.

(b) Frame with hinged beam. Moments are generated in the columns only.

$0.5Ph$

(c) Tabletop frame. Moments and axial forces are generated in both beams and columns.

$0.5Ph$

$0.5Ph$

(d) Fully fixed frame. Moments and axial forces are generated in both beams and columns. These moments, however, are numerically smaller those in the frames in (b) and (c).

$0.225Ph$

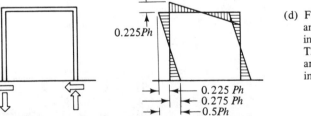

$0.225\ Ph$
$0.275\ Ph$
$0.5Ph$

FIGURE 9-16 Effects of end conditions on moments in single bay rigid frames.

mum moments differ as do distributions. Figure 9-16 also shows structural responses based on the moment diagrams obtained.

A comparison of the moment diagrams indicates that higher moments are developed in some frames than others. Comparing the frame in Figure 9-16(b) with that in Figure 9-16(c) (the *table top* frame) reveals that while maximum moments are the same in the columns in both frames, the former has no moments developed in the beam at all, while the latter does. Differences also exist in member axial forces. From a design viewpoint these differences mean that the tabletop frame generally requires more total material to support the lateral load in space than does the first structure and is thus less preferable with regard to this criterion. Because of the need to assume full base fixity at the column, however, the first structure requires more substantial foundations than does the tabletop frame, with its pinned base connections. The maximum moments developed in the frame shown in Figure 9-16(d), which has rigid joints and fixed base connections, are less than those developed in either of the initial two structures. It is interesting to note that the absolute sum of the positive and negative column moments in this frame numerically equals the magnitude of the column moments present in the first two structures (members are, however, designed for either positive or negative moment values and not their sum). The same total moment is present in all cases, but its distribution is changed by the type of member end conditions present.

When all factors, including vertical loads, are considered, the fully rigid frame shown in Figure 9-16(d) usually displays the most advantages in terms of structural efficiency. In cases where the design of the foundation is a problem, however, an approach using pinned-base connections such as illustrated in Figure 9-16(c) may be the best overall solution. The moments induced by differential settlements in a frame having pinned-base connections are less than those induced in a fully rigid frame. Also, the foundation for a pinned-base frame need not be designed to provide moment resistance. Horizontal thrusts associated with vertical loads are also usually smaller in a pinned-base frame than in a fixed-base one. Still the advantages of full rigidity in terms of minimizing moments and reducing deflections often overshadow those advantages of a pinned-base frame. Each specific design must be evaluated in its own context to see which approach proves most desirable.

9.4.5 General Considerations

As mentioned earlier, special attention should be paid to whether or not a frame is actually the best structural solution possible for use in a given context. As Figure 9-16 indicates, an assembly using pinned connections and full diagonal bracing (a simple form of truss) is really quite preferable to any of the frame systems shown in terms of structural efficiency (as defined by the relative amount of material needed to support a given loading in space). Using a frame structure makes doubtful sense if other approaches can be used.

In many cases where a frame is still considered an appropriate or necessary structure because of its openness, it is possible to use a *braced-frame* approach wherein a basic frame is reinforced with diagonals in bays where the diagonals can be safely placed without interfering in building functions (see Chapter 14). Typical locations are around elevator shafts and other continuous vertical bays that are normally enclosed anyhow. Use of diagonal bracing greatly reduces the moments in the frame caused by lateral loads. Consequently, members can be reduced in size.

QUESTIONS

9-1. Discuss in detail the historical development of quantitative methods of analyzing framed structures. Consult your library.

9-2. Find a frame building under construction in your local area. Draw a sketch of a typical beam-and-column connection.

9-3. A single bay frame of the type generally illustrated in Figure 9-4 carries a horizontal load of 3000 lb acting at the upper left joint. Assume that $h = 12$ ft and $L = 22$ ft. Draw shear and moment diagrams. Indicate numerical values.

9-4. A single bay frame of the type generally illustrated in Figure 9-5 carries a horizontal load of 5000 lb acting at the upper left joint. Assume that $h = 15$ ft and $L = 25$ ft. Draw shear and moment diagrams. Indicate numerical values.

9-5. Draw a sketch that diagrammatically illustrates possible member-size variations (along the lines of the sketches in Figure 9-14 or 9-15) for the frame analyzed in Question 9-3.

9-6. Consider the two single bay structures shown in Figure 9-16(c) and (d), respectively. Assuming that both structures are identical in all respects, except for member end conditions, and carry identical loads, which would you expect to deflect more horizontally? Why?

9-7. Repeat Question 9-6 for the frames shown in Figure 9-16(b) and (c), respectively.

chapter 10

PLATE AND GRID STRUCTURES

10.1 INTRODUCTION

Plates are rigid planar structures typically made of monolithic material whose depths are small with respect to their other dimensions. A multidirectional dispersal of applied loads characterizes the way loads are carried to supports in plate structures. The advent of modern reinforced concrete has made the plate among the most common of all building elements.

Plates can be supported along their entire boundaries or only at selected points (e.g. columns) or with some mixture of continuous and point supports. Support conditions can be simple or fixed. The variety of support conditions possible makes the plate a highly versatile structural element.

While strictly speaking, a plate is made of a relatively homogeneous solid material exhibiting similar properties in all directions, there are also several other types of structures whose general structural behavior is analogous to that of a plate (Figure 10-1). The *space frame* (actually a space truss), which is composed of short rigid elements triangulated in three dimensions and assembled to form a large rigid planar surface structure of relatively thin thickness, is one such structure. The *grid structure* is another. Plane grid structures are typically made up of a series of intersecting long, rigid linear elements such as beams or trusses, with parallel upper and lower chords. Joints at points of intersection are rigid. The basic shear and moment distributions in

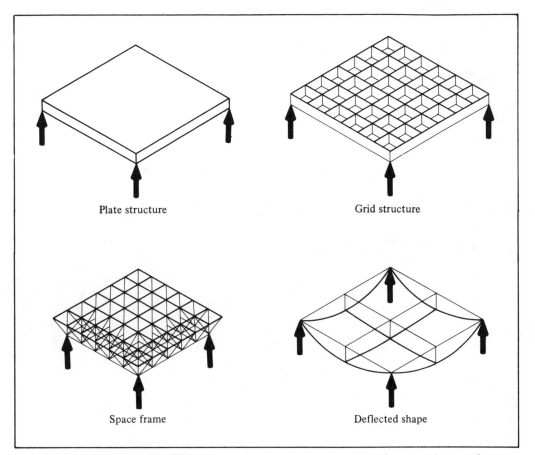

Plate structure

Grid structure

Space frame

Deflected shape

FIGURE 10-1 Plate, grid, and space-frame structures. The general type of curvatures and external moments induced in plate, grid, and space-frame structures are basically similar if loads and general dimensions are similar. The exact way in which each structure provides balancing internal resisting moments and the specifics of behavior, however, is different.

structures of these types are not unlike those present in comparably dimensioned monolithic plates. There are also, however, some basic differences which will be explored later in this chapter.

In general, coarse-grained grids tend to be better in supporting a series of concentrated loads while plates and space frames (with many small members) tend to be preferable for carrying uniformly distributed loads.

10.2 GENERAL PRINCIPLES

10.2.1 Grids

It is easiest to understand the structural behavior of plates by first looking at some simple grids. Consider the simple crossed-beam system supported on four sides shown in Figure 10-2(a). It is evident that as long as both beams are exactly identical, the load will be equally dispersed along both beams (i.e., each beam will pick up one-half of

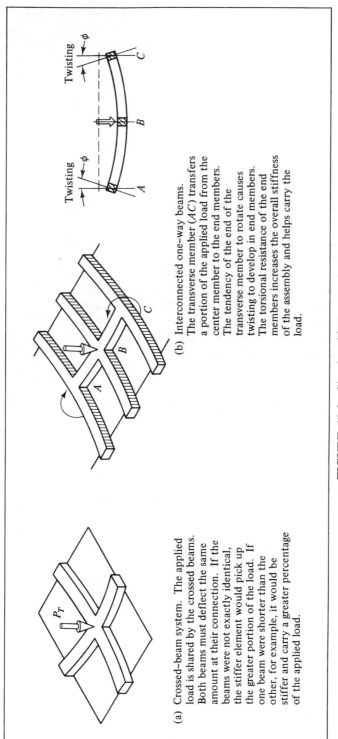

(a) Crossed-beam system. The applied load is shared by the crossed beams. Both beams must deflect the same amount at their connection. If the beams were not exactly identical, the stiffer element would pick up the greater portion of the load. If one beam were shorter than the other, for example, it would be stiffer and carry a greater percentage of the applied load.

(b) Interconnected one-way beams. The transverse member (AC) transfers a portion of the applied load from the center member to the end members. The tendency of the end of the transverse member to rotate causes twisting to develop in end members. The torsional resistance of the end members increases the overall stiffness of the assembly and helps carry the load.

FIGURE 10-2 Simple grid structures.

the total load and transfer it to its supports). If the beams are not identical, a greater portion of the load will be carried by the stiffer member. If, for example, the beams were of unequal length, the shorter member would carry a greater percentage of the load than the longer member because it is stiffer. Both members would have to deflect equally at their intersection because they are connected. For both beams to deflect equally, it is necessary that a greater force be applied to the shorter beam than to the longer one to achieve the same deflection. Therefore, the shorter element picks up more of the applied load. This point will be returned to in the following section on analysis. The primary point herein is that the relative amount of load carried in mutually perpendicular directions in a grid system is dependent on the physical properties and dimensions of the grid elements.

Consider another simple grid, in this case a one-way spanning system shown in Figure 10-2(b). As the center beam deflects downward under the action of the load, the transverse member transfers some of the load to the adjacent longitudinal elements. Just by looking at the geometry of the probable deflected shapes, it is possible to see that the center longitudinal member carries a greater portion of the load than do the outside members (the center member is bent more, which means that the internal moments are higher, which in turn means that the percentage of load supported is greater). All the grid elements, however, share in carrying the load. In a simple beam system, only the member beneath the load would carry the load.

Another interesting aspect of the one-way grid shown is the twisting induced in the exterior members by the transverse member. As the transverse member deflects, its ends tend to rotate. This tendency to rotate causes torsion to develop in the exterior members. At the same time these members provide a torsional resistance to the end rotations of the transverse member. The transverse member is, in effect, stiffened by the torsional restraint offered by the exterior members and has end conditions somewhere between a fixed end and a simply supported end. The center longitudinal member is also stiffened by the restraint offered by the transverse member, consequently reducing its deflections. The net effect is that a fractional part of the load is eventually transferred to the supports by a twisting action. All the grid elements participate more in carrying the load because of this twisting action. The stiffness of the whole grid is also thereby increased.

In a more complex grid, such as is illustrated in Figure 10-3, both two-way action and twisting occur. All elements participate in carrying the loads to the supports through a combination of bending and twisting. Note that if the beams were simply crossed and nonrigidly attached at intersection points, the bending rotation of one member would not cause twisting in the other. The consequent loss in overall rigidity due to the loss in torsional resistance associated with the twisting action would cause greater deflections to occur in a nonrigidly connected system than in a rigidly interconnected grid.

10.2.2 Plates

A plate structure behaves in much the same way as the grid structure described above except that the various actions described take place continuously through the slab rather than only at points of interconnection.

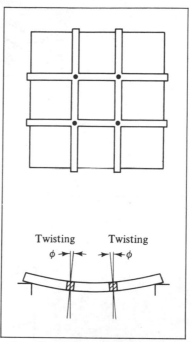

FIGURE 10-3 Behavior of complex grid structures. Twisting is developed in all members as a consequence of the way the structure deforms. The associated torsional resistance of members increases the stiffness of the grid.

Consider the simple one-way plate shown in Figure 10-4. The plate can be *imagined* as a series of adjacent beam strips of unit width interconnected along their lengths. As an applied load is picked up by one beam strip, it tends to deflect downward. The interconnected adjacent strips, however, offer resistance to this tendency, thereby picking up part of the applied load. A series of shear forces are developed at the interface between adjacent strips. Twisting related to these shears is also caused in the adjacent strips. As beam strips farther and farther from the strip under the load are considered, twisting and shear forces are reduced since more and more of the load becomes transferred to the supports by the longitudinal action of the strips.

That the relative percentage of the load which is carried by beam strips reduces away from the load is also evident by looking at the deflected shape of the structure and noting that longitudinal curvatures tend to decrease toward the edges of the plate. Internal moments must consequently also decrease. Figure 10-4 diagrammatically illustrates the way reactions and internal moments are distributed in the plate. It is common to express moments in plates in terms of a moment per unit length m (e.g. ft-lb/ft or kN-m/m). The reactions shown in the diagram are expressed in terms of a force per unit length.

The sum of the reactions must total the applied load acting in the vertical direction and the sum of the internal resisting moments distributed across the plate at a section must equal the total external applied moment. These observations follow from basic equilibrium considerations.

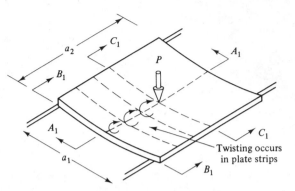

(a) Deflected shape of plate. Plate curvatures are greatest under the load.

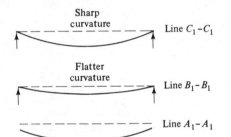

Plate curvatures

Sharp curvature

Line C_1-C_1

Flatter curvature

Line B_1-B_1

Line A_1-A_1

(b) From basic equilibrium considerations it is evident that the structure must provide a total internal resisting moment at a section which balances the external moment present at the same section. If the internal moments were assumed to be uniformly distributed, then the average moment per unit width of plate would be given by: $m = Pa_1/4 \div a_2$.

Uniform reaction distribution (assumed)

$m = Pa_1/4a_2$

Uniform internal moment distribution (assumed)

$M_{total} = Pa_1/4$

(c) An inspection of the deflected shape of the plate indicates that the internal moments cannot be correctly assumed to be uniformly distributed but are greater where curvatures are sharp and smaller where the plate is flatter. Some internal moments are high than the average value illustrated in (b) and some are lower. Their total sum of all moments per unit width, however, still must equal the applied moment.

Actual reaction distribution

$M_{total} = Pa_1/4$

Actual internal moment distribution

FIGURE 10-4 One-way plate structure.

Assuming that the plate shown had a span of a_1, and a width of a_2, the internal moment per unit length in the plate at a transverse section at midspan would be the external moment divided by the plate width $[m = (Pa_1/4)/a_2]$ if it was assumed that the resisting moments were uniformly distributed across this section. As previously argued, however, these moments *cannot* be uniform but vary from a maximum at the middle of the plate to a minimum at each edge. Still, thinking in terms of average moments is a useful way of establishing an average reference line about which actual maximum moment values deviate to a greater or lesser degree. Note that using an average value for design purposes, however, is *not* consevative.

The behavior of a plate resting on other types of supports is not unlike that described in connection with the plate resting on parallel supports. In general, the efficiency of plate structures can be increased by increasing the amount of edge support provided. Thus, a square plate resting on four continuous edge supports is stiffer than one resting on parallel edge supports with two free edges. The behavior of a continuously supported plate is similar to that of the simply supported plate, but all the internal actions previously described occur in two directions (perpendicular to each other) rather than one. Plates of this type will be discussed in more detail in the following section.

A critical aspect of plate structures not yet mentioned concerns their behavior with respect to shear forces. When a large plate rests on discrete point supports (e.g., columns), a local punch through type of failure due to excessively high shear stresses is possible. A way to prevent this is to increase the plate thickness. This can be done over the entire surface or locally around the column tops.

10.3 ANALYSIS OF GRIDS AND PLATES

10.3.1 Grid Structures

The key to analyzing grid structures is to recognize that at each point of connection in a crossed-beam system that a state of deflection compatibility must exist. Consider, for example, the simple crossed-beam system discussed in the last section and illustrated in Figure 10-5. Assuming that the beams are rigidly connected means that both undergo an identical deflection due to the load. By equating deflection expressions appropriate for each beam, it is possible to determine the relative percentage of the load carried by each element. Let P_A be defined as the percentage of the total load (P_T) carried by member A and P_B that carried by member B. By equating deflection expressions for each member because of the compatibility requirement, we obtain

$$\Delta_A = \Delta_B$$

$$\frac{P_A L_A^3}{48 E_A I_A} = \frac{P_B L_B^3}{48 E_B I_B}$$

$$\frac{P_A}{P_B} = \left(\frac{L_B}{L_A}\right)^3 \left(\frac{E_A I_A}{E_B I_B}\right)$$

If the members are identical in all respects except their lengths, the above expression

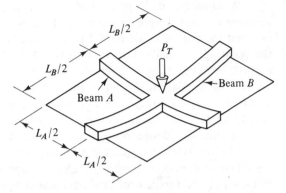

(a) Basic structure.

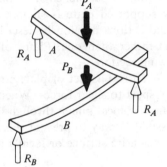

(b) Each beam carries a portion
    of the total load. $P_T = P_A + P_B$.

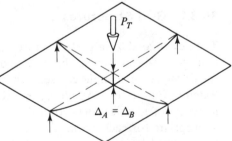

(c) P_A and P_B can be found by
    equating deflection expressions
    since $\Delta_A = \Delta_B$.

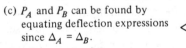

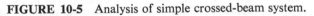

FIGURE 10-5 Analysis of simple crossed-beam system.

becomes

$$\frac{P_A}{P_B} = \left(\frac{L_B}{L_A}\right)^3$$

For members of equal length it is evident that $P_A = P_B = P_T/2$. If $L_B = 2L_A$, then

$$\frac{P_A}{P_B} = \left(\frac{2L_A}{L_A}\right)^3 = 8$$

$$P_A = 8P_B$$

Thus, the shorter (more rigid) beam picks up eight times the load that the longer beams does. Noting that $P_A + P_B = P_T$ it is evident that $P_A = 8P_T/9$, $P_B = P_T/9$. Associated beam moments are $M_A = 4P_T L_A/18$ and $M_B = P_T L_A/18$. Thus, moments in the short span are four times as large as those in the long member.

The analysis of more complex grids with multiple crossed beams proceeds in a fashion similar to that described above. Deflections at the intersections of the various beams can be equated. An analytical difficulty arises, however, because of the fact that there are multiple points at which deflections must be compatible and that the interactive reaction forces between members at one point contribute to the deflections at another. Invariably, for a complex grid a number of equations are generated that must be solved simultaneously—a task for computers. The reader is referred to more advanced texts for rigorous solutions to complex grids.

A study of the results of such analyses reveals what should be expected. If the cross members are of different lengths, the shorter, more rigid members pick up the predominate share of the applied load. The more rectangular the grid becomes, the less load is carried by the longer members. For very long, narrow bay dimensions, the longitudinal ribs can simply become nothing more than dead weight and are of limited value as structural elements, except as stiffeners.

10.3.2 Plate Structures

SIMPLE PLATES ON COLUMNS. While plates can be analyzed as continuous grids, it is perhaps more useful to look at them in terms of how different types of plates provide an internal set of shears and moments which balance the external applied shears and moments.

Consider a plate that is simply supported on four columns [see Figure 10-6(a)]. The probable deflected shape of the structure is shown in Figure 10-6(b). As is evident, the curvatures in the plate are highest in the plate strips nearest the free edges of the plate and less toward the middle. This implies that the internal moments in the plate generated to balance the external moments due to the applied forces are greater at the plate edges (with respect to the bending behavior of the plate between the columns) than are the moments in the middle. Since the plate is deformed into a doubly curved shape by the load, it is evident that moments are developed in multidirections rather than in one direction as is normally the case in a single curved element such as a beam.

To find the magnitudes of the internal moments developed, the equilibrium of a section of the plate can be considered. Any segment of the plate can be considered, since it is fundamental that any portion of any structure be in equilibrium. If the moments need to be known along a line D–E–F, a section is passed through that line and the equilibrium of the left or right plate segment considered. In doing this it is convenient to view the plate from the side as illustrated in Figure 10-6(c), in which case the plate appears as a simply supported beam carrying with a uniformly distributed load per unit length. If the width of the plate is denoted by a_1, the span a_2, and the *load per unit area* acting on the plate surface by w', then the uniformly distributed *load per unit length* that appears on the *analogous* simple beam is given by $w'a_1$. The reactions of the analgous beam on either end are simply the reactive forces in the columns (which by inspection must be $w'a_1a_2/4$). The external bending moment at

(a) Deflected shape of plate.

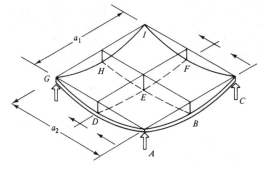

Unit width – Internal
moments are expressed
in terms of moments
per unit width of plate

m m

Bending in plate
element at E

Bending in plate
element at D

(b) Plate curvatures.

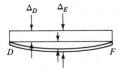

(c) Beam analogy (equivalent to looking
at plate from the side). The total
external moment along line $D-E-F$
is equivalent to the analogous simple
beam amount present.

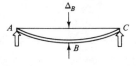

R_G, R_I R_A, R_C

$\leftarrow$ a $\rightarrow$

$M_T = (w'a_1)(a_2)^2/8$
If $a_1 = a_2$, $M_T = w'a^3/8$

(d) The plate provides a total internal
resisting moment that exactly balances
the external moment present. The
internal moment is actually distributed
along line $D-E-F$. Maximum internal
moments occur where plate curvatures
are the greatest.

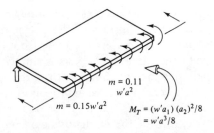

$m = 0.11$
$w'a^2$

$m = 0.15w'a^2$

$M_T = (w'a_1)(a_2)^2/8$
$= w'a^3/8$

Moment variation along
ling $D-E-F$

$m_{average} = 0.125w'a^2$

FIGURE 10-6 Square plate simply supported on four columns (uniformly
distributed load, w').

midspan for this analogous structure is given by the simple beam moment in the form of $wL^2/8$ where w is expressed as a force per unit length. In the case of the plate, $w'a_1$ is analogous to w and a_2 to L. The *total* moment present is consequently $M_T = w'a_1a_2^2/8$ or $M_T = 0.125w'a_1a_2^2$, where w' is expressed as a *force per unit area*. This is the same result that would have been obtained by summing moments about the line *DEF*, a procedure described in Section 2.2.5. If the plate is square, $a_1 = a_2$ and $M_T = 0.125w'a^3$.

The simplified analysis above is a useful one since it is evident that the external applied moment along line *DEF* is, in total, equal to $0.125w'a^3$. The plate must, therefore, provide a total internal resisting moment of $0.125w'a^3$. In a beam, this moment would have been provided by a single discrete couple formed by the compressive and tension stress fields internally developed in the beam in response to the external moment. In a plate the total resisting moment is provided by a continuous line of resisting moments in couples across the width of the plate whose sum effect is to yield a total resisting moment of $0.125w'a^3$. The equivalent discrete moment in the analogous beam has been, so to speak, simply spread out across the width of the plate.

If the total resisting moment that the plate provides is known, the next step is to find how this moment is distributed across the section. *If* the resisting moment was assumed to be uniformly distributed across the section, the resisting *moment per unit of width* (m) would simply be the total moment divided by the width of the section (i.e., $m = 0.125w'a^3/a = 0.125w'a^2$, where m is the internal moment per unit width across the plate width at the section considered).

The argument made in the opening of this section, however, indicates that the total moment is *not* uniformly distributed but is greater at plate edges than in the middle. Quantitatively assessing the exact distribution, however, is fairly complex and beyond the scope of this book. It is obvious that maximum and minimum values vary about the average moment value of $m = 0.125w'a^2$ discussed above. Figure 10-6(d) shows the results of a more exact analysis. There it is seen that the maximum moment per unit width occurs at the plate edge ($m = 0.15w'a^2$) and the minimum at the center of the plate ($m = 0.11w'a^2$). Similar results would be obtained for an analysis about line *B–E–H* since the structure is symmetrical.

Note that the maximum moment developed in the plate occurs at the free edge about an axis perpendicular to that edge. The moment about an axis parallel to the free edge at the same point is zero. At the plate midpoint, however, a moment of $m = 0.11w'a^2$ is developed about both axes.

The analysis discussed above highlights some very interesting points. One is that maximum moments occur not at the midpoint of the plate, where one might normally expect them by virtue of a beam analogy, but at the midspan of the edges. The midpoint of this plate is, for example, a good place to put a hole in the plate if this must be done to accommodate some other building element.

Another interesting point is that this particular plate must provide the same *total* internal resisting moment as an analogous beam. When the plate is designed, it is therefore evident that the use of a plate will not necessarily result in the saving of material as compared to an analogous beam. The plate will, however, be shallower. The much heralded advantages of two-way plate action in terms of material savings are associated with other types of support conditions, not the one described and analyzed herein.

PLATES SIMPLY SUPPORTED ON CONTINUOUS EDGES. Consider a plate similar to the one previously analyzed but supported continuously on all four edges by simple edge supports (e.g., walls) rather than being supported on columns. An inspection of the probable deflected shape of the structure reveals a radically different type of behavior under load. Maximum curvatures occur at the midpoint of the plate and decrease toward its edges. Internal moments can be expected to vary accordingly.

To find the magnitudes of these internal moments, a method of analysis similar to that used for the column-supported plate can be adopted. In this case, however, the first step of finding the plate reactions is not so easy. Considering the way the plate deflects, it can be seen that the reactions are not uniformly distributed but are at a maximum at the center of each line support and then decrease toward the corners [see Figure 10-7(b)]. A curious aspect of the plate reactions stems from the tendency of the corners of the plate to curl upward under the action of the vertical load. If the corners are to be restrained from curling upward, downward reactive forces must exist at the corners. The total sum of the upward non-uniformly-distributed reactions and the downward corner reactions must total the value of the externally applied load acting downward—a fact obvious from basic equilibrium considerations. A quantitative assessment of the way the reactions vary is beyond the scope of this text, but their general form is illustrated in Figure 10-7(b).

By looking at the plate from the side and considering a beam analogy, it can be seen that the total external moment present along a midspan section is less than the moment existant in the previously analyzed plate on columns. The total moment in the plate on the columns was equivalent to the previously encountered $wL^2/8$ moment value, or in terms of the plate dimensions and unit area loads, $w'a^3/8$. This value is associated with concentrated reactions at either end of the structure. By looking at the derivation of the expression, i.e., $M_{L/2} = (wL/2)(L/2) - (wL/2)(L/4) = wL^2/8$, where the first term represents the effects of the end reactions, it can be seen that the dominant moment-producing effect is caused by the moments associated with the reactions and not the external loads. In the plate resting on four continuous supports, the comparable moment caused by the reactions must be less than that associated with point supports because a greater percentage of the reaction forces are nearer the moment center. Consequently, the net external midspan moment must also be reduced, since the moment from the load remains constant. Since external moments are reduced, internal resisting moments are also less. They are considerably reduced from those present in the plate supported on four columns. A shallower plate can therefore be used.

All in all, continuously supporting a plate is far more preferable to using point supports. When continuous supports are used, the true benefit of two-way action can be achieved.

PLATES WITH CONTINUOUS FIXED-EDGE SUPPORTS. Figure 10-8 illustrates a plate similar to that analyzed above except that the edges are now completely fixed. The moments in the plate are even further reduced by fixing the plate ends. This is a very advantageous type of support condition for plates.

(a) Deflected shape of plate.

Corners tend to
curl upward if
not restrained

(b) Plate reactions. The downward
forces shown on the corners keep
the corners from curling up.

(c) Beam analogy (equivalent to
looking at the plate from the
side). The external moment
developed is less than the
comparable $wL^2/8$ for a simple
beam because of the different
reaction distribution. The total
internal resisting moment is
likewise smaller.

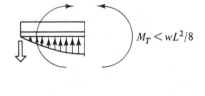

$M_T < wL^2/8$

(d) Plate moments.

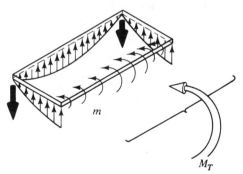

m

M_T

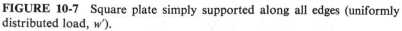

FIGURE 10-7 Square plate simply supported along all edges (uniformly
distributed load, w').

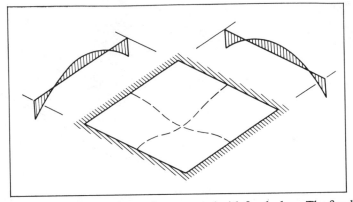

FIGURE 10-8 Plate continuously supported with fixed edges. The fixed-edge conditions cause plate moments to be relatively low.

PLATE ON BEAMS SUPPORTED BY COLUMNS. Figure 10-9(a) illustrates a variant of the first plate analyzed in which a plate is supported by continuous beams, which are in turn supported by columns. This is an interesting type of structure finding wide application in buildings. Assuming that the plate is merely simply supported on the beams gives the plate a set of end conditions somewhere between the situation where the plate is supported only by columns and the situation where the plate is supported by continuous edge supports. If the beams are extremely stiff, the support conditions of the plate approach those of continuous edge supports and similar moments are developed in the plate. If the beams are very flexible, however, little edge support is provided and the plate behaves more as if it were simply resting on four columns. Higher plate moments are accordingly developed than in the previous case. The relative stiffnesses of the edge beams therefore crucially affect the magnitude of the moments developed in the plate.

An interesting facet of this type of structure can be seen by passing a transverse section through midspan. By basic equilibrium considerations, it can be seen that the plate and beam structure must provide an internal resisting moment that exactly balances the external moment at the section. As was the case before when four point supports were used, the moment of interest is $0.125w'a^3$. If the beams were highly flexible ($I_b \rightarrow 0$), the sum of the internal moments that are developed in the plate is also equal to $0.125w'a^3$, as described before. The beams therefore would not provide any of the moment resistance. If, on the other hand, it were somehow possible to have the plate of minimum stiffness ($I_p \rightarrow 0$) and not capable of providing moment resistance, the beams would *have* to provide the entire resistance. Each of the two beams would therefore develop a moment of $0.125w'a^3/2$. The point herein is that the structure must provide a fixed requisite moment resistance, but the exact way in which this occurs is dependent on the nature of the structure used. Moments are never dispensed with, but merely redistributed.

The consequence of using stiffer beams is that plate moments are reduced and the plate can be made shallower. Alternatively, a stiffer plate can be used and the beams reduced in depth or eliminated (the plate, however, must be made thicker than in the previous case).

(a) Plate on walls. This
    is a favorable support
    condition for plates.

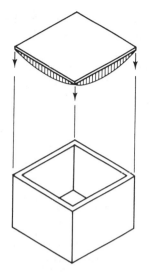

(b) Two–way beam and slab system.
    Plate on beams. If the beams are
    very stiff, the plate will behave
    as if it were supported on walls.
    If the beams are highly flexible,
    then the plate behaves as if it
    were supported on four columns.

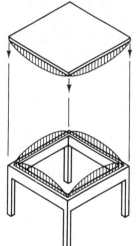

(c) The total internal resisting
    moment provided by the
    beam and slab system must
    equal tht total applied external
    moment.

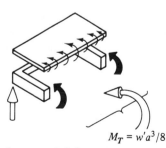

$$M_T = w'a^3/8$$

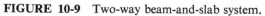

FIGURE 10-9 Two-way beam-and-slab system.

BAY PROPORTIONS: EFFECTS ON MOMENTS. In the section on analysis of grids it was noted that the stiffer portion of a grid structure, which is often the short-span direction, usually carries the greatest percentage of the applied load. The same phenomenon is true in rigid planar structures.

Consider the rectangular plate (continuously supported) shown in Figure 10-10. As its deflected shape indicates, only the plate strips in the short direction are experiencing significant curvatures. The longitudinal strips are virtually undeformed and simply translate downward as a unit. The implication is that only the strips in the short direction are providing an internal moment resistance and are participating significantly in carrying the applied load. The long span strips are almost simply riding along.

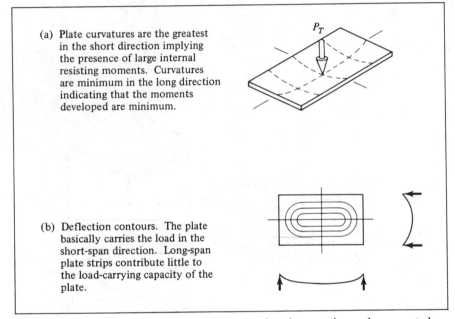

(a) Plate curvatures are the greatest in the short direction implying the presence of large internal resisting moments. Curvatures are minimum in the long direction indicating that the moments developed are minimum.

(b) Deflection contours. The plate basically carries the load in the short-span direction. Long-span plate strips contribute little to the load-carrying capacity of the plate.

FIGURE 10-10 Deformations in a rectangular plate continuously supported along edges. As plates become more rectangular, the two-way action decreases. The plate behaves more like a one-way system spanning in the short direction.

The more rectangular a plate becomes, the more it behaves as a one-way rather than a two-way system. It is as if the plate were supported only along its long parallel edges. Consequently, none of the advantages associated with the two-way action of continuously supported plates are present.

If the *length/width bay dimensions* of a plate exceed a ratio of about 1.5, the plate essentially behaves as a one-way rather than as a two-way system.

OTHER TYPES OF PLATES. There are, of course, a wide variety of other plate types not yet considered. Because of the importance of end conditions, each type of plate must be treated separately. The reader is referred to more advanced texts on the subject. Reliance should not always be placed on finding matnematical solutions to peculiarly supported plates, since often such solutions do not exist. A good way of getting at least a general feeling for the way any plate behaves is to sketch its deflected shape, something that with practice can be done for virtually any type of plate (see Figure 10-11). These sketches will at least identify regions of critical positive and negative moment and thus will be useful in many situations (such as the preliminary design of reinforced concrete plates, where it is necessary to know whether reinforcing steel goes on the top or bottom of the plate).

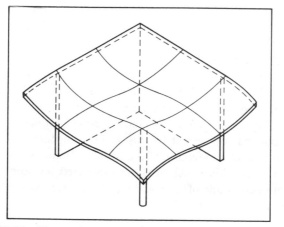

FIGURE 10-11 Unusual support conditions. The basic behavior of plates having unusual support conditions can be at least qualitatively understood by drawing and studying the probable deflected shape of the structure (i.e., noting regions of tension and compression).

COMPARISONS AMONG PLATES, GRIDS, AND SPACE FRAMES. As mentioned earlier, the general types of curvatures and moments induced in plates, grids, or planar space frames of comparable dimensions and carrying equivalent loads are similar. The exact way each structure provides balancing internal moments, however, differs. In a plate structure the internal moment is provided by a continuous line of couples formed by the compression and tension forces developed in the upper and lower surfaces of the plate. In a grid the resisting couples are concentrated in the beams. In terms of the moments developed in a comparable plate and normally expressed in terms of a moment per unit length, the moment in a grid beam is very approximately given by the beam spacing times the average moment per unit length at that location (all the equivalent plate moments in a region are, so to speak, concentrated in a single beam). This approximation, however, can lead to unconservative results in coarse-grained grids, since grids and plates differ in many significant ways (particularly the relative contribution of the torsional stiffness of elements in carrying the applied loads). The approximation described above improves as the grid mesh becomes finer and finer (it begins to approach a continuous plate surface). A similar situation exists for space frame structures except that the resisting couples are provided by a trusslike, rather than a beamlike, fashion.

Since these relations exist, the general moment distributions discussed earlier in connection with plates can also be used to describe approximately the behavior of grids and planar space frames.

EFFECTS OF SHEAR FORCES.   While the discussion has thus far dealt exclusively with bending in plates, shear is also present and is often the dominant factor influencing the design of the plate.

Shear forces are highest in plates on columns or other discrete supports. A punch-through type of failure can occur in plates around such points becuase of the high shear stresses present. The entire reactive force in a column, for example, must be distributed in the form of shear forces in the plate around the interface of the column and plate.

The plate area resisting the external shear can be found by considering potential shear failure lines. A reinforced concrete plate, for example, tends to fail in a manner illustrated in Figure 10-12. The diagonal crack pattern is caused by diagonal tension cracks associated with the shear stresses present (see Section 6.4.10). The area of plate providing the resistance to punch-through failure is therefore this same crack surface. Note that the extent of this surface depends critically on the thickness of the plate and the circumference of the column. Punch-through shear failures are most common in either thin plates or those supported on pointed or small columns. The approximate magnitude of the shear stresses present is given by $f_v = V/A_p$, where A_p is the plate area in shear.

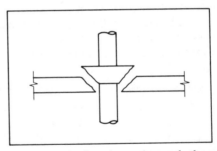

FIGURE 10-12   Shear forces in flat plates. A punch-through type of failure can result from excessively high shear forces. The failure shown is typical of reinforced concrete.

The use of thick plates and/or large-diameter columns (or columns with capitals) reduces the shear stresses present in a plate. Shear stresses are not as critical in plates continuously supported since the area in shear is relatively large.

In rigid planar structures, such as the space frame, that are composed of discrete members, the shear forces around column tops are manifested as very high compressive or tension forces in diagonal members (see Chapter 4). Consequently, these members have to be specially designed (Figure 10-13). Alternatively, increasing the number of elements picking up the shear force (by increasing the size of the column top) decreases the forces in the individual members.

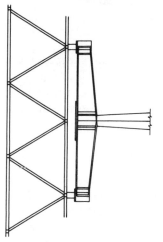

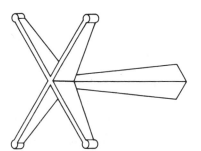

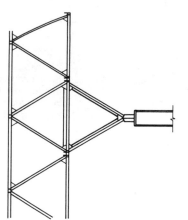

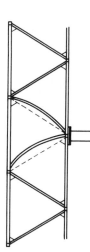

(a) The transfer of shear from a column into the structure can cause high local forces in members.

(b) Forces in members can be reduced by increasing the number of members into which the shear force is transfered. This is done by increasing the bearing area of the support.

FIGURE 10-13 Shear forces around columns in space frames.

10.4 DESIGN OF PLATES AND GRIDS

10.4.1 One-Way Systems

Sizing any type of one-way spanning structure is basically a process of first determining the shears and moments present and then using the techniques for establishing the size of members in bending that were discussed in Chapter 6. Techniques discussed there for increasing the load-carrying capacity of a member in bending by distributing material about the cross section in particular ways (primarily by increasing depths) are also appropriate for plate structures.

RIBBED PLATES. The common ribbed plate results from an application of the principles mentioned above (see Figure 10-14). If the connecting slab is relatively stiff, the whole assembly functions as a one-way plate rather than a series of parallel beams. For design purposes, however, it is still common to treat the structure as a series of parallel T beams in the longitudinal direction. The transverse slab is treated as a one-way plate continuous over the beams.

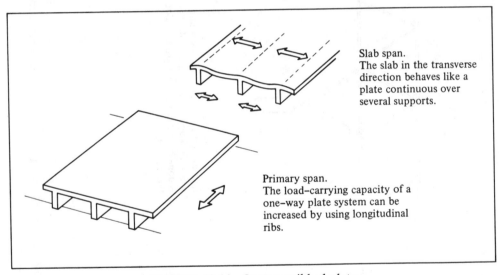

Slab span.
The slab in the transverse direction behaves like a plate continuous over several supports.

Primary span.
The load–carrying capacity of a one-way plate system can be increased by using longitudinal ribs.

FIGURE 10-14 One-way ribbed plates.

FOLDED PLATES. The stiffness of a one-way plate structure can be dramatically increased by not using a planar surface at all (ribbed or otherwise), but by radically deforming the whole plate surface in a way such that structural depths are greatly increased. The result is a form of plate with a unique cross section which can be envisioned as a series of thin, deep elements joined together along their boundaries (see Figure 10-15). Structures of this type, commonly called *folded plates*, are typically used as roofs and hence primarily carry uniformly distributed loads which impinge on the surfaces of the individual plates. A characteristic of folded plate structures is that individual plate elements are long with respect to their width. An underlying design principle is that of moving as much material as possible away from the middle plane of the structure. Thin, deep members reflect this principle, and have a high load-

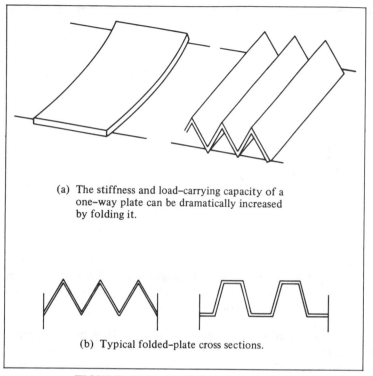

(a) The stiffness and load–carrying capacity of a one–way plate can be dramatically increased by folding it.

(b) Typical folded–plate cross sections.

FIGURE 10-15 Folded-plate structures.

carrying capacity, but only if lateral buckling of the compression zone can be prevented (see the diagram illustrating lateral buckling in Figure 6-29). By arranging individual plates as illustrated in Figure 10-15, the compression zone in each individual plate is inherently braced by the adjacent plate. Lateral buckling is thus not a problem, and the full potential load-carrying capacity of the plates can be realized. The angle formed between adjacent plates must be relatively sharp for this bracing action to occur.

The actual way typical surface loads are carried to supports in folded plate structures is best envisioned by considering two types of beam action—transverse and longitudinal (Figures 10-16 and 10-17). First consider the transverse action. In the same way that one plate braces another against possible buckling in the latter's out-of-plane direction, it can also be regarded as providing a continuous edge support for the adjacent plate. An individual plate in a continuing series of folded plates can consequently be thought of as a one-way slab whose span is the width of the plate. Surface loads are transferred to adjacent folds through *transverse beam action*. If a rigid connection between plate elements is used, a typical transverse strip behaves like a continuous beam supported at several points (the folds). Once the surface loads are carried to the folds, they are transferred through the *longitudinal beam action* of the plates acting in their own deep planes to the supports (i.e., the reactions of the transverse beam strips have components in the in-plane direction of each of the adjacent plates at a fold). For preliminary design purposes the structure can thus be treated as a

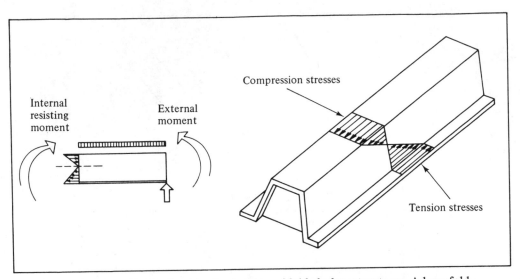

FIGURE 10-16 Longitudinal action of folded-plate structures. A long folded plate structure braced by transverse stiffeners behaves basically like a beam of unique cross section.

series of beams of unique cross section in the long direction and one-way slabs in the transverse direction.

It is crucial that the angle formed between plates be maintained constant. A transverse splaying of the type illustrated in Figure 10-17(a) can occur which will substantially reduce the load-carrying capacity of the structure. This phenomenon can also occur in end plates of a continuous series of plates. To prevent this transverse splaying, stiffener plates of the type illustrated in Figure 10-17(b) are often used at the ends of folded plate structures and, in very long span structures, at interim locations. These stiffeners not only prevent splaying but also provide additional lateral braces for individual plates. If no such stiffeners are used, the rigidity of the joints between plates must be relied on to prevent this splaying. In this case, relatively high bending moments are produced in the transverse direction at the joints which typically require that the thickness of the plate be increased beyond what would be required in a structure using stiffeners.

The general load-carrying actions described above assume that all the plates deflect similarly under load. This is generally true for interior plates. It is highly important that the plates be made to deflect equally, since not doing so would detrimentally affect the structure in the transverse direction. Unequal plate deflections would be equivalent in effect to unequal support settlements in a continuous beam (see Section 8.3.3) and produce undesirable bending moments in the transverse direction. End plates often deflect more than interior plates. Thus, they are often stiffened (see Figure 10-18) by the addition of boundary beams. The stiffening plates discussed earlier also serve to equalize plate deflections.

The efficiency of a folded-plate structure can be increased by placing as much material as possible away from the neutral axis in a way similar to that which was done in beam design. Consequently, plates of the type shown in Figure 10-16 tend to

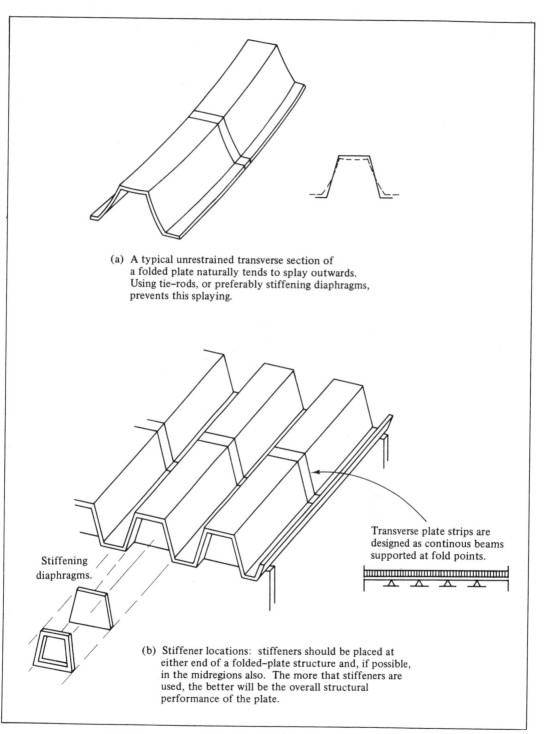

(a) A typical unrestrained transverse section of a folded plate naturally tends to splay outwards. Using tie–rods, or preferably stiffening diaphragms, prevents this splaying.

Transverse plate strips are designed as continous beams supported at fold points.

Stiffening diaphragms.

(b) Stiffener locations: stiffeners should be placed at either end of a folded–plate structure and, if possible, in the midregions also. The more that stiffeners are used, the better will be the overall structural performance of the plate.

FIGURE 10-17 Transverse action of folded-plate structures.

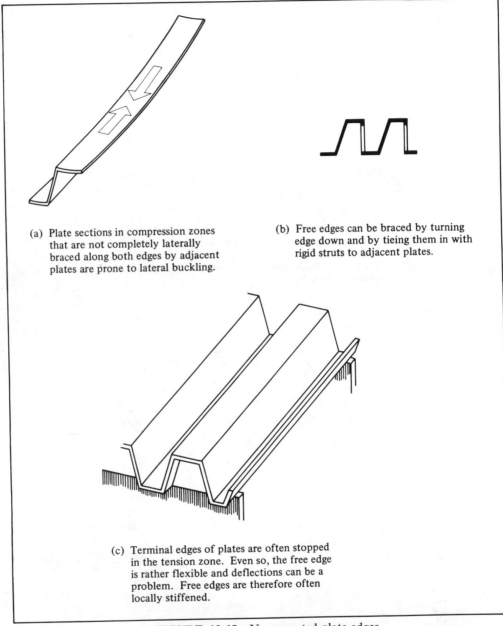

(a) Plate sections in compression zones that are not completely laterally braced along both edges by adjacent plates are prone to lateral buckling.

(b) Free edges can be braced by turning edge down and by tieing them in with rigid struts to adjacent plates.

(c) Terminal edges of plates are often stopped in the tension zone. Even so, the free edge is rather flexible and deflections can be a problem. Free edges are therefore often locally stiffened.

FIGURE 10-18 Unsupported plate edges.

be more efficient than the type shown in Figure 10-15. When reinforced concrete is used, the use of a flattened area at the base of a plate allows plenty of room for the placement of tension-reinforcing steel.

Care must be taken to assure that individual plates do self-brace each other to prevent lateral buckling. The open configuration shown in Figure 10-18, while in a sense a folded plate, is not very efficient, since the top plate is in compression and not laterally braced on one edge. The edge can, however, be locally stiffened.

Anytime a folded plate terminates, special attention should be paid to stiffening the free edge. Terminating a plate in the compression zone as described above is particularly undesirable. Many plates, therefore, are terminated so that the free edge is in the tension rather than compression zone (see Figure 10-18). Even these edges, however, should be stiffened to prevent undesirable deflections.

Folded plate structures can be made out of virtually any material. Reinforced concrete is a widely used material. By specially reinforcing the separating joint, adjoining plates can be rigidly connected along their edges. Reinforced concrete folded plates can also be post-tensioned to further increase their spanning capabilities (see Figure 10-19).

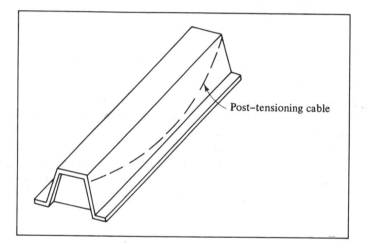

Post–tensioning cable

FIGURE 10-19   Post-tensioning folded plates. Post-tensioning can be effectively used in reinforced concrete plates. The depth of the plates allows a post-tensioning cable to be deeply draped for efficient action.

Steel is also used for folded plates, albeit less frequently than concrete. Rigid joints between adjacent plates can be obtained by welding or other special connections.

Wood, most usually in the form of plywood sheets, is extensively used. Rigid joints between adjacent plates are hard to achieve, however, and special care should be devoted to preventing lateral splaying of plates by using stiffening diaphragms or some other mechanism.

10.4.2 Two-Way Systems

The process of sizing members for the shears and moments present in a two-way structure is as for any member in bending. There are, however, several steps that can be taken which initially minimize the bending present. There are also other critical variables which affect the design of a plate structure and influence the appropriateness of a system selected. These factors will be considered in the following section.

SUPPORT CONDITIONS. A primary design objective is always to minimize the bending moments present in a structure. An effective way of doing this in rigid planar structures is to operate on the support conditions used since, as discussed in the previous section, the moments present in these structures are critically dependent on the type of support conditions present.

Generally speaking, a design that provides as much continuous support to a rigid planar structure as possible will be one in which moments in the structure are minimized. Structural depths can thereby be reduced. Thus, continuous edge supports are preferable to column supports. Fixed supports also result in smaller plate moments than simple supports (see Figure 10-8).

Techniques such as using overhangs described for reducing moments in beams are also applicable to planar structures in bending. Cantilevering a portion of the structure tends to reduce the maximum design moment present.

When the support system is basically repetitive in nature (e.g., a series of column bays), the use of a continuous rigid structure leads to lower design moments than does the use of a series of discrete, simply supported plates. The use of a continuous surface causes the portion of the plate over a single bay to behave more like a plate with fixed ends than does a simply supported one (see Figure 10-20).

BAY PROPORTIONS: EFFECT ON CHOICE OF STRUCTURE. As noted in the section on analyzing rigid planar structures, the less square or more rectangular a supporting bay becomes, the less the supported planar structure behaves like a two-way system and the more it behaves like a one-way system acting in the short-span direction. From a design viewpoint, the consequence of this phenomenon is that bays should be designed to be as dimensionally symmetrical as possible if two-way action is desired. Only if bay dimensions form a ratio of between 1:1 and 1:1.5 does two-way action obtain.

Using what appears to be a two-way structure, such as a space frame, over a long rectangular bay does not really accomplish very much. The structure will behave like a one-way system in the short direction. Members in the short-span direction would carry most of the load. Longitudinal members would simply ride along contributing little, except adding some stiffening. Therefore, a system deliberately designed and intended to be a one-way spanning system may as well be used. Less total material could be used to support the load in space. A system of planar trusses placed in the short-span direction in the case shown in Figure 10-10 could be a preferable solution to a space frame which is two-way in a visual sense only. The same holds true for ribbed reinforced concrete plates or grid systems.

TYPES OF LOADS: EFFECT ON CHOICE OF STRUCTURE. A fundamental reason for using a rigid planar structure is quite often a functional one. These structures can provide usable horizontal floor planes. As such, the loads they typically carry are usually uniformly distributed. When this is the case, either plates, grids with small meshes, or space frames can be used relatively efficiently. These same structures are, however, relatively inefficient when called upon to support large concentrated loads. Surface structures work best supporting surface loads. Large concentrated loads can, however, be fairly effectively carried by coarse-grained grid structures in which the crossed beams intersect beneath the concentrated loads.

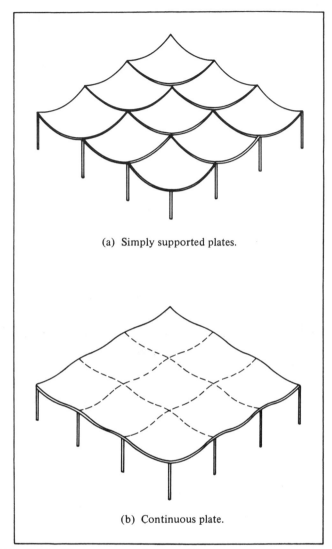

(a) Simply supported plates.

(b) Continuous plate.

FIGURE 10-20 Plates on a column grid. A continuous plate surface is generally preferable to a series of simply supported plates in that design moments are reduced and rigidity increased by the continuity.

An additional point is that many current, commercially available types of space frame structures, such as those that are typically made of cold formed light steel elements, are better suited to roof rather than to floor loads, since roof loads tend to be somewhat lighter. These structures do not work particularly well for floor loads. This is not a consequence of the basic structural approach but of the specific way they are constructed.

SPAN RANGE. There is a wide variation in spans possible. Rigid plates made of reinforced concrete can economically span anywhere from 15 to 60 ft (4.5 to 18 m) or so. Grids and space frames can effectively span higher distances, depending on exactly how they are made. Commercially available space frames, for example, can easily span up to 100 ft (30 m), depending on the member sizes and support conditions used. Much higher spans are even possible with specially designed structures.

For short spans of the type often encountered in buildings [e.g. 15 to 30 ft (4.5 to 9 m)], the construction complexities of grids and space frames often render them less attractive than simple reinforced concrete plates. Chapter 14 discusses the influence of span in more detail.

10.4.3 Design of Reinforced Concrete Plates

Since reinforced concrete is so extensively used to construct plates, it is well worth highlighting design possibilities that are unique to this material.

Reinforced concrete that is poured in place is a highly useful material for making plates for a number of reasons. Concrete, for example, is inherently a two-way material if properly reinforced. Reinforced concrete plates are relatively easy to construct. It is also very easy to make special features, such as increased plate thicknesses at critical points, using reinforced concrete. Continuity is very easy to achieve.

This section will only briefly cover the more important design considerations. The reader is referred to Chapters 14 and 15, where reinforced concrete plates are treated in more detail.

PLATE MOMENTS AND PLACEMENT OF REINFORCING STEEL. The thickness of a reinforced concrete plate and the amount and location of reinforcing steel used in a constant-depth plate or slab is critically dependent on the magnitude and distribution of moments in the plate. Figure 10-21(a) illustrates the deformations present in continuous plate resting on columns. Reinforcing steel must be placed in all tension regions. This requirement results in a placement of reinforcing steel of the type illustrated in Figure 10-21(b). Since moments are continuous, reinforcing steel must be closely spaced. It is thus common to use a series of parallel bars. Since moments are multidirectional, mutually perpendicular sets of bars are used. These bars can be deformed as shown so that the steel is properly located in tension regions or discontinuous shorter linear pieces that are lapped can be similarly placed.

While the design moments in the plate (on which plate thicknesses and steel sizes would depend) can be found from techniques discussed in the section on analysis, some characteristics peculiar to reinforced concrete cause a slightly different design approach to be taken. As noted previously, the total (positive plus negative) moments in a slab supported on columns in given by $M_T = w'a^3/8 = 0.125w'a^3$. It would therefore be thought that the total amount of steel needed would be in response to this value. If a more detailed study of the *plastic behavior* of the plate is made, however, it can be seen that designing for the $0.125w'a^3$ moment is more conservative than necessary (see Section 6.3.9 for a discussion of the idea of plastic behavior in members). Consequently, an empirically substantiated total design moment of $M = 0.09w'a^3$ is often used, which is about 25% less than the elastic moment value.[1]

Since construction considerations generally preclude placing steel sizes which continuously vary with varying moments, a simpler approach is taken to the moment

[1]See, for example, P. M. Ferguson, *Reinforced Concrete Fundamentals*, 3rd ed., John Wiley & Sons, Inc., New York, 1973.

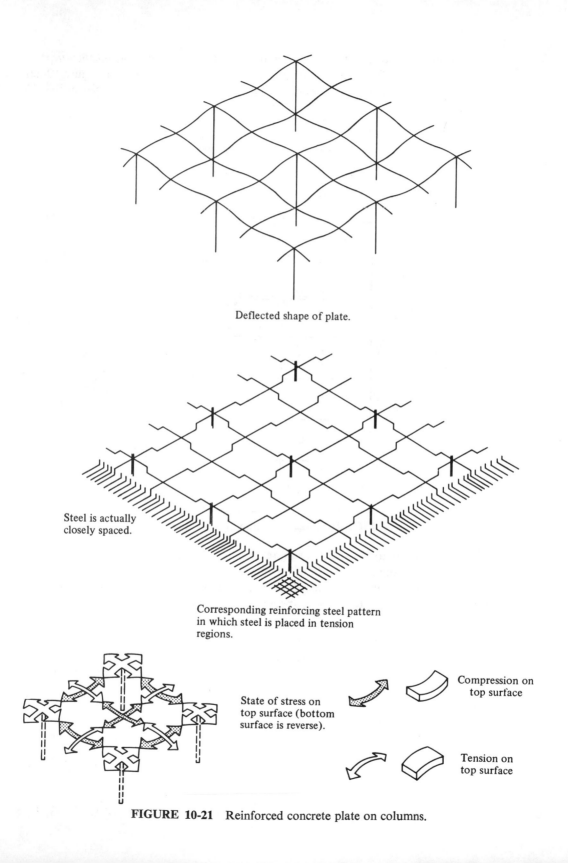

Deflected shape of plate.

Steel is actually
closely spaced.

Corresponding reinforcing steel pattern
in which steel is placed in tension
regions.

State of stress on
top surface (bottom
surface is reverse).

Compression on
top surface

Tension on
top surface

FIGURE 10-21   Reinforced concrete plate on columns.

distribution. A plate is usually considered divided into *column* and *middle strips*, in which moments per foot are assumed to be constant within a strip (see Figure 10-22). The total moment of $M = 0.09w'a^3$ is usually assumed to be apportioned in the following way:

$$\text{column strip} - \text{positive moment} = 22\%$$
$$\text{column strip} - \text{negative moment} = 46\%$$
$$\text{middle strip} - \text{positive moment} = 16\%$$
$$\text{middle strip} - \text{negative moment} = 16\%$$

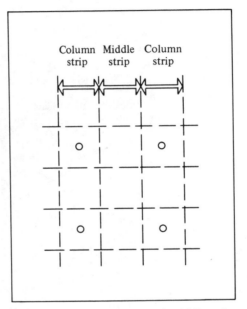

FIGURE 10-22 Column and middle strips.

These values are good for a square or nearly square interior panel of constant thickness. The proportion of the total moment assigned to a particular column or middle strip is then uniformly distributed across that strip. Design moments are consequently expressed in terms of a specific moment per unit width of slab for each strip.

EFFECTS OF SPAN. For low-span ranges [e.g., about 15 to 22 ft (5 to 7 m)], common building loads result in moments that can be handled by relatively thin plates [e.g., on the order of 5 to 10 in. (13 to 26 cm). As spans increase, moments become greater and plate thicknesses must increase. A consequence of using thicker plates, however, is that the dead weight of the structure also dramatically increases. For this reason, plates are often hollowed out to reduce the dead load while at the same time maintaining an appreciable structural depth. One such system, called a *waffle slab*, is shown in Figure 10-23.

Another approach to increased spans is to use what is commonly referred to as

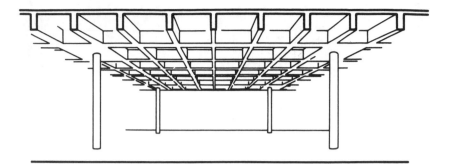

(a) General view. A plate of constant thickness has hollows formed into it to reduce weight. A two–way ribbed structure results.

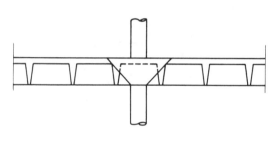

(b) Section. To increase the shear capacity of the structure the plate is not hollowed adjacent to column heads.

(c) The overall spanning capacity of the slab can be increased by not hollowing out lines between columns. Two-way beam-and-slab action is thus obtained.

FIGURE 10-23   Reinforced concrete waffle slab.

a *two-way beam-and-slab system* (see Figure 10-24). Since the beams are relatively stiff and form a continuous edge support for the plate, the plate can be made relatively thin because of the favorable support conditions, which decrease the moments present in the plate.

Both the waffle and two-way beam-and-slab systems can span relatively long distances. A system exhibiting desirable properties of both systems is obtained by not hollowing out the waffle slab between columns. A surrounding stiff-beam system is thereby created. Spans longer than that associated with the basic waffle slab are possible. The systems can be post-tensioned to even further increase their span capabilities. These systems are discussed more in detail in Chapters 14 and 15, to which the reader is referred.

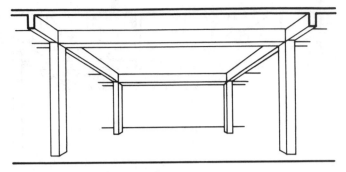

(a) General view.

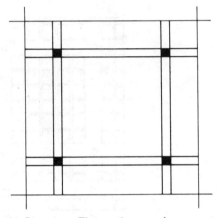

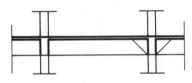

(b) Plan view. The continuous edge support provided to the slab by the beams gives the slab a favorable support condition whereby plate moments are reduced.

(c) Shear is not a major problem in this type of system because of the presence of the beams which have a large shear capacity.

FIGURE 10-24 Two-way beam and slab.

EFFECTS OF SHEAR ON PLATE DESIGN. The punch-through type of failure potentially possible because of high shear forces is often a crucial design issue. As mentioned earlier, the magnitude of the shear stresses present depend directly on the magnitude of the shear force and the plate area in shear. The latter depends on the thickness of the plate and the length of failure plane possible (which depends on the size of the support). If shear stresses are too high, the design options are to use special steel reinforcement in the overstressed regions (called *shear heads*) or to increase the plate area in shear.

Increasing the plate area in shear can be easily done by increasing the plate thickness. This may prove uneconomical to do for the whole plate if not needed for moment considerations. The plate can be locally thickened at critical points, however, through the use of drop panels [see Figure 10-25(a)]. Alternatively, the plate area in shear can be increased by increasing the support size. This can also be done locally

 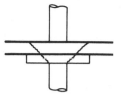

(a) Column with drop panel.   (b) Column with drop panel   (c) The plate area in shear can be
                                   and capital.   increased (thus decreasing shear
                                                                   stresses) by using a drop panel
                                                                   to locally increase the plate
                                                                   thickness.

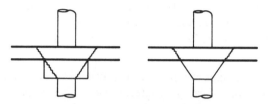 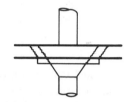

(d) Column capitals. The use of column capitals   (e) Drop panels and capitals. Both drop
    increases the length of the failure plane, thus   panels and capitals can be used
    increasing the area of the plate in shear.   simultaneously to increase the shear
    Material below the diagonal shear failure plane   capacity of the structure in cases
    is inactive and can be eliminated.   where shear forces are extremely high.

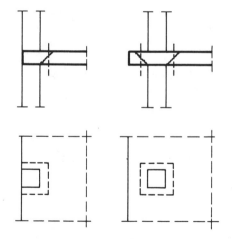

(f) Insetting columns. At free edges of plates
    the shear capacity of the system is reduced
    because of the reduced length of the plate
    thickness in shear. Insetting the column by
    the amount shown increases the shear capacity
    of the system. This type of edge condition is
    commonly used in plate structures.

FIGURE 10-25 Designing for shear in reinforced concrete plates.

through the use of column capitals. The larger the capital, the greater is the plate area in shear. Column capitals can be any shape, but due to the natural tendency to fail in diagonal shear, capitals are often given a sloped shape (simply because material below the 45° line is inactive).

Both drop panels and column capitals can be simultaneously used. The shear capacity of such a system is quite high and is used in special cases where loads are particularly high (e.g., warehouses).

Reinforced concrete slabs that do not use column capitals are usually referred to simply as plates. Those using capitals and drop panels are usually referred to as flat slabs.

Shear can also be a problem in waffle slabs but can be handled quite easily. The area adjacent to a column top is simply not hollowed out. A built-in column capital is thus created (see Figure 10-23). In a two-way beam-and-slab system, the beams framing into the columns pick up most of the shear. Since their areas are relatively large and reinforcement is easy, shear is usually not a great problem in this type of system (see Figure 10-24).

EFFECTS OF LATERAL LOADS. Plates or slabs cast monolithically with columns naturally tend to form frame structures capable of carrying lateral loads. Their capacity to carry lateral loads depends to a great extent on the thickness of the horizontal structure at the column interfaces (deep structures have larger possible moment arms for increased resistance). A thin plate-and-column system can carry lateral loads, but its capacity is limited. The use of column capitals or drop panels increases the system capacity. Waffles and two-way beam-and-slab systems can support substantial lateral loads. The two-way beam-and-slab system forms a natural frame system and is often

FIGURE 10-26 Ribbed plate structure reflecting isostatic lines.

used where lateral load-carrying capacity is important. The reader is referred to Chapters 14 and 15 for a more in-depth coverage of this topic.

SPECIAL PLATES. A detailed analysis of the elastic stress distributions present in a plate reveals the presence of lines of *principal stresses*. These lines, often called *isostatics*, are directions along which the torsional shear stresses are zero. Some designers have devised plates that are ribbed in a manner intended to reflect these isostatic lines (see Figure 10-26). These structures are, of course, expensive to construct but their designers claim that these expenses are not excessive and that material savings compensate for any added construction cost. These structures are undoubtedly interesting but are curious from the viewpoint of classical theory in that they represent the physical manifestation of a model of a stress distribution in an elastic material that is actually constructed using an inelastic material (concrete). That the ribs are indeed lines of principal stress is not argued, but a self-fulfilling prophecy may be present in placing stiffer ribs along these lines.

QUESTIONS

10-1. For the simple crossed-beam system shown in Figure 10-5, draw a graph indicating the relative percentage of the load carried by each crossed beam as the relative lengths of the members vary. Repeat for the bending moments present in each member.

10-2. What are the maximum positive moments present in a square plate that is simply supported at its corners by four columns and which carries a uniformly distributed load of 80 lb/ft². Assume that the plate is 6 in. thick and measures 15 ft by 15 ft. Assume that the unit weight of reinforced concrete is 150 lb/ft³.
Answer: 5231 ft-lb/ft.

10-3. Build a simple folded cardboard model of the folded plate illustrated in Figure 10-18. Do not include transverse stiffeners. Load the model with large flat weights at midspan. Note the behavior of the structure under load and measure the load that causes collapse. Cut out cardboard stiffeners that exactly fit into the end of the original plate configuration and glue them into place. Again load the structure and measure the load required to cause failure. Cut out additional stiffeners and place them in the interior of the structure (e.g., at third points) and repeat. Discuss.

chapter 11

MEMBRANE STRUCTURES

11.1 INTRODUCTION

A *membrane* is a thin flexible surface structure that carries loads primarily through the development of tension stresses. The soap bubble is the classic example used to illustrate both what a membrane is and how it behaves. Like their flexible-line counterparts, membrane surfaces tend to adapt their shapes to the way they are loaded. They are also very sensitive to the aerodynamic effects of wind, which can cause *fluttering* to occur (see the discussion of this phenomenon in Chapter 5). Consequently, most membranes that are used in buildings are stabilized in some way so that their basic shape is retained under a variety of loadings.

There are several basic ways of stabilizing a membrane. An inner rigid supporting framework, for example, could be used. Of special interest in this chapter, however, is stabilization through the prestressing of the membrane surface. This can be accomplished through either the application of an external force which pulls the membrane taut or through some sort of internal pressurization when the membrane is volume-enclosing.

An example of prestressing though the application of an external force are some types of tents—an age-old type of structure. Many tents, however, do not have surfaces that are actually appreciably tensioned by supporting guy cables and thus tend to move under loads. While capable of withstanding normal wind loads, for example, many tent surfaces flutter due to the aerodynamic effects of extreme winds and are not of direct value as permanent structures (although of obvious great value as temporary structures). It is possible, however, to prestress a membrane by external jacking forces sufficiently so that it remains completely taut under all anticipated loads (see Figure 11-1). Usually such membranes are stressed in perpendicular directions all across their surfaces. Many roof structures of this type have been built.

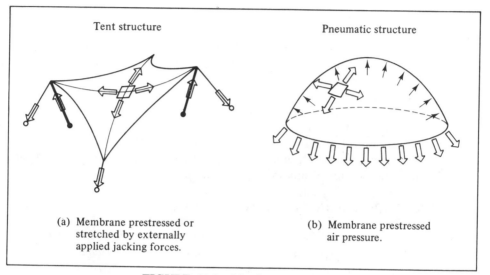

Tent structure   Pneumatic structure

(a) Membrane prestressed or stretched by externally applied jacking forces.

(b) Membrane prestressed air pressure.

FIGURE 11-1 Membrane structures.

Stabilizing a membrane through internal pressurization is possible when the membranes enclose a volume. A class of membranes typically called *pneumatic structures* obtain their stability in this way. While pneumatic structures as applied to architecture are mostly of fairly recent origin, the basic technology of pneumatics has long been known to man. The water skin, for example, is a type of pneumatic structure known to man for a long time. Bubbles have aroused man's curiousity for at least as long. The more immediate genesis of the pneumatic structures currently used in the building fields, however, undoubtedly lies in the balloons and airships that have graced our skies in the recent past.

Building applications of pneumatics are fairly recent. An English engineer by the name of William Lanchester attempted to apply the balloon principle in a patent for a field hospital in 1917. In 1922 the Oasis Theatre in Paris sported a pneumatic hollow roof structure that was rolled into place when it rained. A lot of research work on pneumatics occurred in World War II because of the military value of different types of pneumatic structures. Widespread use of air-supported structures began in 1946 with their application as *radomes*, which housed large radar antenna. The performance of these radomes in harsh arctic climates paved the way for commercial applications. Pneumatics are now part of our common building vocabulary.

11.2 GENERAL PRINCIPLES

11.2.1 Basic Actions

The basic load-carrying mechanism in a membrane structure is one of tension. A membrane carrying a load normal to its surface tends to deform into a three-dimensional curve (depending on exact loading and support conditions) and carries the load by in-plane tension forces that are developed in the surface of the membrane. The load-carrying action is similar to that present in a crossed-cable system. In addition to the tension stresses, a type of tangential shearing stress is developed within the membrane structure which is associated with the twist normally present in the curved surface that also helps carry the applied load. The two-way tension field action and the tangential shearing action result, if the membrane is stabilized, in a surface structure that can carry loads in a funicular manner as long as the loads applied do not actually cause compressive stresses, which are manifested by folds in the membrane, to develop.

In line with the above, it follows that the fundamental principle underlying the design of membrane structures is simply that the surface must be maintained in a state of tension under any loading condition.

11.2.2 Pneumatic Structures

There are several different types of membrane structures in use that obtain their stabilization through internal pressurization. The term *pneumatic structure* is usually used to describe the particular subset of pressurized constructions that find applications in building. Most of these structures use air as the pressurization medium. Other gases (and even liquids), however, could potentially be used.

There are two primary classes of pneumatic structures—air-supported structures and air-inflated structures. An *air-supported structure* consists of a single membrane (enclosing a functionally useful space) which is supported by a small internal pressure differential. The internal volume of building air is consequently at a pressure higher than atmospheric. An *air-inflated structure* is supported by pressurized air contained within inflated building elements. The internal volume of building air remains at atmospheric pressure. Hybrid or combination approaches are also possible.

In both types of structures, the air pressure induces tensile stresses in the membrane. External forces acting on the membrane cause a relaxation of some of these tensile stresses. The internal pressure must be high enough under the action of any possible applied loading to prevent compressive stresses, which are manifested as folds, from appearing. Complete stability is achieved only when the whole membrane remains in tension. The high initial tension stress induced by the pressurization, however, must not exceed the allowable stress of the material.

The air-supported structure relies on the maintenance of an internal air pressure only slightly higher than atmospheric to maintain tension in its membrane [Figure 11-2(a)]. The actual internal pressure involved is actually rather small and usually causes no discomfort to building occupants. If the external loads were uniform, for example, an equal internal pressure would support the loads directly. The membrane

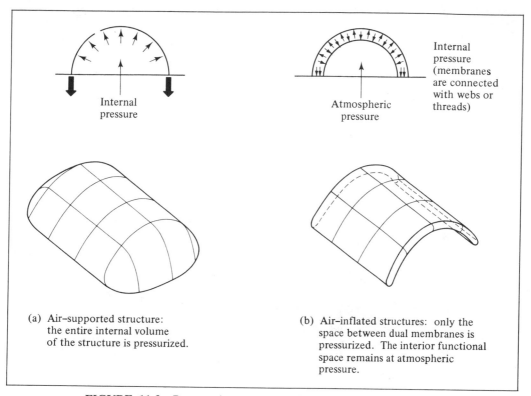

(a) Air–supported structure: the entire internal volume of the structure is pressurized.

(b) Air-inflated structures: only the space between dual membranes is pressurized. The interior functional space remains at atmospheric pressure.

FIGURE 11-2 Pneumatic structures: air-supported and air-inflated forms.

would consequently serve merely as a separator. Since external loads are usually relatively small, the internal pressures required can be similarly small. The correct image of an air supported structure is more one of a low-pressure air bag than of a high-pressure automobile tire.

An air-inflated structure has a different load-supporting mechanism [Figure 11-2(b)]. Pressurized air is used to inflate shapes (e.g., arches, walls, columns) which are used to form the building enclosure. There are two primary types of air-inflated structures in common use: the inflated rib structure and the dual wall structure (Figure 11-3). A ribbed structure is made of a series of inflated tubes, usually arched, which form a space enclosure. Dual wall structures consist of a parallel membrane system. The space between the membranes is pressurized. The membranes are held together by connecting threads and diaphragms. Air-inflated structures generally tend to require a higher degree of pressurization to achieve stability than do air-supported structures. This is because the internal pressurization cannot be directly used to balance the external load but must be used to form shapes which carry loads in a more traditional manner. It is fairly obvious, for example, that a thin air-inflated plate structure of the type illustrated in Figure 11-4 is far more susceptible to collapse due to a type of folding or buckling in the top surface when inflation pressures are low than is a comparable air-supported membrane. Only by using a high internal pressure in the inflated plate can the load-carrying capacities of the systems be made equal.

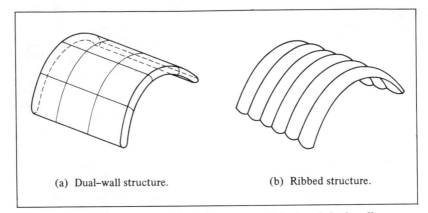

(a) Dual–wall structure. (b) Ribbed structure.

FIGURE 11-3 Air-inflated structures: ribbed and dual-wall.

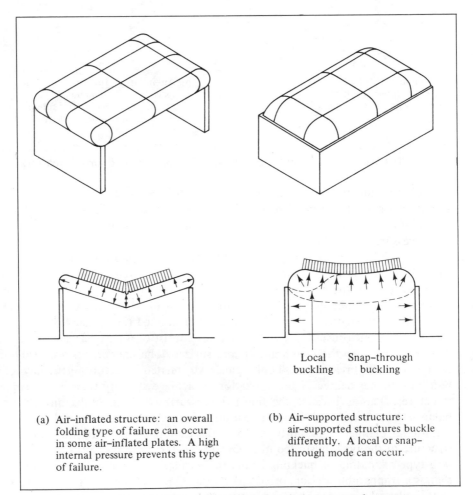

Local   Snap–through
buckling   buckling

(a) Air-inflated structure: an overall folding type of failure can occur in some air–inflated plates. A high internal pressure prevents this type of failure.

(b) Air–supported structure: air-supported structures buckle differently. A local or snap–through mode can occur.

FIGURE 11-4 Plate forms: air-inflated and air-supported structures.

All advantages, however, do not lie strictly with an air-supported structure. An inflated plate provides a certain degree of freedom in the design of the functional space it covers. Clearly, the air-supported system must come equipped with air locks and other paraphernalia in order that the space be usable. The air-supported system must also have an edge anchoring system that ties the membrane firmly to the ground and prevents leakage. The air-inflated plate can be used with simpler supports.

In general, air-supported structures are capable of greater spans than are air-inflated structures. Occasionally, hybrid structures are used that combined features of both air-supported and air-inflated systems. The spanning potential of air-supported structures are achieved while the dual wall system provides added insulation and safety against possible collapse.

11.2.3 Tent and Net Structures

For surfaces that obtain their stability through externally applied prestress forces, application of the design principle that surface tensions must be maintained generally means that prestress forces must be large and/or curvatures in surfaces must be maintained that are fairly sharp. Large flat areas in membranes are to be generally avoided since enormous prestressing forces would be required to maintain the areas when normal loads are applied to them. It will be recalled that as the sag of a cable approaches zero, cable thrusts increase indefinitely. Conversely, an indefinitely large prestress force is required to maintain a cable in a zero sag configuration under an applied load. The same general phenomenon is true in stressed skin structures.

Flat areas can and must be avoided by paying close attention to the exact geometry of the shell surface. Models are often useful for these studies during preliminary design stages.

The amount of prestress force that is applied for stabilization must not create membrane stresses in excess of the capacity of the material used. For large-span structures, it is often necessary that the membrane actually consist of a *net* of closely spaced steel cables which are capable of carrying the relatively high prestress forces involved.

A critical design issue in stressed skin membranes is their edge or support conditions. The use of point supports, such as masts, for example, induces extremely high local stresses in the membrane at the point of interface. Special provisions must be made to relieve these stresses at points such as this. One technique is using an eyelet of the type illustrated in Figure 11-13(a). These and other techniques will be discussed in greater detail in Section 11.4.2.

11.3 ANALYSIS AND DESIGN OF PNEUMATIC STRUCTURES

11.3.1 Air-Supported Structures

TYPES OF EXTERNAL LOADING CONDITIONS.   Most of the points raised in Chapter 3 which discussed loads on buildings are applicable to pneumatic structures as well. There are, however, some unique conditions in connection with loads on pneumatic structures that are of special interest. Several types of loads can act on an air-supported structure. Snow can accumulate and cause a downward-acting load. If the structure is a spherical segment that is nearly a hemisphere, snow can accumulate only near the crown of the structure since it will slide off elsewhere. In low-profile sphere segments, snow can cover the whole roof. Rarely does snow accumulate to any great depth, however, because of the heat losses that normally occur through the membrane itself when the interior volume is heated.

Major concentrated loads, which induce very high local stresses, should be avoided at all costs and other structures selected when such forces exist. Minor concentrated loads such as a person walking on the roof, however, rarely cause much trouble.

Wind loads are a major problem. As previously discussed in Chapter 3, the forces due to wind on a shape such as a sphere are farily complex and consist of both suction and tension forces. Figures 11-5 and 11-6 diagram these forces. Whether or not a shell experiences simply pressure forces or both pressure and suction forces depends on the aspect angle defining the extent of the surface used.

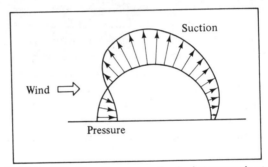

FIGURE 11-5 Wind loadings on spherical pneumatic structures.

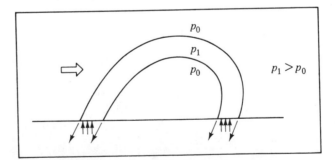

FIGURE 11-6 Effect of wind on a dual-walled pneumatic structure.

MEMBRANE STRESSES DUE TO INTERNAL PRESSURE. The in-plane forces in a membrane induced by an internal pressure are dependent on the dimensions and geometrical shape of the membrane as well as the magnitude of the internal pressure present.

The forces in a simple membrane shape are easy to determine. First consider a planar ring acted upon by an internal pressure per unit area p, directed radially outwards (see Figure 11-7). The tension forces developed in the ring can be found by passing a section through the midline of the ring and considering translatory equilib-

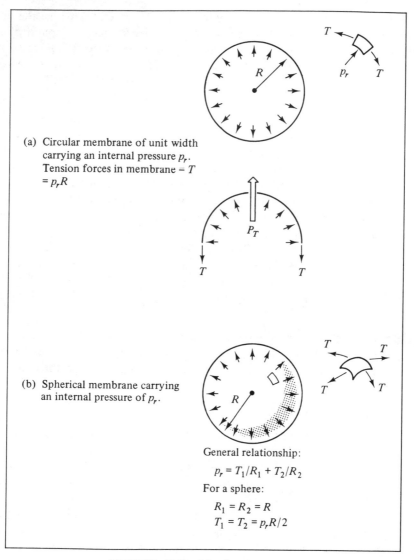

(a) Circular membrane of unit width carrying an internal pressure p_r. Tension forces in membrane = T = $p_r R$

(b) Spherical membrane carrying an internal pressure of p_r.

General relationship:

$$p_r = T_1/R_1 + T_2/R_2$$

For a sphere:

$$R_1 = R_2 = R$$
$$T_1 = T_2 = p_r R/2$$

FIGURE 11-7 Membrane stresses.

rium. The total translatory effect of the pressure p_r is simply the sum of the aligned components of the pressure forces acting on the surface in the direction considered. This total effect is exactly equivalent to the internal pressure acting over the projected area of the ring. The resisting total internal forces is simply given by $2T$ where T is the force developed in the ring. Consequently, $2T = p_r(2R)$ or $p_r = T/R$ or $T = p_r R$. If the shape is actually a surface curved in two directions, the analogous expression is given by $p_r = T_1/R_1 + T_2/R_2$, where r_1 and r_2 are the radii of curvatures involved and T_1 and T_2 are in-plane forces that are perpendicular to one another.

The general expression $p_r = T_1/R_1 + T_2/R_2$ is an invaluable one in analyzing membranes. For a cylindrical surface, $R_1 = R$ and $R_2 \to \infty$, hence $T = p_r R$. For a sphere, $R_1 = R_2 = R$ and consequently $T = p_r R/2$. This is the internal force, expressed in terms of a *force per unit length of membrane*, in a spherical shape carrying an internal pressure p_r. Internal membrane forces can be converted to stresses by taking into account the membrane thickness (i.e., $f = T/tL$, where L is a unit length). These expressions will also be used later in the analysis of shells.

INTERNAL PRESSURES REQUIRED.   The internal pressure required must be sufficient to keep the membrane surface from folding no matter what combination of applied loads exist. Different points on the surface of the sphere must be checked to find what loading combination produces a maximum radial component. The point on the surface having the largest of all radial components then becomes the value to be balanced by internal pressurization. The magnitude of the resultant pressurization, however, must be such that the tensile stresses anywhere in the membrane induced by the pressurization never exceed the allowable stress of the material fabric used. The stresses in the fabric under all possible loading conditions, including those due to internal pressurization, should be checked after an internal pressurization is decided upon since partial loadings may produce some unusual interactive effects.

By and large, the internal pressures required will usually be found to be surprisingly small. Internal pressures on the order of only 100 mm of water pressure are not uncommon.

SUPPORT CONDITIONS.   How an air-supported structure meets the ground is a critical design issue. There is, for example, the problem of ensuring an airtight seal. From a structural rather than detailing viewpoint, however, the dominant problem is that the structure exerts large uplift and, depending on the exact shape of the structure, horizontal forces on its supports.

Consider the air-supported structure shown in Figure 11-8. The structure evidently exerts a total uplift force of $p_r A_t$, where A_t is the projected area of the structure (equivalent to the ground plane covered). Uplift forces of $T_v = p_r A_t/L$, where L is the circumference of the base ring and T_v is expressed in terms of a force per unit length of support, are therefore developed at the support. Note that for a sphere segment, $T_v = p_r \pi (R \sin \phi)^2 / 2\pi R \sin \phi$ and $T = T_v/\sin \phi = p_r R/2$ as before. In addition, horizontal forces of $T_v/\tan \phi$ are also developed at the support. For low-profile shapes, the horizontal forces are inwardly directed. In some external load cases, other forces may be developed as well (e.g., uplift forces due to wind).

The foundation must be designed to resist both the vertical uplift and horizontal forces caused by the membrane in order to anchor it to the ground. For larger structures an often used device for resisting these forces is some form of a base containment ring. For a low-profile structure, the horizontal components of the membrane reactions are inwardly directed. Hence, a containment ring would be in compression.

Containment rings for structures that are sphere segments are circular rings. For other membrane shapes the design objective of minimizing bending dictates that the containment ring shape be funicular for the loading. This may lead to unusually shaped rings if membrane shapes are unusual.

In very large span air-supported structures where a cable-net type of membrane must be used, the junctions between the cables and the ring can be particular problem areas because of the high concentrated loads involved. Special care must be taken to

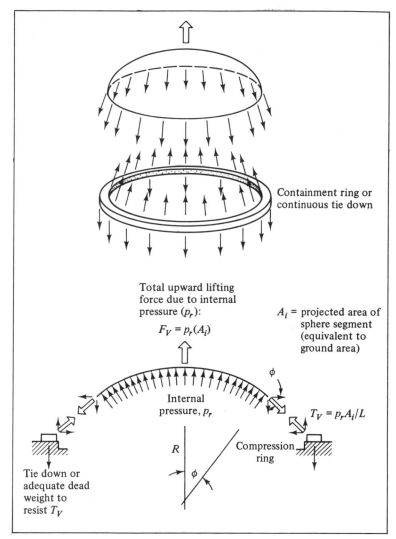

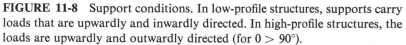

Total upward lifting
force due to internal
pressure (p_r):

A_i = projected area of
sphere segment
(equivalent to
ground area)

$$F_V = p_r(A_i)$$

Containment ring or
continuous tie down

Internal
pressure, p_r

ϕ

$$T_V = p_r A_i / L$$

Compression
ring

R

ϕ

Tie down or
adequate dead
weight to
resist T_V

FIGURE 11-8   Support conditions. In low-profile structures, supports carry
loads that are upwardly and inwardly directed. In high-profile structures, the
loads are upwardly and outwardly directed (for $0 > 90°$).

ensure that excessive local stresses are not developed at such points. Ring shapes may
be affected if a system of primary and secondary cables are used, in which case ring
loads can no longer be characterized as nearly uniform.

PROFILE.   One of the most interesting design problems associated with a pneumatic
air-supported structure is deciding on the profile of the structure. Consider a sphere
that has a radius R_M that is cut off to cover a ground area A_t (see Figure 11-9). Consider another sphere of a much larger radius, R_N, that is also cut off such that it covers
exactly the same ground area A_t. Assuming that each must have an internal pressurization of p in order to support the external loads, it is evident that for a given p the

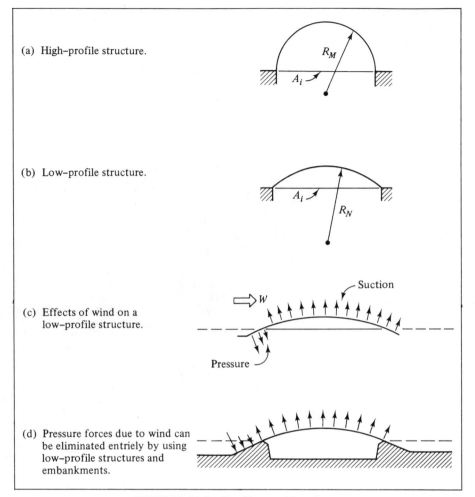

(a) High–profile structure.

(b) Low–profile structure.

(c) Effects of wind on a
low–profile structure.

Suction

Pressure

(d) Pressure forces due to wind can
be eliminated entriely by using
low–profile structures and
embankments.

FIGURE 11-9 Profile considerations.

membrane stresses in the low-profile sphere segment of larger radius ($T_N = pR_N/2$) are much larger than those in the high-profile membrane with a smaller radius ($T_M = pR_M/2$) (i.e., since $R_N > R_M$, $T_N > T_M$). Since the choice of membrane is strongly affected by the magnitude of the force present, it would therefore seem that high-profile, small-radius sphere segments are preferable to low-profile large-radius sphere segments. In addition, low-profile structures develop larger horizontal reactive forces and thus necessitate larger compression rings.

There are, however, trade-offs involved. One is that a larger building volume is enclosed in a high-profile structure than in low-profile structures, and there is thus a greater demand placed on mechanical systems intended to ensure thermal comfort conditions for occupants.

Another trade-off is that the external forces due to wind effects can be utilized to advantage in low-profile structures. By cutting off the sphere segment at the point diagrammed in Figure 11-9(c), a condition can be created where the effect of wind is one of suction only. Pressure forces tending to cause folds to develop are thus avoided. Wind forces literally tend to hold up the roof rather than force it down. Design internal pressurization values are thus favorably affected.

There are examples of both high- and low-profile air-supported structures existant. Most very long span structures are low-profile and use cable-net membranes. Small-span structures are more often high-profile structures.

11.3.2 Air-Inflated Structures

Structures using air-inflated elements carry external loads to the ground in a much more traditional way than do air-supported structures. Common elements, such as beams, columns or arches, are made rigid by high internal pressurization.

Consider the air-inflated beam shown in Figure 11-10 that is inflated with a

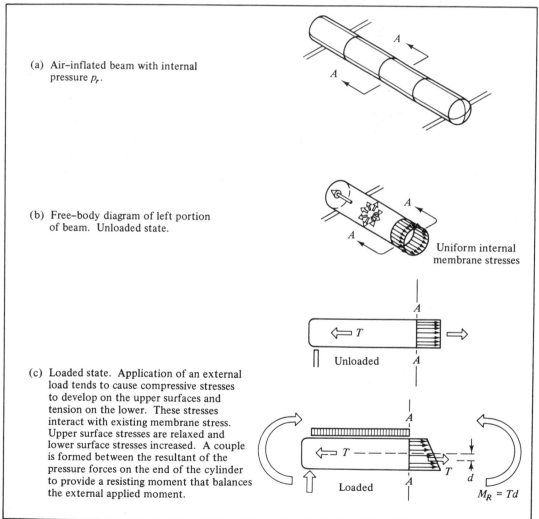

(a) Air–inflated beam with internal pressure p_r.

(b) Free–body diagram of left portion of beam. Unloaded state.

Uniform internal membrane stresses

(c) Loaded state. Application of an external load tends to cause compressive stresses to develop on the upper surfaces and tension on the lower. These stresses interact with existing membrane stress. Upper surface stresses are relaxed and lower surface stresses increased. A couple is formed between the resultant of the pressure forces on the end of the cylinder to provide a resisting moment that balances the external applied moment.

$M_R = Td$

FIGURE 11-10 Air-inflated beam: general load-carrying behavior.

pressure of p_r. Uniform longitudinal membrane tensile stresses due to the pressure are developed along its length when the structure is unloaded. Application of an external load tends to cause compression stresses to develop along the upper surface and tension stresses along the lower in much the same way as occurs in a beam made of a rigid material. These load-induced stresses interact with those caused by the pressurization. The consequence is that tensile stresses originally present along the upper surface are reduced and those along the lower surface increased. Clearly, the internal pressurization must be of such magnitude that no compressive stresses, which would be manifested as folds, are developed along the top surface. The resultant internal stress distribution and the resultant of the pressure force couple together to develop an internal resisting moment which balances the applied moment. The load-carrying mechanism is thus quite similar to a prestressed beam.

Other shapes also depend on internal pressure for rigidity. A double-walled sphere, for example, acts quite like a rigid thin shell. As such, it is particularly sensitive to buckling that might be induced by external loads.

The rigidity of all air-inflated structures largely depend on the degree of internal pressurization. Pressurization is usually higher in air-inflated than in air-supported structures.

11.3.3 Other Considerations

A common concern when pneumatic structures are used is what happens if the membrane is punctured. In this connection it is useful to compare two pneumatics, a common balloon and an air-supported structure.

When a balloon is punctured, it obviously seems to literally explode. In actuality, the puncture causes a crack to develop which propagates very, very rapidly. One reason for the rapid crack propagation is that the very high degree of internal pressurization present in the balloon induces a very high state of stress in the membrane. A lot of energy is stored in the highly stressed membrane. As a crack begins to develop, all this stored energy is available to help the crack propagate. In such structures, often called *high-energy systems*, crack propagation is consequently extremely rapid. In contrast, a typical air-supported building pneumatic is a *low-energy system*. Internal pressurization is relatively small, as are the stresses induced in the membrane, as compared to a balloon. There is relatively little energy stored in the membrane. Consequently, a crack initiated by a puncture does not propagate rapidly, since there is little stored energy in the membrane to help the crack propagate (Figure 11-11). Some tearing may occur, but the puncture effects remain essentially localized.

The relatively low pressure present in building pneumatics is also such that the pressure loss associated with holes is often not initially significant. The mechanical system used to inflate the structure often has sufficient capacity (or can be easily designed to have) to maintain the shape of the membrane even when minor punctures occur.

In any event, even a severely punctured roof will still tend to settle relatively gradually rather than instantaneously, thus allowing evacuation. Consternation will invariably be caused, but the danger is not as life-threatening as is often envisioned.

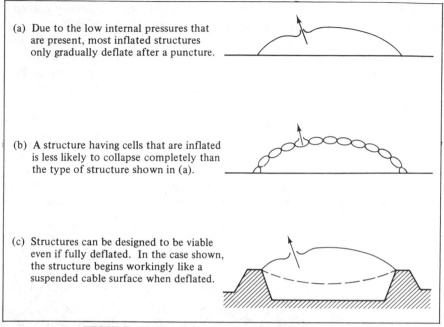

(a) Due to the low internal pressures that are present, most inflated structures only gradually deflate after a puncture.

(b) A structure having cells that are inflated is less likely to collapse completely than the type of structure shown in (a).

(c) Structures can be designed to be viable even if fully deflated. In the case shown, the structure begins workingly like a suspended cable surface when deflated.

FIGURE 11-11   Punctures in air-inflated structures.

For complete safety in the event of a loss of pressure, pneumatics can be designed to act as suspended roofs (admittedly unstabilized ones). Supports are typically elevated so that clearance exists in the event of deflation.

Another important problem in connection with pneumatics is the choice of membrane material, since many tend to degrade with time due to the ultraviolet effects of the sun. This problem however, is solvable by correct choice of material.

11.4 ANALYSIS AND DESIGN OF NET AND TENT STRUCTURES

A detailed analysis of how prestressed membrane net and tent structures behave is fairly complex and beyond the scope of this book. Some of the important underlying principles, however, have already been covered in Chapter 5, since many prestressed membrane structures are basically crossed-cable systems. This section will discuss a few of the more salient design problems.

11.4.1 Curvatures

As noted earlier, it is imperative that relatively flat areas be avoided in membrane surfaces, since extremely high prestress forces would be required to maintain the shape of such areas under the action of external loading. Sharp surface curvatures are preferable.

Flat areas can generally be avoided by paying close attention to the geometry of the surface. Transition zones between positively and negatively curved surfaces can be particularly problematic and should be specially considered.

Curvatures can be assured by carefully controlling the placement of high points and low points. Two high points must always be separated by a low point, and vice versa. This principle is evident in the four-point structure shown in Figure 11-12(a). The two high points (*A* and *C*) are separated by a point that is relatively lower (*B*). The same point (*B*) is, however, high relative to the other two points (*D* and *E*). A continuous skin connecting points *A*, *C*, *D*, and *E* is therefore doubly curved. If the height differential between *A* and *C* and *D* and *E* is reasonable, the surface will have a relatively sharp curvature. Obviously, if all points are near the same level, the surface will be relatively flat.

The same principle of separating highs with lows and vice versa can be applied to more complex or more extensive membranes, as is illustrated in Figure 11-12(b).

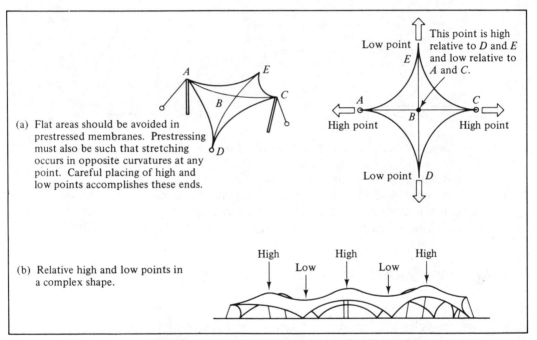

(a) Flat areas should be avoided in prestressed membranes. Prestressing must also be such that stretching occurs in opposite curvatures at any point. Careful placing of high and low points accomplishes these ends.

(b) Relative high and low points in a complex shape.

FIGURE 11-12 Curvatures in pretensioned membranes.

11.4.2 Support Conditions

Most prestressed membrane structures are usually supported by a series discrete point supports. Primary high points are usually formed by using large compression masts. Masts of this type are designed basically as large columns that are almost invariably pin-ended. Primary low points are usually ground connections. Large horizontal and vertical uplift forces usually exist at ground foundations, since the prestressing force in the membrane is obtained by stretching the membrane, through a jacking procedure, between these tie-down points and high points. Since very high concentrated forces are present, the tie-down of ground connection points represents a major design problem.

The number and placement of support points generally determines the amount of force present in a mast or ground connection. The more supports, the less is the load present on any single point support. Care must be taken, however, not to use so many supports as to not be able to develop sufficient curvatures in the membrane.

An exception to the primary point-support type of condition for a membrane often occurs internally in the structure at the free edge of the prestressed membrane (see Figure 11-13). Edge cables are often used which stiffen the free edge and allow a more uniform tension field to be developed in the membrane itself (by essentially pulling all along the free edge of the surface). Forces in these edge cables are usually fairly high.

Since the local membrane stresses that would be generated where a surface drapes over a high point would be very great, eyelets of the type illustrated in Figure

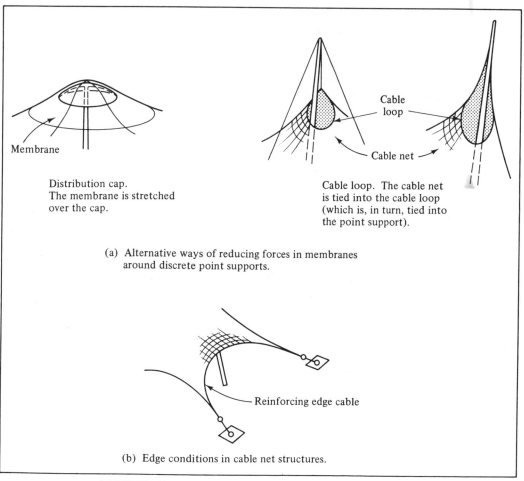

Distribution cap.
The membrane is stretched over the cap.

Cable loop

Cable net

Cable loop. The cable net is tied into the cable loop (which is, in turn, tied into the point support).

Membrane

(a) Alternative ways of reducing forces in membranes around discrete point supports.

Reinforcing edge cable

(b) Edge conditions in cable net structures.

FIGURE 11-13 Boundary conditions in pretensioned membranes.

11-13 are used to reduce membrane stresses by opening the surface at the point and using a cable ring. The membrane is tied into the ring which distributes internal forces more evenly than a pure point support. The ring is, in turn, connected to the mast.

11.4.3 Other Considerations

While not a difficulty from a purely theoretical point of view, there are enormous difficulties involved in actually making doubly curved surfaces. Since such surfaces are typically not developable, the use of naturally flat material sheets to make the surface poses problems. Usually, the surface is made of smaller non-uniformly-shaped strips that are specially cut. Determining the shapes of these strips is a task unto itself. Analogous problems arise when a net of crossed cables are used such as would be needed for longer-span structures. The small grid shapes formed are, of course, not uniformly shaped squares or rectangles but highly variable shapes. Establishing individual cable lengths and cross-connection points is consequently a major task, as is devising an enclosure system to be supported by these cables.

While these problems are, of course, solvable, they do tend to put a different perspective on the apparent freedom in choosing shapes that is often associated with using prestressed membranes. In many cases the choice of form is derived by model studies using nice extensible membranes, such as a rubber sheet, that freely deform in any direction. The making of such shapes from materials that are naturally not freely extensible is a fundamentally inconsistent but necessary fact. The resultant freedom of expression is thus a visual one only and contradicted by the way such structures are constructed.

QUESTIONS

11-1. A spherical balloon has a radius of 12 in. and a thickness of 0.05 in. The balloon is blown up with an internal pressure of 100 lb/in.². What are the membrane stresses developed in the surface of the skin of the balloon?

11-2. How internally pressurized pneumatic structures meet the ground is a critical design and construction issue. Sketch at least one acceptable type of ground connection detail. Consult your local library.

11-3. Using literature from your library, sketch several different ways stretched skin or cable-net structure are connected to cable masts.

11-4. How prestressed tent structures are tied down to the ground is a critical design and construction issue. Sketch several different ground anchorage systems capable of carrying tension loads. Consult your local library.

chapter 12

SHELL STRUCTURES

12.1 INTRODUCTION

A *shell* is a thin, rigid, three-dimensional structural form taken by the enclosure of a volume bounded by a curved surface. A shell surface may assume virtually any shape. Common forms include rotational surfaces generated by the rotation of a curve about an axis (e.g., spherical, elliptical, conical, and parabolic surfaces); translational surfaces generated by sliding one plane curve over another plane curve (e.g., cylindrical and elliptic paraboloid surfaces); ruled surfaces generated by sliding the two ends of a line segment on two individual plane curves (e.g., conoids and hyperbolic paraboloid surfaces); and a wide variety of complex surfaces formed by various combinations of rotational, translational, and ruled surfaces (Figure 12-1). A shell can assume any of these forms. The shape used, however, need not be restricted to those easily described in mathematical terms. Free-form shapes may prove a viable solution to many structural problems. Construction considerations, however, may limit the range of form options.

Loads applied to shell surfaces are carried to the ground by the development of compressive, tensile, and shear stresses acting in the in-plane direction of he surface. The thinness of the surface does not allow the development of appreciable bending resistance. Thin shell structures are uniquely suited to carrying distributed loads and

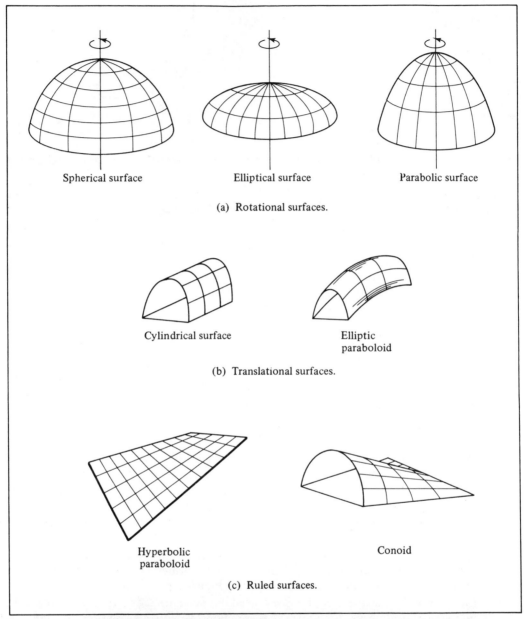

Spherical surface   Elliptical surface   Parabolic surface

(a) Rotational surfaces.

Cylindrical surface   Elliptic
                       paraboloid

(b) Translational surfaces.

Hyperbolic   Conoid
paraboloid

(c) Ruled surfaces.

FIGURE 12-1 Example of different types of continuous shell surfaces.

find wide application as roof structures in buildings. They are, however, unsuited to carrying concentrated loads.

As a consequence of their carrying loads by in-plane forces (primarily tension and compression), shell structures can be very thin and still span great distances. Span/thickness ratios of 400 or 500 are not uncommon [e.g., 3 in. (8 cm) thicknesses are possible for domes spanning 100 to 125 ft (30 to 38 m)]. Shells of this thickness, however, are a relatively recent structural innovation made possible by the development of new materials such as reinforced concrete, which is uniquely appropriate for making shell surfaces. Older three-dimensional shapes, such as masonry domes, are considerably thicker relative to their span and cannot be exactly characterized as carrying loads by in-plane axial or shear stresses (more bending stresses exist), even though an approximation of this type is good for conceptualizing the behavior of such structures.

Three-dimensional forms may also be made of assemblies of short, rigid bars (Figure 12-2). These structures are not, strictly speaking, shell structures since they are not surface elements. Still their structural behavior can be conceptualized as being quite similar to continuous surface shells in which the stresses normally present in a continuous surface are concentrated into individual members. Structures of this type were first extensively used in the last century. The *Schwedler dome*, consisting of an irregular mesh of hinged bars, for example, was introduced by Schwedler in Berlin in

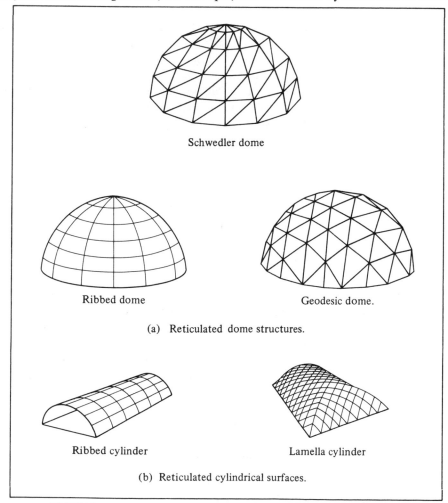

Schwedler dome

Ribbed dome

Geodesic dome.

(a) Reticulated dome structures.

Ribbed cylinder

Lamella cylinder

(b) Reticulated cylindrical surfaces.

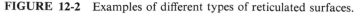

FIGURE 12-2 Examples of different types of reticulated surfaces.

1863 when he designed a dome with a span of 132 ft (40 m). Other more recent structures have bars placed on curves generated by the medians and parallels of surfaces of revolution. Some of the largest domes in the world follow this latter scheme.

To minimize the construction difficulties involved in having to use bars of different length to create the shell surface, a facet present in the bar structures mentioned above, some systems have been developed with the use of equal-length bars as a goal. The most widely publicized are the *geodesic domes* associated with Buckminster Fuller. Since the surface of a sphere is not developable, the number of identical repetitive patterns into which the surface can be divided is limited. The spherical icosohedron, for example, consists of 20 equilateral triangles formed by the connecting the points of crossing of great circles encompassing the sphere. Such geometrical considerations form the point of departure for much of Fuller's work. Interpreting the structural behavior of structures of this type with reference to relatively meaningless terms emanating from proponents of the Fuller world, however, such as "energetic synergetic geometry" should be done carefully, if at all, since many of the properties attributed to these structures are doubtful at best. The structural advantages often claimed are not necessarily greater than other forms of recticulated domes.

Shapes other than surfaces of revolution can also be made using bar elements. These include typical ribbed barrel roofs and the *Lamella roof* which is made of a skew grid of archlike forms composed of discrete elements. The latter form was developed largely in connection with wood as the building material, but concrete and steel have been used as well. Remarkably large spans can be obtained with lameller systems.

12.2 GENERAL PRINCIPLES

12.2.1 Membrane Action

A good way to envision the behavior of any shell surface under the action of a load is to think of it as analogous to a membrane, a surface element so thin that only tension forces can be developed (see chapter 11). A soap bubble or thin sheet of rubber are examples of membranes. A membrane carrying a load normal to its surface deforms into a three-dimensional curve and carries the load by in-plane tension forces that are developed in the surface of the membrane. The load-carrying action is similar to that present in a crossed-cable system. The basic load-carrying mechanism of a rigid shell of a similar geometry is analogous to those produced in an inverted membrane. Of primary importance is the existance of two sets of internal forces in the surface of a membrane that act in perpendicular directions. Also of importance is the existence of a type of tangential shearing stress which is developed within the membrane surface (which is associated with the twist normally present in the surface), which also helps carry the applied load.

12.2.2 Shell Structures Having Rotational Surfaces

The existence of two sets of forces in separate directions within the shell surface tends to make the shell act in a fashion similar to a two-way plate structure. The shear forces between adjacent plate strips in a planar plate structure that were shown to contribute to the load-carrying capacity of the plate are present in shell structures as well.

It is these two characteristics, the development of shear forces and two sets of axial forces rather than one, that characterize the difference between the structural behavior of a shell and that of a series of arch shapes of a similar general geometry that are rotated about a point to form a similar shape. In an arch no bending is present if the arch is funicularly shaped for the applied full-loading condition. If the loading is changed, however, to partial loading, substantial bending is developed in the arch.

In a shell in-plane forces in the meridional direction called *meridional forces* (see Figure 12-3) are induced under full loading that are not unlike those in the analogous arch. Under a partial-loading condition, however, the action of the shell differs remarkably from that of an analogous arch, in that no bending is developed because of the other set of forces that act in the hoop direction, which are also developed in the shell. These *hoop forces* act perpendicularly to the meridional forces. The hoop forces restrain the out-of-plane movement of the meridional strips in the shell that is tended to be caused by the partial loading (bending under partial loading in an arch is accompanied by a movement of this type). In a shell the restraint offered by the hoop forces causes no bending to be developed in the meridional direction (or in the hoop direction for that matter). As a consequence, a shell can carry variations in loads by the development of in-plane stresses only. The plate shears mentioned earlier also contribute to this capacity.

The variations in load patterns involved, however, must be gradual transitions (e.g., a gentle change from full to partial loads) for bending not to be developed. Sharp discontinuities in load patterns (e.g., concentrated loads) do cause local bending to occur. While loads of this type would produce very high bending stresses in arches, the hoop forces in shells again come into play and cause any bending induced to die out rapidly. Thus, any unusual forces in an arch, say those perhaps caused by edge disturbances associated with the supports, cause bending that is propagated throughout the entire arch, while the effects of analogous forces in shells remain more localized.

The shell is thus a unique structure. It can be said to act funicularly for many different types of loads even if its shape is not exactly funicular. In the example dis-

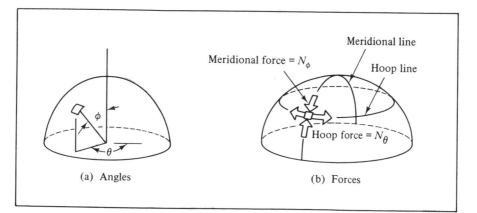

(a) Angles (b) Forces

FIGURE 12-3 In-plane axial forces in a thin spherical shell.

cussed above, the funicular shape for an arch carrying a uniform load would be parabolic. A spherical shell which has a surface form that is not parabolic will also carry loads by in-plane forces. In this case, however, hoop forces will be set up even in the full-loading case, since the shape is not exactly funicular.

The meridional forces in a shell under full vertical loading are always compressive (by analogy with the action of an arch). The hoop forces, however, may be in either tension or compression, depending on the location in the shell considered (see Figure 12-4). In a semicircular shell, or simply one with a high rise, there is a tendency for lower meridional strips to deform in an outward direction. Hoop forces containing this movement are therefore in tension. Near the top of the same shell, the meridional strips tend to deform inwardly. Hoop forces resisting this movement are therefore in compression.

The stresses associated with meridional and hoop forces are generally fairly

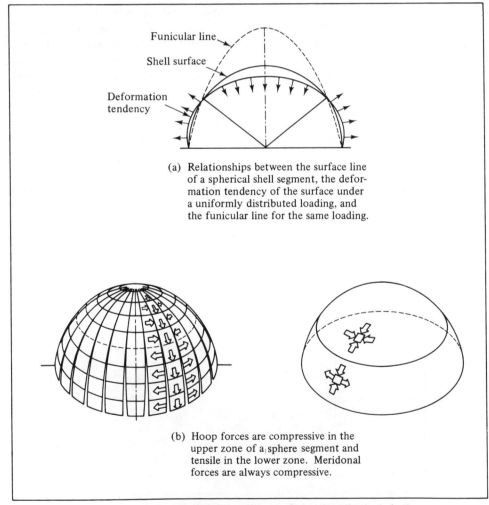

(a) Relationships between the surface line of a spherical shell segment, the deformation tendency of the surface under a uniformly distributed loading, and the funicular line for the same loading.

(b) Hoop forces are compressive in the upper zone of a sphere segment and tensile in the lower zone. Meridonal forces are always compressive.

FIGURE 12-4 Meridional and hoop forces in spherical shells.

small for a uniform loading condition. Point loads, however, can cause very high stresses and should be avoided on shell surfaces.

A major design consideration in a shell of revolution is the nature of the boundary or support conditions. In much the same way that buttresses or tie-rods must be used to contain the horizontal thrusts of arches, some device must be used to absorb the horizontal thrusts associated with the meridional in-plane forces in the shell at the lower edge of the shell. In a dome, for example, a circular buttress system could be used. Alternatively, a planar circular ring, called a *tension ring*, could be used to encircle the base of the dome and contain the outward components of the meridional forces. Since the latter are always in compression, their horizontal components are always outwardly directed at the shell base. The containment ring is therefore always in tension. If a hole were cut out of the shell at its crown, however, the same meridional force components would be inwardly directed. A ring used to absorb these forces would therefore be in compression.

Holes in shell surfaces of the type mentioned above are possible but should generally be avoided since they disrupt the continuity and hence efficiency of the shell surface. If holes are used, the shell must be specially reinforced around hole edges.

Another critical design feature of a shell is its degree of curvature. In shells of low profile, or those whose surfaces are relatively flat, the shell surface can buckle inwardly—a disastrous type of failure. The buckling is an instability type of failure similar to that which occurs in long slender columns. The buckling can either be local over a portion of the surface or the whole surface can pop through. Buckling can be prevented by using shell surfaces of sharp curvature.

12.2.3   Shell Structures Having Translational Surfaces

The behavior of structural forms that are defined by translational surfaces is critically influenced by the relative proportions of the shell and its support conditions.

Consider the cylindrical surface supported on walls illustrated in Figure 12-5(a). In this structure, commonly called a *vault*, the surface behaves much like a series of parallel arches as long as the supporting walls can provide the necessary reactions. If the surface is rigid (e.g., made out of reinforced concrete), the surface also exhibits plate action (i.e., shearing forces are developed between adjacent transverse strips), which is useful for carrying nonuniform loads. The same general type of action occurs when such a surface is supported on very stiff beams. The beams, in turn, transfer the loads to the supports by bending.

The behavior of a very short shell may differ appreciably from that described immediately above if transverse end stiffeners are used. The surface loads may be transferred directly to the end stiffeners through longitudinal plate action.

As the shell structure becomes very much longer relative to its transverse span, a completely different type of action begins, particularly if edge beams are either not used or highly flexible ones are used. Any type of supporting edge beam, it should be noted, naturally tends to become more flexible as its length increases. The cylindrical surface will again *tend* to start behaving like an arch in the transverse direction. Flexible edge beams (or no edge beams) do not, however, provide any appreciable resistance to the horizontal thrusts involved. The consequence is that no archlike

(a) Vaults: vaults are supported
continuously along their
longitudinal edges. Internal
transverse forces are characterized
by an archlike action.

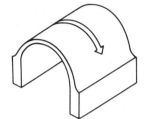

(b) Short shell with stiff edge
beams: the edge beams serve
basically the same function as a
wall in a vault if they are
sufficiently stiff. An archlike
action transfers surface loads to
the beams, which carry the loads
by bending to the supports.

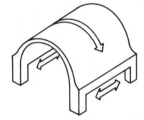

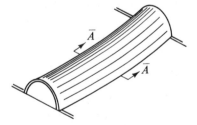

(c) Long barrel shells: if the shell does
not have stiff edge beams, archlike
action cannot develop in the transverse
direction. Instead, loads are carried by
a longitudinal bending action which is
similar to that present in beams.

Internal
stress
distribution

Internal
resisting
moment

External
moment

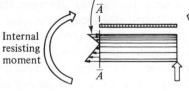

FIGURE 12-5 Cylindrical shells.

action is ever actually exhibited in this direction. Indeed, if no edge beams are present, the longitudinal free edges tend to deflect inwardly rather than outward under a full loading. A different type of load-carrying mechanism must therefore be present. Structures of this type are called *barrel* shells. The principal action in a long cylindrical barrel shell is in the longitudinal rather than the transverse direction. A type of longitudinal bending occurs that is very analogous to that which occurs in either simple beams or folded plates. Compressive stresses in the longitudinal direction are developed near the crown of the curved surface and tension stresses in the lower part.

The analogy with folded plate structures is a useful one since many of the same design principles are present. Transverse stiffeners, for example, are useful in increasing the load-carrying capacity of a barrel shell. The more that stiffeners are used, or if the barrel shell considered is one of a series of adjacent shells, the beamlike behavior becomes more pronounced and beam analysis techniques yield more accurate results. Barrel shells whose lengths are about three times (or more) their transverse spans exhibit this longitudinal type of behavior.

12.2.4 Shell Structures Having Ruled Surfaces

Ruled surfaces can be very complex to analyze. In general, the behavior of such shells may be envisioned by looking at the nature of the curvatures formed by the straight-line generators. If the edge conditions are of the type which can offer restraint (foundations or very stiff edge beams), an archlike action will exist in regions of convex curvature and a cablelike action in regions of concave curvature. The presence of compression or tension forces in the surface depends on which action is present. When surfaces get very flat due to reduced curvatures, a plate action in which bending dominates may be present (necessitating increased plate thicknesses). If shell edges are not supported, a beam behavior may be present.

A ruled surface made by translating two ends of a straight line over two parallel but twisted straight lines (rather than more complex curves) is illustrated in Figure 12-6. Interestingly enough, the shape formed can also be described as a translational surface generated by translating a concave parabola over a convex one. In this type of structure an archlike action will be present in the direction of convex curvature and a cablelike action in the perpendicular direction which is concavely curved. The stress field in the plate is thus compressive in one direction and tensile in the perpendicular direction. These directions are at 45° to the original straight-line generators.

12.3 ANALYSIS AND DESIGN OF SPHERICAL SHELLS

12.3.1 Meridional Forces

The internal forces and stresses in axisymmetric shells uniformly loaded can be found quite easily through application of the basic equations of equilibrium. A dome will be analyzed in detail as an example.

Consider the dome segment illustrated in Figure 12-7. Assume that the loading is a uniform gravity load distributed on the surface of the shell (e.g., the shell's own dead weight and insulative or protective coverings). If the total of all such loads acting

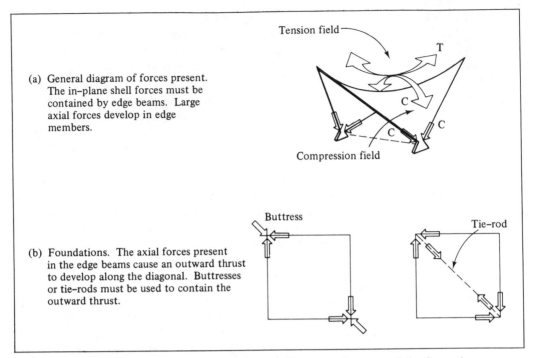

(a) General diagram of forces present. The in-plane shell forces must be contained by edge beams. Large axial forces develop in edge members.

Tension field

Compression field

(b) Foundations. The axial forces present in the edge beams cause an outward thrust to develop along the diagonal. Buttresses or tie-rods must be used to contain the outward thrust.

Buttress

Tie-rod

FIGURE 12-6 Hyperbolic paraboloid. The general nature of the forces in many shells can be ascertained by looking at the types of shell curvatures present. Concave curvatures generally indicate a type of cablelike action (a tension field) and convex curvatures an archlike action (a compression field).

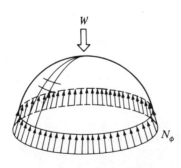

(a) Meridional forces acting in the plane of a shell.

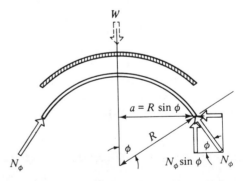

(b) The sum of the vertical components of the in-plane meridional forces developed internally in the shell equals the external vertical live and dead load.

FIGURE 12-7 Meridional forces in a spherical shell.

downward is denoted as W and the meridional in-plane internal force per unit length present in the shell surface as N_ϕ, a consideration of equilibrium yields the following:

$\sum F_Y = 0$:

$$W = (N_\phi \sin \phi)(2\pi a)$$

where ϕ is the angle defining the shell cutoff and a is the instantaneous planar radius of the sphere at that point. The N_ϕ forces in the shell are in-plane compressive forces developed in the shell at the horizontal section defined by ϕ. The vertical components of these forces (assumed uniform around the periphery of the shell) are simply $N_\phi \sin \phi$. Since the N_ϕ forces are expressed as a force per unit length (e.g., lb/ft or kN/m) along the section, the total upward force associated with all the continuous N_ϕ forces is simply the instaneous circumference of the shell at that point (which is given by $2\pi a$) multiplied by $N_\phi \sin \phi$ (i.e., a total length times a force per unit length yields a total force). This upward force must be of a magnitude to exactly balance the downward force, hence the expression $W = N_\phi \sin \phi(2\pi a)$. This expression can be rewritten in terms of the actual radius of the sphere by noting that $a = R \sin \phi$. Thus, $W = N_\phi \sin \phi(2\pi R \sin \phi)$. Solving for N_ϕ, we obtain

$$N_\phi = \frac{W}{2\pi R \sin^2 \phi}$$

If the total load acting downward (W), is determined, the internal forces in the shell can thus be found directly. Since these internal forces are expressed in terms of a force per unit length, actual internal stresses expressed in terms of a force per unit area (e.g., lb/in.^{2} or kN/mm^2) can be found by dividing by the shell thickness. Thus, $f_\phi = N_\phi/tL$, where L is a unit length and N_ϕ is the force per unit length.

An expression can be easily derived which directly incorporates the load acting downward. If the load per unit area of shell surface acting downward is denoted by w, equilibrium in the vertical direction yields:

$\sum F_Y = 0$:

$$-\int_{\phi_1}^{\phi_2} w(2\pi R \sin \phi)R \, d\phi + N_\phi \sin \phi(2\pi R \sin \phi) = 0$$

where ϕ_1 and ϕ_2 define the segment of shell considered. The term on the left thus defines W. For $\phi_1 = 0$,

$$N_\phi = \frac{Rw}{1 + \cos\phi}$$

This expression is actually identical to $N_\phi = W/2\pi R \sin^2 \phi$. Either expression defines the meridional forces present at a horizontal section.

12.3.2 Hoop Forces

Hoop forces, typically denoted as N_θ and also expressed in terms of a force per unit length, can be found by considering equilibrium in the transverse direction. Alternatively, use can be made of the results of the membrane analysis discussed in Section

11.2. There it was found that the in-plane forces in a membrane which act perpendicularly to one another are related by the general expression: $p_r = T_1/r_1 + T_2/r_2$. This expression is immensely valuable in the analysis of shells, since by using it the hoop forces (N_θ) can be related to the meridional forces (N_ϕ).

Since the load being studied actually acts downward rather than radially outward, however, the external force expression must be adjusted. The radial component of the downward load is given by the following: $p_r = w \cos \phi$. The expression relating hoop to meridional forces then becomes

$$p_r = \frac{T_1}{r_1} + \frac{T_2}{r_2}$$

or

$$w \cos \phi = \frac{N_\phi}{r_1} + \frac{N_\theta}{r_2}$$

or

$$N_\theta = r_2(w \cos \phi) - \left(\frac{r_2}{r_1}\right) N_\phi$$

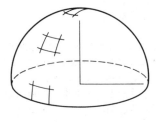

(a) Shell structure

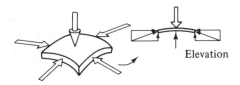

Elevation

(b) Free–body diagram of shell element at top of shell. All internal in–plane forces must be in compression for the element to be in equilibrium in the vertical direction.

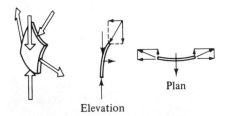

Plan

Elevation

(c) Free–body diagram of shell element at $\phi = 90°$. The outward component of the compressive meridional force must be balanced by forces developed in the hoop direction (which must be tensile). Hoop forces thus vary from compression at the top of the shell to tension at the bottom depending on the geometry of the element and applied loads.

FIGURE 12-8 Hoop forces in spherical shells.

in a sphere, $r_1 = r_2 = R$, and substituting the expression previously found for N_ϕ, we have

$$N_\theta = Rw\left(-\frac{1}{1 + \cos\phi} + \cos\phi\right)$$

The preceding is a simple expression for hoop forces in terms of the radiius (R) of the sphere and the downward load (w). Figure 12-8 illustrates hoop forces in shells.

EXAMPLE

Given a dome having a spherical radius of 100 ft (30.48 m) and a thickness of 4 in. (100 mm) that has an aspect angle of 45°, determine the meridional and hoop forces at the base of the shell for a loading of 100 lb/ft² (4788 N/m²).

Solution:

Meridional forces:

$$N_\phi = \frac{Rw}{1 + \cos\phi}$$

$$= \frac{(100 \text{ ft})(100 \text{ lb/ft}^2)}{1 + 0.707}$$

$$= 5850 \text{ lb/ft in compression}$$

$$= \frac{(30.48 \text{ m})(4788 \text{ N/m}^2)}{1.707} = 85{,}494 \text{ N/m}$$

Meridional stresses:

$$f_\phi = \frac{N_\phi}{Lt} = \frac{5850 \text{ lb/ft}}{(12 \text{ in./ft})(4 \text{ in.})}$$

$$= 122 \text{ lb/in.}^2 \text{ in compression}$$

$$= \frac{85{,}494 \text{ N/m}}{(1000 \text{ mm/m})(100 \text{ mm})} = 0.85 \text{ N/mm}^2$$

Hoop forces:

$$N_\theta = Rw\left(-\frac{1}{1 + \cos\phi} + \cos\phi\right)$$

$$= (100 \text{ ft})(100 \text{ lb/ft}^2)\left[-\left(\frac{1}{1.707}\right) + 0.707\right]$$

$$= 1{,}212 \text{ lb/ft in compression}$$

$$= (30.48 \text{ m})(4788 \text{ N/m}^2)\left[-\left(\frac{1}{1.707}\right) + 0.707\right] = 17{,}684 \text{ N/m}$$

Hoop stresses:

$$f_\theta = \frac{N_\theta}{Lt} = \frac{1{,}212 \text{ lb/ft}}{(12 \text{ in./ft})(4 \text{ in.})}$$

$$= 25.3 \text{ lb/in.}^2 \text{ in compression}$$

$$= \frac{17{,}684 \text{ N/m}}{(1000 \text{ mm/m})(100 \text{ mm})} = 0.177 \text{ N/mm}^2$$

The stresses in the sphere at this point are thus extremely low. This is characteristic of most shell structures.

12.3.3 Distribution of Forces

The distribution of meridional and hoop forces can be found by simply plotting the equations for the two forces (Figure 12-9). As is evident, the meridional forces are always in compression while the hoop forces undergo a transition at an angle of 51° 49' as measured from the perpendicular. Shells cut off above this only develop compressive stresses in their surfaces while deeper shells can develop tension stresses in the hoop direction. The magnitude of the stresses, however, always remains relatively low.

An interesting way of looking at the overall behavior of a dome and ring assembly is illustrated in Figure 12-10. As in other structures, the external moment at a section must be balanced by an internal resisting moment (in this case provided by a couple formed between the hoop and ring forces). Thinking of the structure in these terms also helps explain the development of tension hoop stresses in a dome.

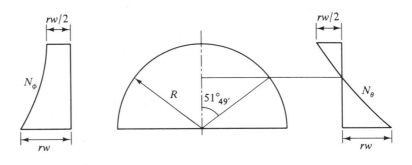

(a) Meridional forces.　　　　　　　　　　　　　　(b) Hoop forces.

FIGURE 12-9　Stress distribution in a spherical dome carrying a uniformly distributed load along the surface of the sphere.

12.3.4 Concentrated Forces

The reason why concentrated loads should be avoided on shells can readily be seen by analyzing the meridional forces present under such a loading. The general expression found before was $N_\phi = W/2\pi R \sin^2 \phi$, where W was the total load acting downward. For a shell carrying a concentrated load, P, the expression becomes $N_\phi = P/2\pi R \sin^2 \phi$. If the load is applied at $\phi = 0$ (the crown of the shell), a situation arises where directly beneath the load, the stresses become indefinitely large in the shell (i.e., as $\phi \rightarrow 0$, $\sin \phi \rightarrow 0$, and $N_\phi \rightarrow \infty$). Obviously a failure would occur if the shell surface could offer no bending resistance and the load was truly characterized as a point force. In any event, such forces should be avoided on shell surfaces.

12.3.5 Support Conditions

As with all structures, the nature of a dome's support conditions critically influence its behavior and design. Ideally, the supports should be such that they do not tend to cause any bending to be developed in the shell surface. Fixed edge conditions should thus always be avoided. One possible solution is illustrated in Figure 12-11(b), where

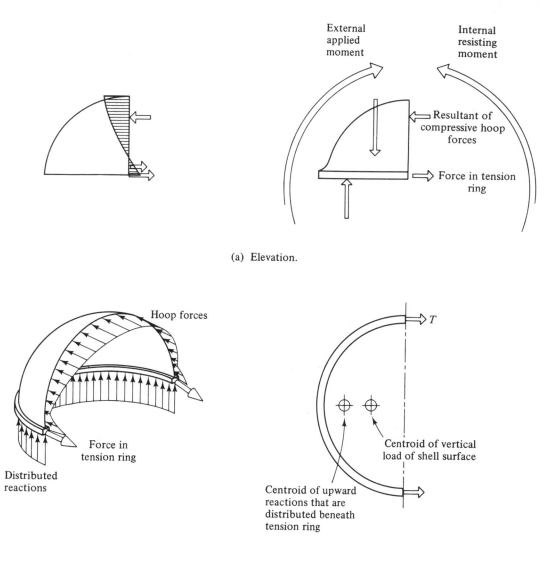

(a) Elevation.

(b) Free-body diagram.

(c) Plan.

FIGURE 12-10 Basic shell behavior.

the shell is pinned along its peripheral edge. Unlike the arch, however, the presence of hoop forces causes the shell to deform naturally in the out-of-plane direction. Restraining this deformation by a pinned connection would be equivalent to applying forces to the shell's edge, thus inducing bending. A peripheral roller support, as illustrated in Figure 12-11(c), is preferable since no restraint exists in the out-of-plane direction. Such supports are, however, extremely difficult to build for shells. In addition, slight angular changes due to movement still cause some bending to be induced (although less than in the case of fixed or pinned conditions).

For reasons of practical construction considerations, some bending is often

(a) Fixed–edge conditions induce undesirable bending in the shell surface.

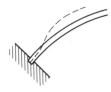

(b) Pinned edges also tend to restrain the shell surface from free movement.

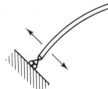

(c) Roller edges are desirable in that they allow movements to occur more freely but are very difficult to construct.

FIGURE 12-11 Shell support conditions.

allowed to develop at the shell edge for the sake of using easy-to-construct edge and foundation conditions. The shell is simply locally stiffened (usually by increasing its thickness) around the edge and specially reinforced for bending.

A primary consideration in the actual foundation design is how to absorb the horizontal thrusts associated with the outwardly directed components of the in-plane meridional forces. A system of continuous buttresses can be used (Figure 12-12). Such buttresses have been used in many buildings, particularly in older masonry domes.

Another way is to contain the thrusts via a tension ring. The tension ring is simply a planar ring against which the outward thrusts push, thus causing tension to develop in the ring. The magnitude of the outward thrusts at the springing of the shell are simply given by $N_\phi \cos \phi$. These outward thrusts per unit length of tension ring cause a tension force of $T = (N_\phi \cos \phi)a$ to be developed in the ring, where a is the radius of the planar ring.

A tension ring completely absorbs all the horizontal thrusts involved. When resting on the ground, it also provides a continuous footing for transferring the vertical reaction components to the ground. Alternatively, the ring can be supported on other elements (e.g., columns), which then receive vertical loads only.

The use of a tension ring does, however, induce bending in the shell surface where the ring and shell intersect. These bending moments are largely due to deformation incompatibilities that exist between the ring and shell. Since the ring is always in tension, it always wants to expand outward. The hoop deformations in the shell, however, may be compressive (thus, the shell edge wants to deform inwardly) depending on the N_ϕ and N_θ forces. In any event it is highly unusual for the deformation tenden-

(a) Buttresses. The vertical and horizontal components of the meridional forces can be supported by buttresses. The buttresses must be capable of containing the large outward thrusts involved.

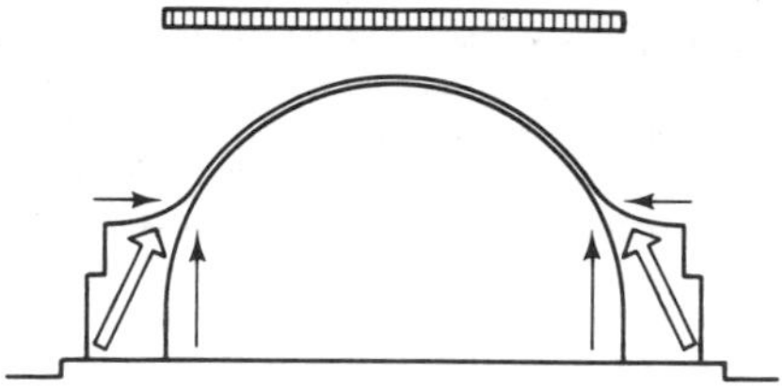

(b) Tension rings. A continuous tension ring can be used to contain the horizontal thrusts. Only downward forces are then exerted on the ground.

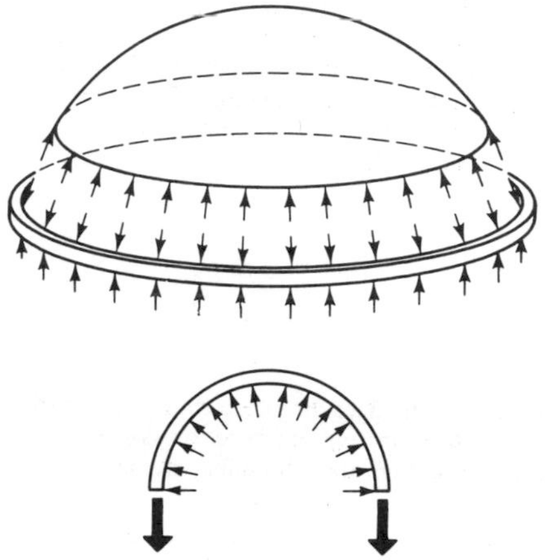

(c) If tension rings are used, they must be continuous around the shell base; otherwise, they are useless and severe stresses will develop in the shell.

(d) Shells using tension rings can be elevated on columns since below the ring only vertical forces need to be transmitted to the ground. Shells without tension rings would need a buttressing system.

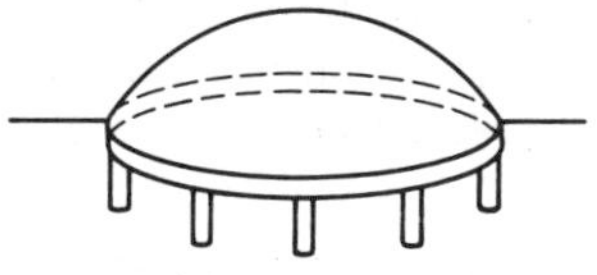

FIGURE 12-12   Spherical shell support conditions.

cies to be exactly similar. Since the elements must be joined, the edge beam restricts the free movement of the shell surface, and thus bending is induced in the shell's edge. As mentioned before, however, this bending rapidly dies out in shells, so the bulk of the shell's surface is generally unaffected. The shell edge is locally stiffened and reinforced for bending (Figure 12-13).

The problem with deformation incompatibilities can become a generator of

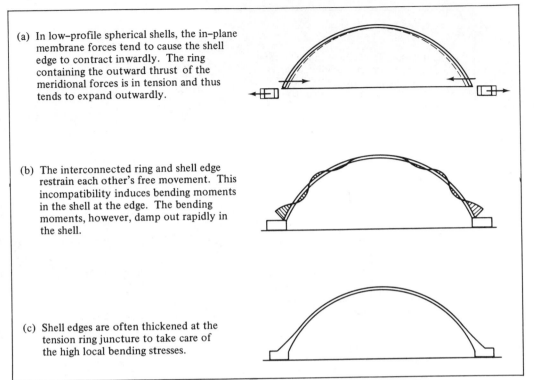

(a) In low–profile spherical shells, the in–plane membrane forces tend to cause the shell edge to contract inwardly. The ring containing the outward thrust of the meridional forces is in tension and thus tends to expand outwardly.

(b) The interconnected ring and shell edge restrain each other's free movement. This incompatibility induces bending moments in the shell at the edge. The bending moments, however, damp out rapidly in the shell.

(c) Shell edges are often thickened at the tension ring juncture to take care of the high local bending stresses.

FIGURE 12-13   Edge disturbances in spherical shells.

design approaches intended to minimize undesirable consequences. One effective method is to utilize post-tensioning as a way of controlling deformations. A support ring normally in tension, for example, can be post-tensioned so that compressive forces (hence compressive deformations compatible with those in the shell in the hoop direction) initially exist in the ring. The outward thrusts of the dome would cause a relaxation of the compressive forces (and an increase in the tension in the post-tensioning wires). If the initial amount of post-tensioning is controlled, final ring deformations can be controlled and thus deformation incompatibilities with the shell minimized. The shell surface itself can also be post-tensioned in the hoop direction for added control of shell forces and deformations.

12.3.6   Other Considerations

There are a host of factors other than those already discussed that must be taken into account in the design of shells. Critical among these additional factors is the need to ensure that a shell does not fail prematurely in a *buckling* or surface instability mode. When the curvature of the shell's surface is relatively flat, snap-through or local buckling can be a severe problem (see Figure 12-14). As with the analogous phenomenon in long columns, instability can occur at very low stress levels. It can be prevented by using surfaces of sharp curvature. The need to maintain sharp curvatures, however, effectively hampers the use of very large span but low-profile shells, in which cases

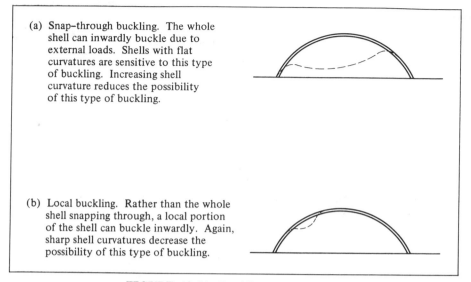

(a) Snap–through buckling. The whole shell can inwardly buckle due to external loads. Shells with flat curvatures are sensitive to this type of buckling. Increasing shell curvature reduces the possibility of this type of buckling.

(b) Local buckling. Rather than the whole shell snapping through, a local portion of the shell can buckle inwardly. Again, sharp shell curvatures decrease the possibility of this type of buckling.

FIGURE 12-14 Buckling in thin shells.

there is relatively little curvature present. The problem is also extremely severe in recticulated shells made of small rigid linear elements (e.g., geodesic domes). More advanced texts discuss the shell buckling problem and propose measures for predicting if instability is a problem in a given design context.

Another major concern is that shells must withstand loads other than those which act vertically. Figure 12-15 illustrates the stress trajectories in a spherical dome due to wind forces. Usually wind forces are not too critical in the design of shell structures. Earthquake forces that also act laterally, however, can pose extremely serious design problems. When such loadings are possible, special care must be taken with the design of the support conditions present.

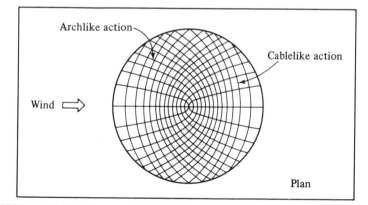

FIGURE 12-15 Stress trajectories in a sperical dome due to wind forces.

QUESTIONS

12-1. A rigid shell having a spherical radius of 300 ft is cut off at an angle of ϕ of 51°49′. Assume that the shell carries a distributed live load of 50 lb/ft². What are the in-plane forces in the shell at $\phi = 0°$? At $\phi = 51°49′$? What are the associated f_ϕ and f_θ stresses? Ignore the dead load of the shell itself.

12-2. A rigid reinforced concrete shell having a spherical radius of 200 ft is cut off at an angle ϕ of 35°. The shell thickness is 3 in. Assume that the unit weight of the shell material is 150 lb/ft³ and that the shell carries a live load of 60 lb/ft². Draw force-distribution diagrams of the type illustrated in Figure 12-9. Indicate numerical values. What is the maximum compressive stress f_ϕ developed in the shell surface? What is the maximum tension stress f_θ?

12-3. Assume that a tension ring is used in conjunction with the shell described in Question 12-1. What is the magnitude of the force developed in the ring?

part III

PRINCIPLES
OF STRUCTURAL
DESIGN

The chapters in Part III discuss principles that are important in designing structures in a building context. The structural design process is largely heuristic in nature and closely linked to the design of the remainder of the building. The following chapters discuss structural· design issues as they generally arise in the various stages of design development of a building.

Chapter 13 discusses structural design considerations as they arise during the very early design phases of a building, where the designer is still in the stages of manipulating functional spaces in buildings and establishing its basic character. Following this stage are periods of design development during which time the basic strategy adopted is given substance. Chapters 14, 15, and 16 address issues that arise as the structural design is developed in detail. Reference is made to the chapters in Part II, however, for specifics regarding actual member design (e.g., sizing a beam for a given span and loading).

The usual design process, it should be noted, is actually far more involved than is reflected in the way the chapters in Part III are organized. Usually, the process is iterative in nature and often involves following several structural options through to a fairly detailed design stage rather than just one. During this process, structural requirements for a particular design strategy might also dictate that other building characteristics be modified, or vice versa. An interactive design approach is normal. The structural system finally adopted is usually one that works well both on its own terms and with respect to other architectural objectives.

Many of the design suggestions made in the following chapters are presented in a polemical way. Some of the principles espoused can be demonstrated as being valid generalizations. Others are more subjective and spring less logically from observations of built structures. The reader is encouraged to take issue with the polemicisms presented since doing so is an excellent way of coming to grips with the design process, which, despite efforts and allegations to the contrary, still retains components which are highly subjective in nature.

chapter 13

STRUCTURAL GRIDS AND PATTERNS

13.1 INTRODUCTION

A study of existing buildings reveals the rather obvious fact that there are often strong and easily identifiable patterns present in the way buildings are functionally organized and in the structural systems used. The patterns formed by the structural configurations are intimately related to those of the functional organizations. Each affects the other. The adoption of a particular functional configuration for a building often determines to a large extent the structure used, and vice versa. For this reason the process of designing a structure is implicitly linked to the process of designing the overall building. Neither process can really exist without the other.

During very early design stages relatively little is initially known about what patterns will be present in a building. One way of interpreting design activities during this early stage is that the designer is manipulating the programmatic functions of the building with the objective of ordering them into a form such that when a physical building fabric is developed, the result will achieve the desired design goals or aspirations. During this process the designer often either adopts a pattern organization to give order to the building or encounters reasons that give rise to generalized patterns of both space usage and structure. The cellular nature of multiunit housing is an obvious example of the latter. The privacy requirements for individual dwelling units naturally lead to a repetitive pattern of space modules related to the size of a typical dwelling unit. In a typical building, the forces that give rise to patterns are often more a consequence of the functional programming requirements of the building or of construction considerations as they are of purely structural motivations. The fact that there are associated structural advantages to using patterns, however, is still a contributing factor, particularly in large-span or large-load situations.

The following sections will briefly expand on different aspects of the pattern nature of building structures. Before getting involved in looking at structural patterns in detail, however, one fundamental point initially addressed in the first chapter should be emphasized. This point is that patterns are to a considerable extent arrived at by simple abstractions from trends toward uniformity. Patterns are forms, but not necessarily structures. A structure is a conception at a much higher plane of abstraction than a pattern. The concept of structure necessarily involves the interrelationships between the elements comprising a form. Patterns are more limited in concept. This chapter, however, will focus strictly on this more limited view of structure.

13.2 BASIC UNIT AGGREGATION PATTERNS

Many different types of structural patterns are possible. In most buildings there is usually a repetitive geometrical pattern or grid present in both the vertical support system, whether it is formed of load-bearing walls, columns, or some combination of the two, and in the horizontal spanning system. Patterns are also typically present in how elements at different elevational levels relate to one another.

The following sections focus on buildings of this type which are composed of repetitive bays or other forms of cellular units. The geometries and dimensions of individual units and the way these units aggregate in such buildings are invariably dependent on the programmatic requirements of the buildings (Figure 13-1). These same dimensions and geometries, in turn, strongly influence the types of structural systems most appropriate for use.

In order to establish a basis for later discussions on the structural implications of different pattern types, it is useful first to review briefly some basic characteristics of pattern types themselves. Figure 13-2 illustrates, in a simplistic way, how some

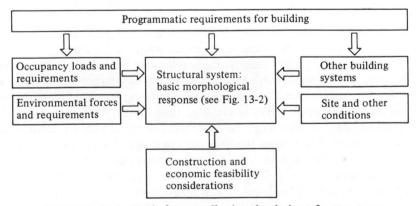

FIGURE 13-1 Basic factors affecting the design of a structure.

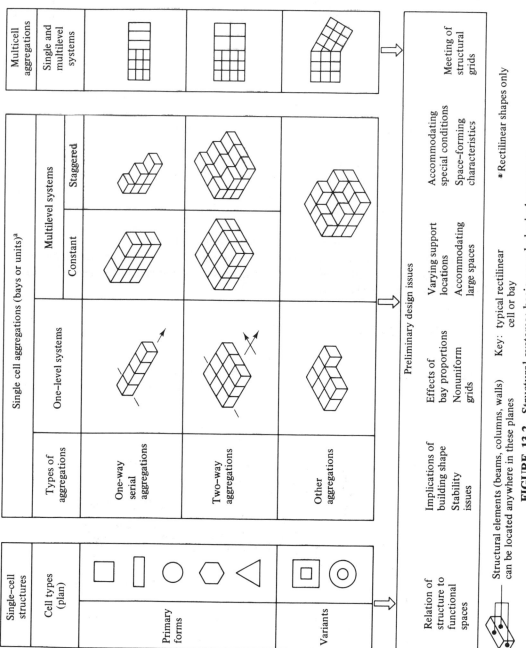

FIGURE 13-2 Structural systems: basic morphological responses.

439

common building configurations can be characterized in terms of their basic aggregation patterns. Obviously many buildings (e.g., gymnasiums) are actually *single-cell* structures. Most buildings, however, are composed of a large number of *aggregated bays* or units. Conceptualizing building forms in terms of aggregation patterns is, of course, only one way of characterizing configurations. It is, however, a useful one from a structural viewpoint.

As is evident from the limited sampling shown, there are a huge number of different configurations which are possible. It can certainly be expected that different structural responses may be appropriate for different configurations. Obvious examples can be drawn with respect to simple linear aggregations. Lateral loads would be the dominant design loading in a tall vertical stacking, while vertical loads would probably be the dominant consideration in an aggregation in the horizontal plane. Consequently, different structural responses could be expected to be appropriate for each of these cases (see Chapter 14). Even with respect to patterns only in the horizontal plane, it could likewise be expected that certain responses might be appropriate for one way serial aggregations and others for two-way serial aggregations (see Section 13.7).

Following the introduction of several examples to clarify the notion of structural patterns or grids in buildings, the remainder of this chapter will explore the design issues mentioned above—and others as well—involved in the use of different types of structural grids or patterns.

13.3 COMMON STRUCTURAL PATTERNS

Several common types of structural patterns are evident in the buildings illustrated in Figures 13-3, 13-4, 13-5, and 13-6. The first figure shows a housing scheme in which a serial aggregation pattern is used to arrange the basic units. The structural pattern

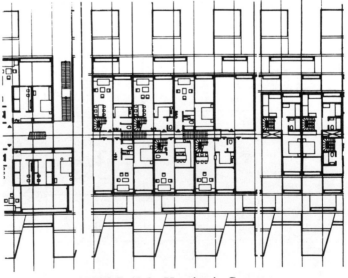

FIGURE 13-3 Housing in Germany.

part III / PRINCIPLES OF STRUCTURAL DESIGN

itself consists primarily of a series of parallel lines. The parallel lines could consist of a series of load-bearing walls or a column-and-beam system placed in the same pattern locations (the horizontal spanning system could be the same for either type of vertical support approach). It is the basic grid pattern that is of first importance. The selection of a specific type of vertical support system from among the several types available often is, or can be, a secondary decision for this building type.

A common square grid of discrete vertical support elements is illustrated in the plan shown in Figure 13-4. A two-way aggregation (staggered) approach is taken. In this plan the pattern is uniform throughout the building except near the left edge of the building and in the auditorium area. Grids are often altered in zones where something special occurs, such as an auditorium, or at boundaries. Some of the reasons for these alterations will be explored in later sections.

A more complex tartan grid of discrete vertical support elements is illustrated in the plan shown in Figure 13-5. This building is an often used example of how the

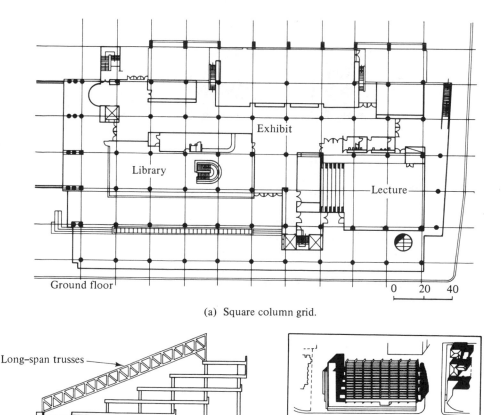

(a) Square column grid.

(b) Section.   (c) Plan.

FIGURE 13-4 Gund Hall, Harvard University, Cambridge, Mass.

chapter 13 / STRUCTURAL GRIDS AND PATTERNS   **441**

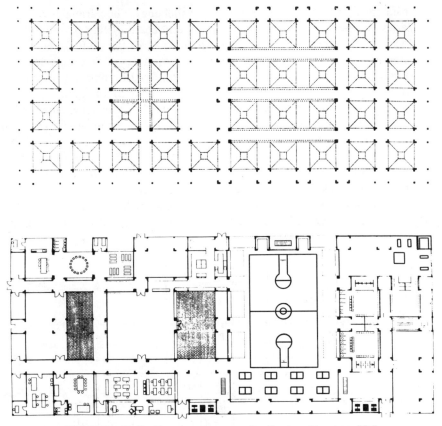

FIGURE 13-5 Trenton Community Center, Trenton, N.J.

structural pattern of a building can be related to and reinforce how a building is functionally zoned. In this building the basic pattern is again altered where a special event, the gymnasium, occurs.

Other geometries (e.g., hexagonal or triangular) can and do find use as structural patterns. Still, by and large, when the forces that normally give rise to patterns in buildings are considered in relation to structural and constructional issues, the consequence is that rectilinear patterns tend to be more generally useful than other configurations and find more widespread application.

Figure 13-6 illustrates a multicell building that contains *several* types of structural patterns instead of just one primary type. In the back part of the building, a double beam-and-column system is used, while in the front part a more traditional single beam-and-column system is employed. A third system, consisting of radial arches, is used over the additorium complex. Transition structures of yet other types exist in the interstitial spaces between the different primary structural patterns. The exact reason for this multiplicity of patterns stems largely from the different functional zones present in the building. The following sections will explore this point in more detail. At this point it is more important to note that this building brings up the highly important structural design question of whether it is more preferable to attempt find-

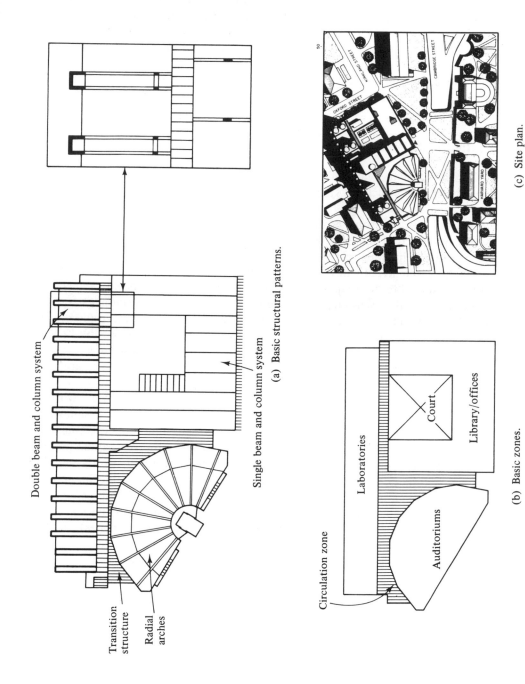

Double beam and column system

Single beam and column system

(a) Basic structural patterns.

Transition structure

Radial arches

Circulation zone

Laboratories

Court

Library/offices

Auditoriums

(b) Basic zones.

OXFORD STREET

KIRKLAND STREET

CAMBRIDGE STREET

HARVARD YARD

50

(c) Site plan.

FIGURE 13-6 Beam-and-column patterns: Science Center, Harvard University, Cambridge, Mass. (Also see Figure 13-22.)

443

ing a single generalized pattern into which all the varied functional activities of a building are fitted, albeit occasionally with minor alterations or deformations, or, alternatively, to attempt to adopt specific structural systems in response to the specific conditions present, and thereby potentially ending up with a single building containing multiple patterns. Figures 13-4 and 13-5 are illustrations of the former strategy, while Figure 13-6 illustrates the latter. The scale of the building and the variance in the types of spaces required are clearly important considerations. There are, however, structural and constructional considerations as well.

13.4 PATTERN MANIPULATIONS

One of the first activities normally undertaken by a designer in the process of developing an appropriate structural system for a building is to try fitting various types of common structural patterns to the proposed building. As a way of clarifying both the process and illustrating some of the difficulties involved, a hypothetical building will be briefly examined. The intent will be largely to establish a context for the topics discussed in the remaining sections of this chapter.

Consider the hypothetical building plan shown in Figure 13-7(a), which is arbitrarily assumed to be composed of a series of modular spaces having an arrangement and individual dimensions responsive to the programmatic requirements of the building. Assume that the critical modular dimensions are denoted by a_1 and a_2. The length of a_1 or a_2 is assumed to be the minimum clear span dimension possible because of functional reasons. Vertical supports thus cannot be placed closer together than this dimension. They can, however, be spaced farther apart since a_1 and a_2 are minimum and not a maximum spans.

For many typical buildings, it is common to begin the design process with the placement of elements intended to support horizontal members. In early design stages it is normal to think in terms of patterns of only point and line supports. *Point supports* are obviously columns. The concept of a *line support*, however, can be usefully extended from the obvious load bearing wall to frame or post-and-beam assemblies in which a continuous length of beams forms the line support. Clearly, the line support could also represent any type of eroded wall system [see Figure 13-7(b)].

It is evident that either one-or two-way horizontal spanning systems composed of any of the basic horizontal spanning elements (e.g., beams, trusses, plates) could be used in conjunction with the different vertical support patterns indicated.

For the hypothetical building shown, it should be obvious that if no design strategy other than that of random generation is used, a huge number of different types of structural patterns could be generated by different combinations of vertical and horizontal spanning systems. There is nothing that says, however, that all are equally viable. Nonetheless, as yet there is no basis for real choice. Any of the solutions shown in Figure 13-7(c)–(h), for example, could be adopted and perhaps be made to work. Obviously, they form radically different buildings.

It is clear that the process of pattern generation without the adoption of appropriate criteria for choice is limited in value. Unfortunately, exact criteria for structural choice rarely exist in the abstract but develop with, and in response to, the design

Critical functional
dimensions: a_1 and a_2

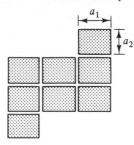

(a) Basic functional
modules.

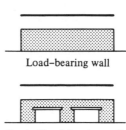

Load-bearing wall

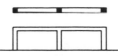

Eroded load-bearing wall

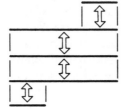

Framed system

One-way
spanning system

Two-way
spanning system

(b) Vertical and horizontal support systems.

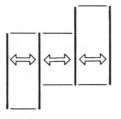

(c) One-way system.

(d) One-way system.

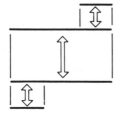

(e) One-way system.

(f) Two-way system.

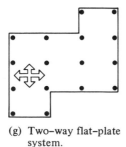

(g) Two-way flat-plate
system.

(h) One-way system.

FIGURE 13-7 Vertical support patterns. Most buildings can be fitted with
many different types of structural patterns that correspond to different structural approaches.

development of the remainder of the building. At this stage the basis for choice must relate to an almost intuitive assessment of the basic reasonableness of the various solutions. It is doubtful, for example, if the solutions that are more chaotic or less orderly than others will prove viable in the long run. While possible, for example, the solution indicated in Figure 13-7(h) would probably prove more difficult to construct and hence more costly to build than some of the other solutions (even though this actually depends on the scale of the building). Simple patterns usually prove superior to complex ones. As an aside, it is interesting to note that usually when a building can be framed in many different ways with no one solution clearly better than others, the building will probably never prove to be problematical (from a structural point of view) as the design develops, no matter what option is selected.

While exact criteria for system choice are often elusive during early design stages, it is possible to begin identifying some of the basic issues involved and to begin developing strategies for pattern development that are more cogent than random generation, which will potentially lead to viable and efficient structural systems.

Is, for example, a design strategy that invariably employs a one-on-one fit between the vertical support system and the modular spaces of the building (which thus minimizes clear spans at the expense of having to utilize numerous vertical elements) intrinsically better or worse than a stategy in which larger horizontal spans and correspondingly fewer column elements are characteristically employed.

Another critical issue is that analysis and experience demonstrates that associated with each different vertical support pattern considered is a range of horizontal spanning systems, each having its own pattern characteristics, that are most appropriate. Hence selecting a vertical support system typically implies something about the horizontal spanning system. Knowledge of the interactive nature of this relationship is valuable in immediately assessing the viability of a proposed combination of vertical and horizontal systems.

Another point of concern is that of whether unique forces, such as associated with earthquakes, would cause some patterns to be more preferable than others. These issues and others which are important in establishing viable structural patterns early in the design process will be discussed in following sections and chapters.

13.5 STABILITY ISSUES

While a subject to be more extensively discussed in Chapter 14, it should be noted immediately that the exact types of loads present strongly influence the choice of an appropriate structural pattern.

Of fundamental concern is that the structure must be organized such that sufficient lateral-load-carrying mechanisms exist to render the resultant assembly stable under any conceivable type of lateral loading condition. Some systems, such as two-way beam-and-slab frame constructions of reinforced concrete, normally inherently provide such mechanisms. Other systems, however, such as bar joists used with load-bearing masonry walls, do not, and special attention should be paid to their stability.

For common square or rectangular grids, the stability requirement usually means that frames, shear walls, or diagonal braces be placed in a series of mutually perpendicular planes placed throughout the building. Pattern implications are obviously present. The pattern illustrated in Figure 13-8(f), for example, is obviously problematical with respect to loads acting perpendicular to the planes of the shear walls. Those illustrated in Figure 13-8(b)–(e) are intrinsically more viable.

In early design stages it is often usually adequate to note only the basic lateral force resistance strategy and to identify its pattern implications. It is not always necessary to decide immediately whether a frame, shear wall, or diagonal bracing action will be used to carry the lateral loads. Appropriate locations of planes for these elements, however, must be provided. Chapter 14 addresses these issues in greater detail.

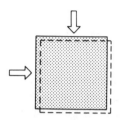

(a) Lateral loads: forces can act in any direction.

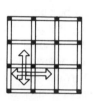

(b) Lateral bracing mechanisms throughout plan (e.g., two–way beam and slab system).

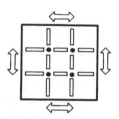

(c) Lateral bracing mechanisms around periphery only.

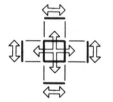

(d) Plan with symmetrically placed shear walls.

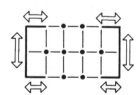

(e) Plan with end bracing.

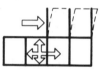

(f) Plan with problematical shear wall arrangement in upper portion (for flooring systems not acting like rigid diaphragms).

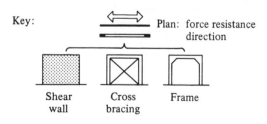

Key:

Plan: force resistance direction

Shear wall   Cross bracing   Frame

Typical lateral resisting mechanisms

FIGURE 13-8 The lateral stability of any structure under any type of loading must be assured by the correct placement of lateral force-resisting mechanisms.

13.6 RELATION OF STRUCTURE TO FUNCTIONAL SPACES

The type, pattern, and scale of the vertical support system present in a building greatly influences the type of horizontal spanning system used, and vice versa. The support geometry is, in turn, intimately related to critical dimensions derived from considering the intended functional usage of the building.

Consider the simple hypothetical building shown in Figure 13-9, in which the critical modular dimension related to the function of the building is denoted by a, which is the minimum clear span dimension. This critical modular dimension can be small in terms of absolute dimensions, as in housing, or large as might be the case in an athletic facility.

Given a building that can be characterized in this way, it is evident that a wide variety of structural approaches are possible, depending on how the structure relates to the functional grid. Figure 13-9 illustrates several possibilities. A primary initial structural design objective is to determine the degree of fit which is possible or preferable between the pattern formed by the vertical support system and the critical functional dimension a. The selection of a degree of fit determines the magnitude of the horizontal span that the structural system must provide and, consequently, has an important bearing on the system selected for use. Figure 13-9(a) illustrates several one-on-one fits where the primary structural unit that is used repetitively corresponds directly with a. The other figures illustrate looser fits where the structural unit is a multiple of a. If a defines the minimum clear span possible due to functional reasons then either one-on-one fits or looser fits are possible. It is obvious, of course, that other functional requirements (e.g., circulation) may preclude one or more of these approaches.

Considering only the degree of fit, however, it is evident that if the fit possible is a one-on-one and a is not a very small dimension, the adoption of a looser fit is not usually justified in purely structural terms. A general structural objective is always to minimize bending and this is not done by using spans of unnecessary length. Obviously if the a dimension is very small [e.g., less than 10 ft (3 m)], however, it would probably be preferable to use a multiple of a for a structural span since spanning less than the minimum capacity of any typical system is not economic.

In some very special situations involving even relatively large a distances, going to a greater than one-on-one fit (even when such a fit is possible) may still be feasible and desirable. With respect to Figure 13-9, for example, it may prove more advantageous to use horizontal spanning elements of a length of $2a$ rather than a if the elements used operate more effectively at a longer span than a shorter one because of construction ease, cost, or some other reason. This is sometimes the case with precast concrete elements (see Section 14.2).

Resolution of the question of which of the approaches illustrated in Figure 13-9 is preferable from a structural viewpoint largely hinges on the issues to be discussed in Chapter 14 (e.g., span length, loading), the obvious question of whether the wall elements showing in some of the figures are at all possible from a functional viewpoint (if walls are needed for functional reasons anyway, they might very effectively be used as load-bearing elements), and the proportions of the bays themselves (the importance

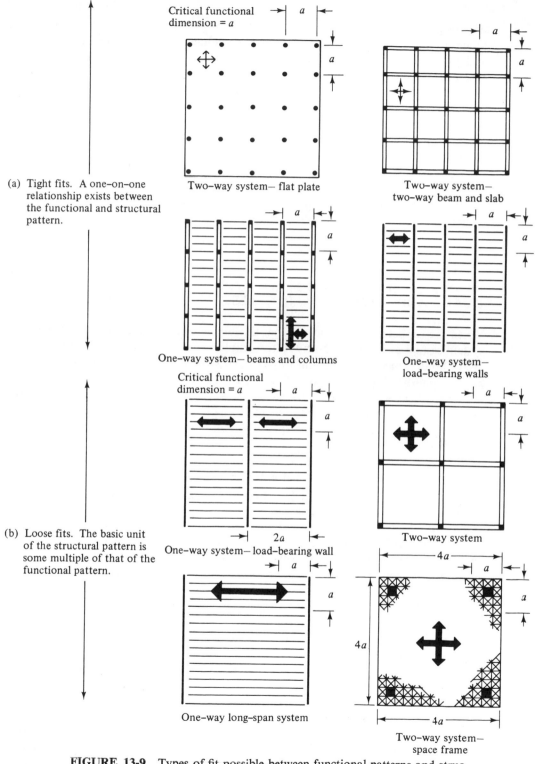

FIGURE 13-9 Types of fit possible between functional patterns and structural patterns.

of bay proportions will be discussed in a following section). With respect to span alone, Figure 13-10 illustrates some typical economic spans for several different one- and two-way structural systems. Chapters 14 and 15 explore the relation between specific structural systems and effective span ranges in greater detail.

Other factors that might affect the degree of fit used are the foundation conditions present. Under some conditions building loads are best distributed over the entire site so that the load present at any discrete point is reduced. The structural implication is that a fine-grained grid system should be employed and one-on-one fits employed. If, on the other hand, unique conditions exist of the type where some site locations can carry enormous loads and other locations virtually no loads, it is obvious that the structural pattern should respond to these conditions by having vertical elements concentrated at high-capacity points and by using long-span horizontal elements to bridge over low-capacity points. A rough-grained pattern to the vertical support system often results.

13.7 EFFECTS OF BAY PROPORTIONS: SQUARE AND RECTANGULAR GRIDS

When the support grid is square, either a two-way or a one-way horizontal spanning system can be used, with two-way systems often proving more appropriate when spans are of modest or medium length. In particular, where the support system is a square grid of load-bearing walls (see Figure 13-11), the full benefit of two-way action (see Chapter 10) can be obtained, which makes such systems very desirable. One-way systems would not take advantage of the support capability of walls parallel to the spanning elements and would thus typically require more structural depth and volume than a two-way system.

Even when a support grid consists of square modules, it is still necessary to examine more closely the overall pattern of the grid before opting for a two-way rather than one-way system. When square grid modules form one-way serial aggregation patterns, many of the structural advantages of two-way action are not easily achieved because continuity for the horizontal spanning system is directly possible only in one direction rather than two. Construction considerations might also cause one-way elements to be preferred for linear patterns. When the square support grid forms a field in both directions, however, continuity is easily achieved in both directions and two-way systems can quite efficiently be used.

Another issue involved in the choice between one- and two-way systems for square grids is that of the absolute dimensions present between support points. As will be discussed in the following chapters, there are a wide variety of two-way structural systems available for use in low- and intermediate-span ranges [e.g., 15 to 40 ft (5 to 13 m)] and either type of system could potentially be used. When square bay dimensions are very long, however, one-way systems are often used for square grids simply because there are not many long-span two-way constructional systems available. For spans up to around 60 ft (18 m) or so, there are still several two-way concrete systems commonly available. For longer spans there are fewer. For roofs in longer span ranges, the two-way steel space frame system is useful but is generally not appropriate

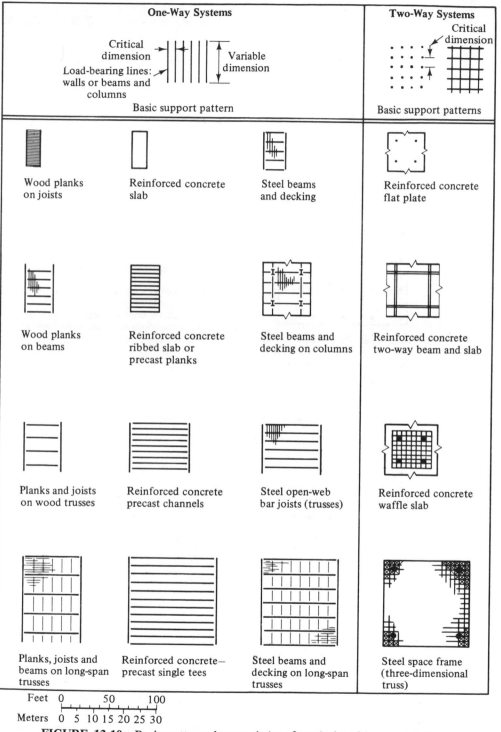

FIGURE 13-10 Basic pattern characteristics of vertical and horizontal elements of different planar structural systems. Spans shown are typical (see Chapters 14 and 15).

(a) Wall supports—very effective use can be made of two-way action. One-way systems are usually less desirable since the load-carrying capability of one set of walls cannot be utilized.

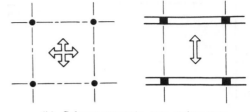

(b) Column supports—one- or two-way systems are possible.

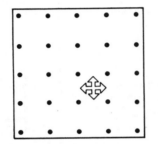

(c) Two-way serial aggregations: when the square support grid extends in both directions, a two-way system is particularly appropriate.

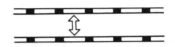

(d) One-way serial aggregations: when a square support grid forms a linear pattern, one-way systems are often more appropriate than two-way systems.

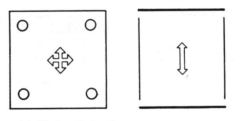

(e) Single cell structures: span limitations of available systems often make one-way systems more appropriate than two-way systems when bay dimensions are very large.

FIGURE 13-11 Vertical support systems: square patterns.

for heavier floor loads. Specially constructed two-way grid structures of either steel or reinforced concrete can be used but such systems are relatively expensive. Thus, for large square bays, a one-way approach is often used, since individual elements in the system can be designed easily to meet the unique load and span requirements. In actuality, however, large square bays are not frequently encountered in practice.

Whenever the support system is markedly rectangular (see Figure 13-12), it is probable that a one-way horizontal spanning system would prove preferable to a

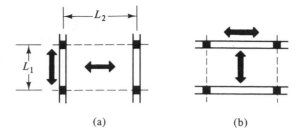

FIGURE 13-12 Vertical support systems: rectangular patterns. One-way systems are usually preferred when the support system forms a rectangular pattern. Two-way structural action is not obtainable unless the ratio of long to short bay dimensions is less than about 1.5.

two-way system. As discussed earlier in Chapter 10, two-way structural action is not realized when bay dimensions are not similar. The structure begins behaving like a one-way system when bay dimensions fall outside the range 1:1 to 1:1.5.

If a column supported bay is rectangular and one-way spanning elements are to be used, the basic question is whether to arrange the spanning elements as indicated in Figure 13-12(a) or (b). In the case shown in Figure 13-12(a), the long-span beams carry only uniformly distributed loads, while the short-span elements serve as collector beams carrying the reactions from the long-span members. The reverse is true of the approach shown in Figure 13-12(b). There is nothing that a priori suggests that the two approaches both will require the same amount of structural material to support the same load in space or that structural depths will be the same. Indeed, it can be expected that one or the other approach would prove preferable with respect to either or both of these criteria for a given set of dimensions and loadings. The question of which way to arrange primary spanning elements is a highly interesting one from a theoretical viewpoint. Unfortunately, the question is not one that can easily be answered in the abstract. For a specific case, however, involving known dimensions, loads, and materials, the two framing approaches can be compared by designing them and then comparing resultant volumes and depths. For example, if $L_1 = 30$ ft (9 m) and $L_2 = 80$ ft (24 m) and reinforced concrete is to be used, a possible approach would be to use precast concrete single tees for the long-span direction (these elements are good primarily for distributed loads only) and two specially designed poured-in-place concrete beams for use as collectors in the shorter direction. If the alternative framing approach were used, it would be necessary to design and construct unique long-span collector beams (although smaller, prefabricated plank elements could be used in the short direction). The structural volumes and depths of those two systems could be evaluated after the members are sized. Costs could be assessed as could other important considerations, such as the interaction of the structure with other building components. A statement of preferability could then be made depending on the relative importance assigned the different criteria.

There is no hard and fast answer to which approach for framing rectangular bays is preferable. Often, but not always, the longer the absolute dimension of the long span (L_2) becomes, the more it is usually both desirable and necessary to opt for the approach shown in Figure 13-12(a), in which the longest elements carry distributed loads only and do not act as collectors. The collector is kept as short as possible. When absolute dimensions are small (and thus collector depths are not excessive) and when both the collectors and the surface load-transfer elements can be off-the-shelf items instead of uniquely constructed ones, the other approach often proves preferable.

13.8 MEETING OF STRUCTURAL GRIDS

In many common cases more than one generalized structural pattern is used in a building. The reasons for this multiplicity of patterns vary. Quite often there is a wide variation in the programmatic requirements of the building such that a variety of minimum clear span dimensions exist. One or more standard functional patterns may be adopted to respond to this variation instead of seeking a single pattern that responds in a compromise way to the varying requirements present. In other cases physical constraints, such as nonuniform foundation conditions, may dictate using different structural grids in different areas. When more than one generalized structural pattern is used in a building, a basic structural design issue is how these patterns meet. Intersection points always call for unique treatment or special elements.

Consider the two structural grid systems shown in Figure 13-13(a). If these two systems randomly intersected, it is evident that the structure in the region of the intersection cannot be characterized as typical of either general grid. If a beam were used along boundary line *B*, for example, the beam would be loaded in a way that is uniquely different from either those present in either the grid to the left or right. It would therefore have to be designed differently. Additional columns may likewise be needed. While randomly intersecting grids are certainly possible, other strategies are possible as well which may prove preferable.

Figure 13-13 illustrates several basic approaches to pattern intersections. Other approaches exist as well. One basic approach illustrated in Figure 13-13(b) and (c) would be to use a strategy of alignment in which the smaller grid is designed as either directly related to, or as a subdivision of, the larger one. This alignment reduces the misfit between the systems and does not require additional vertical supports.

Another general approach is to use some sort of mediating system between the two general systems. One type of mediator would simply be a strip of space itself, in which case the whole problem is avoided simply by bypassing the two general systems [see Figure 13-13(d)]. Another type of mediator would be a third structural system of some sort. This third system must necessarily reflect or be adaptable to the characteristics that are common to both of the general systems or be extremely flexible in some other way. A structural grid that is much smaller than either of the primary systems (but a logical subdivision of each), for example, is often used to join primary systems [see Figure 13-13(e)]. Smaller-grained systems of this sort are often, but need not be, made of different materials than the primary systems. The primary systems, for example, might be constructed of poured-in-place reinforced concrete, while the smaller-grained mediating system might be a light steel system. Mixing of materials in this controlled way is usually quite acceptable even in purely economic terms. Intermixing materials within primary systems, however, is usually more difficult. An advan-

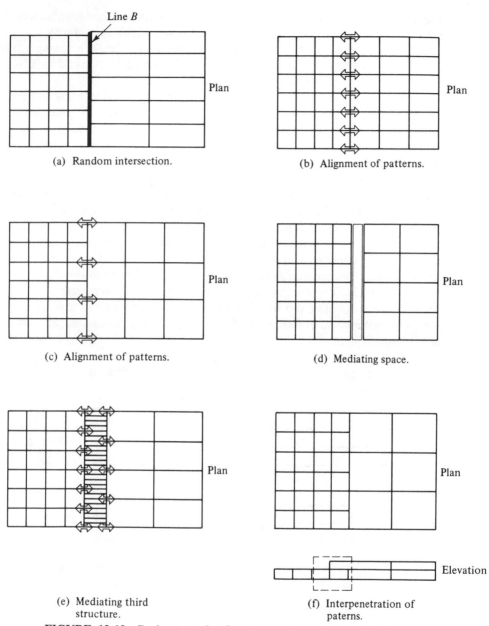

(a) Random intersection.

(b) Alignment of patterns.

(c) Alignment of patterns.

(d) Mediating space.

(e) Mediating third structure.

(f) Interpenetration of paterns.

FIGURE 13-13 Basic strategies for the meeting of generalized structural patterns.

tage of using a small-grained mediating system is that the designer has greater freedom with placing the primary grids. Primary grid lines, for example, do not have to be aligned. The building illustrated in Figure 13-6 uses a small-grained mediating structure of the type described. Load-bearing walls are also often used as third element meditors, particularly when the primary systems are not aligned, as illustrated in

Figure 13-13(a). A load-bearing wall line can be construed, however, as a continuous grid having an infinitesimal grid spacing and worked with accordingly.

Other approaches to intersecting patterns also exist; for example, an inter-penetration strategy of the type illustrated in Figure 13-13(f) could be adopted as a variant of the basic alignment approach.

Pattern intersection can also be a problem with respect to using one grid on top of another. This situation quite often arises when it is desired to place a roof system having one grid geometry on top of an internal system having a different geometry. In such cases, a strategy of alignment is often used wherein vertical support points common to both systems are used [see Figure 13-14(a)]. An alignment system is used in the building illustrated in Figure 13-4. Occasionally, a bypassing type of approach is used wherein the vertical supports for the upper grid are maintained as different from those of the lower grid [see Figure 13-14(b)]. The systems are thus independent of one another. The former alignment strategy is commonly used when the overlapping grid patterns are fairly similar to one another. The latter bypassing strategy is frequently used when the upper grid structure (e.g., the roof) is relatively long span as compared to the lower, of simple overall geometry, and has a different internal pattern than the lower structures. The overlapping structures need not necessarily be of the same material, although they can be. When the overlapping grid patterns are very similar, the same materials are often used. When they are very dissimilar, different materials are frequently used.

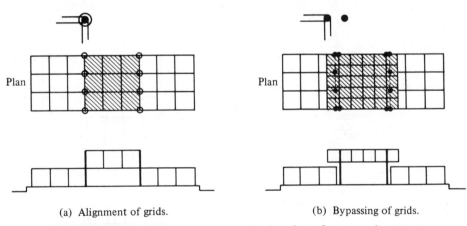

(a) Alignment of grids. (b) Bypassing of grids.

FIGURE 13-14 Basic strategies for the layering of structural patterns.

13.9 IMPLICATIONS OF BUILDING SHAPE

The overall geometry and scale of a building is quite often a major design determinate for the structural system to be used. Very tall, slender high-rise buildings, which are subject to high lateral loads, for example, commonly call for different structural approaches than do wide, low-rise buildings. Likewise, certain building geometries require special structures when unique loadings, such as caused by earthquakes, are present. Structural responses in situations of this type are covered in Chapter 14. In this section the subjects of interest are the more commonly encountered geometric situations in more typical buildings that influence the choice of structures.

An obvious case in point is a circular plan, which typically calls for either a radial or circumferential system of structural elements or a naturally spherical dome or pneumatic structure (see Figure 13-15). Triangular or other special shapes call for equally obvious responses.

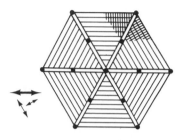

(a) One–way radial beam–and–column system for a hexagonal or circular configuration.

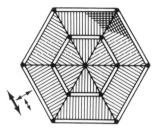

(b) One–way circumferential beam–and–column system plan for hexagonal or circular configuration.

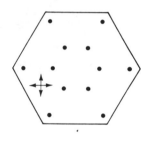

(c) Two–way flat–plate system (without beams) for a hexagonal or circular configuration.

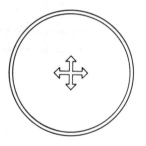

(d) Three–dimensional shell or pneumatic structure for a circular configuration.

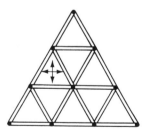

(e) Two–way slab–and–beam system for a triangular configuration.

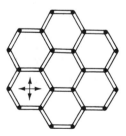

(f) Two–way slab–and–beam system for a hexagonal pattern.

FIGURE 13-15 Influence of special building geometries on structural approach. Typical structures for circular, hexagonal, and triangular geometries.

In addition to the more obvious cases, however, there are numerous conditions that more subtly affect the choice of structure. Whenever it is necessary to bend, distort, or do something else unusual to a structural grid because of an overall building geometry, there are certain to be structural implications. Even simply terminating a grid, for example, usually demands special end treatments.

Consider the plan shown in Figure 13-16(a), in which the whole building literally

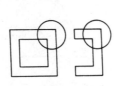

(a) Typical corner conditions.

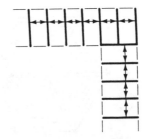

(b) One–way systems: turning corners with one–way systems is usually awkward.

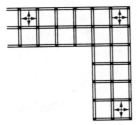

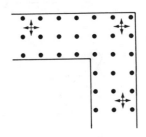

(c) Two–way systems: turning corners with two–way systems usually presents no problem.

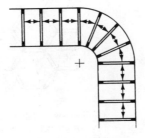

(d) Bent–axis approach.

(e) Transistion structure.

FIGURE 13-16 Influence of building morphology on structural approach. Basic strategies for turning corners.

turns a corner. Under some circumstances, structuring the corner region can pose some problems. Assume that the functional modules in the building are basically rectilinear, as illustrated in Figure 13-16(b). This is a very common pattern in many building types, such as housing, where each module would represent a dwelling unit. As is evident, this basic type of pattern lends itself well to a structural system that uses parallel beams or load-bearing walls along the depth of the building and a planar one-way system of some sort that spans horizontally between these elements. The only difficulty in using such a system is at the corner, where the one-way system cannot be oriented in two directions at once. It would thus be necessary to adopt an approach of the general type shown in Figure 13-16(b). While such an approach is obviously possible, it certainly lacks elegance. The basic difficulty, of course, is not really a structural one but one associated with the basic rectilinear pattern and the corner turning requirement. As a consequence, the structure most logical for the basic parts of the building becomes awkward at the corners.

Consider another type of functional pattern, which consists of aggregated squares [see Figure 13-16(c)]. In this case the basic pattern lends itself well to a two-way structural system supported by columns at grid intersection lines. In this case it is evident that no difficulty at all exists in structuring the corner region. The two-way system is naturally capable of turning corners because of its symmetry in both geometry and structural behavior. It would consequently seem that, whenever L-shaped conditions exist in a building, square rather than rectangular functional and structural grids should be used. A rule of this sort is somewhat simplistic, however, since square grids may not accommodate building functional requirements as well as rectangular ones (housing is a case in point). In addition, there are other structuring possibilities that may prove effective.

Several other ways a structural pattern can turn a corner are illustrated in Figure 13-16(d) and (e). Obviously, the axes of the basic grid can simply be *bent* around the corner in a radial fashion. While possible in a great many cases, building functional requirements may preclude this approach in others. If a bent axis system is adopted, one-way structural systems using prefabricated elements usually, but not always, prove difficult to use. There are construction difficulties involved in having to form a variable length spanning surface in the bent region when materials such as steel or timber are used since member lengths are not constant. A poured-in-place one-way reinforced concrete system using radial beams and simple one-way slabs, however, is quite appropriate. As will be discussed more in Chapters 14 and 15, however, span lengths are limited with this system. The two-way flat plate is another natural system for use in a bent-axis region. Again, spans are limited. In general, the bent-axis system is fairly difficult to structure when span distances between grid lines are large.

Another way of handling the corner condition is illustrated in Figure 13-16(e). In this case the primary system is simply terminated at either boundary of the corner and a *transition* structure is used. This transition structure invariably has two-way characteristics and often has a relatively small grain. One of the strategies discussed in Section 13.8 in the meeting of patterns must be used at the intersection between the primary grid structure and the transition structure. The transition structure can be of the same materials as the parent structure, or it can be a different material altogether. This approach is somewhat of an overkill if the building is generally small and the

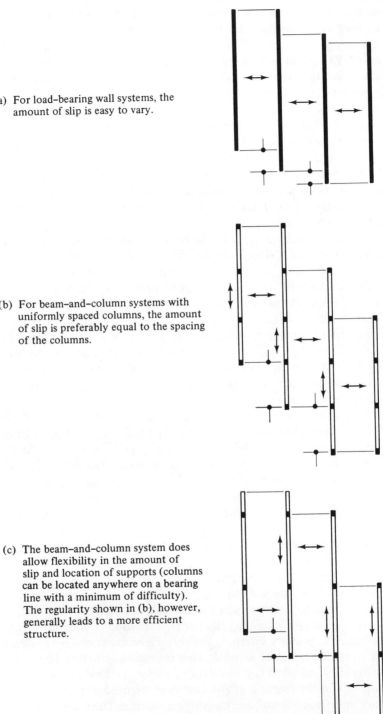

(a) For load–bearing wall systems, the amount of slip is easy to vary.

(b) For beam–and–column systems with uniformly spaced columns, the amount of slip is preferably equal to the spacing of the columns.

(c) The beam–and–column system does allow flexibility in the amount of slip and location of supports (columns can be located anywhere on a bearing line with a minimum of difficulty). The regularity shown in (b), however, generally leads to a more efficient structure.

FIGURE 13-17 Some basic strategies for slipped-unit buildings.

absolute depth of the building is relatively shallow; it is simply not worth the trouble. In large buildings that are relatively deep, the transition-structure approach works quite well.

Another commonly encountered building geometry involves units that are slipped with respect to one another. For buildings of this type, one-way structural systems usually prove more appropriate than two-way systems. One-way systems involving load-bearing walls, for example, can adapt to very small increments of slip [see Figure 13-17(a)]. If a beam-and-column system is used instead of load-bearing walls, the degree of slip most preferably corresponds to the normal repetitive distance between columns [see Figure 13-17(b)]. This same type of system does allow, however, for the fairly easy shifting of columns along the bearing line. Indeed, column spacings can be different along different bearing lines (particularly if beam spans are small), which makes this approach extremely versatile. This is not to say, of course, that such manipulations are at all desirable from a structural or constructional viewpoint. Still, such flexibility is occasionally highly important from a functional viewpoint. Two-way systems, other than the flat plate of limited span, do not really have this flexibility.

13.10 NONUNIFORM GRIDS

In situations where the pattern of the vertical support system is irregular, some structural systems are more attractive than others. Totally irregular grids mitigate against the effective or economic use of systems involving a number of highly repetitive and typically off-site-produced, modular units (e.g., precast concrete elements). For full advantage to be gained from the construction process possibilities inherent in such systems, they must be used in an almost brutally repetitive way. Poured-in-place concrete systems are usually preferable in cases involving pattern irregularities, since anomalies can be more easily handled (Figure 13-18). In a like vein, wood joist systems made on the site are preferable to prefabricated stressed skin panels unless the latter can be used in a repetitive way over a regular vertical support system.

Irregularity in the vertical support grid can most easily be accommodated when support points are relatively closely spaced. As horizontal spans become greater, so do the structural requirements for horizontal members. For maximum spans of around 16 to 20 ft (5 to 6 m), the reinforced concrete flat plate system, discussed in more detail in Chapter 15, can accommodate many irregularities in support placement. For

FIGURE 13-18 For irregular grids, structural systems made on-site (e.g., poured-in-place concrete) are usually preferable to those utilizing relatively large off-site-produced elements.

(a) Separation strategy: simply separating the large volume from the more general system allows efficient structural responses for each system to be developed independently.

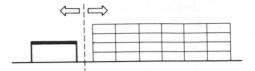

(b) Embedment strategy — low placement of volume: placing the large volume at low levels requires the use of either large transfer members or long span members (where not required by functional considerations). Either alternative is usually structurally possible but economically dubious.

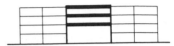

(c) Embedment strategy — roof level placement of volume: placing the large space at roof level does not require the use of either large load–transfer members or a potentially uneconomical loose fit as dictated by a low placement of the volume.

FIGURE 13-19 Strategies for accommodating large volumes in a small grained system.

longer spans, where this system is not appropriate, the rigid orthogonality associated with longer-span systems dictates a more orderly vertical support arrangement.

13.11 ACCOMMODATING LARGE SPACES

In most buildings the bulk of the functional requirements are handled through using generalized grids. In some situations, however, programmatic requirements exist that give rise to one or more special spaces that are very different in scale than other cases. An example is a school where there might be a large number of similar-size small spaces (classrooms) and a few very large spaces (gymnasiums, auditoriums, or cafeterias). Handling both types of spaces can present problems.

Figure 13-19 illustrates several basic ways of including a large volume in a more extensive cellular grid of smaller grain. A time-honored solution is to simply *separate* the two systems completely and connect them only along edges or with simple linkages of some sort. This allows the most appropriate structure to be chosen for each system independently of the other. A long-span, one-way, steel truss system could be selected for the large space, for example, and something else chosen for the smaller-grained system. One of the strategies discussed in Section 13.8 on the meeting of different patterns would have to be used at the interface between the systems. The auditorium complex in the building shown in Figure 13-6 is an example of how a large volume can be simply separated out and given its own structural system.

A completely opposite approach is also often used wherein the large space is *embedded* into the finer-grained system. Two situations are of interest here—one where the largest space is placed such that much of the finer-grained system is above the space, and the second reverse case where the large space is located completely above the finer-grained system.

When the large space is embedded at a low level, a number of structural problems arise. There are two basic structural options available [see Figure 13-19(b)]. One is to use a one-on-one structural fit to the smaller-grained system and to use large transfer members to pick up the loads from the closely spaced vertical supports above the open span space and carry them to the vertical supports at the edges of the large space. Such a strategy is used in the auditorium area in the building shown in Figure 13-4. The second basic option is to adopt the span of the larger space as the repetitively used dimension and use a series of large-span members everywhere, even in the smaller-grained functional areas. A loose-fit approach would thus be used.

Both approaches are possible, but both have their problems. In the first case where transfer members are used, it is evident that these members must of necessity carry very heavy loads and thus be large members that have to be uniquely designed and constructed to carry the loads involved (i.e., off-the-shelf members will usually not work). These members are obviously problematic. The remainder of the structure in the fine-grained area can work quite well. The alternative of using long-span elements everywhere is that loose fits of the type present in the fine-grained area are inherently noneconomic. Spanning a large distance when a small one could be spanned instead usually does not lead to economic structures. Using special systems (e.g., precast and prestressed reinforced concrete tee shapes), designed specially for long spans, may tend to offset this inherent disadvantage somewhat, but it remains an influential factor.

Note that most of the difficulties mentioned above can be circumvented by placing the large span space on top of the smaller grid system. Then the long-span elements carry roof loads only, which are relatively light, and any of a variety of systems could work. The design of the finer-grained system is not appreciably affected by the presence of the large space above it. One of the strategies for the overlapping of generalized patterns discussed in Section 13.8, however, must be adopted. A difficulty with this approach is not a structural one but a functional one, in that large spaces are usually more appropriately placed lower in a building because of access reasons—an important factor in view of the large number of people often using such a space.

13.12 ACCOMMODATING SPECIAL CONDITIONS

There invariably arise a host of special conditions in a building to which the structure must respond. These conditions usually represent particularized events related to building functions such as circulation (e.g., stairwells) or to building service systems (e.g., heating, ventilating, and air conditioning).

There are several basic strategies for accommodating building service system elements that run horizontally. One is to use a one-way spanning system that naturally provides a space for parallel runs (see Figure 13-20). Except with trusses, service

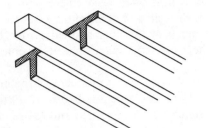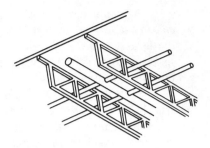

(a) One–way spanning systems easily accommodate service elements that run parallel to main members. Vertical service risers are often buried in walls or chases.

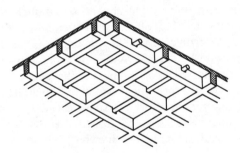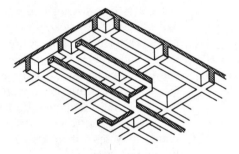

(b) Penetration approach: for minor service elements the basic structural fabric often can simply be penetrated.

(c) Passing of systems approach: when the horizontal service system is complex or contains large elements, an approach wherein structural and service elements are placed on nonconflicting planes is often very effective.

FIGURE 13-20 Accommodating horizontal service elements.

elements that run perpendicular to the spanning direction are obviously more difficult. It is possible to locally deform or penetrate the basic structural fabric to accommodate such instances. This is most often done when the elements to be accommodated are few in number and/or small in dimension. Most structural systems can be designed to accommodate minor penetrations in horizontal members.

The penetration strategy can also be used to accommodate horizontal elements in two-way structural systems [see Figure 13-20(b)].

Some structural systems, such as trusses, can easily accommodate horizontal service elements, since they can simply pass through web members. Excepting such systems, it is generally true that whenever penetrations become numerous or large, the basic structural fabric can be compromised to the extent that the structure is more correctly characterized as a series of ad hoc solutions than as a system. When this occurs, it is usually preferable to switch to a different strategy for accommodating the special conditions.

Another strategy for accommodating horizontal service elements in either one or two-way systems is simply to pass them beneath the primary structural system (leaving it intact) and either leave them exposed or enclose them with a dropped chase or ceiling. This approach is highly efficient from a construction viewpoint. Total building heights, however, are disadvantageously increased.

Most structural systems can be designed to accommodate minor penetrations for vertical service and functional elements without difficulty (Figure 13-21). More major penetrations, such as for stairways, must often be accompanied by special local framing. Still the basic structural pattern is not affected to any great degree.

One way of handling situations where numerous vertical penetrations would ordinarily be required is to group the special functions into clusters and then to design the structure to accommodate this cluster. This is often done for vertical elements, in which case the elements are grouped into core units that fit into individual structural bays. The structure is simply eliminated or specially treated in this bay. The advantage

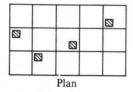

Plan

(a) Vertical service elements — penetration approach: for randomly distributed minor vertical elements the basic structural fabric can often be simply penetrated.

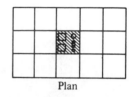

Plan

(b) Vertical service elements and special features — clustered passing approach: when vertical penetrating elements are large or complex, an approach wherein such elements are deliberately clustered into groups which in turn bypass the basic structure is often utilized.

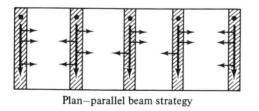

Plan—parallel beam strategy

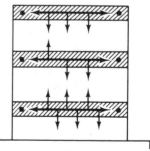

Section—interstitial floor strategy

(c) Horizontal and vertical elements — provision of regular interstitial zones: it is possible to provide regularly spaced zones for vertical and/or horizontal service elements. Doing so is most appropriate in cases where the service elements must be distributed throughout the building and are relatively large or complex.

FIGURE 13-21 Strategies for accommodating building service systems and special features.

of doing this is that the general structural fabric in other areas is undisturbed and can be designed to function efficiently in its own terms. This is a common approach in many high-rise structures.

Quite often the cluster approach does not work well with the functional requirements of the service elements. In many instances these elements have to be distributed throughout the building as in the case of some types of heating, ventilating, and air-conditioning systems, or laboratory piping systems. Under some conditions a struc-

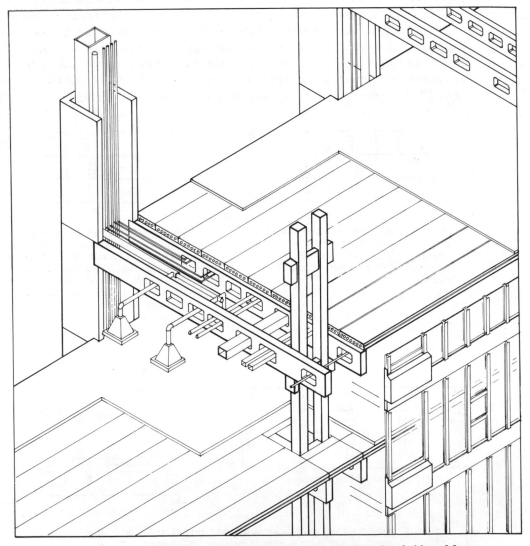

FIGURE 13-22 Science Center, Harvard University, Cambridge, Massachusetts. The building is made with precast concrete elements. The part of the building housing laboratories uses a double beam and column system which facilitates the distribution of mechanical services by providing regularly spaced interstitial zones between beams [see also Figures 13-6 and 13-21(c)].

tural system that provides spaces for locating these elements at fixed intervals may prove effective. Such a system is shown in Figure 13-21(c). The interstitial spaces between the parallel beams and double columns is reserved for horizontal and vertical mechanical runs. The science center illustrated in Figures 13-6 and 13-22 has such interstitial spaces for mechanical service elements in the back part of the building where the double beam-and-column system is used. Mechanically intensive physics and chemistry laboratories are located in this part of the building. The front part of the building houses only libraries and offices, functions that are much less mechanically intensive. A single beam-and-column system without significant provisions made for mechanical equipment is accordingly used.

A conceptual difficulty involved in using structural systems that provide fixed intervals for the provision of mechanical services is that there is nothing that a priori suggests that the rhythm of the mechanical system, in terms of where it is located to best provide its services, is at all the same as that of the structural system. Forcing the mechanical system into a rigid pattern related to the structural system may not prove most efficient from the viewpoint of the overall building design, since the design of *both* the structural and mechanical systems might well be compromised. Whether this is true or not, however, depends on the specific building program involved as well as many other design variables.

Interstitial zones that provide service *layers* between floors can also be introduced if the intensity of service or flexibility demands so warrant. Service elements can be dropped from interstitial zones into floors below or rise to floors above. Changes

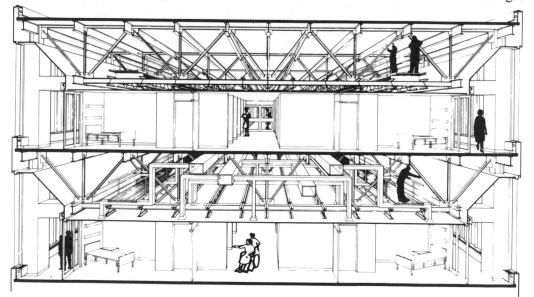

FIGURE 13-23 Woodhull Medical and Mental Health Center, Brooklyn, N.Y. Walk-through interstitial levels are provided to house mechanical elements servicing floors. Mechanical elements can be modified and maintained without disruption to hospital activities. Deep one-way trusses are used to form the interstitial layers [see also Figure 13-21(c)]. (Drawing Courtesy of Kallmann and McKinnell, Russo and Sonder, Associated Architects.)

chapter 13 / STRUCTURAL GRIDS AND PATTERNS

and modifications can be made with minimum disruption to functional activities. A structural solution that takes advantage of the full depth of the interstitial zone is often, but not necessarily, employed (see Figure 13-23).

In general, highly integrated and expensive approaches of the type described are justified only in cases when the mechanical system in a building is extremely complex and extensive, as might be the case in a hospital, where they can work well. In buildings of low mechanical intensity, such as churches, such integrated approaches are usually not economically justifiable, since distributed mechanical elements are few and small. The expensive interstitial spaces would thus simply not be utilitized to the extent needed to justify their existence.

13.13 VARYING SUPPORT LOCATIONS TO REDUCE BENDING MOMENTS

It was pointed out in earlier chapters on beams and frames that manipulating support locations can potentially lead to improved structural efficiency by reducing the bending moments present. Some of these manipulations have strong implications on the structural patterning used. In many cases, once the structural pattern is fixed, such manipulations are impossible. A classic way of reducing design bending moments in

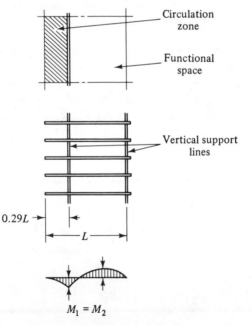

FIGURE 13-24 Optimizing support locations. In many cases a fit between the vertical support pattern and building use pattern can be achieved that works well from both structural and building use viewpoints.

beams or trusses, for example, is that of using end cantilevers (see Sections 4.4.10 and 6.4.6). In the overall process of selecting a structural pattern in a building it might be quite possible to introduce cantilevers into the building. The rules of thumb of $\frac{1}{4}L$ and $\frac{1}{3}L$ discussed in Chapter 6 that define optimum cantilevers for homogeneous members, for example, often lead to absolute cantilever dimensions that serve quite well as overhangs covering circulation routes. The pattern implications of using cantilevers are, of course, significant (Figure 13-24). Other examples with continuous beams and frames could also be noted.

13.14 SPACE-FORMING CHARACTERISTICS

A subjective consideration of unquestioned importance from a larger architectural viewpoint is that each of the systems illustrated have different spatial implications. The spaces formed by the bearing-wall system [Figure 13-25(a)] are one-directional (or linear) and have a strong planar quality which is imparted by the vertical and horizontal enclosure elements.

The spaces formed by the beam-and-column system [Figure 13-25(b)] are also primarily one-directional, owing to the organization of the collector beams, but have in addition a secondary axis in the transverse direction. These characteristics would be significantly influenced by the shape and orientation of the columns. Rectangular columns with a long axis in the direction of the beams would further emphasize the linearity of the space, while placing the long axis in the other direction would emphasize the secondary axis more. Round columns are a neutral statement.

The spaces formed by the flat-plate system [Figure 13-25(c)] are two-directional, since neither of the axes is dominant. The spaces formed are relatively neutral. Round columns or square columns do not alter this two-way directionality. Rectangular columns would start causing one or the other axis to become more dominant. The spaces formed by the two-way beam-and-slab system [Figure 13-25(d)] are less neutral, with the beams even further emphasizing the two-way directionality of the spaces.

The space formed underneath a very large, completely neutral surface (such as a flat plane) is not directional. This assumes that any directionality that the support system might impart does not influence the observer. The space frame illustrated in Figure 13-25(e) would provide this type of space (the structural member spacing is sufficiently small as to cause the lower surface to appear as a neutral plane).

While other structural arrangements may yield slightly different spatial properties, they are usually some variant or combination of the general types discussed above.

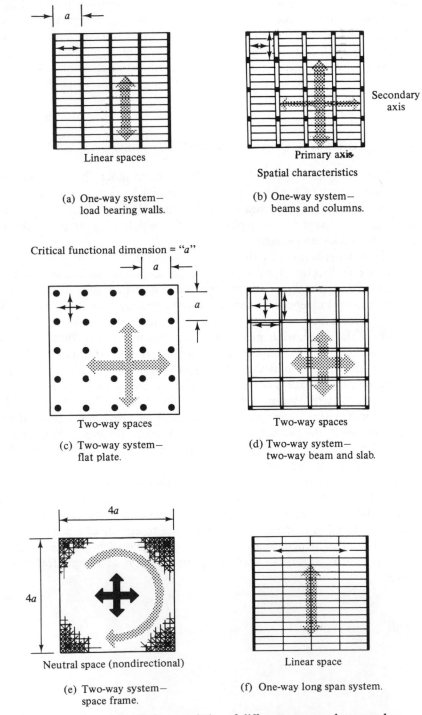

Critical functional dimension = *a*

a

Linear spaces

(a) One-way system—
load bearing walls.

Spatial characteristics

Secondary
axis

Primary axis

(b) One-way system—
beams and columns.

Critical functional dimension = "*a*"

a

a

Two-way spaces

(c) Two-way system—
flat plate.

Two-way spaces

(d) Two-way system—
two-way beam and slab.

4*a*

4*a*

Neutral space (nondirectional)

(e) Two-way system—
space frame.

Linear space

(f) One-way long span system.

FIGURE 13-25 Spatial characteristics of different structural approaches.

QUESTIONS

13-1. Visit four different large multilevel buildings in your local area. Draw a sketch to scale of the vertical support pattern present in each building (ground level). Diagrammatically indicate whether the vertical support elements are columns or load-bearing walls. Identify whether the horizontal spanning elements are one-way or two-way systems.

13-2. For the same buildings used in Question 13-1, identify how the general vertical support pattern is affected by or locally responds to the presence of other vertical building service elements, if at all. Are mechanical elements clustered into zones, for example, or do they simply penetrate the structural fabric? How are vertical circulation zones handled? Illustrate your answers with diagrams.

13-3. Speculate on the reasons why square and rectangular vertical support patterns are more typically encountered in common buildings than are other possible patterns (e.g., hexagonal). Group your responses into two categories:
(a) Reasons that stem from construction or structural considerations.
(b) Reasons that stem from programmatic or architectural considerations.

13-4. Identify two buildings in your local area or from the literature which basically have fine-grained vertical support patterns but include one or more larger volume spaces as well. Identify what structural strategy is used to include these large spaces (e.g., are they embedded or separated), and what the are structural implications involved? Illustrate your answers with diagrams.

chapter 14

STRUCTURAL
DESIGN ISSUES

14.1 INTRODUCTION

Usually, it is possible to generate one or more general structural strategies by simultaneously manipulating functional and structural patterns in the ways described in
Chapter 13. Since several strategies are typically possible (e.g., different degrees of fit),
it is necessary to further develop the different strategies by giving them physical real-
ities so that they can be rationally compared. Typically, the most promising strategy is
first developed in detail, next it is analyzed to measure how closely design objectives
(e.g., objectives linked to the architectural design of the building, efficiency, serviceability, construction ease, costs, etc.) are met, and then it is either modified or rejected,
depending on the results of the analysis. This process is often repeated several times,
with the intent of continuously developing an alternative so that it better and better
meets design objectives. Normally, this procedure is then repeated for several other
general strategies so that comparisons can be made.

The general procedure described above is obviously not an easy or a simple
process. Neither does it assure that of the many structural solutions possible for a
given situation the one which, in an absolute sense, is actually most preferable is
among the postulated design alternatives selected for study. One of the fundamental
tasks of the designer is to explore as many alternatives as feasible with the intent of
assuring that potentially preferable solutions are not simply overlooked. Similarly, it

is the task of the designer to weed out as soon as possible many alternatives that in the long run would not prove feasible, so that energies can be more fruitfully utilized in exploring potentially appropriate solutions.

A necessary prelude to the development of specific design alternatives for a given purpose is the often difficult task of defining exactly the scope, limits, and role of the structure to be designed. This is analogous to the step in the analysis process where the extent of the structure considered for analysis must be defined and then modeled. In a design context, however, this step is more complex, since many decisions about the building that may affect the structure are usually still open. Adoption of specific solutions for some elements such as walls, for example, may influence the design of the structure. An attitude that walls are to serve only as enclosure elements and serve no structural function (even as a lateral stability device, for example) leads to a very different set of potential structural solutions than does an attitude that walls can serve dual functions. Which strategy is appropriate depends largely on the building program. The latter strategy might be quite reasonable for housing, but unreasonable for office buildings, while the former might be quite reasonable for office buildings and doubtful for housing. Only a detailed consideration of building program requirements will tell.

Even once basic attitudes about the role of various buildings have been established, there are still an enormous number of factors that must be considered in selecting a structural system for detailed development. Some are readily apparent. Span length must be a crucial determinate, since not all systems operate efficiently over the same span ranges. The influence of span on the choice of structure will be explored in a following section. Span length, however, is not the only or necessarily the primary determinant of structural response, as is often assumed. The nature and magnitude of the forces a structure must carry is of corresponding importance. Indeed, in a design context the question of how a yet to-be designed structure will resist the lateral forces that it necessarily must carry in addition to vertical loads is probably *the* crucial one. This issue will be explored in Section 14.3.

Once a specific structural response has been tentatively selected for a given situation, the next step is a twofold one: exploring its pattern implications in detail (along the lines suggested in Chapter 13, but in a less abstract way) and developing it in further detail so that all the implications of its use will be understood. The latter step will be discussed more in Chapter 15.

14.2 VERTICAL LOADS: EFFECTS ON THE DESIGN OF STRUCTURES

14.2.1 Types of Loads

Whether a loading is distributed or concentrated is a crucial factor in selecting an appropriate structural system for a given condition. There are a number of structural systems available that are particularly suited for light, uniformly distributed floor or roof loads. These structures are typically surface-forming in nature and are usually made up of a series of closely spaced similar elements or involve mass-produced components. Included are wood joist systems, steel deck systems, long-span precast concrete planks, precast channels, open-web steel joists, and reinforced concrete flat plates.

When uniformly distributed floor loads are present, every effort is usually made to make certain that the surface elements that first pick up the loads operate at their maximum efficiency. Therefore, the spacing of more specially designed members supporting these surface elements is often based on the most efficient span length for the supported surface elements rather than on some other criterion. The sizes and spacings of wide-flange members supporting a steel floor deck, for example, are often based on optimizing the decking used. The design process thus usually begins with the smallest and lightest (but most numerous) members first taking the load and proceeds to supporting beams, and then to the columns, instead of vice versa.

Surface-forming elements are generally inappropriate when large concentrated loads are present. In most cases, members carrying large concentrated loads must be uniquely designed for the particular situation at hand. They are most often linear one-way spanning elements. Included are steel or timber trusses having unique bar configurations, plate girders, and post-tensioned reinforced concrete beams. Most of these structures are typically handcrafted. More attention is usually paid to designing the structure in response to the force distributions within it.

14.2.2 Span Lengths

BASIC PRINCIPLES. Span length is unquestionably a crucial determinant in selecting a structural response for a given situation. Some structural systems are quite appropriate for certain span ranges and not for others.

The importance of the structural span is evident from noting that design moments for uniformly distributed loads are proportional to the square of the length of the span. Doubling a span length, for example, increases design moments by a factor of 4; quadrupling span lengths increases design moments by a factor of 16. Member sizes, of course, depend closely on the magnitude of the design moment present. The appropriateness of a particular structural option is dependent on this factor as well. For this reason it is useful to review briefly how different structural options provide an internal resisting moment to balance the external applied moment.

Figure 14-1 illustrates how several different structural options provide an internal resisting moment to balance the external applied moment. Note that in all cases the basic mechanism is the same. A couple is formed between compression and tension zones whose magnitude exactly equals the applied moment. For a given applied moment the magnitude of the internal forces or stresses developed in the compression and tension zones depends directly on the magnitude of the moment arm present. The deeper the structure, the greater the moment arm and the less are the tension and compression forces or stresses present.

The process of designing an appropriate structure for a given span range is directly linked to applying the principles discussed above. The sensitivity of the design moment to the span is critical. In low spans the design moments are low, and any of the basic structural options shown in Figure 14-1 are possible. As spans lengthen, however, design moments increase so rapidly that some of the options shown become

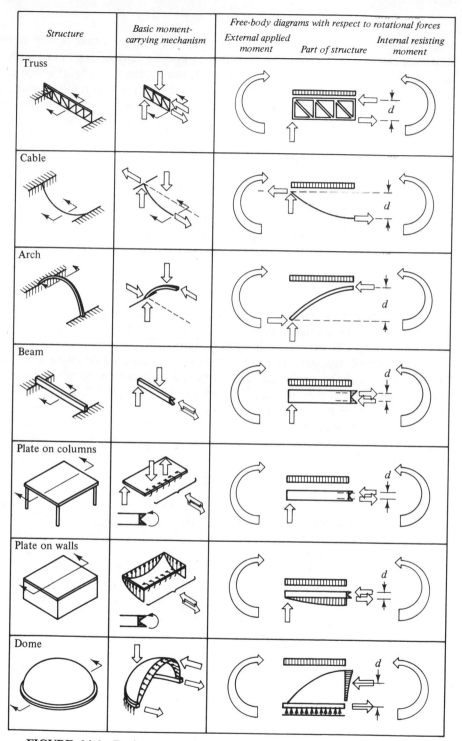

FIGURE 14-1 Basic moment-carrying mechanisms in different types of structures. In all cases the structure provides an internal force couple which generates an internal resisting moment which equilibrates the applied external moment at the same section. As the depth of the internal moment arm increases, the associated internal forces decrease, and vice versa.

less feasible. Constant-depth members such as beams, for example, are normally relatively shallow, and consequently increases in span lengths are linked with large increases in the magnitudes of the tensile and compressive stress fields which provide the resisting couple. Since the depths of such members is inherently limited, these increases cannot entirely be controlled or compensated for by increasing the moment arm of the resisting couple or by other means (e.g., increasing flange areas). Consequently, members of this type are not capable of extremely large spans, since a point is reached where the internal compression and tension forces become too large to be handled efficiently. Deflection control also becomes a governing consideration. Obviously, if structural depths were *always* increased in response to the increased design moments associated with longer spans, internal forces could be kept at a reasonable level. This is essentially what the shaped truss, cable, or arch does. These structures are usually relatively deep and consequently inherently provide a very large internal moment arm. The forces in the resisting couple can thus be fairly small with the structure still capable of providing a very large internal resisting moment. Large spans are thus possible.

TYPICAL RELATIONS BETWEEN SPAN AND STRUCTURAL TYPE.   Figure 14-2 illustrates the approximate span ranges for a variety or common horizontal spanning systems suitable for either floors or roofs. The *maximum* spans indicated do not represent maximum *possible* spans (since most of the systems indicated could be made to span farther than indicated) but represent spans that are not to be considered unduly unusual, even though they are longer than typically encountered. The minimum limitations are intended to represent a system's lower feasibility range based on construction or economic considerations. There is obviously nothing sacred or fixed about any of the precise span figures shown. They are intended to give a feeling for relationships between structural systems and spans and nothing more. More detailed span-range charts for specific materials are contained in Chapter 15.

LONG-SPAN STRUCTURES.   In many ways the selection of an appropriate structure for use in a given context becomes easier if the structural requirements, in terms of span, are very large. As noted above, a characteristic of appropriate long-span systems is that their structural depths are relatively large in comparison with their spans. A linked characteristic is that the structures are usually shaped in some way. Consequently, appropriate structural systems for long-span applications typically include shaped trusses, arches, cables, nets, pneumatics, and shells. Exceptions, to be sure, exist. There are many successful examples of long-span constant-depth trusses, space frames, and other structures. Rarely do these flat systems, however, match the efficiency of the simply conceived cable, arch, or shaped truss. Interestingly enough, the internal structural actions of these shaped structures are less complex than other elements, such as beams, which do not have long-span capabilities.

As a consequence of the geometry of common shaped structures, such structures are primarily useful only for roofs in buildings or other situations where structural shapes can be allowed to vary. Rarely are such systems useful for providing the planar surfaces that are needed elsewhere in buildings for obvious functional reasons.

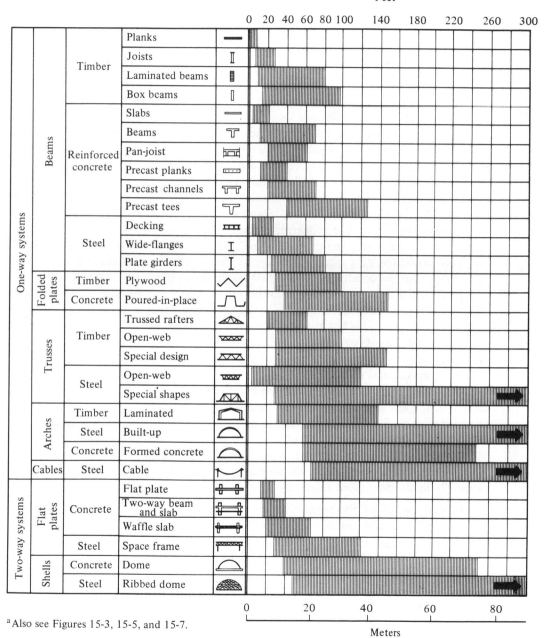

Span range[a]
Feet

FIGURE 14-2 Approximate span ranges of different systems.

[a] Also see Figures 15-3, 15-5, and 15-7.

477

INTERMEDIATE- AND LOW-SPAN SYSTEMS. As span or load-carrying requirements decrease in a building from the dramatic to the more commonplace, and if the structure must provide a rigid planar surface, the choice of structure becomes more difficult since more systems are possible in low- or intermediate-span ranges. There are a whole host of different systems made of different materials that are potentially applicable.

In the span ranges normally encountered in most buildings, it is evident from inspecting Figure 14-2 that a number of systems are potentially appropriate for use. For spans on the order of 16 to 20 ft (5 to 6 m), for example, either one- or two-way systems of all major types are potentially competitive. Which of the systems is most appropriate cannot easily be determined without a separate analysis consisting of designing alternatives for the specific span and load condition considered and making a cost analysis. Usually, the lightest or least involved construction type appropriate for a given span which is capable of carrying the design load is the most preferable. For example, in reinforced concrete, for a span of 20 ft (6 m) and for relatively light, uniformly distributed loads, the flat plate would probably prove the most preferable of all two-way concrete systems. The other reinforced concrete systems (e.g., waffle slabs) are possible but are more substantial and have higher span and load capacities by virtue of the way they are constructed than is needed. From a structural viewpoint there is no incentive to go to the trouble and expense of creating the special formwork required to construct these more complex systems when a simple flat plate would work just as well for the conditions at hand. With increases in span or load, however, the flat plate begins losing its viability and these other systems become more appropriate.

In low-span ranges, the deeper-shaped structures capable of very long spans are still structurally possible but are often not used because of construction considerations (reflected as high costs). A cable system could be used to span 15 ft (5 m), for example, but using such a system when a series of simple wooden joists would work just as well does not make good sense.

The versatility of the truss in terms of both its shape and the way it is made make it one of the few structural elements useful for the full range of short, intermediate, and long spans. Specific characteristics and geometries of trusses for these extreme span ranges, of course, vary. Short- and intermediate-span trusses are typically used closely spaced and in highly repetitive ways—as is commonly done with typical bar joist systems. Often such elements are mass produced. Long-span trusses are usually specially made and placed relatively far apart.

ONE- AND TWO-WAY SYSTEMS. In low-span ranges, [e.g., 15 to 35 ft (4.5 to 10.6 m)] a whole range of both one- and two-way systems is potentially appropriate. As spans begin to increase beyond a certain point around 50 to 60 ft (15 to 18 m), however, flat two-way concrete systems typically begin to lose some of their construction viability (although still possible) and one-way spanning elements are frequently found to be more appropriate. For roof systems carrying light distributed loads, closely spaced open-web steel joists are often highly efficient in this span range. Precast and prestressed reinforced concrete members are also extensively used.

As spans increase to 80 to 100 ft (24 to 30 m), systems are often predominantly made of one-way elements. Steel is frequently used, although some prestressed con-

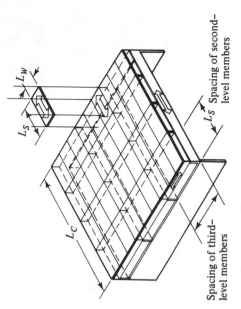

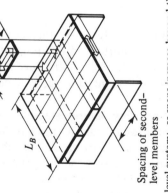

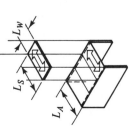

(a) One-level system: for short spans, the surface elements can be used directly.

(b) Two-level system: as spans increase beyond the reach of the short-span surface elements, a second level of supporting one-way elements can be used. These elements are usually placed at the optimum economic span for the supported surface elements, and are then specially designed to carry the loads involved. Otherwise off-the-shelf second-level members are used and are so spaced as to be loaded to their most efficient capacity. Surface elements most appropriate for the selected spacing are then chosen.

(c) Three-level system: as spans continue to increase, it may be desirable to go to a three-level system. In this case the surface elements are supported at their maximum economic spans by second-level elements. These elements are in turn supported by fewer and larger third-order elements. The spacing of the third-order elements is often chosen on the basis of what spans are most economic for the second-level elements. Trade-offs in the design of the second and third-level elements, however, often occur.

FIGURE 14-3 Planar structures assembled from repetitively used short-span one-way rigid surface elements (e.g., decking, planks, precast plates) and one-way rigid line elements (e.g., beams or trusses). Special long-span systems could, of course, be used which would not necessitate the multilevel heirarchial approaches indicated above. Which general strategy is most preferable can only be determined by detailed analysis.

crete members can very efficiently still span this far. Of the flat two-way systems, the steel space frame has this span capability. It is typically most appropriate for lighter roof loads rather than floor loads. Specially designed two-way systems, such as a grid approach could, of course, be used. Such systems are, however, usually quite expensive.

HIERARCHICAL SYSTEMS. A consequence of increasing spans in those types of one-way structural systems that are hierarchical in nature and use constant-depth members is that there is often an increase in the number of levels used with increasing spans (see Figure 14-3). Short spans often use one-level arrangements while larger spans use two- or three-level arrangements. Higher levels are possible but rare. This principle is, of course, not universally applicable since there are a number of special elements (such as prestressed concrete tees) that can very efficiently span large distances directly. The principle is more applicable to systems that initially use minor surface-forming elements (e.g., planks or steel decking) to initially pick up a uniformly distributed load. As a consequence of meeting the principle of optimally supporting elements of this type (see Section 14.2.1) a multilayer system often results (see Figure 14-3). Layered systems can be demonstrated to be quite efficient minimum volume solutions and are extensively used.

14.3 LATERAL FORCES: EFFECTS ON THE DESIGN OF STRUCTURES

14.3.1 Basic Design Issues

In designing structural systems, the way lateral stability is achieved is an issue of fundamental importance. The issue is important in buildings of any height but absolutely crucial in high-rise construction. The way a structure resists lateral forces not only influences the design of vertical elements but, as will be seen, the horizontal spanning elements as well.

The basic mechanisms for assuring lateral stability (shear walls, diagonal bracing, and frame action obtained through rigid joints) have already been discussed to some extent. All these mechanisms serve to create stiff shear planes capable of carrying lateral loads. In low- or medium-rise structures, the adoption of any one of these approaches will provide sufficient lateral resistance. It is obvious, however, that a sufficient number of shear planes must be used to ensure lateral stability. In some building types doing this is very easy. In housing, for example, the cellular nature of the building type lends itself well to using shear walls or diagonal bracing in the interface area between adjacent units. A frame could also be used in these locations but, as noted in Chapter 9, frames tend to be less efficient than shear walls or diagonal bracing as a lateral load-carrying device. In housing the use of shear walls or diagonal bracing in the interfaces between units poses no functional problem and hence are preferably used. In many other building types, however, the barriers formed by shear walls or diagonal bracing systems create functional problems and cannot be freely used. Frames or a combination of frames and shear walls or diagonal bracing are consequently used. Intermediate- and low-rise office buildings, for example, quite often employ a basic frame throughout the building which is further stiffened by the placement of shear walls or diagonal bracing at building ends or around service cores. Usually, placing shear walls or diagonal bracing in these locations poses no functional problems.

In rectangular buildings the greatest problem with lateral forces is in the short direction of the building, although stability must be assured in both directions (see Figure 14-4). Sometimes one type of lateral load-carrying mechanism is used in one direction and another in the other direction (primarily for functional reasons). In simple steel buildings, for example, the more efficient mechanisms (e.g., shear walls or diagonal bracing, are often used in the shorter direction). Stability in the long direction is achieved either by similar means or by frame action (see Figure 14-5). In a steel building this approach dictates how specific structural elements are organized. In relatively low buildings, wide-flange elements are usually organized such that about their strong axis they function as part of the frame in the longitudinal direction, and about the weak axes they function as part of the diagonal bracing system in the short direction. Since the diagonal bracing acts in trusslike fashion, bending is minimum in a member in the short direction, so it is okay to have the weak axis oriented in this way. In the long direction, however, lateral loads are carried by frame action involving high bending in the members. Consequently, organizing the strong axis of the member in this direction makes sense.

When the narrow dimension of a steel building is very small relative to the building height (indicating a severe problem with lateral loads in the short direction) and the building is adequately braced in the long direction, a combination frame and diagonal bracing system can be used to carry loads in the short direction. In this event, wide-flange members are organized such that their strong-axes contribute to frame action in the short direction. The frame and diagonals supplement each other yielding a total system of increased load-carrying capacity.

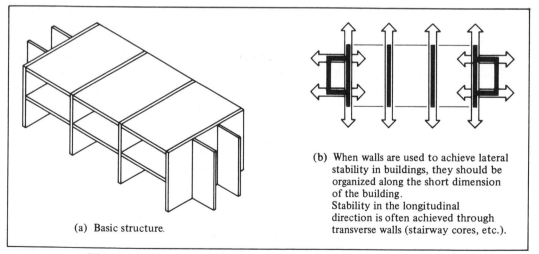

(a) Basic structure.

(b) When walls are used to achieve lateral stability in buildings, they should be organized along the short dimension of the building.
Stability in the longitudinal direction is often achieved through transverse walls (stairway cores, etc.).

FIGURE 14-4 Organization of shear planes and/or bearing walls in low- to medium-rise structure.

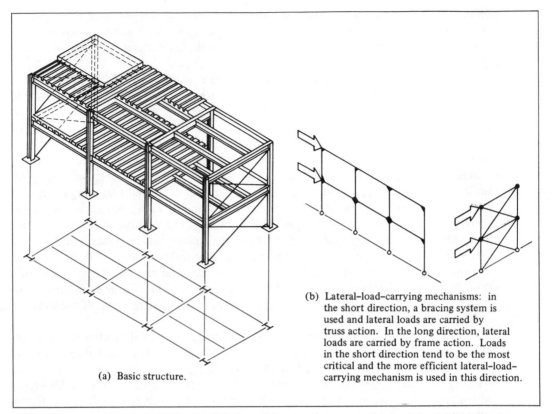

(b) Lateral–load–carrying mechanisms: in
    the short direction, a bracing system is
    used and lateral loads are carried by
    truss action. In the long direction, lateral
    loads are carried by frame action. Loads
    in the short direction tend to be the most
    critical and the more efficient lateral–load–
    carrying mechanism is used in this direction.

(a) Basic structure.

FIGURE 14-5 Possible lateral-load resistance mechanisms in a simple steel structure.

The same principles described above (placement of strong and weak axes) also apply to columns in reinforced concrete buildings. Placing stiffer elements (e.g., shear walls in shorter building directions is also a generally applicable principle.

If shear planes are to be effectively used in conjunction with other vertical planes not having any significant ability to carry lateral loads, such as a pin-connected steel beam and column system, floor planes must be designed to serve as rigid horizontal *diaphragms* which act as thin horizontal beam elements spanning between shear planes (Figure 14-6). This diaphragm action serves to transfer lateral loads from interior non-load-carrying vertical planes to the load-carrying shear planes. When reinforced concrete systems are used, making floors serve as rigid diaphragms is typically no problem. With steel special care should be taken to assure that diaphragm action does indeed occur. Depending on the spacing of the shear planes, it may be necessary to impart some lateral load-carrying capacity to interior vertical planes as well.

Anytime a structure is specially stiffened with shear walls or diagonal bracing systems, care should be taken to place these elements symmetrically. If this is not done, highly undesirable *torsional* effects can develop (see Figure 14-7) due to the fact that the center of rigidity of the building becomes noncoincident with the centroid of the applied lateral load. Placing elements symmetrically is particularly important in

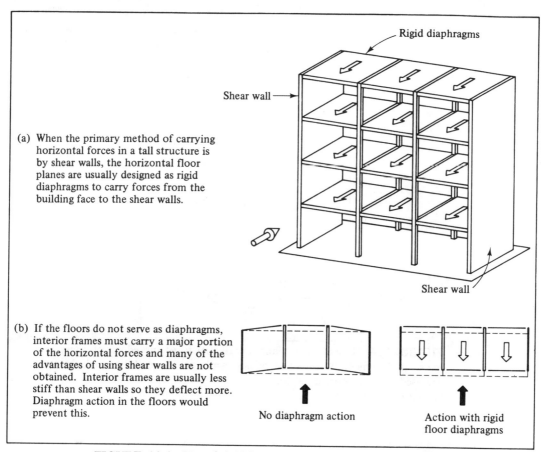

(a) When the primary method of carrying horizontal forces in a tall structure is by shear walls, the horizontal floor planes are usually designed as rigid diaphragms to carry forces from the building face to the shear walls.

Rigid diaphragms

Shear wall

Shear wall

(b) If the floors do not serve as diaphragms, interior frames must carry a major portion of the horizontal forces and many of the advantages of using shear walls are not obtained. Interior frames are usually less stiff than shear walls so they deflect more. Diaphragm action in the floors would prevent this.

No diaphragm action

Action with rigid floor diaphragms

FIGURE 14-6 Use of rigid floor diaphragms in framed buildings relying on shear walls to carry horizontal forces.

very tall buildings or when earthquake hazards are present. In either case, huge lateral loads capable of inducing high torsional forces are possible. Structures can be designed to resist these torsional effects, but a very high cost premium is paid, to the extent that rarely is nonsymmetrical placement of stiff elements justified.

14.3.2 Influence on Member and Connection Design

Of interest here is that the adoption of one or more of the lateral load-carrying mechanisms discussed above influences the design and/or selection of individual members and their connections. Figure 14-8 illustrates some basic implications.

   If frame action alone were being relied on to assure stability, it is preferable to employ a structural system where rigid joints can be readily obtained. If shear planes or diagonal bracing is used, the requirement of joint rigidity is not crucial. Herein arise implications about the design and selection of individual beam and column elements. Not all systems available readily lend themselves to achieving moment-resisting connections, while in other systems joint rigidity can be easily accomplished.

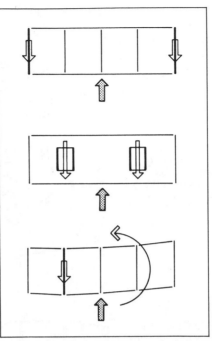

FIGURE 14-7 Use of shear planes in tall structures and effects of nonsymmetrical placement. When shear planes (walls or diagonally braced planes) are used to carry horizontal forces in a tall structure, the planes should be arranged symmetrically to avoid undesirable torsional effects.

In selecting systems, it is important to make the best fit between the intrinsic nature of the individual structural elements available and their most appropriate end conditions and the demands placed on them. Precast concrete elements, for example, are quite naturally suited to situations where joint rigidity is not required (thus implying the existence of shear planes or diagonal bracing in the building). While it is true that joint rigidity in precast members is possible, it is difficult. On the other hand, if frame action alone is being used to assure the stability of the whole building, then a poured-in-place reinforced concrete system might be quite appropriate, since joint rigidity can be achieved quite easily. Indeed, obtaining a pinned connection is what would be difficult. Steel systems lend themselves well to either type of approach. Timber connections are most naturally pinned (rigid joints are difficult to achieve).

Not always does the choice of a method for achieving lateral stability precede the choice of a horizontal spanning system. In cases where the horizontal span is extremely long or loads are unique, the nature of the horizontal element may dictate the type of lateral stability approach used. If spans are on the order of 100 ft (33 m), for example, and precast single tees seem appropriate for the vertical loads, the provision of lateral stability through some other technique than joint rigidity is mandated, since the type and scale of the horizontal element, and the difficulty of getting rigid connections, make the idea of frame action unfeasible. In general, the longer the

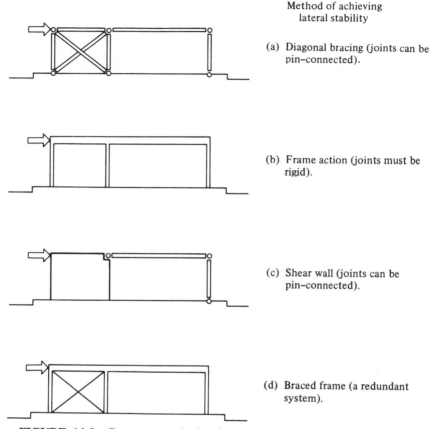

(a) Diagonal bracing (joints can be pin–connected).

(b) Frame action (joints must be rigid).

(c) Shear wall (joints can be pin–connected).

(d) Braced frame (a redundant system).

FIGURE 14-8 Common methods of resisting lateral forces: implications on connection type.

horizontal spans are, the less likely is frame action to be an appropriate mechanism for achieving lateral stability (Figure 14-9).

Often the unique characteristics of one structural element influence the selection of another. When masonry load-bearing walls are used in low-rise buildings, the walls serve as shear planes in resisting forces in the lateral direction. It is obviously best to organize them along the short dimension of the building. Lateral stability along the long dimension can be achieved (depending on building height) through specially designed stair cores or interior walls placed transverse to bearing walls. The use of masonry walls implies that horizontal spanning elements should be simply supported since masonry walls, unless specially reinforced, cannot carry moments. Thus, attempting to use rigid connections at the end of horizontal elements (which would induce moments into the wall) would be counterproductive.

The descriptions of the systems described in more detail in the following chapter should be reviewed with special attention paid to the type of joint conditions implied by a particular approach. The discussion in Section 3.3.2 is also relevant.

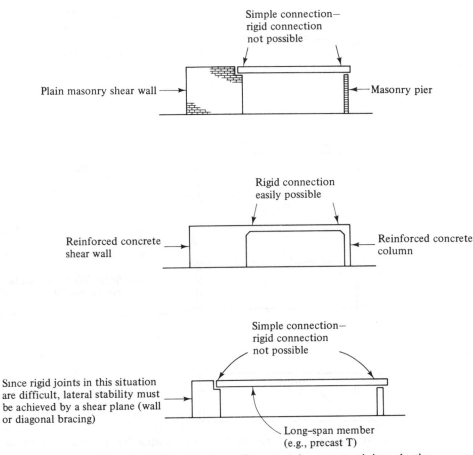

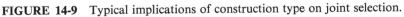

Plain masonry shear wall

Simple connection—
rigid connection
not possible

Masonry pier

Reinforced concrete
shear wall

Rigid connection
easily possible

Reinforced concrete
column

Since rigid joints in this situation
are difficult, lateral stability must
be achieved by a shear plane (wall
or diagonal bracing)

Simple connection—
rigid connection
not possible

Long–span member
(e.g., precast T)

FIGURE 14-9 Typical implications of construction type on joint selection.

14.3.3 Multistory Construction

STRENGTH DESIGN.   As buildings increase in height, the importance of responding to lateral forces caused by either wind or earthquakes becomes an increasingly important structural design consideration.

Of the basic approaches to stability, frame action is probably the least efficient way of achieving lateral stability and is preferably used only when lateral forces are not excessive—as in low- to medium-rise buildings. Typical steel framed buildings that carry loads exclusively by frame action, for example, do so efficiently only up to 10 stories or so. After that a significant premium is paid in terms of excessive material used if a framed structure is used.

Even in low- to medium-rise structures where frame action may still be appropriate, there are differences in lateral load-carrying capacities among the different types of framed systems possible. Steel systems are unique in that they can be designed to be responsive to virtually any situation. Reinforced concrete systems, on the other hand, must be used more carefully. As will be discussed in more detail in the following chapter, for example, systems such as the flat plate and column assembly have a much smaller lateral-load-carrying capacity than other poured-in-place concrete systems. The flat-plate spanning element is uniquely suited to carrying the limited moments induced by relatively light floor loads and *not* the moments generated at the interfaces with columns by large lateral forces. The plate is too thin to carry such moments effectively. Systems with deeper horizontal members, such as the two-way beam-and-slab system or the waffle system, more effectively provide frame action in both directions. When flat plates are used, their lateral-load-carrying capacity is often supplemented by some other mechanism. In common rectangular apartment buildings, for example, which often use flat-plate construction, end walls are often turned into shear planes, which in turn largely carry the later loads. The flat-plate system can then be designed primarily in response to vertical loads. The enclosure around elevator cores is also often specially designed to serve this same function.

As heights increase from the low to intermediate range, it is preferable either to begin supplementing *any* type of frame system used with additional lateral bracing mechanisms, such as diagonal braces around elevator cores, or to adopt a radically different structural approach.

To get a feeling for what different types of structural approaches might be possible for *very* tall structures, it is useful to first briefly review some fundamental principles of the way a tall structure carries lateral loads. Most high-rise structures tend to be relatively tall and slender. Under the action of lateral forces they tend to act like vertical cantilever members. The lateral loads tend to produce an overturning moment which must be balanced by an internal resisting moment provided by the structure. This internal resisting moment is typically provided by couples formed between forces developed in vertical members (see Figure 14-10). If the building is very slender, the small moment arm present between the forces in the vertical members means that very high forces must be developed to provide the internal resisting moment. Buildings of similar heights with wider bases and generally less slender proportions could provide the same internal resisting moment with smaller forces developed in vertical members because of the increased size of the internal moment arm— obviously an advantageous consequence.

The view of a high-rise building as a vertical cantilever is useful in that in addition to implying something about appropriate building proportions, it also can help inform other design responses. In Chapter 6 it was noted that the most efficient use of material in bending is obtained by locating the greatest amount of material as far as practically possible from the neutral axis of the section. The overall moment of inertia of the section for a given amount of material is increased by doing this, as is the resistance to bending provided. The same principle held true for truss design. In a high-rise building of a given proportion, application of this same principle means that the greatest amount of material should be located in the outer, rather than the inner, vertical elements (see Figure 14-11). Many efficient high-rise structures are conceived in this way. The exterior column and beam assemblies are designed to provide a very stiff ring, or *tube*, capable of carrying lateral loads from any direction. Exterior columns are typically very closely spaced to one another. Spandrel beams are usually rigidly connected to columns to assure that the whole outer assembly acts in an integral way—like a stiff surface element. Although the outer frame assembly can and do carry gravity loads and act like frames in the horizontal direction, their primary

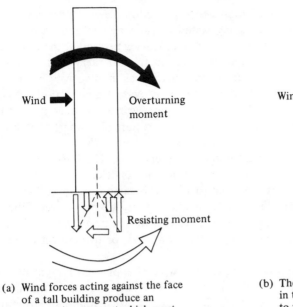

Wind ►   Overturning
moment

Resisting moment

(a) Wind forces acting against the face
of a tall building produce an
overturning moment which must
be balanced by an internal resisting
moment provided by forces developed
in column members. The structure
acts much like a vertical cantiliver beam.

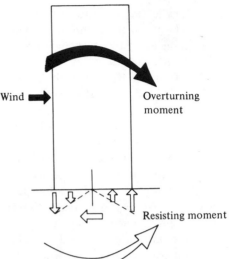

Wind ►   Overturning
moment

Resisting moment

(b) The magnitude of the forces developed
in the columns are generally proportional
to the moment arm separating the columns.
Smaller column forces are developed in
wider buildings than in those of more
slender proportions.

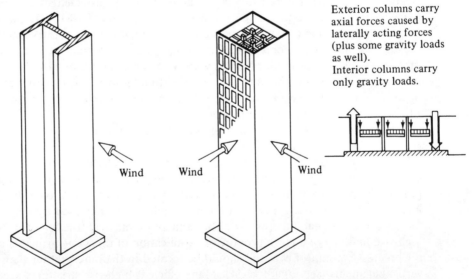

Exterior columns carry
axial forces caused by
laterally acting forces
(plus some gravity loads
as well).
Interior columns carry
only gravity loads.

Wind   Wind   Wind

(c) For very tall structures, an efficient structure
often results from conceiving of the building
as a laterally loaded cantilever beam and
distributing material to achieve a large moment
of inertia.

(d) Many structures have been built on the principle
described in (c). The structure must, however, be
responsive to lateral forces in all directions. A
tube type of structure results in which the exterior
columns provide all the resisting moment to overturning
lateral forces. Interior columns are designed to carry
only gravity loads.

FIGURE 14-10 Very tall buildings can usefully be conceived of as vertical
cantilever beams.

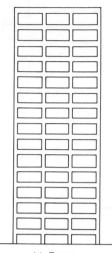
(a) Frame.

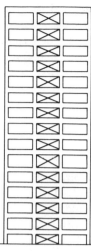
(b) Braced frame.

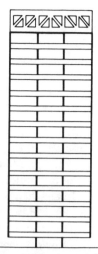
(c) Suspended structure, reinforced concrete core.

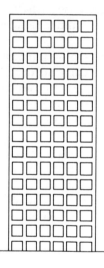

(d) Tube structure. The exterior columns are closely spaced. Horizontal spandrel beams are rigidly connected to columns to form an exterior tube which carries all lateral forces and some gravity forces. Interior columns carry only vertical forces.

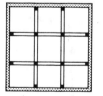

(e) Trussed frame. The lateral stiffness of very tall structures can be markedly increased by superimposing major diagonal bracing members (either a frame or tube structure). Lateral forces are again taken by the exterior trusswork and frame. Interior columns carry only vertical forces.

FIGURE 14-11 Typical structural approaches in high-rise construction.

function is to carry forces generated by the overturning moments associated with lateral loads. Interior columns are designed to carry gravity loads only and are consequently typically smaller than exterior columns. Special attention must be paid to designing a floor system that will make the overall resultant structural assembly behave like a unit in carrying lateral loads. The stiffness of a tube of this type can be even further increased by the addition of large cross-bracing on the outside faces of the structure.

DEFLECTION AND MOTION CONTROL. The structural responses for multistory buildings discussed thus far have all been strength-oriented. Equally important, however, are deflection and motion considerations associated with the dynamic effects of winds. A typical tall building, for example, sways under the buffeting action of nonconstant winds. Even when winds are steady-state, however, a dynamic behavior is still present. As the wind blows against a building, the building bends over and slightly changes shape. The exact magnitude and distribution of wind forces thus also changes slightly. Increases or decreases in forces due to this phenomenon, coupled with any buffeting action of the wind that might be present, cause the building to oscillate. Typically there is an average deflection in the direction of the wind force about which the building oscillates. The magnitude and frequency of these oscillations depend both on the characteristics of the impinging wind forces and the stiffness and the mass-distribution characteristics of the building itself.

These deflections and oscillations are highly important. Excessive deflections can impair the functioning of other building elements (e.g., building service systems) even though the structure itself is unharmed. Oscillations can cause extreme discomfort to building occupants. Human beings do not sense absolute eflections if they occur slowly, but they do sense the accelerations associated with rapid oscillations. A type of motion sickness can consequently occur. What constitutes an acceptable level of acceleration is not an easy criterion to establish rationally, although some values have been suggested by investigators. In addition, analyzing a building to predict what motions actually occur in a given circumstance is also difficult. The most common procedure is to model the structure as an assembly of springs (having elasticities derived from a study of the actual load-deformation characteristics of the structure) and concentrated masses representing building weights. The vibrations a model of this type undergoes when dynamically loaded can be used to predict the response of the real structure.

Thinking about a multistory structure as a spring–mass assembly is also useful in formulating design responses for how to deal with excessive motions or deflections. Clearly, one obvious way to control deflections is to increase the *stiffness* of the structure (not necessarily its strength) until wind-induced oscillations are held to an acceptable level. Increasing member sizes, using diagonal braces or other shear planes, and redistributing material placement are all ways of increasing stiffness.

Another way of controlling motions, particularly accelerations, is to build into the building physical *damping mechanisms* (these dampers are conceptually similar to door-closing devices that cause a door to close slowly and evenly when opened and released). Damping devices are typically installed at joints between beams and columns. As movements begin to occur, these devices absorb energy and damp the motion.

Excessive motions can also be controlled by installing a very large mass of material mounted on rollers and attached to the building structure with dashpot or damping devices. This device, often called a *tuned mass damper*, is placed in the upper floors of a tall building. As motion begins to occur, the inertial tendency of the mass is to remain at rest. Hence, building movements cause the damping devices to accentuate and begin absorbing energy, thus damping out motions.

The use of damping devices of the type described above is usually restricted to very large buildings. They have, however, been successfully used in several situations. More commonly the stiffness of the structure is controlled. A very rough rule of thumb often used to limit both accelerations and deflections when dynamic analyses are not appropriate or feasible is to use a static wind-load analysis and limit maximum steady-state deflections to a particular value, say $h/500$, where h is the height of the building in feet, and to vary the stiffness of the structure until this criterion is met. This approach is much easier than a dynamic analysis but is dubious at best. Still the approach has historically proven useful if not always a sure bet.

14.3.4 Earthquake Design Considerations

GENERAL PRINCIPLES. As discussed in Chapter 3, earthquakes tend to produce forces on structures that are primarily lateral in character. Most of the principles discussed in the preceding sections are therefore as applicable to the design of structures for earthquake resistance as they are for wind resistance. The extremely pronounced dynamic character of earthquake forces, however, make the design problem even more complex. There still exists some controversy among architects and engineers involved in designing structures with respect to earthquake forces as to what constitutes good design strategies for buildings in earthquake-prone regions. This section will briefly explore some of the primary issues involved and alternative strategies that are possible.

There are several basic types of hazards associated with earthquakes. These include *surface fault ruptures, ground shaking, ground failures*, and *tsunamis* (sea waves generated by earthquakes). While all these effects are important, this text will focus primarily on ground-shaking effects. The reader is referred elsewhere for discussions of other phenomena.

As a way of getting into some of the design issues involved with respect to seismic design for ground shaking, it is useful to review briefly the behavior of a structure subjected to an earthquake (also see Section 3.2.4). Figure 14-12 illustrates characteristic vibration modes for a tall building. As the figure illustrates, the ground acceleration can cause all floors to move in the same direction, or it can cause different floors to accelerate in different directions in a whiplash type of movement. The whiplash movement can occur because of the tendency for an elastic building to spring back to the vertical after its base has been initially translated and accelerated. Because of inertial tendencies, the upper mass of the building would lag behind the base movement. As the upper mass begins springing back to the vertical after the translation has

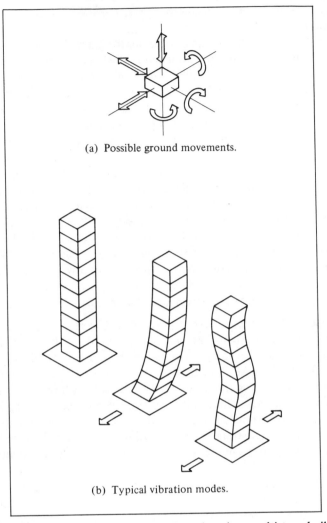

(a) Possible ground movements.

(b) Typical vibration modes.

FIGURE 14-12 Earthquake-induced motions in a multistory building.

occurred, it begins gaining momentum. Because of the momentum gained, the upper mass can actually swing past the vertical. If during this process the ground movement is changed such that accelerations and translations are reversed, very complex deformations can occur because of the inertial tendencies of the building masses to continue the movement already started, while at the same time countermovements are beginning. There is a lag as the effects of the reversed ground movement work their way up the building.

Associated with these complex accelerations are extremely high inertial forces that pose significant structural design problems. In addition to the forces normally expected, unique forces can also be induced in a structure because of unusual building proportions or features. Twisting, for example, can be induced because of nonsymmetrical placement of building masses. These effects will be explored in the following section.

GENERAL DESIGN AND PLANNING CONSIDERATIONS. Probably the single most important design principle with respect to ensuring that a building performs well during an earthquake is that of making certain that both the general masses present in the building (e.g., floors, roofs, etc.) and the stiffening lateral force-resisting mechanisms present in the building (e.g., shear walls, braced frames, etc.) are symmetrically located with respect to one another. Lateral forces associated with earthquakes are, of course, inertial in character and consequently related to the masses of different building elements. Nonsymmetrical placement of building masses and elements that resist the forces associated with these masses can lead to highly undesirable torsional effects in the building, which can be extremely destructive. This is the same issue that was generally addressed earlier in connection with wind forces (see Figure 14-7) but is of even more fundamental importance with respect to earthquake forces.

The principle noted above has significant implications for the overall building shape adopted, since both the mass distribution and placement of lateral force resistance mechanisms is strongly influenced by the shape of the building. An *L*-shaped building, for example, clearly has a nonsymmetrical mass distribution and normally a nonsymmetrical placement of stiffening elements as well. Destructive torsional forces develop in such shapes because of the reasons discussed above. Another way of understanding why the configuration is undesirable is to conceive of the building as made up of two separate masses (each leg of the *L*), each of which tends to vibrate in its own natural frequency. Since the stiffnesses of the two units differ, their natural periods also differ—a condition that sets up a deflection incompatibility at the interface of the two masses. The consequence is that failures often occur at interface locations. The problem can be somewhat alleviated by separating the building into symmetrical units connected by a *seismic joint*, which physically separates the masses and allows free vibratory movement to independently occur in each (the device is conceptually similar to an expansion joint).

From an earthquake-design viewpoint, preferred building shapes are those of simple plane geometry (e.g., squares, circles); L shapes, T shapes, H shapes, or other unusual shapes are invariably difficult to structure when earthquake hazards are present. Seismic joints can be used to advantage in most such configurations. Difficulties can arise, however, even in buildings that are seemingly symmetrical in plan if *stiffening* elements are nonsymmetrically placed. This is a condition that often arises when elevator cores or other natural stiffening elements are unusually located [see Figure 14-13(b)].

The U shape found in many common commercial buildings is another example of an apparently symmetrical building subject to undesirable torsion effects. Such buildings should preferably have their open faces stiffened by the insertion of frames. Stiff roof diaphragms should also be used.

Some completely symmetrical buildings can be susceptable to torsion because of nonsymmetrical placements of occupancy loads—a condition that frequently arises in warehouses. This situation can be prevented somewhat by careful design.

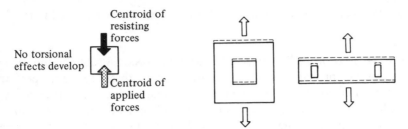

Centroid of
resisting
forces

No torsional
effects develop

Centroid of
applied
forces

(a) Symmetrical structures do not experience exceptionally high torsional forces and are
hence preferred to nonsymmetrical structures.

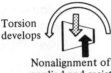

Torsion
develops

Nonalignment of
applied and resisting
forces

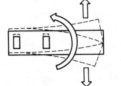

Off–center
stiffening elements
(e.g., elevator cores)

Open–ended
bearing wall
structure

Off–center
loading

(b) Structures that are nonsymmetrical because of either their basic configuration or the
nonsymmetrical placement of lateral–load–resisting elements typically experience high
torsional forces which are very destructive. Nonsymmetrically–placed masses can also
lead to similar torsional effects.

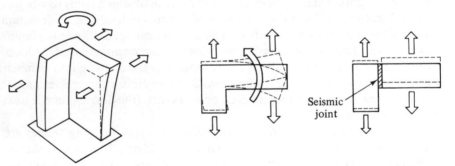

Seismic
joint

(c) Nonsymmetrical configurations with reentrant corners (e.g., L– or H–shaped buildings)
are particularly susceptible to destructive torsional effects. Primary damage often occurs
at the reentrant corners. Allowing separate building masses to vibrate independently by
using seismic separator joints that allow free movement to occur generally improves
structural performance.

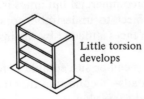

Little torsion
develops

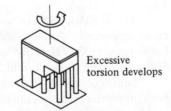

Excessive
torsion develops

(d) Buildings that are nonsymmetrical in the vertical direction also experience destructive
torsional effects. Discontinuous shear walls are particularly problematical.

FIGURE 14-13 General planning considerations: symmetry versus non-
symmetry.

The principle of imparting symmetrical stiffness and massing characteristics to a building also holds true for a building in its vertical dimensions as well as in plan. If the stiffness characteristics of different building elements change with building height, for example, nonsymmetries can develop that lead to destructive torsional effects at different levels of the building. Buildings with discontinuous shear walls can be particularly problematical.

There are other considerations that affect a building's sensitivity to earthquake movements. Buildings that are extremely elongated, for example, are preferably avoided even if they are symmetrically organized (Figure 14-14). The longer a building is in plan, the greater is the possibility that opposite ends of the building will be subjected to different relative ground movements that do not act in the same manner. This would tend to cause the building to be torn apart. Designing to prevent this tearing apart is very difficult. It is often necessary to separate extremely long buildings into series of adjacent volumes (through the use of seismic joints) that can move freely with respect to one another.

Another basic consideration is that tall slender buildings are less able to resist efficiently the overturning moments associated with earthquake forces than are lower, wider buildings (see Figure 14-15).

Buildings that are constructed adjacent to one another should be adequately separated so that each can freely vibrate in its natural mode without touching each other. Otherwise, severe damage can occur because of *pounding* effects (Figure 14-16).

While forces associated with lateral movements are of primary design importance, vertically acting accelerations due to earthquake motions can also cause trouble. Cantilevers, for example, are particularly susceptible to failure because of vertical movements (see Figure 14-17). In such events, failure often also follows in interior members because of the moment redistributions that occur following the dropping off of cantilevers.

All the considerations noted above generally point to the fact that from the

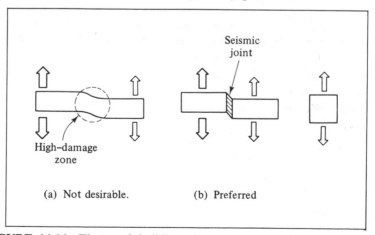

FIGURE 14-14 Elongated buildings are more susceptible to destructive forces associated with differences in ground movements along the length of the building than are more compact shapes. Long buildings can be subdivided by using seismic joints.

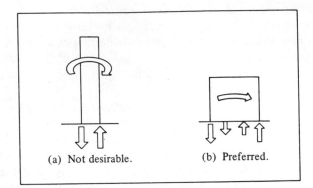

(a) Not desirable.   (b) Preferred.

FIGURE 14-15 Relatively slender buildings are less able to efficiently resist overturning moments caused by earthquakes.

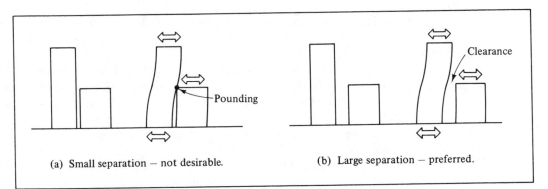

(a) Small separation — not desirable.   (b) Large separation — preferred.

FIGURE 14-16 Adjacent buildings should be adequately separated so that buildings do not pound against each other during seismic events.

viewpoint of designing for earthquake resistance, buildings that are very symmetrically organized, of relatively compact proportions, and which have few or small overhangs or other protrusions are preferable to very nonsymmetrically organized buildings with exaggerated proportions. Other considerations, it should be noted, such as the specific site and soil conditions present, may also strongly affect the type of building configuration that is preferred.

GENERAL CHARACTERISTICS OF EARTHQUAKE-RESISTANT STRUCTURES. Structures that are *continuous* in nature and more or less uniformly distributed throughout a building generally perform well when subjected to earthquakes. The primary reason for this is that the earthquake resistance of a structure is significantly dependent on its ability to absorb the energy input associated with ground motions. Pin-connected structures, such as traditional post and beam assemblies, are far less capable of absorbing energy than are comparable continuous structures (e.g., frames with monolithic joints). The formation of plastic hinges in framed structures (which *must* precede their collapse) requires a significant energy input. Continuous structures, therefore, are quite effectively used in buildings in earthquake-hazard zones (Figure 14-18). Either steel or poured-in-place reinforced concrete framed structures designed to be

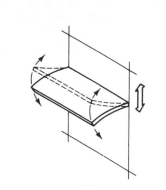

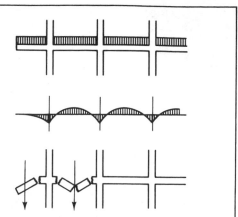

(a) Movements in cantilever induced by vertical accelerations: cantilevers can literally snap off due to moment failures associated with these movements.

(b) Failure of cantilevers often leads to other failures because of the moment redistributions that occur. In the example above, midspan positive moments increase if the overhang is eliminated, which leads to a failure of the interior member.

FIGURE 14-17 Importance of the vertical components of earthquake ground motions.

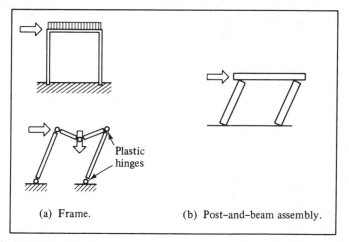

(a) Frame.   (b) Post–and–beam assembly.

FIGURE 14-18 Continuous structures are generally preferable to pin-connected ones. The necessary formation of plastic hinges in continuous structures before collapse can occur absorbs large amounts of energy.

ductile under earthquake forces can be successfully used. Precast concrete structures, with their lack of continuity at connections, are, by a similar token, more difficult to make successfully work.

Even within the category of continuous structures, however, differences exist. Figure 14-19 illustrates the plans of two structures deriving their ability to carry lateral forces from frame action. The first structure, illustrated in Figure 14-19(a),

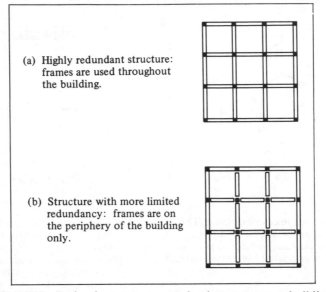

(a) Highly redundant structure: frames are used throughout the building.

(b) Structure with more limited redundancy: frames are on the periphery of the building only.

FIGURE 14-19   Redundant versus nonredundant structures: buildings with highly redundant structural systems in which multiple load paths exist for carrying earthquake forces to the ground usually perform better than less redundant structures during seismic events.

with frames distributed throughout the structure, is generally preferable to that shown in Figure 14-19(b), which has frames around the periphery only. The first structure shown has a greater *redundancy*, and consequently has greater reserve strength than the second structure. Failure of relatively few members in the outer plane of the structure shown in Figure 14-19(b) may lead to total collapse, while many more members must fail in the more redundant structure. Generally, structures with redundancy are preferable to those without redundancy.

Another general characteristic of viable earthquake-resistant structures is that column-and-beam elements are generally coaxial. Offsets or nonaligned members often present extremely difficult design problems.

Earthquake-resistant structures also typically have floor and roof planes designed as rigid diaphragms capable of transmitting inertial forces to lateral-load-resisting elements through beamlike action (Figure 14-20).

Another aspect of earthquake-resistant structures is that they are designed such that horizontal elements which fail due to earthquake motions do so before any vertical members fail (Figure 14-21). Failure should never occur in vertical members first. The reason for this is simply a life-saving consideration. Horizontal elements in *continuous* structures (e.g., slabs or beams) rarely completely fall down even after experiencing extreme damage and when they do, the collapse is fairly localized. (This is not as true with pin-connected horizontal members.) When columns experience damage, however, complete collapse is imminent. The collapse of a column generally causes other portions of a structure to collapse as well. The effects of a single column collapsing can be extensive.

To assure that horizontal elements fail first, care is usually exercised in the

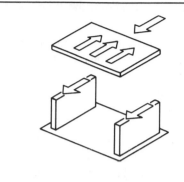

(a) Typical diaphragm action: the horizontal plane acts like a beam in carrying earthquake–induced forces to shear walls or other lateral–load–carrying mechanisms.

(b) If diaphragms are improperly designed, failure can result in floor or roof planes.

FIGURE 14-20 Importance of rigid floor and roof elements: for earthquake-induced inertial forces to be transferred to lateral-load-carrying elements, floor and roof elements must be capable of acting like rigid diaphragms.

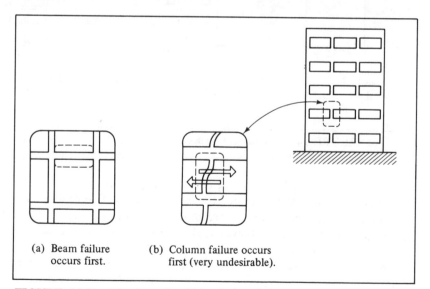

(a) Beam failure occurs first.

(b) Column failure occurs first (very undesirable).

FIGURE 14-21 Members should be designed that such that failure occurs first in horizontal members rather than in vertical members.

design and general proportioning of beam-and-column elements. Extremely deep beams (e.g., spandrels) on light columns are generally best avoided, since experience indicates that such buildings often experience high damage in the light columns which have to pick up all the laterally acting forces by shear and bending. On the other hand, types such as shear wall with a series of spaced small openings tend to perform somewhat better. Good engineering design, it should be noted, can make all types work adequately.

STIFF OR FLEXIBLE STRUCTURES. Of primary importance in seismic design is the *natural period of vibration* of the building under consideration. It has already been noted in Chapter 3 that if the ground input motions have a similar frequency to that of the natural period of the building, then a resonance situation can begin to develop and high destructive forces can be generated. In any event, however, the forces that a building experiences are always directly related to its period even if resonance conditions do not occur. Forces specified in many building codes, for example, often assume the general form discussed in Section 3.3, which implies that "short"-period (high-stiffness) buildings must be designed for larger forces than "long"-period (low-stiffness) buildings (Figure 14-22). As will be discussed below, however, this is *not* the only factor that must be taken into account in choosing to opt for a relatively stiff building (e.g., one using reinforced concrete shear walls) or a relatively flexible (low-stiffness) one (e.g., a steel frame).

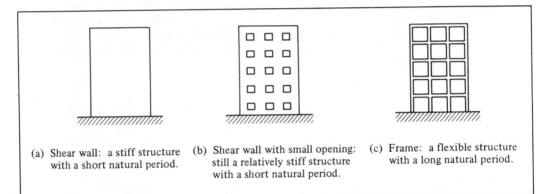

(a) Shear wall: a stiff structure with a short natural period.

(b) Shear wall with small opening: still a relatively stiff structure with a short natural period.

(c) Frame: a flexible structure with a long natural period.

FIGURE 14-22 Different structural responses have widely varying natural periods of vibrations, an important consideration in seismic design.

One important component of the decision has to do with the notion that for ground motions that are close to the natural period of the building itself, the structure will receive maximum punishment because of the tendency toward resonance. If the frequencies are different, however, the structure will experience lower seismic forces (see the discussion on this point in Chapter 3). A very stiff building should seemingly experience lower seismic forces when ground motions have a long period than a flexible building, and vice versa. It would therefore seem that the choice of relative stiffness should be made on the basis of the expected character of the ground motions that might be perceived by the building. The character of these motions, however, are highly difficult to predict. The actual motions perceived by a building, it should be noted, are *strongly* influenced by the nature of the soil conditions present beneath the building and the interaction of the building with the soil. Still in certain circumstances, the general nature of expected ground motions can be predicted to a greater or lesser degree.

Many designers generally advocate for all situations the use of fully flexible buildings in which a relatively flexible structure (e.g., a steel frame) is used. Such buildings are also normally designed such that no actual or potential stiffness is imparted by other building elements (e.g., partitions). Such nonstructural elements are detailed to allow the structure to move freely (a difficult and sometimes nonfeasible task).

Briefly, the primary advantages of a full flexible approach appear to be that such buildings are especially suitable for sites where ground-motion periods are expected to be short. Ductility and continuity are also naturally easy to achieve because steel is typically used extensively. Disadvantages occur when long-period sites are present and resonance situations can develop. Nonstructural elements are also difficult to detail and often experience great damage if not designed properly when earthquakes do occur.

Stiff structures appear advantageous for long-period sites. Nonstructural elements are usually fairly easy to detail in stiff buildings, since structural movements are not large and damage to such elements is limited when earthquakes do occur. Stiff structures, on the other hand, are disadvantageous for short-period sites. For some materials the notion of high stiffness is also not generally compatible with the notion of ductility, which is important from an energy absorption viewpoint. (The presence of some ductility is important even in stiff structures.) By careful design, however, even apparently stiff structures (such as those made of reinforced concrete) can be made to be very ductile through the incorporation of appropriate amounts of steel in proper locations.

On some occasions, mixed systems are often advantageous. The mixed frame/shear wall system shown in Figure 14-23, for example, is often a good system in that for earthquakes of a low magnitude the stiff shear wall takes most of the forces involved, and the whole building responds stiffly, thus limiting nonstructural damage. The more ductile frame, however, provides a large measure of reserve capacity when high-magnitude earthquake forces are present.

MATERIALS. As noted before, fully continuous structures are desirable for use in earthquake-prone regions. Coupled with the idea of continuity is that of ductility. For energy absorption to take place, ductility is essential. Steel is a naturally ductile material and is consequently often used. Poured-in-place reinforced concrete can also be made to have a high degree of ductility by carefully controlling member proportions and the amount and placement of reinforcing steel. Reinforced concrete buildings of this type are often used in earthquake zones. Precast concrete structures, however, are often unsuitable for earthquake zones because of the difficulties involved in achieving a continuous and ductile structure. Such structures are also typically high-mass structures. The high mass contributes to the magnitude of the seismic forces developed and compounds the problem of their use.

Timber can be an extremely good material for use in earthquake regions. It is light in weight and capable of absorbing large amounts of energy when deformed and before collapse. Low-rise wood-framed structures are highly earthquake-resistant and perform well in earthquake regions.

FIGURE 14-23 A combination shear wall/frame structure in which the structure behaves stiffly during low-magnitude earthquakes but still has reserve strength in high-magnitude earthquakes because of the coupled frame.

Materials such as masonry, unless highly reinforced with steel, generally tend not to be suitable for use in earthquake regions. Masonry structures are highly massive but have little ductility. Masonry buildings frequently crack up when subjected to earthquake motions.

NONSTRUCTURAL ELEMENTS. It should be emphasized that many nonstructural elements can significantly influence the dynamic behavior, and hence the seismic forces present, in a building. Depending on how such elements interact with the primary structure, they may alter the natural period of vibration of the structure, thus changing the forces present. They can also affect the distribution of lateral stiffness in a building, which can also have significant consequences.

One of two basic strategies are normally taken with respect to nonstructural elements (Figure 14-24). One is to carefully analyze all elements and include those that contribute to the stiffness of the primary structure in the analysis and design of the primary structure. Contributing elements would have to be carefully detailed to assure that they did indeed contribute to the building stiffness as expected. The second basic

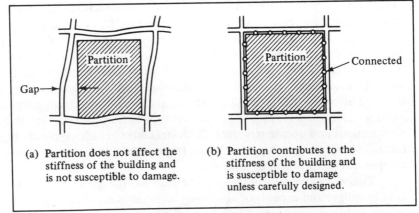

(a) Partition does not affect the stiffness of the building and is not susceptible to damage.

(b) Partition contributes to the stiffness of the building and is susceptible to damage unless carefully designed.

FIGURE 14-24 Nonstructural elements.

strategy is that noted earlier of preventing any of the nonstructural elements from contributing to the stiffness of the structure at all. This is done by attempting to detail connections such that gaps exist between the primary structure and nonstructural elements. The structure can then deform freely. Which of these two approaches is assumed is largely dependent on the attitude the designer takes toward the overall stiffness of the structure and whether flexibility is desired or not.

14.4 CONCENTRATED VERSUS DISTRIBUTED STRUCTURES

When involved in the process of designing structures of even widely differing types, some common issues often arise. One of the most interesting of these issues is whether it is better to use what are perhaps best described as *concentrated* structures or to use *distributed* structures. The descriptive terms refer to alternative design strategies for a given loading condition wherein the first approach is characterized by designing a few very large members to carry the load and the second by designing a greater number of relatively smaller elements. Is it more efficient, for example, to carry a given load with one large column or with two slightly smaller ones? Or is it more efficient to use one large beam or two parallel smaller beams? On a more general scale, is it preferable to use a space frame (with many very small members dispersed throughout the system) or a series of more widely spaced trusses made of relatively larger elements? The basic choice of concentrated versus distributed structural strategies appears over and over again in many different situations.

As would be expected, there is no hard-and-fast answer to this question. It was noted earlier that surface-forming structures (which typically reflect a distributed element structural strategy) are best used when uniformly distributed loads are present and concentrated loads are absent. Similarly, discretized structures were said to respond better to concentrated loads. These principles form one set of guidelines that are useful in formulating initial design strategies. Using distributed structures, such as space frames, in conditions where concentrated loads are present is one of the most typical of all faulty judgments made by novice designers. Curiously, such misuses are often made when the novice designer is attempting to use a system that *looks* structural.

The principles mentioned above, however, do not help resolve questions where the type of load is not a variable and the issue is one of either using a single large member or multiple smaller ones to carry the load. There does seem to be evidence indicating that using a fewer larger elements to carry a given load is often preferable to using a number of smaller elements to carry the same load. This principle is primarily applicable to design situations, such as beam or column design, that involve manipulations with moments of inertia (which does not simply increase linearly with increases in member dimensions). In other situations, such as the design of tension members not involving moments of inertia, the principle is not applicable. Often no differences exist in such situations.

14.5 THE PRINCIPLE OF STRUCTURAL JUSTICE

Implicit in the sections presented earlier on the analysis and design of indeterminate structures is a seemingly curious principle. Roughly stated, the essence of this principle is that in an indeterminate structure *the part of the structure that is the stiffest and most capable of providing resisting forces will end up carrying the greatest portion of the applied load.* In the analysis of continuous beams, for example, it was shown that increasing the member size at supports ("haunching" the beam) also led to an increase in the design moment present, thus in some senses justifying the original increased member size. Moments at other locations were reduced and member sizes consequently made relatively smaller at such places. Analogous phenomena can be shown to exist in the analysis and design of many other indeterminate structures (e.g., a series of parallel columns or tension elements supporting a load or a crossedbeam system). Similarly, it was noted in a previous section that stiffer buildings had to be designed for greater earthquake induced forces than more flexible ones.

This phenomenon, nicely described as the *principle of structural justice*,[1] can be used as a design tool. The designer can, to some extent, control the relative way material is distributed in space in order to achieve some specified goal. A continuous beam can be deliberately haunched at its supports, for example, to provide more headroom or space for mechanical devices at midspan since design moments and thus relative member sizes would be reduced by the haunching. Alternatively, a designer might employ a similar approach of varying the relative stiffness of the structure at different points with the intent of minimizing the total amount of material required to support a given load in space.

An interesting aspect of this principle (which is applicable to indeterminate structures only) is that the design moments obtained as the results of an analysis are dependent on the proportions of the structure originally selected to be analyzed. The design moments obtained apparently lead in turn to member sizes that generally conform to the proportions originally selected and thus seemingly justify their choice, which might have been arbitrary to begin with. A circular sort of reasoning apparently begins to emerge which can certainly provide those readers inclined toward enjoying paradoxes with an interesting case in point. Part of the argument is real and part of it becomes a nonissue as design iterations involving manipulations with member proportions are actually made. Which is which is a point left to inquisitive readers interested in such things to resolve.

14.6 VARYING BOUNDARY CONDITIONS

Throughout the chapters in Part II on specific elements the point was made over and over again that manipulations with support locations or altering boundary conditions in some other way (e.g., fixing the ends of a beam) can lead to more efficient designs. It is not the intent to rehash these subjects here in detail but merely to recall their importance to the reader, since the principle of doing *everything* possible to reduce the bending moments present by whatever means possible is fundamental to the overall

[1]A term first coined, to the author's knowledge, by E. Traum.

process of designing an efficient structure. By paying careful attention to support locations, for example, design bending moments can be reduced with the consequence that smaller members can be used (see Sections 4.3.9 and 6.4.6). Likewise, fixing the ends of members or utilizing continuous members can also lead to reduced design moments (see Sections 8.3 and 10.3.2). These general principles are applicable to virtually all structural types. The principle of providing continuity, for example, can be applied equally to beams, folded plates, barrel shells, and trusses as a way of reducing design moments and thus achieving more efficient structures. There are fewer more powerful tools available to the structural designer than the concious control of a structure's boundary conditions.

14.7 IMPOSED CONSTRAINTS: FIRE SAFETY REQUIREMENTS

A strong determinant of the structural system used in a building are the fire safety requirements imposed by building regulations. Codes typically classify buildings according to construction types, which can be roughly characterized as light, medium, and heavy, according to the construction's degree of fire resistivity. Light construction is typically wood-frame or unprotected metal framing. Medium construction typically uses masonry walls as load-bearing elements. Heavy construction is typically reinforced concrete or protected steel framing.

The main influence of building regulations on the choice of structure is to place restrictions on the type of construction allowed in accordance with the degree of fire hazard present. High-hazard occupancies require heavily fire-resistant construction (typically necessitating reinforced concrete or protected steel structures), while less hazardous occupancies require less substantial construction types. In a similar vein, construction types are coupled with building heights. In low buildings, almost any type of construction is typically allowed unless a high-hazard occupancy is present. As heights increase, the construction used must have a greater degree of inherent fire safety. Buildings using masonry load-bearing walls are typically acceptable up to five or six stories. Above this height reinforced concrete or protected steel structures are generally the only type of construction allowed. It should be noted that any heavier type of construction suitable for more hazardous occupancies or taller buildings is allowable in less restrictive cases, but, of course, not vice versa. A single-family detached house can be built out of reinforced concrete but a high-rise office building cannot be built out of light wood framing. Using a heavier system when a lighter one is allowed, however, is usually not economical.

Coupled with construction types and heights are limitations placed on the maximum floor areas allowed in buildings between specially designed fire division walls. The intent of these walls is to separate the building into compartments and thus restrain the spread of a fire beginning in a compartment to that compartment. In many cases, particularly in low- to medium-rise buildings, the load-bearing walls are designed to be coincident in location with where fire divisions are required so that the same masonry walls serve both functions. This has implications on where load-bearing walls are placed and often on the design and selection of horizontal spanning elements.

QUESTIONS

14-1. Identify at least one multistory building in your local area or that is documented in the literature that uses shear walls for lateral stability. Draw a plan of the building and indicate diagrammatically the locations of these elements. Do they form a pattern capable of resisting loads in all directions? If not, what other load-carrying mechanisms are present to assure complete stability?

14-2. Repeat Question 14-1 using a framed building.

14-3. Repeat Question 14-1 for a diagonally braced building.

14-4. Review the literature and identify at least six different multistory buildings in different height ranges. The height range should vary from about 10 stories or less to some very tall buildings (e.g., the Hancock Building or World Trade Center). Identify the general type of structural system used in each case. Draw to the same scale a schematic diagram of the elevation of each building (emphasizing structural elements).

14-5. Review the literature and identify a major building that was specially designed for earthquake forces. Identify all structural and nonstructural design devices that were incorporated to reduce earthquake hazards. Record your findings with a series of annotated diagrams.

14-6. Review the literature and identify a major building that uses a "tuned mass damper" in its upper floors as a way of controlling lateral motions [consider, for example, the Citicorp Center building in New York (*Architectural Record*, Mid-August issue, 1976)]. Draw diagrams of how the device works.

chapter 15

STRUCTURAL SYSTEMS: CONSTRUCTIONAL APPROACHES

15.1 INTRODUCTION

The design process always involves looking in detail at specific alternatives. Only in this way can the full implications of different alternatives be understood and a rational choice made of the final structure to be used. At this stage the designer is working with specific materials and constructional systems. Knowledge of the range of possible constructional approaches available is an intrinsically important part of the designer's working vocabulary.

This chapter provides a brief descriptive overview of many of the different structural systems found in common use. A highly useful way of characterizing these typical approaches is according to the nature of the primary material used in the structure. Three common materials, wood, reinforced concrete, and steel, will be addressed here. No attempt is made to do anything other than highlight basic characteristics of systems as grouped according to materials. Reference is made to preceding chapters on specific elements for analysis and design techniques.

The following sections also contain some *rule-of-thumb* information for determining approximate sizes of elements. Using rules of thumb is a time-honored way of initially sizing structures that at one time was the only way of utilizing empirical knowledge gained from observing built structures that performed successfully to inform the design of new structures (see Figure 15-1). They are still of interest for the

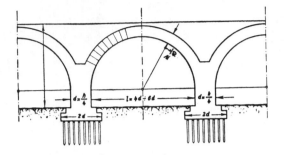

(a) Arch bridge dimensions, according to Leon Battista Alberti.

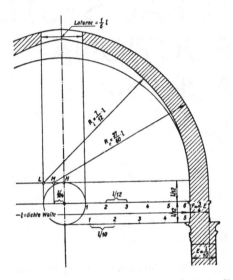

(b) Determining the dimensions of a cupola, according to
    Carlo Fontana.

FIGURE 15-1   Early rules of thumb for sizing structures.

same reason, but should be used with care. Since they were derived by looking at previously built structures their use in design tends to propagate the past state of the art rather than be forward-looking. As such, they inherently tend to preclude innovative structural solutions based on an understanding and application of the theory of structures.

15.2 WOOD CONSTRUCTION

PRIMARY SYSTEMS.   Almost all constructional systems using timber as the base material can be characterized as being composed of linear one-way spanning elements. Hierarchical arrangements are typical. Figure 15-2 illustrates a sampling of the many different types of wood construction systems in common use. Many other systems are, of course, also possible and in use.

LIGHT FRAMING. The familiar light joist system illustrated in Figure 15-2(a) is undoubtedly the most ubiquitous of all wood-framing systems used in current construction. The floor joist system is primarily useful for light occupancy loads which are uniformly distributed and for modest spans, conditions typically found in house construction. Joists are normally simply supported. Rarely are moment-resisting joists used since special connections are required. Usually, the transverse decking is not considered to act integrally with the joists unless special care is taken with connections.

The vertical support system is typically a load-bearing wall of either masonry or closely spaced wood elements (studs) sheathed in plywood. In the latter case, the lateral resistance of the whole structural assembly to horizontal forces is obtained by arranging plywood-sheathed walls to serve as shear planes. Structures of this type are typically restricted in height to three or four stories. This is not so much because of their theoretical load-carrying capacity as it is for fire-safety requirements stipulated in building codes.

Since each element is individually put in place on the job site, a great deal of flexibility is available in the use of the system and in how other building components are integrated with it.

STRESSED SKIN ELEMENTS. *Stressed skin members* are obviously related to the standard joist system [see Figure 15-2(b)]. In these elements, plywood sheathing is affixed to both sides of stringers in a way that assures that the sheathing will act integrally with the stringers in carrying bending. A form of plate is thus achieved.

The rigidity of the complete system is also increased by this integral action. Structural depths are consequently less than in standard joist systems. Stressed skin elements are typically made off-site and put in place as modular units on the site. They are highly useful when they can be used in a highly repetitive way. They can be used in a variety of ways, including being made into long-span folded-plate systems.

BOX BEAMS. The built-up nature of plywood box beams [see Figure 15-2(c)] allows great latitude in the range of span and load conditions that they can be designed to meet. They are highly useful in long-span situations, or where unique loading conditions are present. Box beams can efficiently span greater distances than can homogeneous or laminated beams.

HEAVY TIMBER CONSTRUCTION. Heavy timber beams with transverse planking historically preceded the lighter joist systems [see Figure 15-2(e)]. Laminated timber beams are now often used in lieu of homogeneous members. Such systems can have a much higher load-carrying capacity and span range than joist systems. With laminated beams, for example, relatively long spans are possible, since member depths can be increased almost at will. Elements of this type are normally simply supported, but rigid joints can be achieved through special connections.

The vertical support system is typically either masonry walls or timber columns. The lateral resistance of the whole structural assembly to horizontal forces is typically obtained by using walls as shear planes. Knee braces are also used to obtain lateral stability when columns are used. Gaining lateral stability through moment-resisting joints is possible in low structures but is not frequently done.

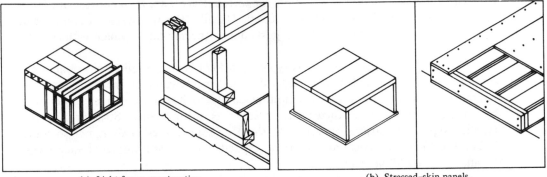

(a) Light frame construction.

(b) Stressed–skin panels.

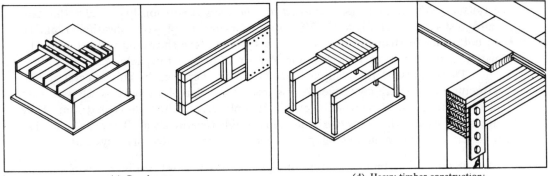

(c) Box beams.

(d) Heavy timber construction: laminated beams.

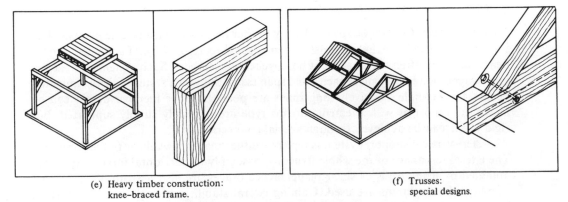

(e) Heavy timber construction: knee-braced frame.

(f) Trusses: special designs.

FIGURE 15-2 Timber construction systems.

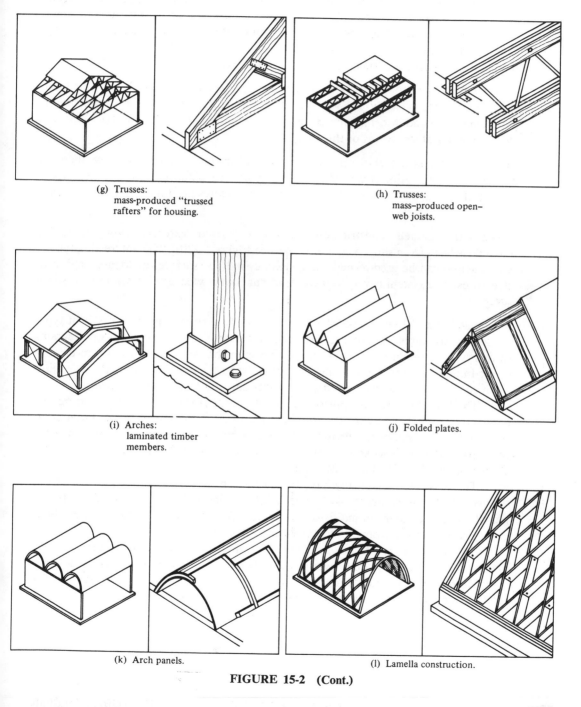

(g) Trusses:
mass-produced "trussed
rafters" for housing.

(h) Trusses:
mass-produced open–
web joists.

(i) Arches:
laminated timber
members.

(j) Folded plates.

(k) Arch panels.

(l) Lamella construction.

FIGURE 15-2   (Cont.)

511

TRUSSES. The timber truss is among the most versatile of all one-way spanning elements, since a wide variation is possible in the configuration and member properties used. Special handcrafted trusses are specially suitable for unique load or span conditions. Mass-produced timber trusses, however, are also available. These are typically used in light-load and modest span situations. The *trussed rafter* illustrated in Figure 15-2(g), for example, is extensively used in roofs of single-family detached housing. The system shown in Figure 15-2(h) is analogous to the steel open-web bar joist and is useful in long-span situations (particularly for roofs).

FOLDED PLATES AND ARCH PANELS. A variety of special flat- or curved-plate structures, which are typically one-way spanning elements useful for roofs, can be constructed from wood. Most involve the use of plywood. Figure 15-2(j) and (k) illustrates two examples of structures of this type.

ARCHES. Standard arch forms can be made from timber. Laminated members are most often used. Almost any shape of arch can be made using laminated wood. Relatively long spans can be obtained. These structures are typically useful as roofs only.

LAMELLAS. Lamella construction is a way of making extensive singly or doubly curved surfaces from short pieces of wood [see Figure 15-2(l)]. This interesting type of construction can be used to make long-span cylindrical surfaces or long-span domed structures. The system is very versatile and has found wide application in roof structures.

MEMBER SIZES. Figure 15-3 illustrates *approximate* span ranges for different types of timber structures. As noted earlier, the "maximum" spans indicated on diagrams of this type represent not maximum possible spans but spans that are not to be considered unduly unusual even if they are longer than typically encountered. The minimum span limitations represent a system's lower-economic-feasibility span.

The figure also shows approximate depth ranges for the different spanning systems. One number represents the usual minimum depth system while the other common maximums. An approximate depth of $L/20$ means, for example, that a member that spans 16 ft (4.9 m) should have a depth of about 16 ft (12 in/ft)/20 = 9.8, or 10 in. [(4.9 m)/20 = 0.24 m = 240 mm].

Timber columns typically have thickness to height (t/h) ratios that range from between 1 : 25 for relatively short and lightly loaded columns to around 1 : 10 for heavily loaded columns in multistory buildings. Walls made of timber elements have comparable t/h ratios, ranging from about 1 : 30 to 1 : 15.

Tables of the type shown in Figure 15-3 and the t/h ratios noted above should always be used with care and never as the only means for selecting and sizing a structure. Rule-of-thumb information should always be used only to develop a feeling for relationships among systems, spans, and depths, and nothing more.

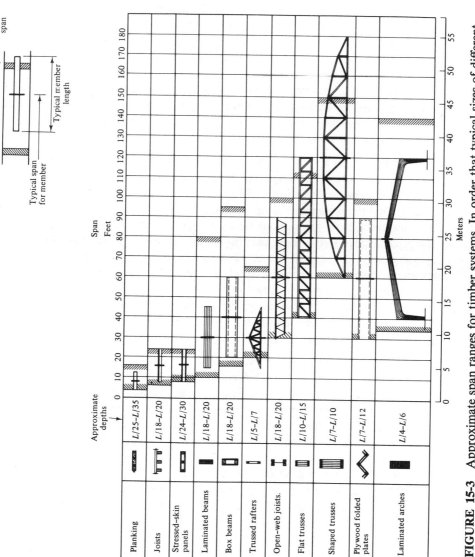

FIGURE 15-3 Approximate span ranges for timber systems. In order that typical sizes of different timber members can be relatively compared, the diagrams shown of the members are scaled to represent typical span lengths for each of the respective elements. The span lengths that are actually possible for each element are noted by the "maximum" and "minimum" span marks.

513

15.3 REINFORCED CONCRETE CONSTRUCTION

PRIMARY SYSTEMS. Reinforced concrete is a highly versatile material from which virtually any type of rigid structural element can be made. Figure 15-4 illustrates some common reinforced concrete systems.

SLABS AND BEAMS. Among the simplest of all reinforced concrete spanning systems is the conventional one-way solid slab [Figure 15-4(a)]. The easy formwork is an attractive feature of this system. These constant-depth systems are particularly suitable for short spans. With longer spans, the dead weight of the solid slab becomes excessive and ribbed slabs become preferable [Figure 15-4(b)].

One-way beam systems with transverse one-way slabs can be used to span relatively long distances (particularly if the beams are post-tensioned) and carry heavy loads. Such systems are relatively deep. Beam spacing is usually determined on the basis of what is most reasonable for the transverse slab.

ONE-WAY PAN JOIST SYSTEM. A one-way pan joist system is ribbed slab constructed by pouring concrete around special forms made of steel or fiberglass [see Figure 15-4(c)]. Transverse beams of virtually any depth can easily be cast in place at the ends of pans so that the system can adapt to a variety of column grids. Longitudinal beams more substantial than the normal ribs can also be cast in easily by varying the pan spacing. The ribbed slab is more suitable for longer spans than a solid slab. With post-tensioning, very long spans can be obtained. The pan joist system tends to be too complex and hence uneconomical for short spans. The vertical support system can be either columns or load-bearing masonry walls.

The ribbed slab and column system is capable of considerable lateral-load resistance, since transverse and longitudinal beams are cast into the floor system monolithically with the columns. Frame action can thus be achieved in both directions.

FLAT-PLATE CONSTRUCTION. The flat plate is a two-way constant-depth reinforced concrete slab system [see Figure 15-4(d)]. It is appropriate for use with light floor and roof loads and relatively short spans. It finds wide application in housing construction. Although a regular column grid is most appropriate, some flexibility is allowed in this respect. Indeed, flat plates are often used where the rigid orthogonality demanded by many other systems on the layout of the vertical supports is not desirable or possible. Spans are, however, limited in comparison with ribbed or beamed systems.

Lower floor-to-ceiling heights are possible with flat-plate construction than with many other systems. Relatively large amounts of reinforced steel, however, are typically required as a consequence of the thinness of the plates used. The governing design factor for flat plates is often the punch-through shear in the plate at the columns. Special steel reinforcement is often used at these points. At plate edges columns are also typically moved in from the free edge to ensure that the interface area between the slab and column remains as large as possible (see Chapter 10).

Lateral stability for the entire plate and column assembly can be a problem. Since the plate and columns are poured monolithically, joint rigidity is achieved which contributes to the lateral resistance of the structure and is sufficient for low buildings. Because of the thinness of the plate element, however, this resistance is limited. For

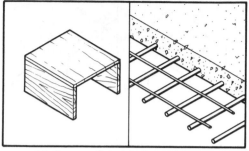

(a) One-way flat plate (poured in place).

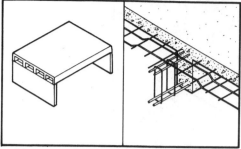

(b) One-way beam and slab system (poured in place).

(c) One-way pan joist system (poured in place).

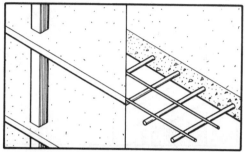

(d) Two-way flat plate (poured in place).

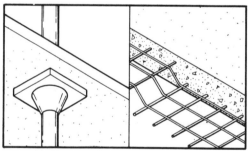

(e) Two-way flat slab (poured in place).

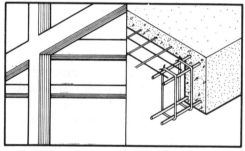

(f) Two-way beam and slab (poured in place).

(g) Two-way waffle slab (poured in place).

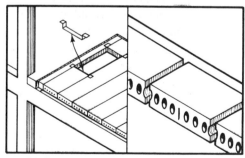

(h) Prestressed long-span planks (precast).

FIGURE 15-4 Reinforced concrete construction systems.

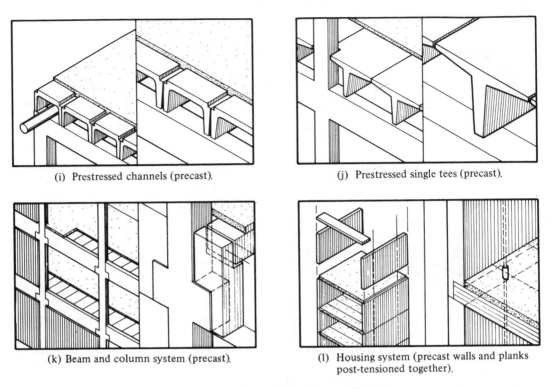

(i) Prestressed channels (precast).

(j) Prestressed single tees (precast).

(k) Beam and column system (precast).

(l) Housing system (precast walls and planks post-tensioned together).

FIGURE 15-4 (Cont.)

tall structures, stability is often achieved through shear walls or stiff poured-in-place core elements in the building such as might be possible around elevators or stairways.

The simple formwork involved is an undoubted virtue of this system. The planar nature of the lower surface also facilitates the design and placement of other building components. The system is quite often used for apartment buildings and dormitories which are comprised of functional spaces that are of limited span and cellular in nature.

FLAT-SLAB CONSTRUCTION. The flat slab is a two-way reinforced concrete system similar to the flat plate except that the area of interface between the plate and columns is increased by the addition of drop panels and/or column capitals [see Figure 15-4(e)]. The drop panels or column capitals serve to reduce the possibility of punch-through shear failure in the slab (see Chapter 10). The system is particularly appropriate for relatively heavy loading conditions (e.g., such as found in warehouses) and is suitable for spans larger than those possible with flat plates.

The capitals and drop panels also contribute to making the slab-and-column assembly more resistant to lateral loads than is the flat-plate system.

TWO-WAY BEAM AND SLAB CONSTRUCTION. A two-way beam and slab system consists of a flat reinforced concrete plate with beams, monolithically cast in place, along the periphery of the plate [Figure 15-4(f)]. The system is good for medium spans and high loading conditions. Large concentrated loads can also be supported if carried directly by the beams. Columns are invariably used for the vertical support system. Since beams and columns are cast monolithically, and there is a substantial interface between these elements, the system naturally forms a frame in two directions. This provides an appreciable lateral-load-carrying capacity and the system can consequently be used for multistory construction with ease.

THE WAFFLE SLAB. The waffle slab is a two-way constant-depth reinforced concrete system having ribs in two directions [see Figure 15-4(g)]. These ribs are formed by the use of special domed pans made of steel or fiberglass. The voids formed by the pans reduce the dead load weight of the structure. Waffle slabs are useful in longer-span situations than flat plates. These slabs can also be post-tensioned to increase their spans.

A thickness of concrete is usually left around column tops (by not using pans in these locations). This solid area serves the same function as drop panels or capitals in a flat slab. The possibility of shear failure is reduced and the moment-resisting capacity of the system is increased (and so is its lateral loadcarrying capacity).

Both the span of the waffle system and its lateral load-carrying capacity can be increased by casting in place beams spanning between columns. This is done quite simply by eliminating the pans along these lines (or spacing them farther apart), adding appropriate reinforcing, and casting a full depth of concrete (see Chapter 10).

CURVED SHAPES. Virtually any singly or doubly curved shape (e.g., a cylinder, dome, etc.) can be made from reinforced concrete. Reinforcing typically consists of a mesh of light steel rods throughout the shell with special additional steel used in localized areas of high internal force. Post-tensioning is commonly used for special elements (e.g., tension rings in domes; see Chapter 12).

PRECAST CONCRETE ELEMENTS. Precast concrete elements are fabricated off-site and transported to the job. They are typically one-way spanning elements that are most often pre-tensioned (see Section 6.4.7). A range of cross-sectional shapes are fabricated which are suitable for a wide variety of load and span conditions. They are most typically appropriate for uniformly distributed occupancy and roof loads and not for concentrated loads or unusually heavy distributed loads. These members are most often simply supported.

Moment connections are possible by using special connections, but are difficult. Large cantilevers are also difficult and must be kept to a minimum. Precast elements are most successfully used when they can be used in a highly repetitive way. Several different types of precast members are discussed below (many other types are also available).

PRECAST PLANKS. Both short-span and long-span planks are available. Short-span planks are used in much the same way as timber planking is used, although appropriate spans are slightly longer. A poured-in-place concrete wearing surface is normally placed on top of the planks. They are often used in conjunction with precast reinforced concrete beams or with steel open-web joists. Long-span planks are available which can span typically between 16 and 34 ft (5 to 11 m), depending on the exact width and depth of the element. These long-span planks are typically prestressed and also cored to reduce dead weights. A poured-in-place concrete wearing surface is placed on top of the planks. This concrete also forms a shear key between adjacent elements so that the resultant structure behaves like a one-way plate [see Figure 15-4(h)]. These planks are most appropriate for light occupancy or roof loads. They are invariably simply supported and are often used with load-bearing walls as the vertical support system (these walls must be either masonry or concrete, not wood stud). They are also frequently used with steel or reinforced concrete beams.

CHANNELS AND DOUBLE TEES. These are ribbed, one-way, precast, and prestressed elements that are suitable for longer spans than planks [Figure 15-4(i)]. They are appropriate for occupancy and roof loads. A poured-in-place concrete wearing surface is placed on top of adjacent members.

SINGLE TEES. These are typically large, precast, and prestressed elements most suitable for relatively long spans. They are rarely used in short-span situations because of the difficulties in erecting them. They are invariably simply supported. They are suitable for fairly heavy occupancy and roof loads. They are, for example, quite often used in parking garages and other buildings having fairly large spans and heavier than usual loads [Figure 15-4(j)].

SPECIAL BUILDING SYSTEMS. A whole host of systems are available for completely forming the entire shell of a building [Figure 15-4(l)]. Systems specially designed for housing are common. Approaches used to date usually fall into one of two categories: (1) those systems which employ off-site-produced linear or planar elements, such as walls or horizontal spanning systems that are assembled on-site (usually by a post-tensioning system) to form volumes; and (2) those systems in which volume-forming assemblies, such as complete boxes, are constructed off-site and then shipped to the site and aggregated. Many of these systems are proprietary in nature.

MEMBER SIZES. Figure 15-5 illustrates typical span ranges and depths for several typical reinforced concrete spanning systems. Reinforced concrete columns typically have thickness to height (t/h) ratios that range from between 1:15 for relatively short and lightly loaded columns to around 1:6 for column heavily loaded columns in multistory buildings. Reinforced concrete load-bearing walls have comparable t/h ratios ranging from about 1:22 to 1:10.

15.4 STEEL CONSTRUCTION

PRIMARY SYSTEMS. Most heavy steel constructional systems are made up of linear one-way spanning elements. A variety of rolled steel sections (e.g., wide flange), are available in various sizes for use. This variety allows a remarkable flexibility in the design of beam-and-column elements. Although simple support connections are typically used when possible, for construction convenience, it is relatively easy to make moment-resisting joints. Frames thus formed are capable of resisting lateral loads. Lateral stability is also often achieved by the use of shear walls or diagonal bracing elements.

BEAMS. Wide-flange shapes are commonly used as horizontal spanning elements [see Figure 15-6(a)]. The span range possible is rather wide. Members are usually simply

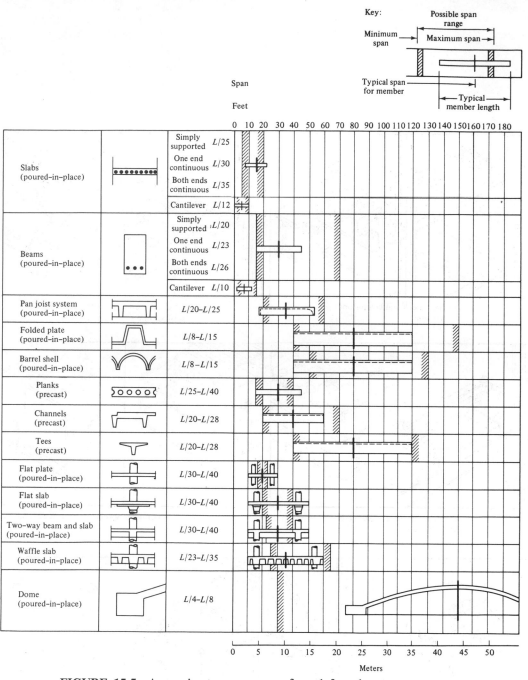

FIGURE 15-5 Approximate span ranges for reinforced concrete systems. In order that typical sizes of different members can be relatively compared, the diagrams shown of the members are scaled to represent typical span lengths for each of the respective elements. The span lengths that are actually possible for each element are noted by the "maximum" and "minimum" span marks.

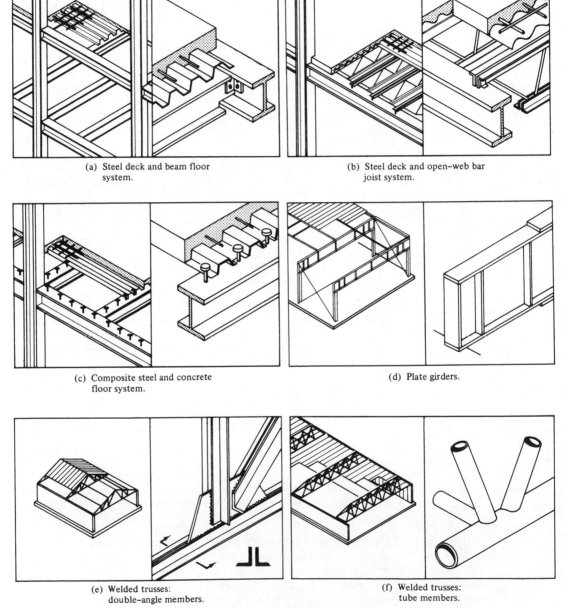

(a) Steel deck and beam floor system.

(b) Steel deck and open-web bar joist system.

(c) Composite steel and concrete floor system.

(d) Plate girders.

(e) Welded trusses: double-angle members.

(f) Welded trusses: tube members.

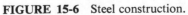

FIGURE 15-6 Steel construction.

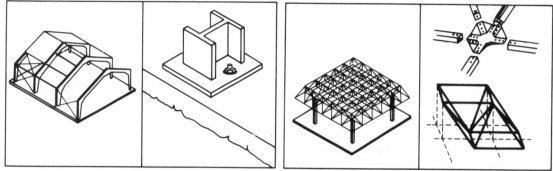

(g) Arches.

(h) Space frame.

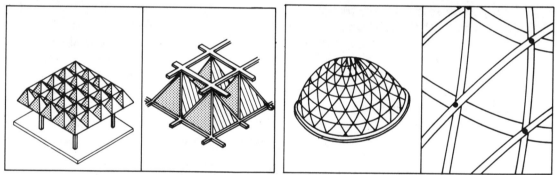

(i) Stressed–skin space frame.

(j) Ribbed dome.

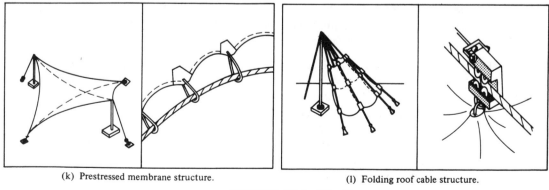

(k) Prestressed membrane structure.

(l) Folding roof cable structure.

FIGURE 15-6   (Cont.)

521

supported unless frame action is needed for stability, in which case moment-resisting connections are used. Other shapes, such as channels, are occasionally used to carry bending but are usually limited to light loads and short spans.

PLATE GIRDERS.   Plate girders are special types of beams. The built-up nature of plate girders [see Figure 15-6(d)] allows great latitude in the range of span and load conditions that they can be designed to meet. They are extremely useful when very high loads must be carried over medium spans. They are often used, for example, as major load-transfer elements carrying column loads over clear spans. For typically encountered distributed floor loads and span ranges, however, they are rarely used, since wide-flange elements are more appropriate and economical.

COMPOSITE CONSTRUCTION.   Many structural systems cannot be easily characterized according to material. The composite beam system illustrated in Figure 15-6(c) is fairly common. In this case the steel is first put in place and then concrete is cast around the shear connectors at the top of the steel beam. The shear connectors cause the steel and concrete to act integrally. A type of T section is thus formed with steel in the tension zone and concrete in the compression zone. If a concrete wearing deck is needed anyway, considerable economies can be achieved by this approach.

TRUSSES AND OPEN-WEB JOISTS.   An endless variety of truss configurations are of course possible. Potential load capacities and span ranges are enormous for uniquely designed or specially made trusses.

Open-web joists, which are mass-produced [see Figure 15-6(b)], can be used for both floor and roof systems. They are often highly economical for intermediate to long-span situations in which relatively light, uniformly distributed loads are involved. The bar joist, for example, is frequently used as a long-span roofing element. Open-web joists are typically simply supported (although it is possible to devise rigid connections) and thus make no direct contribution to the lateral resistance of the whole assembly. Often wide-flange and open-web joists are used in the same system, with the former having moment-resisting joints so that frame action capable of resisting lateral forces can be obtained.

Special heavier-gage open-web joists are available for very long span situations. These are normally appropriate only for light, uniformly distributed roof loadings.

ARCHES.   Rigid arches of virtually any shape can be specially made from steel. Prefabricated arches are available for low to moderate spans. Specially designed arches have been used in very long span situations [e.g., spans on the order of 300 ft (90 m) or more]. Steel arches can be made of solid sections or open-web sections.

SHELLS.   A whole variety of shell forms are possible using steel. The primary problem in using steel to achieve doubly curved surfaces is that of creating the shape by using line elements. In domes, for example, either ribbed or geodesic approaches are possible. Light steel decking of small dimensions is commonly used to form enclosure surfaces.

In some small-span situations, curved steel surfaces can be made by specially pressing steel sheets in a way similar to that done to create singly or doubly curved steel forms in automobile bodies.

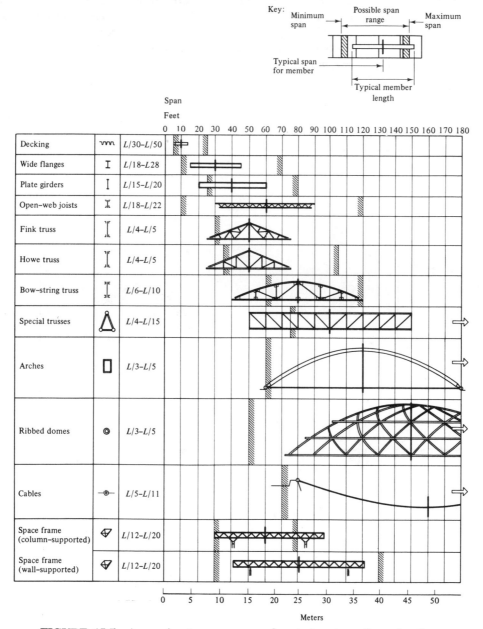

FIGURE 15-7 Approximate span ranges for steel systems. In order that typical sizes of different members can be relatively compared, the diagrams shown of the members are scaled to represent typical span lengths for each of the respective elements. The span lengths that are actually possible for each element are noted by the "maximum" and "minimum" span marks.

CABLE STRUCTURES. Steel is really the only material extensively used to create cable structures. The forms that cable structures can assume vary enormously. Cables can be used to create permanent roofs whose enclosure surfaces are made by either rigid steel planar elements such as decking or by membrane surfaces. The former are usually singly curved shapes while the latter are doubly curved. Movable roofs are also possible using a cable system [see Figure 15-6(l)]. These movable roofs are usually temporary enclosures or semipermanent at best.

MEMBER SIZES. Figure 15-7 illustrates the span ranges and depth/span ratios for several common steel spanning systems. Structural steel columns typically have thickness to height (t/h) ratios that range from between 1:24 to 1:9, depending on column heights and loads.

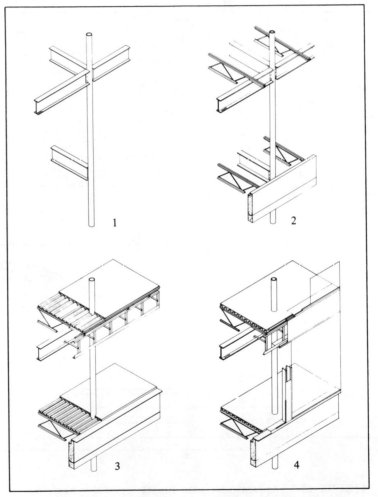

FIGURE 15-8 Willard State Hospital Administration Building, Cortland, N.Y. Werner Seligmann, Architect. Relationship between structure and enclosure fabric. (Drawings by Jeffrey Berg, Carl Gaines, David Holdorf, and Brian Randoll).

THE STRUCTURE IN CONTEXT. While obvious, it should nonetheless be emphasized that the structure of a building is only one aspect of the complete building (Figure 15-8). The selection of an enclosure system, for example, dictates in many cases the type of structural system used and its precise characteristics (and vice-versa). Rarely are structural systems selected in the abstract without consideration of other building systems.

QUESTIONS

15-1. Obtain a set of working drawings for a structural steel building, preferably one in your local area that you can visit. Draw to scale vignettes of the type illustrated in Figure 15-6. A scale of $\frac{3}{4}$ in. $= 1$ ft is suggested for the connection detail drawing and a scale of $\frac{1}{4}$ in. $= 1$ ft is suggested for the overview.

15-2. Repeat Question 15-1, except use a building made of poured-in-place reinforced concrete.

15-3. Repeat Question 15-1, except use building made of precast reinforced concrete elements.

15-4. Repeat Question 15-1, except use a building employing laminated timber elements.

15-5. Visit four different buildings that each have a different type of structural system and measure the dimensions of critical structural elements (e.g., beams, columns, trusses). Convert this information into rule-of-thumb type of information (e.g., L/d ratios).

chapter 16

CONNECTIONS

16.1 INTRODUCTION

How structural members join or meet is often a very critical design issue and one that can under certain circumstances influence the choice of the basic structural system itself, particularly its patterns and materials. The strategies possible in joining structural elements are strongly dependent on the physical properties and geometries of the elements to be joined. This subject has already been considered to some extent in Section 14.3. The following section broadens the discussion.

16.2 BASIC TYPES OF JOINTS

Consider the joining of simple linear rigid members. It is evident that most joints that are prevalently used employ a strategy of either *lapping* the basic elements, *deforming and interlocking* them or *butting* them. Monolithic joints, particularly in reinforced concrete, are also possible.

Most joints using one or another of these basic strategies also commonly make use of some type of *third-element connector*. Bolts, nails, and welds are typical third-element connectors. Third-element connectors may also involve the use of additional pieces (e.g., cover plates). The basic function of such connecting elements or assemblies is to aid in transferring the total load present at the joint from one element to another. A bolt connecting two lapped pieces, for example, transfers forces from one member to another through shear forces developed in the bolt itself. As will be discussed below, however, not all joints necessarily have to employ third-element connectors to serve this function. When used, the connecting elements or devices are typically smaller in terms of absolute dimensions than are the primary elements and are of higher relative strength (e.g., bolts connecting two lapped timber elements). If of the same relative strength, connectors are often of necessity larger so that loads can be transferred safely.

Figure 16-1(a) illustrates several *butt* joints. Third-element connectors are most often used with such joints. Joints can be made either pinned or rigid as discussed previously. In joints of this general type it is usual that either the vertical or horizontal element can be made continuous through the joint, but rarely both. This caveat obviously has significant structural implications, since if continuity is desired both vertically and horizontally, it is necessary to use special rigid joints with at least some of the elements. In steel construction, for example, it is typical to make columns continuous and to frame beams into column sides. If frame action is desired, rigid joints are then made. Butt joints can be advantageously used for unusual horizontal framing patterns. Figure 16-1(a) illustrates how many horizontal members can frame into the same point in a column by use of a third-element connector (the circular shelf). In such cases the column is invariably continuous and the beams discontinuous unless special rigid connections are made.

Figure 16-1(b) illustrates several *lapped* joints. These joints are particularly useful when it is desired that both the connected elements be made continuous through the joint. Third-element connectors are used to join the bypassing pieces. This approach is frequently used in timber construction to achieve continuity. This joinery strategy lends itself well to one-way framing systems. Complex patterns of connected elements are difficult to handle with this type of joint.

Figure 16-1(c) illustrates several joints in which members are *deformed* in order to make the connection. The strategy shown to the right in Figure 16-1(c) is often used with materials that are easily moldable, such as reinforced concrete. Many precast concrete building systems employ such joints. Timber elements can also be deformed to make connections. The ancient mortise-and-tendon joint is such a connection. By and large, these joints employ either simple or pinned connections between elements. It is, however, possible to make rigid connections.

Joints made by deforming and interlocking basic elements sometimes do not need obvious third-element connectors, nor do those which are made by taking advantage of the moldability properties of certain materials (such as reinforced concrete). The latter is considered as a subclass of a deformation strategy. Sometimes the use of a deformed and interlocking strategy means that the member must be made larger at the joint in order to accommodate the deformations. In some cases this increase may imply that the remainder of the member is made relatively larger also so that a constant-size member can be used. Occasionally, third-element connectors are used internally in deformed joints (e.g., in poured-in-place concrete special reinforcing steel is often used at joints).

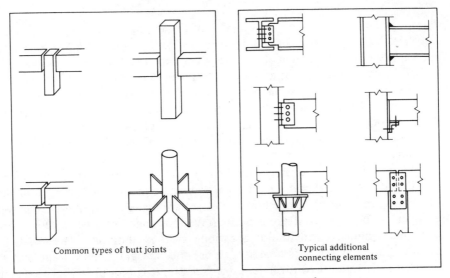

Common types of butt joints

Typical additional connecting elements

(a) Butt joints. Most butt joints require the use of one or more additional connecting elements to effect a load transfer from one element to another.

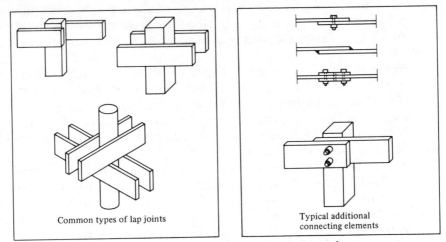

Common types of lap joints

Typical additional connecting elements

(b) Lap joints. Lap joints also typical require the use of one or more connecting elements to effect a load transfer from one element to another.

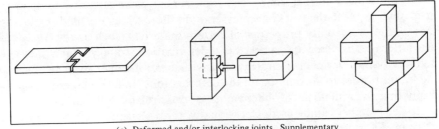

(c) Deformed and/or interlocking joints. Supplementary pieces are also often used but less reliance is placed on the supplementary pieces to transfer loads.

FIGURE 16-1 Basic types of joints.

16.3 BASIC TYPES OF CONNECTORS

16.3.1 General Considerations

The types of third-element connectors that are commonly used in making joints can be characterized as either *point* connectors (e.g., bolts, nails, rivets), line connectors (e.g., welds), or *surface* connectors (e.g., glued surfaces) (see Figure 16-2). The general type of connector used depends largely on the specific physical nature and geometry of the elements to be joined. The joining of large rigid surfaces, for example, generally calls for the use of distributed connectors. Joining two sheets of plywood via a lapping approach by using a single bolt, for example, is not feasible, since very large localized stresses would develop in the plywood around the bolt as it attempts to transfer loads from one sheet to the other. Bearing failures would undoubtedly occur. A preferable approach would be to use a series of distributed smaller bolts such that the amount of load transferred by each bolt is relatively small. Alternatively, a lapped and glued joint is also possible, depending on the loads involved.

The general approach noted above of using a number of smaller distributed-point connectors (or line or surface connectors) is a useful one to consider for virtually any situation. While it is true that single point connectors can effectively be used in many situations (particularly those involving primary members having small dimensions) a distributed approach usually allows for reduced localized stress concentrations in the members joined, which is usually very advantageous or necessary.

Another factor to consider in selecting a joining strategy is the obvious one of whether a pinned or rigid connection is necessary or desirable (Figure 16-3). As discussed elsewhere, each has its relative structural advantages and disadvantages. Single point connections are, of course, clearly most appropriate for making pinned joints, while line and surface connectors lend themselves more to rigid joints. Point connectors, however, can be used to create rigid joints if at least two point connectors are used which are separated in space (e.g., at the top and the bottom of a member) so that a resisting internal moment can be developed. Forces developing in the separated point connections act over the involved moment arm and provide the resisting

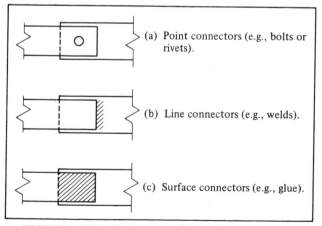

FIGURE 16-2 Basic types of lap joint connectors.

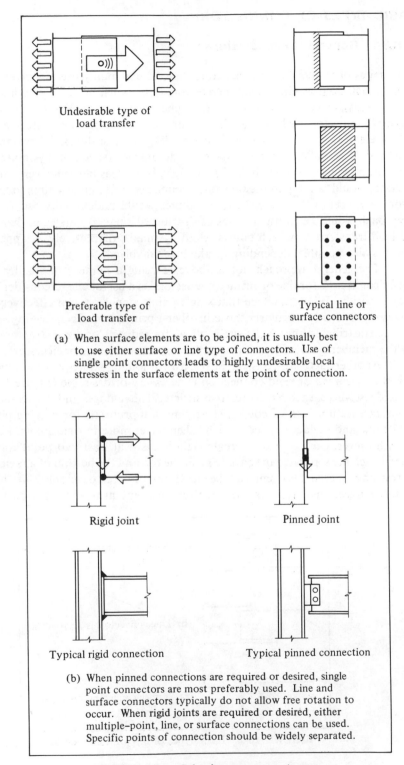

Undesirable type of
load transfer

Preferable type of
load transfer

Typical line or
surface connectors

(a) When surface elements are to be joined, it is usually best
to use either surface or line type of connectors. Use of
single point connectors leads to highly undesirable local
stresses in the surface elements at the point of connection.

Rigid joint

Pinned joint

Typical rigid connection

Typical pinned connection

(b) When pinned connections are required or desired, single
point connectors are most preferably used. Line and
surface connectors typically do not allow free rotation to
occur. When rigid joints are required or desired, either
multiple–point, line, or surface connections can be used.
Specific points of connection should be widely separated.

FIGURE 16-3 Selecting connector types.

moment. The greater the separation in space, the smaller are the forces developed in the connectors for a given moment resistance.

Line or surface connections can be used to approximate a pinned joint if localized in a region near the neutral axes of the connected members and if the extent of the joint is small relative to the size of the primary elements.

16.3.2 Bolts and Rivets

Simple lapped bolted joints rely on the shear capacity of the bolts to effect a load transfer between connected members. Many other types of bolted joints, however, do not depend on the shear capacity of the bolt at all for safety. Bolts are often tightened such that the frictional forces that develop between the compressed lapped plates that are joined are sufficient to effect a load transfer between elements. Obviously, bearing problems are reduced, as are other problematic phenomena. Care must be taken in field operations, however, to assure that bolts are adequately tightened. The following considers the simpler type of joints that rely only on the shear capacity of the bolts or rivets to transfer loads.

Bolts or rivets are usually used in situations involving lap joints. In a typical lap joint (see Figure 16-4) several types of failure are possible: a shear failure in the bolt, a tension failure in the reduced cross section of the loaded member, a crushing or bearing failure in the loaded member caused by excessively high bearing stresses at the interface with the bolt, a shear failure in the loaded member behind the bolt, and a tearing-out type of failure in the loaded member behind the bolt.

For sizing a bolt in a simple lap connection, the most important considerations are shear stresses in the bolt and bearing stresses in the connected members. With respect to shear stresses, it is possible to assume reasonably that shearing stresses are uniformly distributed across the face of the bolt. This assumption is generally valid *if* the depth of the shearing area is small and the distance between shearing forces is also small. This is the type of condition that exists when a bolt is used to connect plates together. Such an approximation is not valid and would lead to unconservative answers in other situations involving large shearing areas and large distances between shear forces—the case in a common beam. If shear stresses can be assumed to be uniformly distributed, then they are simply given by $f_v = P/A$, where P is the applied load and A is the cross-sectional area of the bolt. The allowable load in shear for a bolt in the single shear state illustrated in Figure 16-4(a) is simply $P_v = AF_v$, where A is the cross-sectional area of the bolt and F_v is the allowable stress of the bolt in shear. Note that $P_v = (\pi d^2/4)F_v$.

EXAMPLE

For a bolt in single shear, what bolt diameter is required if the shear force present, P_v, is 4000 lb (17,992 N)? Assume that $F_v = 14{,}000$ lb/in.^{2} (96.5 N/mm^2).

Solution:

Area required:

$$A = \frac{P_v}{F_v} = \frac{4000 \text{ lb}}{14{,}000 \text{ lb/in.}^2} = 0.285 \text{ in.}^2$$

$$= \frac{17{,}992 \text{ N}}{96.5 \text{ N/mm}^2} = 184 \text{ mm}^2$$

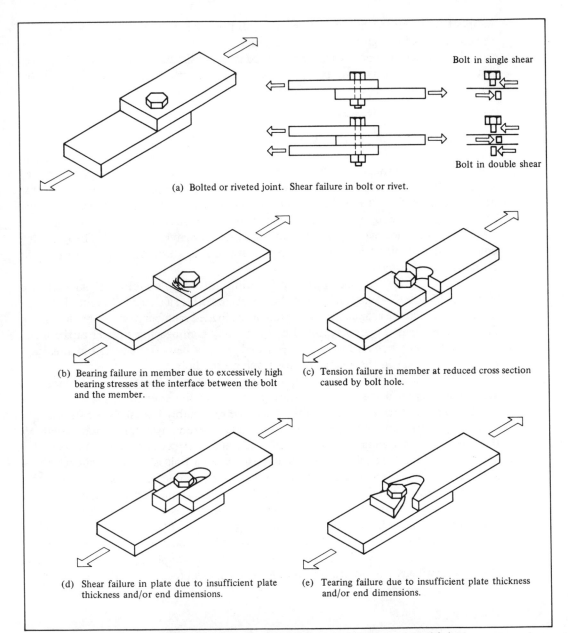

(a) Bolted or riveted joint. Shear failure in bolt or rivet.

Bolt in single shear

Bolt in double shear

(b) Bearing failure in member due to excessively high bearing stresses at the interface between the bolt and the member.

(c) Tension failure in member at reduced cross section caused by bolt hole.

(d) Shear failure in plate due to insufficient plate thickness and/or end dimensions.

(e) Tearing failure due to insufficient plate thickness and/or end dimensions.

FIGURE 16-4 Types of failure in bolted or riveted joints.

Diameter required:

$$\frac{\pi d^2}{4} = 0.285 \text{ in.}^2 \quad \text{or} \quad d = 0.605 \text{ in.}$$

$$= 184 \text{ mm}^2 \quad \text{or} \quad d = 15.3 \text{ mm}$$

The nearest stock size bolt would be used. When a bolt is in the double-shear state illustrated in Figure 16-4(a), two shear planes are present and thus $P_v = 2(\pi d^2/4)F_v$.

Bearing stresses can be found by simply $f_b = P/A_b$, where A_b is the area of the bolt and plate interface. A_b is usually taken as the projected area of the bolt, or $A = td$. Thus, $f_b = P/td$. Bearing stresses often control the design of a joint. Bearing stresses can evidently be reduced by either increasing the thickness of the plate present or by increasing the diamter of the bolt.

Other types of allowable loads for the connected elements can be found similarly by simply considering the area of the probable failure plane (A_p) and noting that $P_{max} = F_p A_p$, where F_p is the allowable stress for the material failure mode present.

When bolts are not symmetrically loaded, care must be taken to include in the load calculations any forces induced by the torque or twisting action of the imbalanced load (see F. L., Singer, *Strength of Materials*, and ed., Harper & Row, Publisher, New York, 1962. Chap. 12).

16.3.3 Welded Joints

Welding is the joining of metals by fusion. High heat melts and fuses metal from a welding rod onto the plates being connected. A continuous and homogenous joint results. There are a huge number of different types of welded joints. Most are, however, a variant of either a butt weld or a fillet weld (see Figure 16-5).

The strength of a unit length of a butt weld in tension is simply equal to the allowable stress of the weld material in tension times the minimum thickness of the weld. For a specified thickness, the length of weld used is directly proportional to the load to be transmitted.

The strength of a fillet weld is dependent on the shear resistance of the weld since failure normally occurs at the throat of the weld. In a 45° weld with a leg thickness, t,

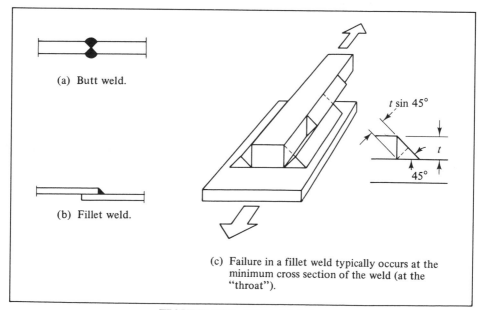

(a) Butt weld.

(b) Fillet weld.

(c) Failure in a fillet weld typically occurs at the minimum cross section of the weld (at the "throat").

FIGURE 16-5 Welded joints.

the shearing area through the throat is $A = Lt \sin 45°$, where L is the length of the weld. Thus, the strength of the weld is given by $P = AF_v = L(0.707t)(F_v)$. A common allowable stress is $F_v = 13{,}600$ lb/in.^{2}. Thus, $P = 9600tL$. In terms of a $\frac{1}{16}$-in. weld 1 in. long, $P = (600 \text{ lb/in.})/(\frac{1}{16} \text{ in})$. Values such as this are useful for quickly estimating required weld lengths and thicknesses to carry given loads. A load of 4800 lb, for example, requires 8 in. of a $\frac{1}{16}$-in. weld, 4 in. of $\frac{1}{8}$-in. weld, or 2 in. of $\frac{1}{4}$-in. weld.

As with bolts, care must be taken to account for additional torque forces induced in weld joints by nonsymmetrically applied forces. Often, loads may look symmetrically applied and welds may look symmetrically placed, as when an angle member is welded to a plate, but the welds may be unequally loaded because the centroid of the member is nonsymmetrical and hence so is the applied load (see Singer, *Strength of Materials*).

QUESTIONS

16-1. Examine and draw diagrams of several different types of structural joints found in buildings in your local area. Classify the joints as lapped, butt, or deformed (or a combination thereof). Are third element connectors present?

16-2. With respect to shear stresses alone, what is the required diameter for a bolt in single shear that transfers a shear force of 6000 lb between two plates? Assume that $F_v = 14{,}000$ lb/in.^{2}.
Answer: $\frac{3}{4}$ in. diameter.

16-3. How many inches of $\frac{1}{8}$-in.-weld are necessary to transfer a shear force of 6000 lb from one plate to another? Assume that $F_v = 13{,}600$ lb/in.^{2}.
Answer: 5 in.

APPENDICES

A1—A13

APPENDIX 1: CONVERSION TABLES

The following table shows the relationship between the system of units currently used in the United States, the conventional metric system, and the *Système International d'Unités* (SI). The SI system is gaining broad acceptance as the preferred system of units.

Current U.S. units	Conventional metric units	SI units
Length		
1 ft	0.3048 m	0.3048 m
1 in.	2.540 cm	25.40 mm
Area		
1 ft²	929.0 cm²	0.09290 m²
1 in.²	6.452 cm²	645.2 mm²
Density		
1 lb/ft³	16.019 kg/m³	16.019 kg/m³
Force		
1 lb	0.4536 kg	4.448 N

Current U.S. units	Conventional metric units	SI units
Moment		
1 ft-lb	0.1383 kg-m	1.356 N·m
1 in.-lb	0.01152 kg-m	0.1130 N·m
Stress		
1 lb/ft² (psf)	4.882 kg/m²	47.88 N/m² = 47.88 Pa
1 lb/in.² (psi)	0.07031 kg/cm²	6895 N/m² = 6.895 kPa
		= 0.006895 N/mm²
		= 0.006895 MPa
Section modulus		
1 in.³	16.39 cm³	16.39 × 10³ mm³
Moment of inertia		
1 in.⁴	41.62 cm⁴	416.2 × 10³ mm⁴

The basic unit of force in the SI system is the *newton* (N). A newton is the force required to give a mass of 1 kilogram due to an acceleration of 1 meter per second per second. Since weight is a force, it is measured in newtons. Since the acceleration due to earth's gravity is 9.807 m/s^2, the weight of a 1 kilogram mass is taken as 9.807 N. Also note that:

1 pascal (Pa)	$= 1 \text{ N/m}^2$
1 kPa	$= 1 \times 10^3 \text{ N/m}^2 = 1 \times 10^{-3} \text{ N/mm}^2$
1 MPa	$= 1 \times 10^6 \text{ N/m}^2 = 1 \text{ N/mm}^2$

APPENDIX 2: CENTROIDS

Of fundamental importance in the study of mechanics are the centers of mass and gravity of bodies and the centroids of geometric figures. The center of mass is that point which represents the mean position of matter in a body. The center of gravity of a body essentially represents the position of the resultant of the earth's gravitational pull on a body and is thus that point in a body from or on which the body could be perfectly balanced or poised in equilibrium. Strictly speaking, the concept of center of gravity is applicable only to bodies having weight. The term *centroid* is used to describe the analogous point in a geometric form such as a line or area. The centroid of an area can be defined as that point at which the entire area may be conceived to be concentrated and have the same moment with respect to any axis as the original distributed area. With respect to this definition, it may be useful to visualize the geometric figure considered, such as a planar area, as a sheet of constant infinitesimal thickness which has a uniform mass per unit area. This makes the analogy with the center of gravity of a body more direct. Thus, the centroid of an area can be visualized as that point on which a geometric figure could be balanced. More precise definitions, however, can be found in almost any basic calculus textbook. A typical formulation is as follows.

For a plane area in the xy plane, the coordinates of the centroid are

$$\bar{x} = \frac{\int_A x \, dA}{A} \qquad \bar{y} = \frac{\int_A y \, dA}{A}$$

These expressions are simply formal statements of the definitions previously cited. The algebraic sum of the moments of each elementary area unit with respect to a given point or line is equal to the moment of total area about the same point or line. The terms $x \, dA$ and $y \, dA$ are often referred to as the *first moments* of the area about the reference area. If the reference axis used actually coincides with the centroidal axis (i.e., $\bar{x} = 0$, $\bar{y} = 0$), it follows that

$$\int_{A_c} x \, dA = 0 \qquad \int_{A_c} y \, dA = 0$$

When an area is symmetrical with respect to an axis, the centroid invariably coincides with the axis of symmetry (see Figure A-2-1). This follows from the fact that the moments of the areas on opposite sides of the axis are equal in magnitude but opposite in sign. For areas having more than one axis of symmetry, the centroid must be a point that lies at the intersection of the axes of symmetry.

Determining the location of the centroid of an area is a straightforward process based on the use of the equations for $\bar{x}$ and $\bar{y}$. Any set of reference axes can be used to find the centroid of an area. The process for finding the centroid of a common figure, a triangle, will be illustrated. The base of the triangle is arbitrarily selected as a reference axis.

EXAMPLE

Find the centroid for the triangle indicated in Figure A-2-1.

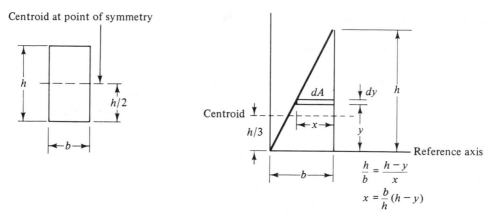

(a) Symmetrical section.   (b) Nonsymmetrical section.

FIGURE A-2-1 Centroids of geometric figures.

Solution:

First set up an expression for use in

$$\bar{y} = \frac{\int_A y \, dA}{A} = \frac{\int_A y \, dA}{\int dA}$$

Thus,

$$dA = x \, dy = \frac{b}{h}(h - y) \, dy$$

Hence,

$$\bar{y} = \frac{\int_0^h y[(b/h)(h - y)] \, dy}{\int_0^h (b/h)(h - y) \, dy} = \frac{1}{3} h$$

Many complex figures, commonly referred to as *composite areas*, can be considered as composed of simpler geometric shapes. If a given area can indeed be divided into parts (whose individual centroidal locations are known), the location of the centroid for the whole composite figure can be found without integration. This is done by obtaining the algebraic sum of the area moments of each individual part and dividing by the total area of the composite figure. This procedure follows from the definitions cited previously. Stated formally:

$$\bar{x} = \frac{\sum A_{x_i}}{A_T} \qquad \bar{y} = \frac{\sum A_{y_i}}{A_T}$$

These expressions are obviously similar in character to $\bar{x} = \int_A x \, dA/A$ and $\bar{y} = \int_A y \, dA/A$. An example of how this expression is applied is given in Chapter 6.

APPENDIX 3: MOMENTS OF INERTIA

A.3.1 General Formulation

One of the most common expressions encountered in the analysis of structures is of the form $\int y^2 \, dA$. This expression is typically termed the moment of inertia (I) of the area with respect to a reference axis. Examined by itself, it has little physical meaning. In the context of analyzing beams, columns, and other elements, however, it is of paramount importance as a general descriptor of the way material in the element is organized or distributed with reference to the cross section of the shape.

The moment of inertia of an area with respect to an axis in the plane of the area can be defined as the sum of the products obtained by multiplying each element of the area by the square of its distance from this axis. Thus, the moments of inertia of an area with respect to the x and y axes are

$$I_x = \int_A y^2 \, dA \qquad I_y = \int_A x^2 \, dA$$

Strictly speaking, the name *moment of inertia* is misleading, since inertia is a property of physical bodies only. Since the above could be written as $I_x = \int_A y(y\,dA)$ and $I_y = \int_A x(x\,dA)$ the term *second moment of the area* is often considered preferable. The first moment of an area, as noted in the previous section, is of the form $x\,dA$ or $y\,dA$; hence, multiplying this term by the distance again yields the second moment of the area. Whereas the first moment of an area about an axis can be either positive or negative, the second moment is always positive. The term "moment of inertia," however, is widely used and is used here.

EXAMPLE

Consider the rectangle shown in Figure A-3-1. Determine the moment of inertia, I, of the area with respect to its centroidal axis (which is, by inspection, midheight). Any other reference axis could be used, but selecting the centroidal axis is most meaningful in the context of analyzing beams or columns since the minimum I value for the section is obtained about this axis.

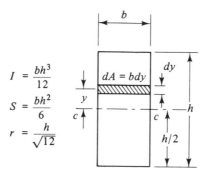

$$I = \frac{bh^3}{12}$$

$$S = \frac{bh^2}{6}$$

$$r = \frac{h}{\sqrt{12}}$$

FIGURE A-3-1 Moment of inertia of a rectangle.

Solution:

$$I_c = \int_A y^2\,dA$$

$$dA = b\,dy$$

so

$$I_c = \int_{-h/2}^{+h/2} y^2(b\,dy) = b\left[\frac{y^3}{3}\right]_{-h/2}^{+h/2}$$

$$I_c = \frac{bh^3}{12}$$

For some shapes, such as circles, it is often more convenient to use polar coordinates rather than cartesian coordinates to find I values.

Table A-3-1 gives moment of inertia values for many common geometric figures.

Table A-3-1. PROPERTIES OF GEOMETRIC SECTIONS

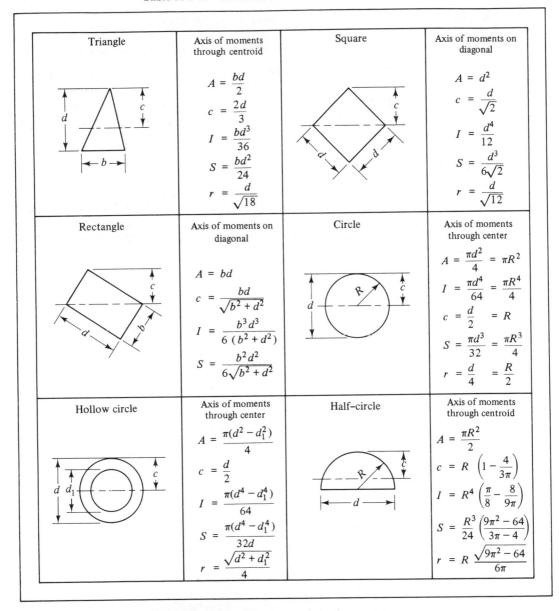

Triangle	Axis of moments through centroid	Square	Axis of moments on diagonal
	$A = \dfrac{bd}{2}$		$A = d^2$
	$c = \dfrac{2d}{3}$		$c = \dfrac{d}{\sqrt{2}}$
	$I = \dfrac{bd^3}{36}$		$I = \dfrac{d^4}{12}$
	$S = \dfrac{bd^2}{24}$		$S = \dfrac{d^3}{6\sqrt{2}}$
	$r = \dfrac{d}{\sqrt{18}}$		$r = \dfrac{d}{\sqrt{12}}$

Rectangle	Axis of moments on diagonal	Circle	Axis of moments through center
	$A = bd$		$A = \dfrac{\pi d^2}{4} = \pi R^2$
	$c = \dfrac{bd}{\sqrt{b^2 + d^2}}$		$I = \dfrac{\pi d^4}{64} = \dfrac{\pi R^4}{4}$
	$I = \dfrac{b^3 d^3}{6(b^2 + d^2)}$		$c = \dfrac{d}{2} = R$
	$S = \dfrac{b^2 d^2}{6\sqrt{b^2 + d^2}}$		$S = \dfrac{\pi d^3}{32} = \dfrac{\pi R^3}{4}$
			$r = \dfrac{d}{4} = \dfrac{R}{2}$

Hollow circle	Axis of moments through center	Half–circle	Axis of moments through centroid
	$A = \dfrac{\pi(d^2 - d_1^2)}{4}$		$A = \dfrac{\pi R^2}{2}$
	$c = \dfrac{d}{2}$		$c = R\left(1 - \dfrac{4}{3\pi}\right)$
	$I = \dfrac{\pi(d^4 - d_1^4)}{64}$		$I = R^4\left(\dfrac{\pi}{8} - \dfrac{8}{9\pi}\right)$
	$S = \dfrac{\pi(d^4 - d_1^4)}{32d}$		$S = \dfrac{R^3}{24}\left(\dfrac{9\pi^2 - 64}{3\pi - 4}\right)$
	$r = \dfrac{\sqrt{d^2 + d_1^2}}{4}$		$r = R\dfrac{\sqrt{9\pi^2 - 64}}{6\pi}$

A.3.2 Parallel-Axis Theorem

If the moment of inertia of a figure with respect to its centroidal axis is known, the moment of inertia with respect to *any* parallel axis may be found quite easily. Doing this is a particularly useful tool in determining the moment of inertia of a composite area.

Consider the area illustrated in Figure A-3-2, which has a centroidal location c and a moment of inertia about its own axis of $\bar{I}$. The moment of inertia of this same area about a parallel axis located a constant distance d from the centroidal axis is given by

$$I_T = \int_A r^2 \, dA = \int_A (y + d)^2 \, dA$$

$$= \int_A y^2 \, dA + 2d \int_A y \, dA + d^2 \int_A dA$$

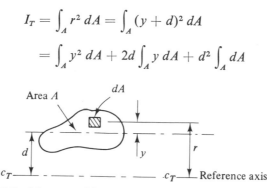

FIGURE A-3-2 Moment of inertia of an area about an axis located a distance d away from the centroid of the area.

The first term, $\int_A y^2 \, dA$, in the expanded expression is the moment of inertia of the area about it own centroidal axis $\bar{I}$. The second term, $2d \int_A y \, dA$, is equal to zero since it involves $\int_A y \, dA$, which is simply the first moment of the area about its own centroidal axis (as noted in the discussion on centroids, this value is identically equal to zero when the reference axis corresponds to the centroid of a figure). The final term, $d^2 \int_A dA$, is more simply $d^2 A$, where A is the total area of the figure. The expression becomes

$$I_T = \bar{I} + Ad^2$$

Thus, the moment of inertia for the area with respect to any axis in the same plane is equal to the moment of inertia of the area about its own axis plus a transfer term composed of the product of the square of the distance between axes and the area of the figure. This implies that *the minimum moment of inertia a figure can have is about its own centroidal axis*. This is why the centroidal axis is often used as reference axis when applied to engineering calculations.

The only time the moments of inertia of individual areas forming a larger composite shape can be added directly is when their individual axes coincide with that of the larger composite shape. In such a situation $d = 0$ for each individual shape. Hence $I = \sum (\bar{I} + Ad^2) = \sum (\bar{I})$.

A.3.3 Negative Areas

In many symmetrical shapes it is often convenient to decompose the figure into what are commonly termed positive and negative areas. A *positive area* adds to the actual area or figure and contributes positively to the moment of inertia of the figure, while a *negative area* produces converse effects. This is merely an alternative way of looking at what has been discussed just previously, but one that provides a different insight into how different areas in a cross section affect the magnitude of the moment of inertia.

EXAMPLE

Determine the moment of inertia of the cross-sectional shape shown in Figure A-3-3. Assume that $b_1 = 10$ in. (254 mm), $h_1 = 10$ in. (254 mm), $b_2 = 8$ in. (203.2 mm) and $h_2 = 6$ in. (152.4 mm). Note that $d_1 = d_2 = 0$, hence $I = \sum (\bar{I})$.

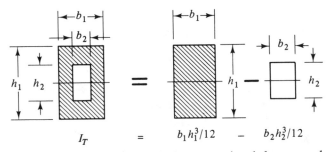

$$I_T \quad = \quad b_1 h_1^3/12 \quad - \quad b_2 h_2^3/12$$

FIGURE A-3-3 Holes in symmetrical cross-sectional shapes can be treated as negative areas having negative moments of inertia.

Solution:

$$I_T = I_{\text{positive}} - I_{\text{negative}}$$

$$= \frac{10 \text{ in.} \times (10 \text{ in.})^3}{12} - \frac{8 \text{ in.} \times (6 \text{ in.})^3}{12}$$

$$= 689 \text{ in.}^4$$

$$= \frac{(254 \text{ mm}) \times (254 \text{ mm})^3}{12} - \frac{(203.2 \text{ mm}) \times (152.4 \text{ mm})^3}{12}$$

$$= 286,900 \times 10^3\text{-mm}^4$$

APPENDIX 4: BENDING STRESSES IN BEAMS

The main issue addressed in this section is the relationship among the external bending moment present at a section in a beam, the properties of the beam, and the bending stresses which are reactively generated at this section in response to the bending moment. The basic principle used in establishing this relationship is that the action of the bending stresses is to produce an internal resisting moment, M_R, to balance the external moment, M_E, present at the section.

Consider the beam shown in Figure A-4-1. As noted in Chapter 6, the effect of bending is to produce deformations in the beam fibers of the type illustrated in Figure A-4-1. It should be noted that while a plane of zero deformations and zero bending stresses (i.e., the *neutral axis*) is known to exist, its exact location is not known a priori. Finding the location of this plane is a necessary prerequisite to further analysis. This surface can be located by considering the equilibrium of the beam in the hori-

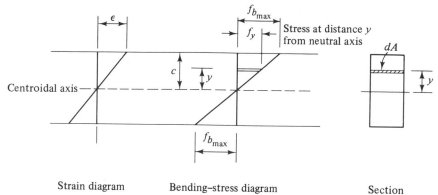

Strain diagram   Bending–stress diagram   Section

FIGURE A-4-1 Bending stresses in beams.

zontal direction. First, we must find the resultant forces of the stresses acting horizontally.

Assume that the maximum stress on a face a distance c from the neutral axis is designated $f_{b_{max}}$. The stress f_y at an arbitrary distance y from the neutral axis can be found through simple proportions [i.e., $f_y/y = f_{b_{max}}/c$ or $f_y = (y/c)(f_{b_{max}})$]. At a level defined by y, the force associated with the stresses f_y is simply $f_y\, dA$. The total force in the horizontal direction produced by the entire stress field is merely $\int f_y\, dA$. We also know that $\sum F_x = 0$. If there are no external horizontal forces acting on the beam, the total force produced by the tensile and compressive components of the whole stress field must total zero. Hence, $\int_A f_y\, dA = 0$. Expressing f_y in terms of the maximum stress, the outer fiber of the beam, we have $(f_{b_{max}}/c)\int_A y\, dA = 0$. Since $(f_{b_{max}}/c)$ cannot be zero, it follows that $\int_A y\, dA = 0$. The quantity $\int y\, da$ is termed the *first moment* of the area of the beam with respect to the neutral axis. Terms of this general form are commonly found in mathematics and are of extreme importance in the study of beams, and, as such, are covered in detail in Appendix 2. There the concept of the *centroid* of an area [i.e., that point on which a geometric figure could be visualized as balancing (another way of saying that $\int_A y\, dA = 0$)] is extensively discussed. For our purpose here, the important point is that $\int_A y\, dA = 0$ defines the centroidal axis of a geometric figure and is readily calculable for any type of beam configuration. In the context of beam analysis, the surfacing of the term $\int_A y\, dA = 0$ means that the neutral axis of a beam corresponds with the centroidal axis of the cross-sectional shape of the beam.

The process of finding the relation between the bending stresses, section properties and the applied bending moment can now begin. It was noted above that the sum of the elemental forces produced by the stresses in the horizontal direction had to equal zero from equilibrium considerations (i.e., $\sum F_x = 0$). Note that for $\sum M = 0$ the sum of the *moments* produced by these elemental forces about the neutral axis must equal the applied external moment. The moment of an elemental force about the neutral axis must equal the applied external moment. The moment of an elemental force about the neutral axis is simply $y(f_y\,dA)$ or $y(f_{b_{max}})(y/c)\,dA$. The sum of the moments of all these elemental forces thus becomes $(f_{b_{max}}/c)\int_A y^2\,dA$. This, then, is the internal resisting moment, M_R, which identically equals the applied external moment, M_E. Thus, $M = (f_{b_{max}}/c)\int_A y^2\,dA$. The term $\int_A y^2\,dA$ is commonly called the *second-moment* of an area when encountered in mathematics, and the *moment of inertia, I,* of an area in an engineering context. In general, it is a descriptor of the way material in a beam is organized or distributed in a cross section, and characterizes the stiffness of the beam as derived from the way material is distributed (*not* from the intrinsic physical properties of the material). The term can be evaluated for any type of cross-sectional shape (see Appendix 3).

The final expression relating the maximum bending stresses ($f_{b_{max}}$), to the properties of the cross section (defined by $I = \int_A y^2\,dA$ evaluated about the neutral or centroidal axis of the cross section), and the moment M present at the section becomes $f_{b_{max}} = Mc/I$. The stress at any location y from the neutral axis is $f_y = My/I$.

APPENDIX 5: SHEARING STRESSES IN BEAMS

As noted in Section 6.3.2, shearing stresses are present in any beam. This section derives relationships among the physical characteristics of a beam, the forces acting on the beam, and the shearing stresses that are developed in the beam as a consequence of these forces. *Horizontal shearing stresses* will be studied first.

Consider an infinitesimal element of a beam as illustrated in Figure A-5-1. As typical in any beam, the bending moment, and consequently the bending stresses, will

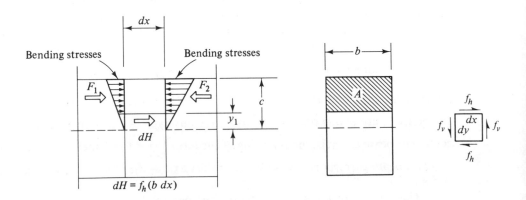

Horizontal shear stresses balance the difference in F_1 and F_2

FIGURE A-5-1 Horizontal shear stresses in beams.

be larger at one section than the other. For equilibrium to obtain in the horizontal direction, a horizontal shear force must be developed as indicated in Figure A-5-1 to balance the difference in force produced by the action of bending. If f_h represents the average shear stress over the differential area of length dx and width b, and if one expresses the bending forces in terms of the bending stresses acting on the beam, then equilibrium considerations in the horizontal direction yield the following:

$$\underbrace{f_h(b\ dx)}_{dH} = \underbrace{\int_{y_1}^{c} \left(\frac{M_2 y}{I}\right) dA}_{F_2} - \underbrace{\int_{y_1}^{c} \left(\frac{M_1 y}{I}\right) dA}_{F_1}$$

or, noting that $M_2 - M_1$, represents the differential change in moment between over the distance dx:

$$f_h = \frac{dM}{Ib\ dx} \int_{y_1}^{c} y\ dA$$

In Section 2.4.5 it was noted that $dM/dx = V$ (the vertical shear force); thus,

$$f_h = \frac{V}{Ib} \int_{y_1}^{c} y\ dA$$

The integral $\int_{y_1}^{c} y\ dA$ is the first moment of the area *above the horizontal section considered* (and where the horizontal shear stresses in the expression above are located) with respect to the neutral axis of the beam. This integral is commonly denoted Q. Hence,

$$f_h = \frac{VQ}{Ib}$$

This is the general expression for horizontal shearing stress in a beam of any cross section. Note that the maximum shear stress occurs at the point where $\int_{y_1}^{c} y\ dA$ is maximum. This occurs when the horizontal section considered coincides with the neutral axis.

In a *rectangular beam*, the distribution of shearing stress can be found by setting up a general expression for the shear stress for a layer a distance y from the neutral axis.

For a rectangular beam:

$$Q = \int_{y_1}^{c} y\ dA = \bar{y}A = \left[b\left(\frac{h}{2} - y\right)\right]\left[y + \frac{1}{2}\left(\frac{h}{2} - y\right)\right]$$

Thus,

$$f = \frac{VQ}{Ib} = \frac{V}{2I}(h^2/4 - y^2)$$

This indicates that the shearing stress is parabolically distributed. As is evident, the maximum shear stress occurs at $y = 0$ (the neutral axis).

At any point in a beam, the horizontal shearing stress is accompanied by an equal vertical shearing stress f_v. This can be demonstrated in a variety of ways. A simple way is to look at the equilibrium of a typical element in a beam (see Figure A-5-1) subjected only to shear. By summing the moments of the forces produced by the stresses acting on the element, we find that

$$f_h(dx\,dz)\,dy - f_v(dy\,dz)(dx) = 0$$

$$f_h = f_v$$

It is this vertical shear stress that actually forms the resisting shear force, $V_R = \int f_v\,dA$, which equilibrates the external applied shear force, V_E. Since $f_h = f_v$, the distribution of the vertical shear on the face of a cross section is described for the horizontal shear stresses. This is the only way of finding the distribution of vertical shearing stresses in a beam.

APPENDIX 6: MOMENT–CURVATURE RELATIONS

This section explores the relation between the moment at a point and the curvature of the member at the same point. That a relation should exist between moments and curvatures should be expected. Moment causes a bowing or bending in a structure. The higher the moment, the greater the bowing. This section quantifies this relationship.

Consider an elemental portion of a member subject to bending (see Figure A-6-1.

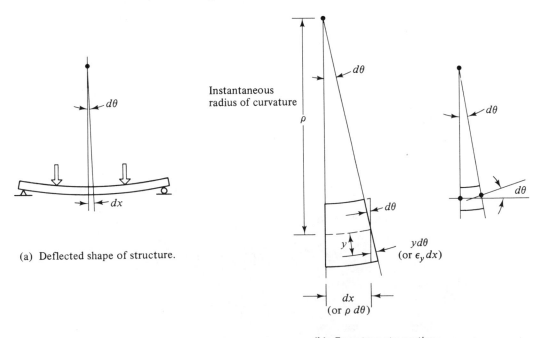

(a) Deflected shape of structure.

(b) Curvature at a section.

FIGURE A-6-1 Moment–curvature relation: $1/\rho = M/EI$.

Assuming that initially plane sections in the member remain plane under the action of bending, the two adjacent planes bounding the element considered undergo a relative rotation ($d\theta$) because of the bending. The fibers at the top of the member are shortened and those at the bottom lengthened. The elongation of a typical fiber at a distance y from the neutral surface (the horizontal plane of zero deformation) may be found by considering the initial undistorted location of the two planes. The elongation is the arc of a circle having a radius y and subtended by the angle $d\theta$, or the elongation $= y\, d\theta$.

If the original undeformed length of the element is a length dx and the strain (deformation per unit length) at y is ϵ_y, then $y\, d\theta = \epsilon_y\, dx$ or $\epsilon_y = y(d\theta/dx)$ or $(d\theta/dx) = \epsilon_y/y$.

From inspecting Figure A-6-1 it can also be seen that the radius of curvature, ρ, for the whole beam and dx are related through the expression $dx = \rho\, d\theta$, since the length dx at the middle surface is the arc of a circle of radius ρ and subtended by $d\theta$. Note that $(d\theta/dx) = 1/\rho$.

Equating the two expressions for $d\theta/dx$,

$$\frac{1}{\rho} = \frac{\epsilon_y}{y} \quad \text{or} \quad \epsilon_y = \frac{y}{\rho}$$

The relation between stress and strain in a homogeneous elastic material is known to be $E = f_y/\epsilon_y$, where E is the modulus of elasticity of the material and f_y is the stress at a point. Alternatively, $\epsilon_y = (1/E)f_y$. From the study of bending stresses we know that $f_y = My/I$. Thus $\epsilon_y = (1/E)(My/I)$, or $y/\rho = (1/E)(My/I)$. Consequently,

$$\frac{1}{\rho} = \frac{M}{EI}$$

This is the *moment–curvature relationship*. The instantaneous radius of curvature (ρ) is directly dependent on the magnitude of the moment (M) in a member and inversely dependent on the product of the modulus of elasticity (E) and moment of inertia (I) for the member. Note that if $M = 0$, then $\rho \rightarrow \infty$, indicating that the member is straight (as it must obviously be under no moment). As the moment increases, the radius of curvature becomes smaller, indicating that the member is being more sharply curved or bowed under the action of the load.

APPENDIX 7: DEFLECTIONS

A.7.1 General Differential Equation

The expression $1/\rho = M/EI$ found in Appendix 6 may be alternatively expressed in terms of the deflection curve of the member. Since the moment (M) in a member typically varies along the length of a member, the corresponding instantaneous radius of curvature will also vary resulting in a deflected shape of nonuniform curvature. If the curve is defined by $y = y(x)$, then from basic calculus it is known that

$$\frac{1}{\rho} = \frac{(d^2y/dx^2)}{[1 + (dy/dx)^2]^{3/2}}$$

The expression dy/dx is the slope of the member at any point. For small deflections, the square of this term is negligible in comparison to other terms. Thus,

$$\frac{1}{\rho} \approx \frac{d^2y}{dx^2}$$

Consequently,

$$\frac{d^2y}{dx^2} = \frac{M}{EI} \quad \text{or} \quad M = \left(\frac{d^2y}{dx^2}\right)EI$$

This is the basic differential equation for the deflection curve of a member subjected to bending.

A.7.2 Deflections: Double-Integration Method

The deflection of a beam at any point can be found by direct application of the result found above ($d^2y/dx^2 = M/EI$).

EXAMPLE

Consider the cantilever beam illustrated in Figure A-7-1, and assume that it is desired to know the deflection at the end of the member. Assume that E and I are constant along the length of the beam.

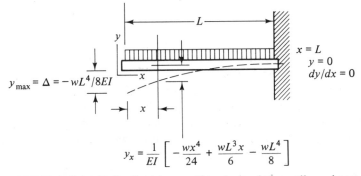

FIGURE A-7-1   Deflections in a uniformly loaded cantilever beam.

Solution:

The moment at x is given by $M_x = -wx^2/2$. Hence,

$$\frac{d^2x}{dy^2} = \frac{1}{EI}\left(-\frac{wx^2}{2}\right)$$

This equation can be integrated once to yield

$$\frac{dx}{dy} = \frac{1}{EI}\left(-\frac{wx^3}{6} + C_1\right)$$

where C_1 is a constant of integration. It can be found by using the boundary condition that the slope of the beam, dy/dx, is zero at $x = L$. C_1 can be found to be $wL^3/6$. Thus,

$$\frac{dy}{dx} = \frac{1}{EI}\left(-\frac{wx^3}{6} + \frac{wL^3}{6}\right)$$

This is the equation for the slope of the beam at any point x. Integrating once more,

$$y = \frac{1}{EI}\left(-\frac{wx^4}{24} + \frac{wL^3x}{6} + C_2\right)$$

where C_2 is the second constant of integration, which can be found by noting that the deflection y at $x = L$ is zero. $C_2 = -wL^4/8$.

$$y = \frac{1}{EI}\left(-\frac{wx^4}{24} + \frac{wL^3x}{6} - \frac{wL^4}{8}\right)$$

This is the basic equation for the deflected shape of the member. The maximum deflection occurs at $x = 0$. Hence,

$$y_{max} = -\frac{wL^4}{8EI}$$

Note that the boundary conditions are extremely important and must be handled with care.

Deflections resultant from any loading condition on any type of member can be evaluated similarly. Also note that if E and I were not constant along the length of the member, they would also have to be expressed as a function of x.

APPENDIX 8: MOMENT–AREA THEOREMS: SLOPES AND DEFLECTIONS

The *moment–area theorems* are a powerful tool in structural analysis. In Appendix 6 on moment–curvature relations, it was noted that with reference to Figure A-6-1 that $d\theta/dx = 1/\rho$ and that $1/\rho = M/EI$. Consequently,

$$\frac{d\theta}{dx} = \frac{M}{EI} \quad \text{or} \quad d\theta = \left(\frac{M}{EI}\right) dx$$

With reference to Figure A-6-1, it can also be seen that $d\theta$ also represents the angular change between two points on the structure. Thus the expression $d\theta = (M/EI)\, dx$ states that the change in slope in a distance dx is equal to $(M/EI)\, dx$. The total change in slope between two points A and B in a member then becomes

$$\theta_B - \theta_A = \int_A^B \left(\frac{M}{EI}\right) dx$$

This is the first of the moment–area theorems: *The change in slope between the tangents to the curve of the beam at two points is identical to the area under the M/EI diagram between those two points.*

The second moment–area theorem relates to deflections. For a structure undergoing the small deflections typically encountered in a real structure, the deviation ($d\Delta$) of one point on a structure with respect to the tangent drawn through another point x distance away is simply $d\Delta = x\,d\theta$. Combining $d\Delta = x\,d\theta$ with $d\theta = (M/EI)\,dx$, we obtain $d\Delta = (Mx/EI)\,dx$. The total change between two points A and B is given by

$$\Delta_B - \Delta_A = \int_A^B \left(\frac{Mx}{EI}\right) dx$$

Or, *the deviation of a tangent to the deflection curve at one point with respect to the tangent at another is identical to the first moment of the area of the M/EI diagram between the two points about the point at which the deviation is sought.*

EXAMPLE

Consider the beam illustrated in Figure A-8-1. Find the slope and deflection at the end of the beam using the moment–area theorems.

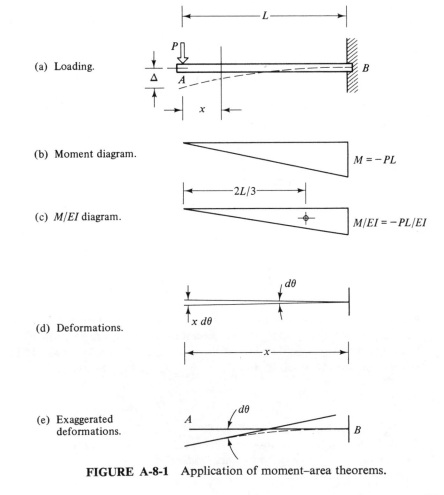

FIGURE A-8-1 Application of moment–area theorems.

Solution:

The first step is to draw the moment diagram. The change in slope from A to B is given by

$$\theta_B - \theta_A = \int_A^B \left(\frac{M}{EI}\right) dx = \frac{1}{EI}\int_0^L (-Px)\, dx$$

$$= -\frac{P}{EI}\left[\frac{x^2}{2}\right]_0^L$$

Since $\theta_B = 0$,

$$\theta_A = +\frac{PL^2}{2EI} \qquad \text{(radians)}$$

In this case the angle found also represents the slope of the free end with respect to the horizontal, since $\theta_B = 0$. The deflection at the end can be found through application of the second moment–area theorem. Noting that the tangent of the member at B is horizontal, the expression for the deviation of point A with respect to the tangent through point B becomes coincidental with the absolute deflection of the beam with respect to the horizontal. Thus,

$$y_{\max} = \Delta_B - \Delta_A = \int_A^B \left(\frac{M}{EI}\right) x\, dx = \frac{1}{EI}\int_0^L (-Px)\, x\, dx$$

$$= -\frac{PL^3}{3EI}$$

Although moment–area expressions can be integrated directly as illustrated above, it is more common to evaluate the area and first moment of the M/EI diagram directly from the moment diagram. Thus,

$$\theta = \left(\frac{PL}{EI}\right)(L)\left(\frac{1}{2}\right) = -\frac{PL^2}{2EI}$$

$$y_{\max} = \Delta_B - \Delta_A = \left(\frac{PL}{EI}\right)(L)\left(\frac{1}{2}\right)\left(\frac{2L}{3}\right) = \frac{PL^3}{3EI}$$

This example is rather simple since it is known that θ_B is horizontal and Δ_B is zero.

EXAMPLE

Find the deflection of the member shown in Figure A-8-2 at midspan.

Solution:

The moment diagram is first drawn. In this case it is evident that the slope of the beam is not zero at either support but must be zero at midspan because of symmetry. The deviation of point C from the tangent to the elastic curve at B is thus equivalent to the total deflection at B. Hence,

$$\Delta_{CB} = \underbrace{\left(\frac{PL}{4EI}\right)\left(\frac{L}{2}\right)\left(\frac{1}{2}\right)}_{(M/EI)\ \text{area}}\underbrace{\left(\frac{2}{3}\right)\left(\frac{L}{2}\right)}_{\text{moment arm about } C}$$

$$= \frac{PL^3}{48EI} = \text{deflection at } b\ (y_b)$$

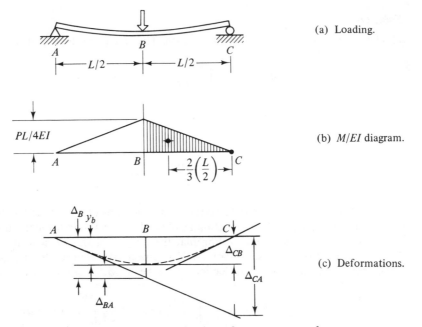

FIGURE A-8-2 Application of moment–area theorems.

For complex loadings when the beam slope at midspan is not zero, it is necessary to adopt the following general procedure (see Figure A-8-2). Δ_{CA} and Δ_{BA} are first calculated by moment-area theorems. Δ_B is next found by proportions. The beam deflection at B then becomes $y_b = \Delta_B - \Delta_{BA}$.

APPENDIX 9: DOUBLE-INTEGRATION METHODS OF ANALYZING INDETERMINATE STRUCTURES

This section addresses a method of analysis which is frankly of little current practical importance as a structural analysis tool. It is, however, of interest in *conceptual* terms as a direct extension and application of some of the formulations previously discussed.

The double-integration method of analysis is based on the moment–curvature relation discussed in Appendix 6. In the example that follows, the moment present at the end of a fixed-ended beam is treated as an unknown in the basic differential equation (see Appendix 7) for the deflection curve of the member. This unknown moment can be solved for by integrating the differential equation and carefully taking into account the appropriate boundary conditions.

EXAMPLE

Consider the fixed-ended beam illustrated in Figure A-9-1. Find the moments developed at the supports.

Solution:

In general

$$EI\frac{d^2y}{dx^2} = M$$

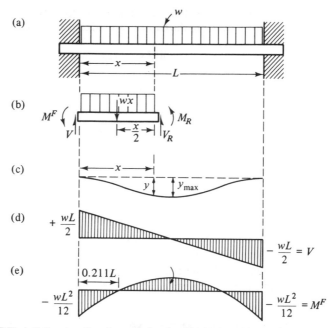

FIGURE A-9-1 Application of the double-integration method of analysis.

For the fixed-ended beam,

$$EI\frac{d^2y}{dx^2} = M^F + \frac{wLx}{2} - \frac{wx^2}{2}$$

Integrating,

$$EI\frac{dy}{dx} = M^F x + \frac{wLx^2}{4} - \frac{wx^3}{6} + C_1$$

Since the slope of the beam is horizontal at the support, $dy/dx = 0$, where $x = 0$; consequently, $C_1 = 0$. Also, $dy/dx = 0$ at $x = L$.

$$0 = M^F L + \frac{wL^3}{4} - \frac{wL^3}{6}$$

$$M^F = -\frac{wL^2}{12}$$

This is the moment developed at the support. Using this moment, the values of the moment present at other points in the beam can be found through equilibrium considerations. The moment at midspan, for example, can be shown to be $M = wL^2/24$. Note that in the example above the moment of inertia, I, of the beam was assumed to be constant. If I was a variable, it would have to be expressed as a function of x and included in the integration. The moment found as a result would no longer be $M = -wL^2/12$, but a different value. This illustrates that moments in continuous beams are not independent of variations in member properties. Variations in the modulus of elasticity, E, would also affect results.

If it were desired to know the deflection at midspan of the member analyzed above, it is only necessary to make use of the moment found at the support.

$$EI\frac{dy}{dx} = -\frac{wL^2x}{12} + \frac{wLx^2}{4} - \frac{wx^3}{6}$$

$$EIy = -\frac{wL^2x^2}{24} + \frac{wLx^3}{12} - \frac{wx^4}{24} + C_2$$

Since $y = 0$ at $x = 0$, $C_2 = 0$.

$$EIy = -\frac{wL^2x^2}{24} + \frac{wLx^3}{12} - \frac{wx^4}{24}$$

By symmetry, the maximum deflection occurs at midspan.

$$\Delta = y_{\max} = -\frac{wL^4}{384EI}$$

While the double-integration method of analysis demonstrated above is conceptually very elegant, its application becomes very cumbersome when symmetrical situations do not exist or if there are a large number of redundancies. For this reason, other methods of analysis are typically utilized.

APPENDIX 10: DEFLECTION METHODS OF ANALYZING INDETERMINATE STRUCTURES

The deflection method of analysis involves isolating one (or more) of the redundant supports, conceptually removing the redundant support and allowing the structure to deflect freely (thereby making the structure statically determinate), determining the deflection at the support point, and calculating the magnitude of the force required to push the structure from its freely deformed condition *back* to its original state. The force required to reestablish the original shape of the structure is equivalent to the reaction normally developed at that point.

Consider the structure shown in Figure A-10-1 and its deflected shape. If the support at A were removed, the structure would become statically determinate and have the deflected shape indicated. The free deflection (Δ_{w_A}) at A due to the load w could be found through methods previously discussed. If an upward force were applied at the free end of the freely deflected structure, it would tend to reduce this deflection. Clearly, a force (R) of some as-yet-unknown magnitude exists that would exactly push the free end back to its original location (before the support was imagined as removed) and thus reestablish the original deflected shape of the member. This force is equivalent to the reaction force normally developed at this point since their effects on the structure are identical.

If the deflection at A due to the force R applied to the free end of the cantilever beam is termed Δ_{R_A} and that associated with the free deflection of a uniformly loaded cantilever beam termed Δ_{w_A}, then $\Delta_{R_A} + \Delta_{w_A} = 0$. This is a deflection compatibility statement. It is based on the fact that deflections resultant from several loads on a member can be superimposed. This statement can be used to find the force R required

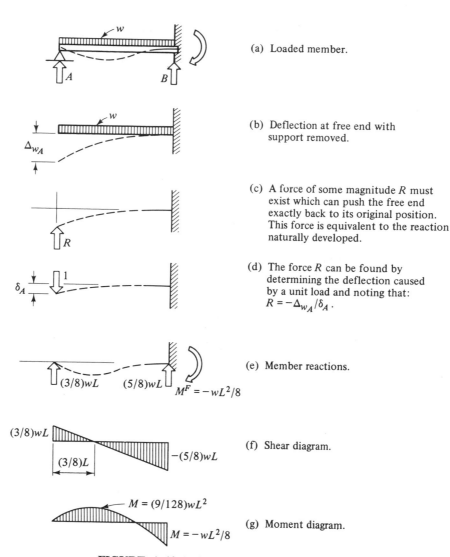

(a) Loaded member.

(b) Deflection at free end with support removed.

(c) A force of some magnitude R must exist which can push the free end exactly back to its original position. This force is equivalent to the reaction naturally developed.

(d) The force R can be found by determining the deflection caused by a unit load and noting that: $R = -\Delta_{w_A}/\delta_A$.

(e) Member reactions.

(f) Shear diagram.

(g) Moment diagram.

FIGURE A-10-1 Deflection method of analysis.

to satisfy the condition that in the actual structure the deflection at A is zero. The easiest way to find this force is to first determine the deflection at A due to a unit load ($P = 1$) applied to the end of the number. If the deflection at A associated with a unit load at A is termed δ_A, the deflection associated with a force R at the same location is simply $R(\delta_A)$. Deflections are linearly dependent on the magnitude of the loads involved as long as the material is in the elastic range and deflections are small.

The original deflection compatibility statement $\Delta_{R_A} + \Delta_{w_A} = 0$ can now be written $R(\delta_A) + \Delta_{w_A} = 0$ or $R = -\Delta_{w_A}/\delta_A$. Thus, the unknown force R can be found by considering the ratio of the deflection under the external load associated with the structure when the support is removed to that associated with the structure under the action of a unit load.

These quantities can be easily found by calculating deflections. Using the methods described earlier, the free-end deflection of a uniformly loaded cantilever beam can be shown to be $\Delta_{w_A} = -wL^4/8EI$. For a cantilever beam carrying a concentrated load of P at its free end, $\Delta = PL^3/3EI$. If $P = 1$ for the unit-load case, then $\delta_A = (1)L^3/3EI$. Thus, $R = -(wL^4/8EI)/[(1)L^3/3EI] = -\frac{3}{8}wL$. This force is equivalent to the reaction at A, since both are associated with a net deflection at A of zero. Knowing this one piece of information, other reactions can be found by $\sum F_y = 0$.

For structures having more than two redundants, the same process can be used but it becomes more complex. Consider the member shown in Figure A-10-2. Two supports must be removed to make the member statically determinate. If B and C are removed, the structure will deflect an amount Δ_B and Δ_C at these points. A unit load can be placed at B and the resultant deflection δ_{BB} at B calculated. This load at B also causes a deflection δ_{BC} at point C. A unit load at C causes a deflection at C of δ_{CC} and a deflection at B of δ_{CB}. With respect to the reactions, a force R_C at C will raise point

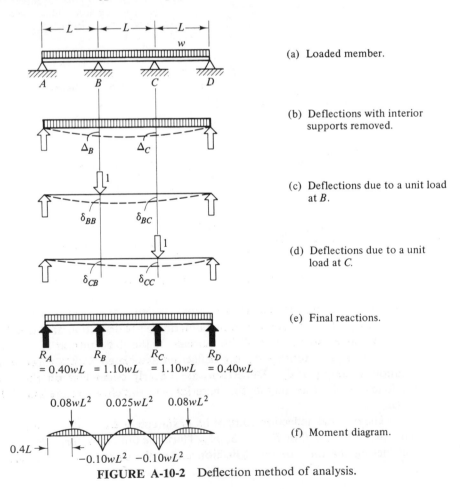

(a) Loaded member.

(b) Deflections with interior supports removed.

(c) Deflections due to a unit load at B.

(d) Deflections due to a unit load at C.

(e) Final reactions.

R_A = 0.40wL R_B = 1.10wL R_C = 1.10wL R_D = 0.40wL

0.08wL^2 0.025wL^2 0.08wL^2

(f) Moment diagram.

0.4L

$-0.10wL^2$ $-0.10wL^2$

FIGURE A-10-2 Deflection method of analysis.

C by an amount $R_C\delta_{CC}$. It will also, however, raise point B by an amount $R_C\delta_{CB}$. Similarly, a force R_B will raise point B by an amount $R_B\delta_{BB}$ and point C by an amount $R_B\delta_{BC}$. Since each affects the other, it is evidently necessary to consider a more involved statement of deflection compatibility at each point. For B and C respectively, the net deflection at each point is

$$\Delta_B + R_B\delta_{BB} + R_C\delta_{CB} = 0$$

$$\Delta_C + R_C\delta_{CC} + R_B\delta_{BC} = 0$$

These equations can be solved simultaneously once appropriate values for the different parameters have been found.

The process can be extended to members over any number of supports. As can be anticipated, however, the number of equations that must be solved simultaneously increases with the increase in number of support points. This can obviously get too involved for solution by hand but can be readily adapted for computer solutions.

APPENDIX 11: OTHER METHODS OF ANALYSIS

The methods of analysis previously discussed are only a sample of the many methods of structural analysis that have been used. Many are also based on variants of the deflection methods discussed above. Others are formulated on entirely different theoretical approaches. Ideas relating to the idea of internal work or to energy form the basis for many powerful and elegant analytical methods. Often used methods of analysis include the three-moment theorem. Castigliano's theorem, column analogy, the slope-deflection method, and moment distribution. The latter two were very widely used prior to the advent of the computer. They are still valuable and elegant.

A look at the example of the beam over two supports discussed in Appendix 10 yields an insight into why many methods were evolved and approaches especially suitable for computer applications developed. As was noted, the more redundants were present in an indeterminate structure, the more equations there are that need to be solved simultaneously for exact results to be obtained. Doing these calculations by hand for more than a few equations is tedious and almost impossible for highly redundant structures. Thus, methods were developed to avoid this process. The moment-distribution approach is the classic example here and deserves special mention.

Moment distribution is often referred to as the *method of successive approximations*, since it involves an iterative process. It is based on a study of the moments and rotations induced by loads at supports and joints and how moments and rotations at several locations are related. The iterative aspect is based on a procedure of fixing a joint in space, determining the moments as if it were a fixed-ended structure, then releasing the joint (i.e., allow rotations to occur) and studying the transferance of moments and rotations to other joints. Moment distribution is a powerful way of analyzing structures and can be used on complex structures involving many redundants. Results can be more or less precise depending on the number of iterations completed. The reader is encouraged to study this technique by reading other texts.

With the advent of the computer most of these methods, including moment distribution, are less used than previously. There are other formulations more appropriate for computer applications. Generally, these can be characterized as either *stiffness methods*, where the problem is addressed in terms of unknown joint displacements, or as *flexibility methods*, where the problem is addressed in terms of the actions of the redundants. Data are put into matrix format. Structures involving almost any number of redundants can be solved through these techniques. These methods, however, tend to be on the black-box side unless the user is fully aware of the details of the analytical techniques used and are thus of little value as a tool for beginning to develop a conceptual understanding of structural behavior.

APPENDIX 12: CRITICAL BUCKLING LOADS FOR COMPRESSION MEMBERS

This section presents a rigorous derivation of the buckling load for a pin-ended column. The critical buckling load is defined to be that axial force that is just sufficient to hold the bar in a slightly deformed configuration. Under the load P the column deflects into a curved shape such that the lateral deflection at a point x from one end of the column is defined by the distance y.

The necessary condition that one end of the bar be able to move axially with respect to the other end in order that the lateral deflection may take place is assumed. The differential equation of the deflection curve is the same as that presented in Appendix 7:

$$EI\frac{d^2y}{dx^2} = M \tag{1}$$

The bending moment in the above expression is merely the moment of the force P times the deflection y at that point. Hence the bending moment is $M = -Py$. Thus,

$$EI\frac{d^2y}{dx^2} = -Py \tag{2}$$

If we set

$$\frac{P}{EI} = k^2 \tag{3}$$

then

$$\frac{d^2y}{dx^2} + k^2y = 0 \tag{4}$$

This differential equation is readily solved by any one of several standard techniques discussed in textbooks on differential equations. The solution is relatively simple. We need merely find a function which when differentiated twice and added to itself (times a constant) is equal to zero. Evidently either $\sin kx$ or $\cos kx$ possesses this property. These terms can be combined so as to form a solution to the above equation.

Thus,

$$y = C \sin kx + D \cos kx \tag{5}$$

It is next necessary to determine the constants C and D. At the left end of the bar, $y = 0$ when $x = 0$. Substituting these values in equation (5), we obtain

$$0 = 0 + D \quad \text{or} \quad D = 0 \tag{6}$$

At the right end of the bar, $y = 0$ when $x = L$. Substituting these values in the expression for y with $D = 0$, we obtain

$$0 = C \sin kL \tag{7}$$

Evidently, either $C = 0$ or $\sin kL = 0$. But if $C = 0$, then y is everywhere zero and we have only the trivial case of a straight bar which is the configuration prior to the occurrence of buckling. Since we are not interested in this solution, we must then take

$$\sin kL = 0 \tag{8}$$

For this to be true, we must have

$$kL = n\pi \text{ radians} \quad (n = 1, 2, 3, \ldots) \tag{9}$$

Substituting $k^2 = P/EI$ in equation (9), we find

$$\left(\sqrt{\frac{P}{EI}} \right) L = n\pi \quad \text{or} \quad P = \frac{n^2 \pi^2 EI}{L^2} \tag{10}$$

The smallest value of this load P evidently occurs when $n = 1$. Then we have the *first mode of buckling*, where the critical load is given by

$$P_{cr} = \frac{\pi^2 EI}{L^2} \tag{11}$$

This is called *Euler's buckling load* for a pin-ended column. The deflection shape corresponding to this load is

$$y = C \sin \sqrt{\frac{P}{EI}} x \tag{12}$$

or

$$y = C \sin \frac{\pi x}{L} \tag{13}$$

Thus, the deflected shape is a sine curve.

The effects of other types of column end connections (e.g., fixed ends) on the buckling load of a column can be determined in a similar manner. The moment expressions, however, and boundary conditions used are different from those presented above.

APPENDIX 13: ELASTIC SECTION MODULUS TABLE (S_x) FOR SELECTED SHAPES USED AS BEAMS[a]

S_x $mm^3 \times 10^3$	S_x $in.^3$	Shape	S_x $mm^3 \times 10^3$	S_x $in.^3$	Shape	S_x $mm^3 \times 10^3$	S_x $in.^3$	Shape
16,882	1030	W 36×280	1,868	114	W 24× 55	395	24.1	C 12× 25
13,718	837	W 36×230	1,836	112	W 14× 74	341	20.8	W 8× 24
10,998	671	W 33×200	1,612	98.4	W 18× 55	339	20.7	C 10× 30
7,343	448	W 33×141	1,598	97.5	W 12× 72	288	17.6	W 12×16.5
5,392	329	W 30×116	1,337	81.6	W 21× 44	279	17.0	W 8× 20
			1,324	80.8	W 16× 50	274	16.7	W 6× 25
4,425	270	W 30× 99	1,121	68.4	W 18× 40	243	14.8	W 12× 14
4,376	267	W 27×102	1,100	67.1	W 10× 60	231	14.1	W 8× 17
3,622	221	W 24× 94	949	57.9	W 18× 35	226	13.8	W 10× 15
3,475	212	W 27× 84	926	56.5	W 16× 36	193	11.8	W 8× 15
3,311	202	W 18×105	774	47.2	W 16× 31	185	11.3	C 9× 15
2,885	176	W 24× 76	762	46.5	C 15× 40	172	10.5	W 10× 11.5
2,885	176	W 14×111	575	35.1	W 14× 26	167	10.2	W 6× 16
2,295	140	W 21× 68	575	35.0	W 10× 33	149	9.09	C 8× 13.75
2,262	138	W 14× 87	474	28.9	W 14× 22	133	8.14	C 8× 11.5
2,147	131	W 14× 84	449	27.4	W 8× 31	128	7.80	W 8× 10
2,131	130	W 24× 61	442	27.0	C 12× 30	127	7.78	C 7× 14.75
2,114	129	W 18× 70	415	25.3	W 12× 22	100	6.08	C 7× 9.8
2,098	128	W 16× 78	398	24.3	W 8× 28	95	5.80	C 6× 13

[a]The beams in the table are arranged according to their relative S_x values in descending order of magnitude for the first entry in each grouping of members. Entries in each grouping are arranged in descending order of S_x for a group. The first entry in a group represents the lightest member and the member having the largest section modulus for all entries in the group. It is thus an efficient and often preferred member.

W, wide-flange shape.
C, channel shape.

Typical designation:

W 21 × 44
 ↑ ↑   └ weight per linear foot, lb
 ↑ └ nominal depth, in.
 └ shape designation

INDEX

Poisson's ratio, 73
Positive moments, 63, 70-71
Post and beam structures (*see also* Beam and
        column systems)
    examples, 10, 16, 24
    horizontal loads, 4, 336
    vertical loads, 4, 10, 335
Post-tensioning, 277-80
Precast concrete elements, 517-19
Pressure coefficient, 92
Pressurized Structures (*see* Pneumatic structures)
Primary structural units:
    definition, 15
    examples, 7, 16
    single volumetric units, 7, 24
Principal stresses, 250
Principle of structural justice, 504
pressure obefficient, 92
Prestressing, 277-80
Proportional limit, 73

Radius of gyration, 295, 301, 306
Reactions:
    calculation of, 53-54, 111-18
    definition, 50
    in typical beams, 111-18
Rectangular grids, 453
Reinforced concrete:
    beams, 257-60
    continuous beams, 331
    flat plate, 395, 469, 474, 514
    flat slab, 516
    precast, 517-18
    prestressing and post-tensioning, 277-80, 517-19
    span ranges, 519
    structures, 514-19
Relative stiffnesses, 348-51
Resisting moments:
    in all structures, 475
    in beams, 225-26, 232, 542
    in cables, 196-99
    in membranes, 409
    in plates, 368, 372-75
    in trusses, 148
    in shells, 429
Resonance, 97, 500
Reticulated surfaces, 417
Ribbed slab, 382
Rigid arches, 12, 182, 213-15
Rigid frames (*see* Frames)
Rigid joints (*see also* Frames):
    models, 104-5
    use in preventing instability, 5, 19, 337
Rigid structures, 8, 10
Rivets, 531-32

Roller joint, 52, 104, 107
Rules of thumb, 507-8, 512-13, 519, 523

S.I. units, 535-36
St. Ouen, 83
Schwedler Dome, 417
Science Center, 443, 466
Section modulus, 261, 265, 268, 271-73, 560
Sections, method of, 140-46
Seismic joints, 493-94
Serviceability, 84
Service systems, 463-67
Shear:
    basic phenomena, 59-61
    distribution, 65
    in beams, 226
    sign conventions, 62-63
    stresses in beams, 227, 239-44, 544-46
Shear and moment:
    basic phenomena, 59-61
    in trusses, 146-48
Shear and moment diagrams, 65-71
    constructing diagrams, 65
    for cantilever structure, 67
    for simply supported beams with concentrated
        loads, 64-66
    for simply supported beams with uniform
        loads, 68-69
    shapes of diagrams, 69
    sign conventions, 62-63
    relations among load, shear, and moment, 71
    relation of moment diagram to deflected shape,
        70
Shear center, 247-48
Shear stresses:
    distribution in beams, 241, 545
    general theory in beams, 544-46
    horizontal, 227, 239-44, 544-46
    in beams, 226-27, 239-44, 544-46
    in bolts or rivets, 531
    in rectangular beams, 242
Shear walls:
    arrangement, 481, 485
    definition, 18
    for resisting lateral forces, 480-83
    use in preventing instability, 18, 19, 447,
        480-83
Shell structures, 415-33
    bar assemblies, 417
    barrel, 423
    buckling, 432
    comparisons with arches, 419
    concentrated forces, 428
    geodesic domes, 418
    hoop forces, 419, 425-27